W9-AQF-326

TOPICS IN POPULATION GENETICS

TOPICS IN POPULATION GENETICS

Bruce Wallace
CORNELL UNIVERSITY

W · W · NORTON & COMPANY · INC ·
NEW YORK

Library of Congress Catalog Card No. 68-22851

Published simultaneously in Canada by
George J. McLeod Limited, Toronto

SBN 393 09813 3

PRINTED IN THE UNITED STATES OF AMERICA

2 3 4 5 6 7 8 9 0

To Miriam, Bruce, and Roberta, whose independent activities during the past year have made the completion of this book not only humanly possible but also financially necessary.

Contents

Preface

Although *Topics in Population Genetics* has been no labor of love, I have enjoyed writing it immensely. I have attempted to present topics much as they might be presented in a series of seminars. Consequently, *Topics* is not a tightly knit, precisely organized textbook. Indeed, some of my own notions have evolved during the preparation of the manuscript. This will be most obvious in the attempt that is made in Chapter 24 to interrelate genetic load, adaptive values, and population size; the ideas presented in that chapter were largely unformed ten chapters earlier.

I anticipate three major criticisms of *Topics in Population Genetics:* first, that important topics have been omitted; second, that important experiments relevant to the topics actually discussed have been neglected; and third, that the generalized mathematical aspects of population genetics have been omitted almost entirely.

In answering these anticipated criticisms, I had best refer to my own experiences and to the use I intend to make of *Topics.* I have found that a tremendous gap separates the content of even an excellent general genetics course and the intellectually interesting aspects of population genetics. Too much time in a course on population genetics must be spent merely acquainting students with elementary observations and the techniques by which these are made; without a feel for the basic tools of the trade, the student can neither grasp the meaning of nor detect the weaknesses of advanced concepts.

For me, *Topics* will not be a textbook whose contents are to be repeated in lectures. Rather, I shall use it to gain time. It will enable me to discuss in class additional topics and additional work in *worthwhile* detail. I can look forward now to an opportunity to cover the ideas and experimental data of (for example) Lerner, Levins, Milkman, Mukai, or Waddington to an extent not possible before. The experimental and theoretical work of a number of ecologists can now be dealt with in reasonable detail.

Topics was never intended to be a text of mathematical genetics. A number of books—those by Falconer, Kempthorne, and Li, for example—already exist; various brochures have informed me that still other excellent ones are about to appear. As a non-mathematician, I have set myself the task of presenting biological information to students with mathematical ability rather than that of impressing biology students with trivial calculations. Nevertheless, I have tried to stress throughout that problems of population or evolutionary genetics are quantitative problems that lend themselves to mathematical analyses.

In concluding these remarks, I should like to thank my friends and former professors, Th. Dobzhansky and John A. Moore, for reading and criticizing the original manuscript; much (but, in their view, probably not enough) of their advice was heeded. The helpful comments of Professor Ken-ichi Kojima are gratefully acknowledged. I would also like to thank Mr. Kenneth Demaree of W. W. Norton, who paled but did not panic when he learned that there are over one hundred tables. I would be remiss if I failed to acknowledge my colleagues at Cornell who have patiently "heard me out" on this or that topic during the past months. I want to express my gratitude to the many persons in the Genetics and Plant Breeding offices at Cornell who have helped with the final typescript. Finally, in a less personal vein but with no less gratitude, I want to acknowledge the long-continued support given my personal research efforts by the U.S. Atomic Energy Commission; many of the observations described in this text were made under their financial aegis.

BRUCE WALLACE

Ithaca, N. Y.

PART I

This section of *Topics in Population Genetics* deals primarily with the Hardy-Weinberg equilibrium—what it is, what it says, how it can be used, and what agencies modify its claims. In order that the theoretical discussion be accompanied by experimental evidence, genetic techniques for studying populations are illustrated in considerable detail. The concept of the "local population" is discussed also, especially in relation to the movement of individuals.

Throughout this book great reliance is placed on data gathered from studies of various species of *Drosophila*; there is scarcely a topic in population genetics to which flies of this genus have not contributed truly basic knowledge. Among the most important species in this respect are those belonging to the subgenus *Sophophora*. As "bearers of wisdom," these flies have surely been well named!

1

ABOUT MODELS

> The existing theories of population genetics will no doubt be simplified and systematized. Many of them will have no more final importance than a good deal of nineteenth-century dynamical theory. This does not mean that they have been a useless exercise of algebraic ingenuity. One must try many possibilities before one reaches even partial truth.
>
> *A Defense of Beanbag Genetics* (Haldane, 1964)

Population genetics is a branch of genetics that deals with the breeding structure and genetic composition of populations. Individuals, the transitory packages that encase the genetic endowment of a population at a given moment, are interesting to the population geneticist only to the extent that their differing reproductive successes mold this endowment. The population geneticist studies the frequencies of different alleles in a population, the interactions of alleles at the same or different loci, and the degree to which such interactions govern gene frequencies themselves. He notes the number of individuals in populations, because these in turn determine the actual range of frequencies with which a given allele can occur and the role chance will play in the transmission of this allele from one generation to the next. He is especially interested in the divergent paths followed by isolated populations, for such divergences are the simplest of all evolutionary changes. Equally interesting are the events which occur when two formerly isolated populations meet once more; from such meetings sometimes come the finishing touches on behavioral differences or habitat preferences that lead to the reproductive isolation of the two groups. Speciation, the splitting of one formerly interbreeding population into two reproductively isolated groups of individuals, involves the perfection of one or another (or an efficient combination of several) isolating mechanisms.

The beauty of population genetics lies not in enormous catalogues of alterations known to have occurred in the genetic composition of various plant and animal species, but in generalizations that allow us to say with remarkable assurance what the genetic composition of a given population is like at a given moment. These generalizations can be viewed as probability statements that refer to the existing situation at a given locus, as statements concerning the distribution pattern of expectations at many loci, or as descriptions of events that are likely to befall a single locus over a long period of time. The beauty of population genetics, in short, lies with its abstractions rather than with descriptions of actual events. Geneticists, as a rule, regard abstract generalization, if not with affection, at least without fear.

Any abstraction involves the use of a model. This chapter is about models. The need for these comments is twofold: Many biologists distrust models; to the extent that their distrust reflects misunderstanding, a short discussion at this point will serve a useful purpose. Furthermore, much of what follows in subsequent chapters is expressed in general—frequently algebraic—terms; these chapters may get a better reception if the usefulness of models is made clear now.

The formulation of generalizations is the goal of all research. Ideal descriptions of events are those which include only pertinent factors and from which everything superfluous has been removed. Thus the shape and size of flasks are not generally included in formulas of chemical reactions. Mathematical equations omit all reference to actual "things." The same search for generalization exists in biology, too; unfortunately, the scope of biology is enormous and its subject matter is exceedingly complex. Indeed, many biologists despair that useful generalized statements exist in their science (see, for example, Sturtevant, 1965, p. 110) and are, at best, prepared to accept only partial generalizations of limited scope.

Living things come in a variety of shapes and sizes—plants and animals, vertebrates and invertebrates, warm-blooded and cold-blooded, flying and crawling, aquatic and terrestrial. The environment by virtue of its physical irregularities, compounded by the added irregularities in the distribution of various forms of life, is extremely complex. Small wonder that a biologist who specializes in the study of one zoological or botanical family or one genus is skeptical of "broad" statements derived from a study of an entirely different organism; differences in the form and habits of organisms are much more obvious than are similarities in the challenges of life which confront them, challenges to which they respond in fundamentally similar ways.

Despite the distrust of models shown by many biologists, every research worker uses them. In effect, models with which we are familiar are no longer regarded as models; on the contrary, models tend to be generalizations which are only poorly understood or with which we disagree. Statistical tests of significance are based on hypothetical models, but there is scarcely a biologist who objects to computing means, variances, errors, or chi-squares for his data whether these quantities are applicable to his problem or not. Most elementary genetics courses now include a discussion of the Hardy-Weinberg equilibrium. This equilibrium is described by the probabilities that two independent events will occur in various combinations given the probability for each alone; despite its abstract nature, many students who study the genetics of populations confidently run through Hardy-Weinberg calculations whether the available data need be subjected to such calculations or whether the calculations are meaningful when made. The Hardy-Weinberg is a familiar and trusted model!

Haldane (1964) has given an eloquent defense of models in a reply to criticisms leveled by Mayr (1963) at the work of early population geneticists. Mayr, noting that genes do not act independently of one another during development, claims that the treatment of genes as independent units in devising mathematical theories of population genetics was in many respects misleading. In fact, looking back from his present-day position, Mayr (1959) asks, "But what, precisely, has been the contribution of this mathematical school to the evolutionary theory?" By "mathematical school" he means the contributions of Fisher, Haldane, and Wright, which began to appear in the early 1920s and ten years later were summarized in three classic works: *The Genetical Theory of Natural Selection* (Fisher, 1930), *The Causes of Evolution* (Haldane, 1932a), and *Evolution in Mendelian Populations* (Wright, 1931).

Haldane's defense of population genetics is based on the following points: First, biologists tend to be overly impressed by the mathematics involved; as we mentioned above, models that are distrusted are frequently those that are poorly understood. Second, he argues that models based on the work of Mendel, Bateson, and Punnett are a great improvement over the earlier statements of Lucretius. Third, he emphasizes the ambiguity of verbal arguments and the improvement that comes from the use of algebraic expressions. Not only is algebraic reasoning exact, but it imposes an exactness of verbal postulates. This, I believe, is the crucial point in the Mayr-Haldane dialogue: Life is complicated, the environment is bewilderingly heterogeneous even for members of a single species, and development itself represents a poorly understood skein of internally regulated, gene-initiated reactions; nevertheless, it does not follow that verbal descriptions based on ill-defined or ambiguous terms are more useful than descriptions that are mathematical in design (if not in symbolism).

A milder criticism of biological models than Mayr's has been expressed by Ford (1964, p. 274); his comments are so lucid that it is worth quoting them: "Indeed I am myself far from convinced that the fitting of field observations to theoretically derived models is a useful technique. It is subject to serious defects of several kinds. In the first place, the required parameters are almost invariably affected by such errors that they could accord with different types of models . . . Furthermore, serious oversimplifications of fact are necessary in order to describe such a model at all . . . Moreover, it does not seem that such models can be made without serious oversimplifications of theory also . . . Whether or not we consider that a comparison of Haldane's (1948) model contributes substantially to our knowledge of the *edda* cline in Shetland, there is no doubt that Kettlewell's field-studies have done so."

Ford is quite right in regarding Kettlewell's analysis (Kettlewell, 1961a and 1961b; Kettlewell and Berry, 1961) of the *edda* cline in *Amathes glareosa* in Shetland as an outstanding piece of work; indeed it is! Without some sort of theoretical model, however, one has no basis for assuming that a second analysis of the same material would agree with the first. Nor would one know on what basis to compare the observations in Shetland with those made in other localities, nor the *edda* cline with clines of other sorts. Actually, without a model, it is *unimportant* what happened to *edda* on Shetland while Kettlewell made his studies; this fine work takes on meaning only in relation to other studies and from the generalizations which emerge from them. It has no meaning standing in splendid isolation.

Andrewartha (1963, p. 181) has also expressed reservations concerning theoretical (conceptual) models. Without such models, he admits, we may not know what experiment to do. He goes on to say, however, that "it is the experiment which provides the raw material for scientific theory. Scientific theory cannot be built directly from the conclusions of conceptual models." As an example of a faulty model, he would probably cite one which begins by assuming a uniform environment; the environment, as he points out, is patchy and irregular in virtually all respects. In reply, I would say that a *useful* model is one which helps us understand a number of similar observations, while a truly *powerful* one serves as a basis for explaining discrepancies between different sets of observations. An accurate mathematical description of the patchiness of the environment would not only vary from species to species (the environment of the American bison is surely different from that of dung beetles that may live underfoot) but, if formulated, would be complicated beyond comprehension. On the contrary, a model based on a uniform environment may be quite accurate for many purposes and, when empirical observations deviate markedly from the model, may serve as a basis for discussing the role of patchiness of the environment in population studies. That this is the role of the powerful model in research can be gathered from the following statement by Sturtevant (1965, p. 51): "One of the striking things about the early *Drosophila* results is that the ratios obtained were, by the standards of the times, very poor . . . with Drosophila such ratios as 3 : 1 or 1 : 1 were rarely closely approximated." The scientific discoveries of the Morgan school of Drosophilists at Columbia were based on conceptual models, not on raw data as such.

The remainder of this chapter will be devoted to a consideration of three models which, although not strikingly genetical in nature, will impinge on the genetic aspects of populations to be discussed in subsequent chapters. The first of these is an attempt to account for the stability in the size of a population over successive generations; the second is a generalization of the Hardy-Weinberg equilibrium—a generalization more interesting for its assumptions than for its conclusions; and, finally, a model based on samples of random digits that illustrates the limitation of genetic mechanisms in recording the evolutionary history of a population. Each of these models in its own way illustrates the uses to which simplified generalizations and mathematical abstractions are put in helping us understand processes occurring within natural populations.

The Self-Regulation of Populations

The first model relates to the numbers of individuals of a given species or, better, to the numbers of individuals of a species inhabiting a given locality. The number of starlings in the Ithaca, New York, area can serve as an example. There are several things we can say about a local population with complete assurance that we shall be correct. First, the number of individuals in the local population is not constant from year to year or from generation to generation. In the case of starlings, there were none anywhere in the United States before 1890, the year they were introduced from Europe. Since then their numbers have increased steadily but with fluctuations large enough so that they have been greater nuisances in certain years than in others.

The second thing we can say about populations is that, despite periodic fluctuations, in the long run they tend to remain constant in size. Following their introduction into the United States (with perhaps a lag of several years), starlings increased in numbers very rapidly. They are no longer increasing at the earlier, enormous rate, however; there has been no spectacular rise in Ithaca, for example, during the past decade. No elaborate argument is required to justify the claim that a species can never increase in numbers geometrically for long; if no other limiting factor intervenes, standing room will be exhausted in a rather small number of generations.

The final statement we can make with certainty is that despite the numbers of individuals actually existing at any one moment, extinction of a population—indeed, of an entire species—is not only possible but, in the long run, is even probable. Most species of the past, lest we have forgotten, have disappeared without issue. The wholesale extinction of numerous dinosaur species followed immediately the period during which they were the dominant animals on earth. The extinction of the passenger pigeon in the United States, and the near extinction of the American bison, remind us that numbers of individuals are no guarantee of a species' continued existence.

The generation by generation continuation of a population can be viewed in two quite different ways. The first of these (a view that I find unrealistic) says simply that each adult female on the average leaves a single adult female offspring as a replacement in a population of constant size (for convenience we neglect males in the discussion). However, this single replacement represents an average and, consequently, a decrease (including extinction) or an increase in population size is fundamentally a chance event. Having increased or decreased in size, according to this view, the population has no inherent tendency to return to its original number; all fluctuations within a given set of environmental conditions are purely chance fluctuations.

The other view, one that has been expounded especially by Nicholson (1955; see, too, Haldane, 1953), regards populations as self-regulating collections of individuals; as numbers of individuals decrease in an otherwise stable environment, the average number of surviving offspring per parental female increases. Conversely, as numbers of individuals increase, the number of surviving offspring per female decreases. A diagrammatic representation of this view of self-regulation is shown in Figure 1-1. On the vertical axis only the point representing one surviving daughter per female is shown. This is the critical point; the remainder of the scale is arbitrary. If in looking at the diagram we think of oysters, the top of the diagram would represent millions of offspring most of which die; in case of fruit flies, the top of the diagram would represent several hundred or a thousand offspring; for many large mammals including man, the top would represent perhaps two dozen. The horizontal axis represents population size; again numbers have not been shown because they vary so much from species to species. Regard-

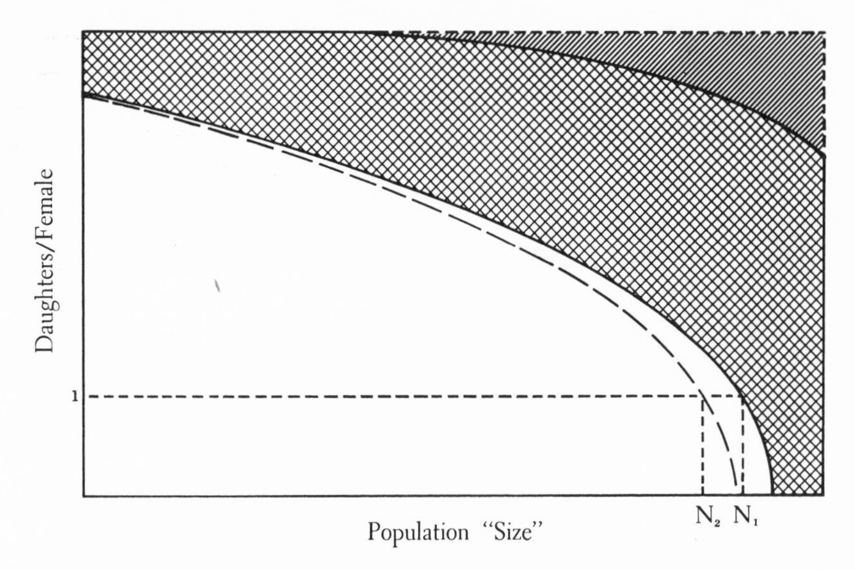

FIGURE 1-1. Conditions that tend to stabilize the number of individuals in a population. The unshaded portion represents the number of adult daughters produced per mother; equilibrium exists when on the average each mother leaves a single replacement. The values N_1 and N_2 show that changes in the severity of the environment are reflected eventually in population size. (Modified from Haldane, 1953.)

less of the species, however, the depicted relationship between number of surviving daughters per female and population size leads to a stable population when the average number of surviving daughters is one (for simplicity we ignore males). A population larger than the critical size will tend to shrink because the number of surviving daughters would be smaller than the number of mothers; conversely, a smaller population would tend to expand because the number of daughters surviving to adulthood would exceed their mothers in number.

An important feature of the model shown in Figure 1-1 is the stability of the population in the face of new environmental hazards. The dashed line in the figure represents a newly imposed hardship that lowers the number of surviving offspring. For the many small mammalian species whose carcasses can be seen on many highways, automobiles would represent such a hazard. Before the advent of cars, the populations of these mammals were stabilized and, consequently, one adult female offspring was left on the average by each mother. In the absence of self-regulation in these populations, the excess number of deaths now caused by automobiles should inevitably force each affected species to extinction. This is a most unlikely possibility, however. Instead of becoming extinct, the population contracts somewhat in numbers until once more each mother leaves one adult daughter. In effect, this type of population control allows the population to shift the causes of death just as money can be shifted from one account to another in a bank. Death is unavoidable as long as individuals are mortal; death before reproductive maturity is unavoidable as long as each pair of parents produces more than two offspring. The causes of these deaths are arbitrary, however, in the sense that if one cause were to be lessened, another would grow to take its place. We shall return to this point in a discussion of genetic deaths in a later chapter.

The Generalized Hardy-Weinberg Equilibrium

The second of the three models is a generalized Hardy-Weinberg equilibrium (Kennedy, 1965). A detailed discussion of the Hardy-Weinberg distribution will be given in Chapter 4. Nevertheless, this distribution is included in nearly all elementary genetics texts, so we can safely refer to it at this time. In effect, the Hardy-Weinberg distribution says that if alleles *A* and *a* occur in male and female gametes with frequencies p and q, then the frequencies of *AA*, *Aa*, and *aa* individuals which arise through random fertilization will be p^2, $2pq$, and q^2. Those who have learned some probability theory will recognize in this equation the following: If the probability that event 1 will be *A* is p, and if the probability that event 2 will be *A* is also p, then the probability that both events are *A* (= *AA*) equals p^2. Similar probability statements can be formulated to account for the expected frequencies of *Aa* and *aa* individuals.

The generalization developed by Kennedy goes as follows: "Consider a large, freely breeding population with s sexes for which the Mendelian model is valid, i.e., the production of a new individual requires the union of s gametes, one from each of the s sexes, such that each gamete carries a single gene for a particular locus." It is not necessary to follow the development of this model in detail; suffice it to say that gene frequencies remain constant from generation to generation, and that a given frequency distribution of zygotic types is regenerated each generation.

I am sure that many biologists will dismiss Kennedy's model as a useless intellectual exercise. Indeed, this is the charge most frequently leveled at an unusual model. (If a model describes a readily recognized situation, it is frequently not called a model. How many biologists would have used Kennedy's terminology "Mendelian model"?) Now, Kennedy's statements are mathematically correct; the model appears useless at first because we know of no "real life" situation that fits it. We do know, though, of organisms such as paramecia in which there are more than two sexes (see Sonneborn, 1947). The discrepancy between the model and life as we know it lies in the stipulated union of *s* gametes in the formation of a new individual; although many species of organisms possess several mating types, offspring arise from the union of *two* gametes, and two gametes *only*. Why? Suddenly we get a glimpse of the problems in communication that a species would encounter if individuals representing many sexes had to assemble for mating; as things now exist, encounters of single males and females require fantastically elaborate auditory, visual, and olfactory signals. We see, too, the new problems that would accompany meiosis in order that each gamete would contain one and only one haploid set of genes. We see problems in the successful development of several different sexes; the sex-determining mechanism would of necessity be an elaborate one. We see that many of the physiological aspects of fertilization that serve to restrict fertilization of an egg to a single sperm are precisely those which would make the utilization of a third gamete exceedingly difficult. In short, the model proposed by Kennedy is mathematically correct. It is a powerful one as well. And because it is powerful, we can ask why the simultaneous union of many gametes should be such an exotic notion. Why is fertilization limited to two gametes? The model makes us ask questions that we might otherwise never have raised.

Populations as Recorders of Information

The final model of the three we are discussing is a modification of one developed by Lewontin (1966) in asking whether nature is probable or capricious. Nature is capricious, according to Lewontin's definition, if repeated samples fail to increase our information regarding a given system. If we have an enormous vat of beans some of which are red and others white, and the two kinds are thoroughly mixed, the analysis of cumulative repeated samples of, say, fifty beans each will gradually lead us to better and better estimates of the true proportions of red and white beans. There is no guarantee that this is so; rather, it is highly probable that large numbers will give more accurate estimates of true frequencies than will small ones.

The key word in the description of samples in the preceding paragraph is "cumulative." Suppose that in recording the results of the repeated samples of beans, our data sheet limits us to a listing of the last ten samples only. Under this scheme, we know only as much as ten samples can tell us; "repeated" samples no longer help us gain better insight regarding the relative frequencies of red and white beans in the vat. Our recording is imperfect, because everything that happened eleven or more samples ago is forgotten. Our data-recording system has no substantial memory.

We can now regard the vat of beans as the source of environmental challenges confronting a population of organisms. The challenge actually encountered by a given generation is that represented by the proportions of red and white beans in a given draw. For simplicity we assume that the true proportions in the vat are 50 red : 50 white. (Lewontin uses random digits instead of proportions in his model; the average size of all digits from 1 through 9 equals 5.) We now say that within our hypothetical population there are two alleles, *A* and *a*, whose initial proportions p_0 and q_0 are 0.50 each. With each draw we alter p in the following way: (1) Calculate the difference between the proportion of red beans and p (negative if p is the larger of the two); (2) multiply the difference by either p or q, whichever is the smaller; and (3) add or subtract (according to sign) this product from the original value of p.

Several features of the above procedure are worth noting: First, the population responds poorly to the environmental challenge if p is close to 0 or 100 percent, because item (2) in the above product is small in these cases. Second, the population responds readily to each draw as long as p retains intermediate (0.30 to 0.70) values. Third, the fate of p depends upon the order of chance events in the individual draws. The same draws taken in reverse order do not necessarily bring a second population to the same final point, or over the same course, as the first; this uniqueness of the historical paths taken by different populations even to the same overall influences is illustrated in Figure 1-2.

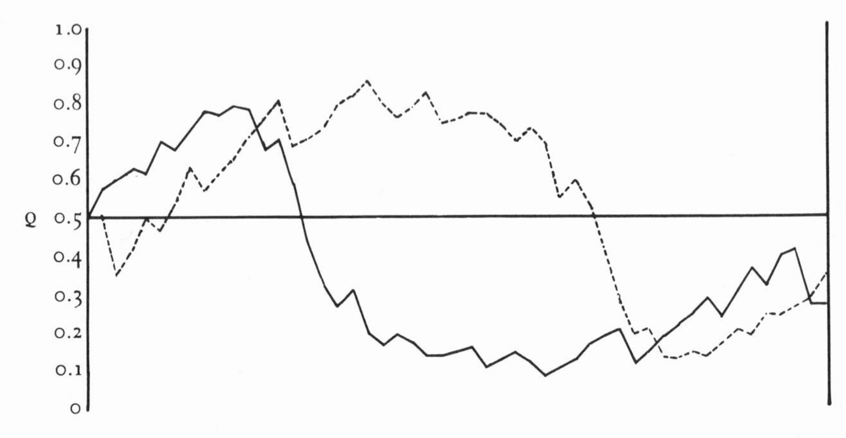

FIGURE 1-2. Gene frequencies (Q) of successive generations in a fluctuating environment. The solid line is a series of frequencies resulting from a random series of environmental conditions; the dashed line is the series of frequencies generated by the identical environments operating in the reverse order. The different courses taken by the two lines illustrate that evolutionary changes are historical changes; they depend upon the historical sequence of environments as well as upon the environments themselves. (After Lewontin, *BioScience, 16:* 25–27, 1966.)

For the present, the importance of this model is that the dependence of a population's current status upon its unique history has been illustrated through the use of a vat of beans or a table of random digits. The model has shown, too, how the past history of a population can determine the magnitude of its reactions to present events. A population in which *p* is nearly 0 or 100 percent does not "record" current observations very well. One point has by now been nearly forgotten. By no stretch of the imagination can a population be equated to a vat of beans or to a table of random digits. Rather, a clever model has served to illustrate some very complex—but very general—interactions between organisms and their environment, and the recording of these interactions in the genetic apparatus of the population.

2

DISPERSION

Population geneticists study populations. Except for those populations that are confined to small cages in the laboratory, most populations do not have sharp boundaries. Local populations, collections of individuals inhabiting the same small locality, generally merge into one another. Individuals continuously enter and leave. Even in the case of plant species where mature individuals do not move, pollen and seeds can be transported over considerable distances. In the present chapter we shall discuss dispersion patterns and, with these in mind, examine the notion of the local population. For example, a person might think that the dispersal mechanisms possessed by various organisms render the idea of a local population useless. We shall see that this is not so. As with much that follows in subsequent chapters, the bulk of the discussion will concern *Drosophila*—the workhorse of population genetics.

It makes no sense, obviously, to ask in connection with dispersal where one fly might go or where a single seed might be blown. A fly—one fly—can go anywhere. It can be carried by plane from one continent to another. Typhoons, hurricanes, or other violent storms can blow individual seeds or small animals such as insects tremendous distances. Migratory birds, as Darwin showed, carry a multitude of small organisms in the mud on their feet. The accidental transport of single individuals may account—and most likely has accounted in the past—for the introduction of species into new geographic areas. It may also account for the occasional introduction of new alleles into an already existing population. In population genetics, however, the study of rare, once in a lifetime (or, once in a geologic age) accidents is unrewarding. Truly rare accidents or unique events are capricious in precisely the sense Lewontin defined the term, and as we used it in Chapter 1. We can admit that such accidents occur. We can admit that some of the distribution patterns, and the compositions of some populations, that we see today must have had accidental beginnings. But we cannot predict where or when such capricious events will recur. Neither accidents nor capricious events lend themselves to investigation.

Rare accidents cannot be studied, but common accidents can be. One cannot predict in advance the movements or the final resting point of an individual fly; one can, however, describe the distribution of movements and the ultimate destinations of many flies. By releasing marked flies and by noting their positions when recaptured, one can describe the probability that a fly released at point x will be found, after a lapse of some time, at a distance y or beyond. Even though such statements are empirical at the moment, they can be made quite precise through the release of large numbers of flies. The behavior of individual flies is no more important in such studies than is the fate of individual atomic nuclei in the analysis of the exponential decay of radioactive elements.

Exceptionally thorough and systematic analyses of the dispersion of *Drosophila pseudoobscura* have been made by Dobzhansky and Wright (1943, 1947). The following account is based upon their data. The analytical procedure used here, however, is extremely simple in comparison with the original one. Consequently, *the original accounts should be consulted* by those seriously interested in the dispersal of organisms. A more detailed account of the present analysis has been reported previously (Wallace, 1966a); Wright (1968), in a critique of the empirical analysis, has pointed out some of its shortcomings.

The first studies of the dispersal of *D. pseudoobscura* were carried out on Mt. San Jacinto in Southern California during the summers of 1941 and 1942. Each experiment consisted of the release of several thousand adult flies (marked by the eye-color mutant, *orange*) at a given point on one day, and the subsequent daily recapture and release of these marked flies at regularly spaced traps. The flies—both marked and unmarked wildtype natives—that were captured were released at the point of capture. In this way the day by day distribution pattern of the released flies can be reconstructed from the numbers of marked flies recaptured at various distances from the point of release on different days. Furthermore, from the changing proportions of wildtype (native) to marked flies, one can estimate the mortality rate of *D. pseudoobscura* in Southern California. All in all, three experiments were carried out in each of which some 3000 marked flies were released and for which recaptures were made in traps set out in a cross-shaped pattern; the results of these experiments have been consolidated in the following analysis.

Our immediate problem is a simple one: Do the recapture data enable us to make a simple yet meaningful statement about the dispersal of flies from a point of release? Yes, they do. We have chosen a simpler model than the original one for describing the observed dispersal, and so our analysis differs from that given originally by Dobzhansky and Wright. In the original analysis, the distribution of recaptured flies is treated as a leptokurtic distribution—that is, as a rather peaked bell-shaped curve centered upon the point of release. Day by day the curve became lower at the peak (point of release) and more flattened as the flies spread farther and farther from the central trap of the cross-shaped experimental field.

Now, the distribution used by Dobzhansky and Wright is only one of many distribution curves that might possibly be used to describe the recapture data. We abandon it in the following discussion in favor of an empirically arrived at distribution that "fits" the data and can be represented as a straight line. According to this model, the logarithm of the number of marked flies recaptured at various distances from the point of release decreases linearly with the square root of distance. The linearity of the distribution makes it an easy one to compare with experimental data; its exponential nature makes it amenable to additional manipulations. In his critique, Wright (1968) suggests that variations in the results of the separate experiments are responsible for the (fortuitous?) linearity of the present analysis.

The consolidated recapture data for *D. pseudoobscura* for the first four days following release are listed in Tables 2-1 through 2-4. The data have been consolidated by adding together the three separate experiments and by pooling as well the four arms of the cross-shaped collection fields. Table 2-5 gives the total recapture data lumped not only through the different experiments and different arms of the collection fields but also through all days (a week or more) during which the experiments lasted; this consolidation yields what we call the "lifetime" dispersal data. Figures 2-1 and 2-2 illustrate the "goodness" of the fit of data from Tables 2-1, 2-2, and 2-5 to curves constructed from them according to our simplified model. Because they represent the consolidation of numerous daily distributions that differ in their dispersion about the point of release, the lifetime recapture data also fit the empirical model (Figure 2-2). This is fortunate because flies taken from a single collection at a given spot represent a cross section of flies of different ages. These will have had different opportunities to move from their places of birth. The reconstruction of probable origins of flies in such a sample can be made most easily from the lifetime dispersal data.

Do all *Drosophila* flies disperse equally rapidly following release? A number of studies have been made using several *Drosophila* species. The first of these (Timofeeff-Ressovsky and Timofeeff-Ressovsky, 1940a, 1940b, 1940c) involved *D. funebris*; probably because of the choice of

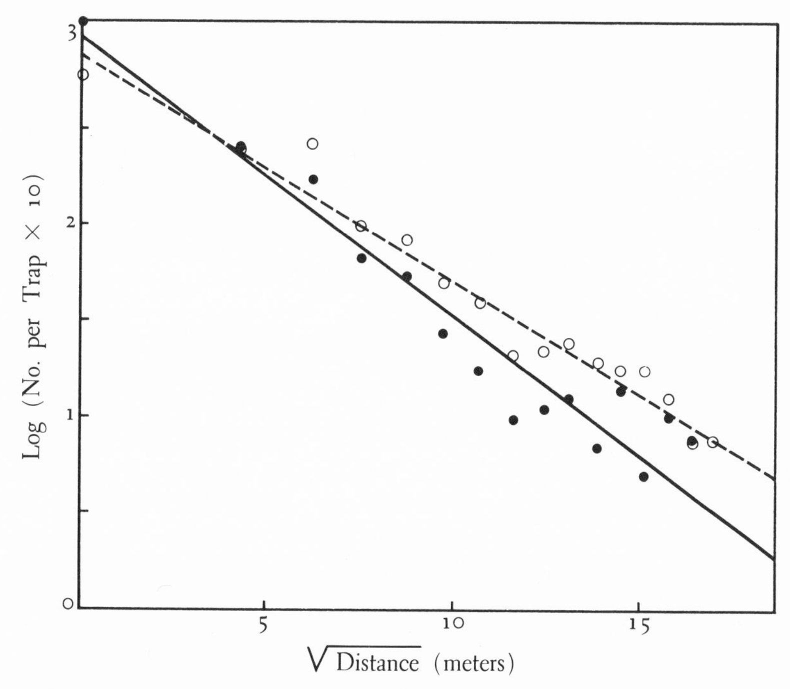

FIGURE 2-1. Recapture of experimentally released *D. pseudoobscura.* Abcissa: square root of distance from point of release. Ordinate: logarithm of the number of recaptured flies. Solid line and closed circles: first day after release. Dashed line and open circles: second day after release. (After Dobzhansky and Wright, 1943; Wallace, 1966a.)

the release point (the center of a village refuse heap), the dispersal observed for these flies was very small. The dispersal *pattern*—except for the actual distances traveled—was the same as that described for *D. pseudoobscura*; the logarithm of numbers of marked flies recaptured declined linearly with the square root of distance from the point of release.

The data reported by Timofeeff-Ressovsky (Table 2-6) include the total numbers of flies recaptured over several days; consequently, these data are comparable to the lifetime dispersal data listed in Table 2-5. In both tables the slope of the regression of the logarithm of the number on square root of distance is given. The slope of the regression is not the most convenient measure for expressing dispersion since it describes the change in the logarithm of the number of flies recaptured as a function of square-root distance—a rather difficult concept to grasp. The reciprocal of the slope, on the other hand, can be used to estimate distances from the points of release (or from any other point, as a matter of fact) at which the numbers of recaptured flies have dropped by

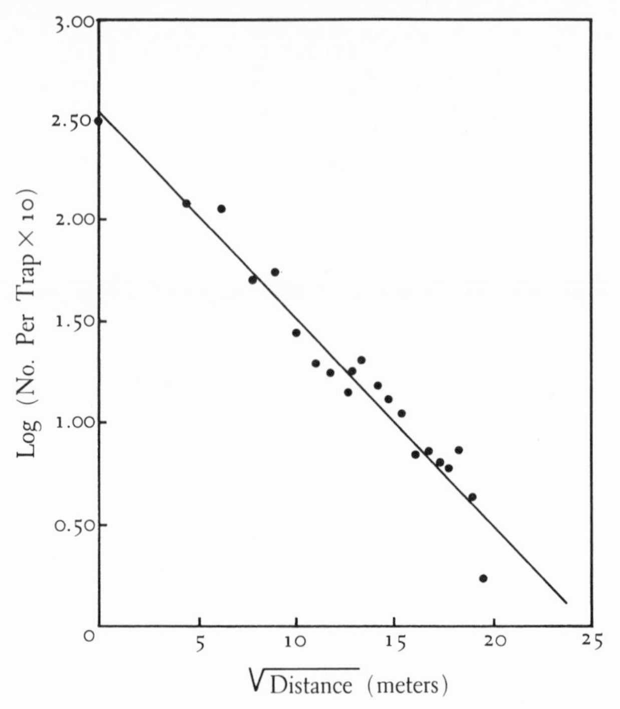

FIGURE 2-2. "Lifetime" dispersal of flies. Circles represent data listed in Table 2-5; the line has been fitted to these data.

90 percent; unit changes on a logarithmic scale reflect tenfold differences in actual numbers. The measure of distance is still given in root distance, but elementary algebra tells us that for $\log N_2 - \log N_1 = 1$, $d_2^{1/2} = d_1^{1/2} + x$, where x is the reciprocal of the slope and d_2 and d_1 are arbitrary distances from the point of release. Squaring both sides of the above equation gives $d_2 = d_1 + 2xd_1^{1/2} + x^2$.

Now, if d_1 is taken as zero (point of release), we see that the total number of recaptured flies has decreased by 90 percent (only 10 percent as many as at the center trap) in approximately 100 meters in the case of *D. pseudoobscura* but in some 5 meters in the case of *D. funebris*. This is indeed a striking difference; the flies released by Timofeeff-Ressovsky scarcely moved from the point of release. Data on a third species, *D. willistoni*, are presented in Table 2-7. Once more the distribution pattern fits the model we have been using. In this case, traps some 15 to 20 meters from the point of release yield only 10 percent as many flies as does the center trap (zero distance from the point of release). Finally, Dobzhansky and Wright (1943) have described a small experiment using *D. melanogaster*. The actual data are not given in

TABLE 2-1

COLLECTION DATA FOR THE FIRST DAY FOLLOWING THE RELEASE OF MARKED DROSOPHILA PSEUDOOBSCURA.

(Based on the first three experiments of Dobzhansky and Wright 1943.) From the left, the columns list (1) distance (meters) of traps from point of release, (2) total number of released flies recaptured, (3) number of traps exposed, (4) average number of flies per trap, (5) weighting factor used in computing regression, (6) log of average ($\times$ 10, to avoid negative values), and (7) square root of distance. Two additional values are listed at the bottom of the table: b, the slope of the regression of the logarithm of the number of flies recaptured on the square root of distance from point of release, and s_b, the error of the slope.

DISTANCE	FLIES	TRAPS	AVERAGE	w_i	LOG(AV. $\times$ 10)	(DISTANCE)$^{1/2}$
0	357	3	119.0	2	3.08	0
20	318	12	26.5	3	2.42	4.47
40	224	12	18.7	3	2.27	6.32
60	85	12	7.1	3	1.85	7.75
80	67	12	5.6	3	1.75	8.94
100	33	12	2.8	3	1.45	10.00
120	21	12	1.8	3	1.26	10.95
140	12	12	1.0	3	1.00	11.83
160	13	12	1.1	3	1.04	12.65
180	15	12	1.3	3	1.11	13.42
200	8	12	0.7	3	0.85	14.14
220	18	12	1.5	3	1.18	14.83
240	2	4	0.5	1	0.70	15.49
260	4	4	1.0	1	1.00	16.13
280	3	4	0.8	1	0.90	16.73
300	0	4	0	1	—	17.32

$b = -0.1447$ $s_b = 0.0115$

their report, but the description of the results suggests that *D. melanogaster* disperses in a manner similar to that of *D. willistoni*; its rate of dispersal is less than that of *D. pseudoobscura* but greater than that of *D. funebris*.

The experiments we have been discussing, we should recall, have not dealt with the movements of members of natural populations. In each experiment, relatively large numbers of flies were released in an already inhabited locality; those which were released bore markings of one sort or another so that they could be distinguished from the natives. Is it possible to get evidence from these studies on the effect overcrowd-

TABLE 2-2

COLLECTION DATA FOR THE SECOND DAY FOLLOWING THE RELEASE OF MARKED DROSOPHILA PSEUDOOBSCURA.

(Based on the first three experiments of Dobzhansky and Wright 1943.) See Table 2-1 for further explanation; column (7) of Table 2-1 has been omitted.

DISTANCE	FLIES	TRAPS	AVERAGE	w_i	LOG(AV. × 10)
0	193	3	64.3	2	2.81
20	288	12	24.0	3	2.38
40	328	12	27.3	3	2.44
60	112	12	9.3	3	1.97
80	106	12	8.8	3	1.94
100	63	12	5.3	3	1.72
120	49	12	4.1	3	1.61
140	25	12	2.1	3	1.32
160	27	12	2.3	3	1.36
180	29	12	2.4	3	1.38
200	25	12	2.1	3	1.32
220	21	12	1.8	3	1.26
240	14	8	1.8	1	1.26
260	10	8	1.3	1	1.11
280	6	8	0.8	1	0.90
300	6	8	0.8	1	0.90

$b = -0.1160 \qquad s_b = 0.0086$

ing might have on dispersal rates? Some information of this sort can be inferred from the data but it is a by-product of the experiment.

Dubinin and Tiniakov (1946), using a technique different from that of Timofeeff-Ressovsky, have suggested that the dispersal rates obtained by him are much too small for *D. funebris.* Dubinin used a naturally occurring inversion as a marker for wildtype flies; Timofeeff used laboratory mutants. Dubinin released his flies in what seems to be a fairly open countryside (Figure 2-3); Timofeeff apparently chose a refuse dump. Dubinin released over 100,000 flies; Timofeeff released only several thousand. Dubinin searched for his marker inversion in adult flies which may or may not have been those actually released (the inversion was not normally present in the experimental locality so that individuals carrying the inversion were either released flies or their descendants); Timofeeff recaptured the released individuals themselves.

About one month after the time of release, Dubinin and Tiniakov found that 10 to 15 percent of all chromosomes carried by flies captured

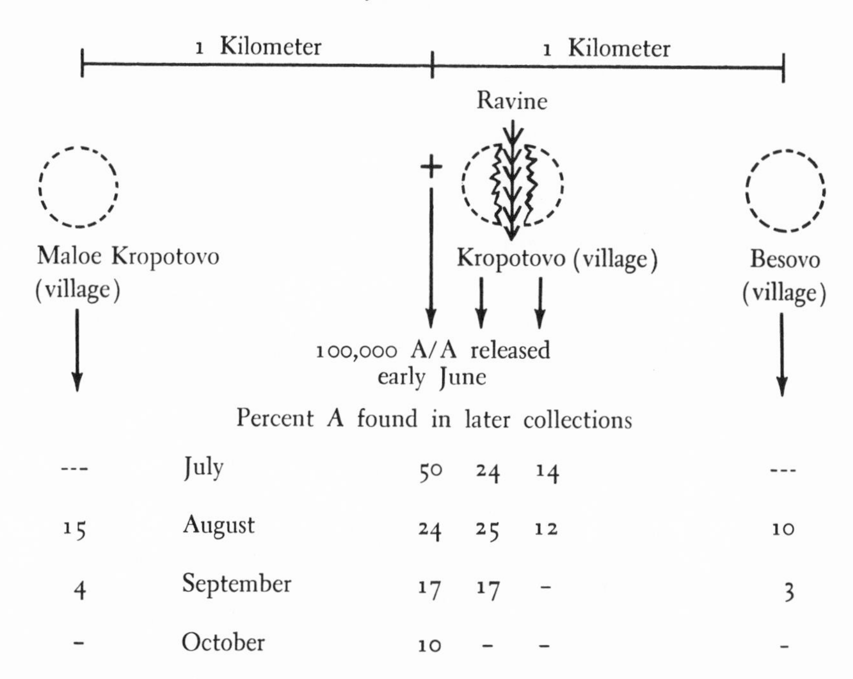

---	July	50	24	14	---
15	August	24	25	12	10
4	September	17	17	-	3
-	October	10	-	-	-

FIGURE 2-3. Dispersal of *D. funebris.* (After Dubinin and Tiniakov, 1946.)

one kilometer from the point of release possessed the inverted gene order that they had chosen as a genetic marker. Clearly, the marker in this case had spread much more rapidly than one would have predicted from Timofeeff's data, according to which the number of flies recaptured some five meters from the point of release was only one tenth the number recaptured at the point of release itself. Perhaps the mutations chosen by Timofeeff as markers diminished the mobility of their carriers. Perhaps, too, there was small incentive to stray from the attractive point of release. Finally, though, 100,000 flies may so overcrowd an area around the point of release that rates of dispersal are greatly exaggerated.

Data on the recapture of flies suggest that 3000 released flies gave exceptionally high initial dispersal rates in *D. pseudoobscura.* Dobzhansky and Wright (1943) found that the released flies scattered more widely on the first than on subsequent days; they also suggest that under generally unfavorable conditions flies might seek an escape by dispersing. Thus, although we may justifiably wonder about the extremely small dispersal found by Timofeeff for *D. funebris,* we can be reasonably sure that, merely because released flies overcrowd the experimental field, the technique of releasing and recapturing flies gives an exaggerated estimate of normal dispersal rate.

TABLE 2-3

COLLECTION DATA FOR THE THIRD DAY FOLLOWING THE RELEASE OF MARKED DROSOPHILA PSEUDOOBSCURA.

(Based on the first three experiments of Dobzhansky and Wright 1943.) See Tables 2-1 and 2-2 for further explanation.

DISTANCE	FLIES	TRAPS	AVERAGE	w_i	LOG(AV. × 10)
0	60	3	20.0	2	2.30
20	183	12	15.3	3	2.18
40	172	12	14.3	3	2.16
60	73	12	6.1	3	1.79
80	87	12	7.3	3	1.86
100	43	12	3.6	3	1.56
120	30	12	2.5	3	1.40
140	39	12	3.3	3	1.52
160	30	12	2.5	3	1.40
180	36	12	3.0	3	1.48
200	27	12	2.3	3	1.36
220	27	12	2.3	3	1.36
240	10	10	1.0	1	1.00
260	1	10	0.1	1	0
280	5	10	0.5	1	0.70
300	5	8	0.6	1	0.78
320	4	6	0.7	1	0.85
340	2	6	0.3	1	0.48
360	1	6	0.2	1	0.30

$b = -0.1031$ $\quad s_b = 0.0120$

The model that we have used to understand the lifetime dispersion of individuals within populations of *Drosophila* is one in which the logarithm of the number of released flies is said to decrease linearly with the square root of distance from the point of release. The model is useful because it enables us to use a straight line rather than a curve in representing experimental data. Nevertheless, flies (not logarithms of numbers of flies) make up a *Drosophila* population and when these flies move, they move certain distances, not square roots of distances. The relation between the model and actual populations may be visualized more easily by transforming the straight line of the model into a curve relating the numbers of flies to distance (rather than the logarithm of numbers of flies to square root distance) as in Figure 2-4. The striking feature of the transformed curve is the extent to which it successfully hugs both axes of the graph. The tall vertical portion reflects the seden-

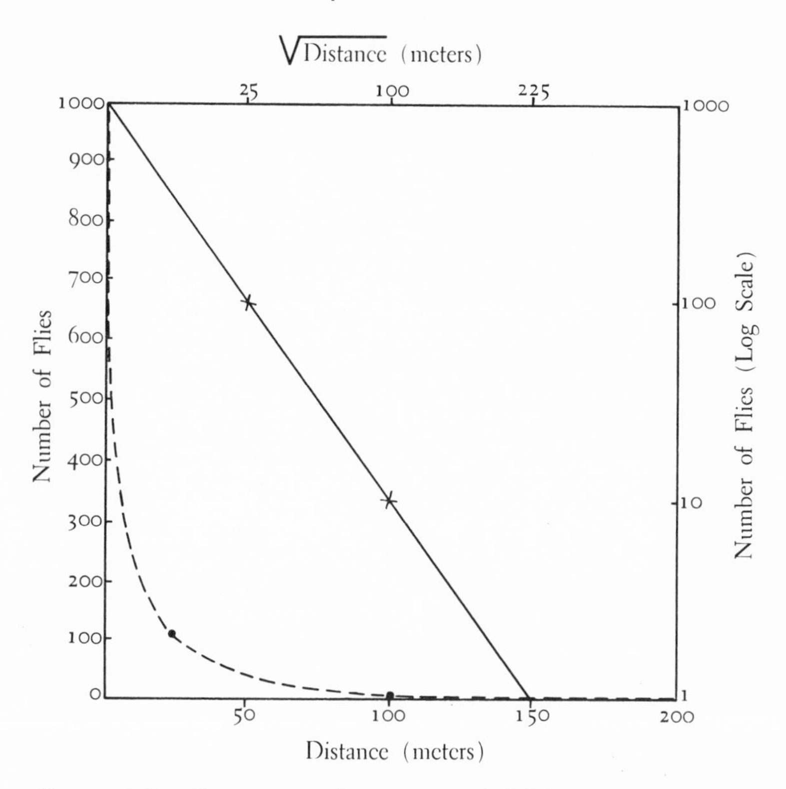

FIGURE 2-4. Comparison of two curves. Solid line: a straight line plotted so that the abcissa represents the square root of distance and the ordinate represents the logarithm of the number of flies. Dashed line: the curve obtained by transforming the abcissa and ordinate into linear measures of distance and number.

tary nature of the bulk of the released flies. The long horizontal tail reflects the great distances to which released flies move *if they move at all.* These two features—the tall spike and the horizontal tail—suggest the following: Within a population of *Drosophila* the majority of all adult individuals found at one spot are "natives" in the sense that they have emerged in the immediate neighborhood of that spot. A few individuals found at any one spot, on the other hand, are immigrants; if an individual is an immigrant, then it is very nearly as likely that he came from a relatively great distance as from a medium one.

Within a steady-state population (or series of adjacent populations) distributed uniformly over a large area, emigrants that leave a local population and migrate a certain distance must be replaced in the long run by an equal number of immigrants that have come from the very same distance. If this were not so, certain spots on the landscape would

TABLE 2-4

COLLECTION DATA FOR THE FOURTH DAY FOLLOWING THE RELEASE OF MARKED DROSOPHILA PSEUDOOBSCURA.

(Based on the first three experiments of Dobzhansky and Wright 1943.) See Tables 2-1 and 2-2 for further explanation.

DISTANCE	FLIES	TRAPS	AVERAGE	w_i	LOG(AV. × 10)
0	46	3	15.3	2	2.18
20	92	12	7.7	3	1.89
40	104	12	8.7	3	1.94
60	51	12	4.3	3	1.63
80	77	12	6.4	3	1.81
100	37	12	3.1	3	1.49
120	25	12	2.1	3	1.32
140	31	12	2.6	3	1.42
160	12	12	1.0	3	1.00
180	27	12	2.3	3	1.36
200	18	12	1.5	3	1.18
220	10	12	0.8	3	0.90
240	14	10	1.4	1	1.15
260	3	10	0.3	1	0.48
280	10	10	1.0	1	1.00
300	4	8	0.5	1	0.70
320	2	6	0.3	1	0.48
340	2	6	0.3	1	0.48
360	2	6	0.3	1	0.48

$b = -0.0940 \qquad s_b = 0.0089$

accumulate more and more individuals at the expense of others, until all individuals would finally be found in these few spots.

Because immigrants must on the average be identical in number and in distance traveled to emigrants under equilibrium conditions, the dispersal data can be used to reconstruct the composition of a local population. Suppose, for example, that I collect one hundred *Drosophila* in a single trap. What can I say regarding the origins of these flies? To answer this question, it is necessary to understand that the dispersal data discussed so far represent cross sections of actual dispersal patterns; flies disperse in a two-dimensional space in all directions, while we have determined numbers that moved certain distances along a single linear transect. To reconstruct the entire dispersal pattern, it is necessary to rotate our curve (preferably the transformed curve of Figure 2-4) around its vertical axis. When we rotate it, it generates a surface resem-

TABLE 2-5

TOTAL COLLECTION DATA FOR THE FIRST THREE EXPERIMENTS OF DOBZHANSKY AND WRIGHT (1943).

(The average numbers of flies captured per day in the most distant traps is exaggerated, because many of these were not set until it seemed likely that flies had spread that far.) These data are referred to in the text as "lifetime" dispersal data because they are equivalent to the expected spatial distribution of flies 1, 2, 3, . . . days old about a given source.

DISTANCE	FLIES	TRAP DAYS	AVERAGE	LOG(AV. × 10)
0	728	23	31.65	2.500
20	993	82	12.11	2.083
40	956	82	11.66	2.067
60	412	83	4.96	1.695
80	464	84	5.52	1.742
100	230	84	2.74	1.438
120	163	84	1.94	1.288
140	148	84	1.76	1.246
160	118	84	1.40	1.146
180	173	84	2.06	1.314
200	125	84	1.49	1.173
220	109	84	1.30	1.114
240	65	58	1.12	1.049
260	40	58	0.69	0.839
280	42	58	0.72	0.857
300	28	44	0.64	0.806
320	15	25	0.60	0.778
340	18	25	0.72	0.857
360	11	25	0.44	0.643
380	1	6	0.17	0.230
		$b = -0.1020$	$s_b = 0.0052$	

bling a sharply peaked witch's hat with an extremely broad, flat brim. The geometric figure obtained by rotating the curve about the vertical axis represents the lifetime dispersal pattern of released flies spreading in all directions. It serves as a basis for reconstructing the origins of flies that inhabit a single locality. The total number of flies that have migrated from a point to a certain distance or beyond must be replaced by an equal number of which have entered the population from that distance or greater ones. This argument has been used to construct the curves shown in Figure 2-5; in this figure we see the proportion of all individuals inhabiting a single spot at a given time which have come to that spot from various distances. The remarkable feature of the figure is what

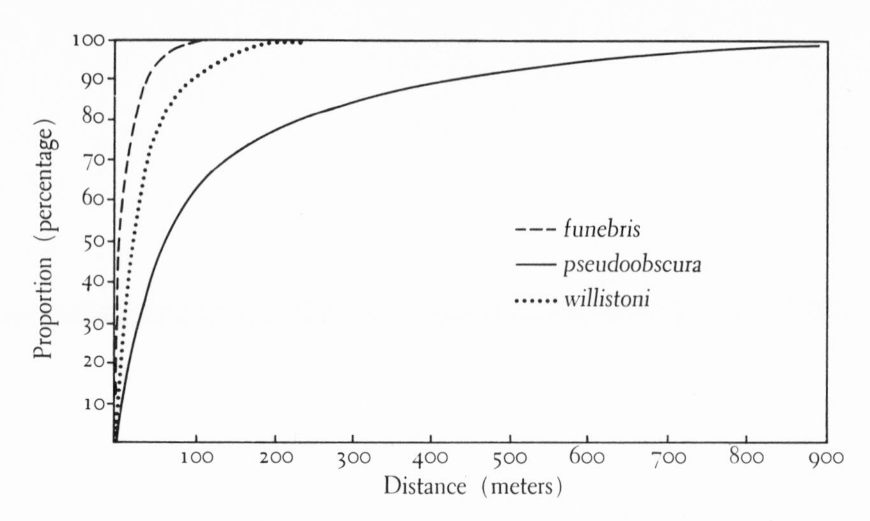

FIGURE 2-5. Reconstruction of the composition of local collections of flies based on dispersal data. The curves represent the proportion of flies in each collection that has arisen within a given distance of the collecting site. (After Wallace, 1966a.)

TABLE 2-6

COLLECTION DATA FOR D. FUNEBRIS TABULATED FROM TIMOFEEFF-RESSOVSKY (1940, FIG. 31).

See Table 2-1 for further explanation. Column 5 has been omitted (no weighting factor was used in analyzing these data). In contrast to Tables 2-1 through 2-4, these data consist of all flies caught throughout the course of the recapture experiment.

DISTANCE	FLIES	TRAPS	AVERAGE	LOG(AV. × 10)	(DISTANCE)$^{1/2}$
12.0	36	4	9.00	1.95	3.46
17.0	27	4	6.75	1.83	4.12
24.0	10	4	2.50	1.40	4.90
26.8	18	8	2.25	1.35	5.18
33.9	5	4	1.25	1.10	5.82
36.0	5	4	1.25	1.10	6.00
37.9	3	8	0.38	0.58	6.16
43.3	8	8	1.00	1.00	6.58
48.0	0	2	0	—	6.93
49.5	3	4	0.75	0.88	7.04
50.9	0	8	0	—	7.14
53.7	1	4	0.25	0.40	7.34

$b = -0.4563 \qquad s_b = 0.0707$

we anticipated above: A substantial proportion of all flies inhabiting a given spot have emerged very near that spot; for example, about 25 percent of *D. pseudoobscura* or 60 percent of *D. willistoni* that are captured at a single trap are likely to have arisen within a radius of some 25 meters. At the other end of the scale, though, the curves flatten

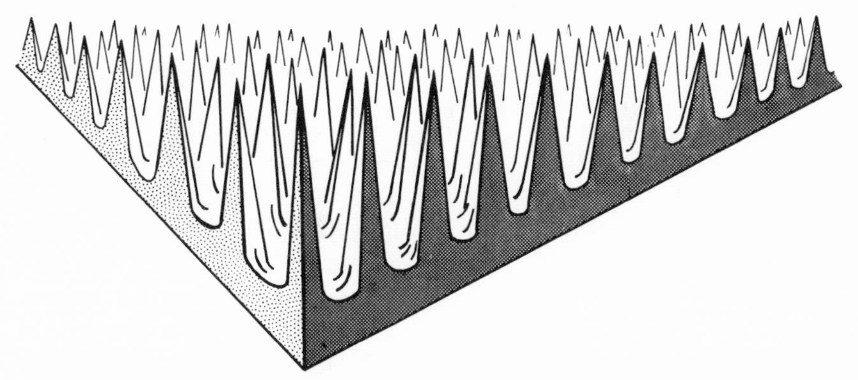

FIGURE 2-6. Attempt to represent the very slight dispersal of flies under normal circumstances and hence to emphasize the local nature of local populations. This figure is referred to in the text as the "bed-of-nails" view of local populations.

TABLE 2-7

COLLECTION DATA FOR D. WILLISTONI FROM BURLA ET AL. (1950).

See Table 2-1 for further explanation; column 5 has been omitted (no weighting factor was used in analyzing these data). In contrast to Tables 2-1 through 2-4, these data consist of flies caught at various times (1 to 3 days) after release.

DISTANCE	FLIES	TRAPS	AVERAGE	LOG(AV. × 10)	(DISTANCE)$^{1/2}$
0	334	4	83.5	2.92	0
10	296	16	18.5	2.27	3.16
20	165	16	10.3	2.01	4.47
30	134	16	8.4	1.92	5.48
40	43	16	2.7	1.43	6.32
50	44	16	2.7	1.44	7.07
60	25	16	1.6	1.20	7.75
70	19	16	1.2	1.08	8.37
80	8	12	0.7	0.85	8.94
90	7	12	0.6	0.78	9.49
100	0	12	—	—	10.00

$b = -0.2306 \qquad s_b = 0.0105$

markedly as they approach 100 percent; although the bulk of any local population may have arisen in the immediate neighborhood, a few individuals have come from elsewhere. Since migrants move great distances nearly as often as lesser distances, it is difficult to say just where they have come from.

A freehand and rather impressionistic representation of local populations is shown in Figure 2-6. A patchy environment would exaggerate rather than diminish the bed-of-nails aspect of the landscape drawn here. In this picture the "nails" represent the natives that make up each local population, while the "board" from which the nails project represents the cosmopolitan individuals whose origins and ultimate destinations are really unknown. Local populations, to an extent that is remarkable indeed, are effectively isolated from one another as a result of dispersion patterns alone.

So far we have concentrated on *Drosophila* populations. What of other organisms? Few have been studied as thoroughly as have those of the genus *Drosophila*. A good review of data on many different organisms has been prepared by Wolfenbarger (1946). Although it is not obvious that all species discussed in this review disperse in a manner compatible with the model we have used, a large number seemingly do disperse as we have described. Bateman (1947a, 1947b, 1947c, 1950) has also dealt with problems of dispersal in considerable detail.

Marriage patterns in human populations can be thought of as indicators of the dispersal of human beings. In the case of *Drosophila* we assume that flies which are found at a given spot are able to mate and, in fact, that they actually do mate. Consequently, if we know something about the birthplaces of flies found in that spot, we assume that we also know something about the mating of flies in reference to points of origin. The simplest statement about this relationship for flies may be made as follows: Here is an individual male that has hatched at a given spot; his potential mates, in reference to points of origin, have frequencies described by our model of dispersal as shown in Figure 2-5. In the case of human populations, data are obtained by an examination of marriage licenses on each of which the birthplaces of the couple are listed. From these one can determine, as Cavalli-Sforza (1959) has done, the probability of marriage as a function of distance between the birthplaces of husband and wife (Figure 2-7). Except for the higher than expected proportion of marriages involving persons whose birthplaces are separated by zero distances (born in the same parish), the data for marriages fit the model we have used for flies extremely well. In the case of the data reported by Cavalli-Sforza, the reciprocal of the slope (distance measured in kilometers) is about 1.75. That is, the probability of occurrence for a marriage is only 10 percent as great if the couple are born some two kilometers apart than if they are born in the same village. People tend to marry persons whom they meet frequently and consequently know well; as a rule, these are persons who live nearby.

If, in a study of local populations of *D. willistoni,* we were to overwhelm our data with the results of matings between flies captured at 20 meters or more from one another, we would be as misled by our results as if we were to study early human populations by limiting our observations to marriages between persons whose birthplaces were at least two kilometers distant. In each case we would discard some 90 percent of all matings (or marriages) and would restrict our attention to the 10 percent that involve individuals from relatively isolated populations. Perhaps this comparison demonstrates as vividly as any that the usual dispersion of individuals does not destroy the very local nature of local populations.

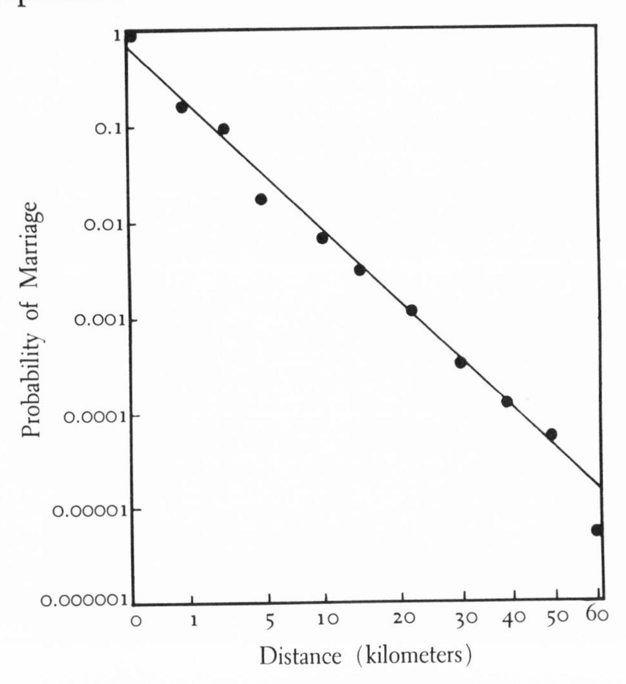

FIGURE 2-7. Matrimonial migration in man. This curve is analogous to that in Figure 2-5; the logarithm of the probability of marriage decreases linearly with the square root of the distance between the birthplaces of the marriage partners. (After Cavalli-Sforza, 1959.)

3

DETECTING AND MEASURING CONCEALED VARIABILITY

Fundamentally, the population geneticist is concerned with gene frequencies, zygotic frequencies, and the comparison of these with some preconceived hypothesis. The hypothesis may state that zygotic ratios are related to gene frequencies as the Hardy-Weinberg equilibrium predicts; the comparison of observed and expected frequencies would be made by means of a chi-square or other appropriate statistical test. Depending upon the results of this comparison, the hypothesis would be either accepted or rejected. Alternatively, the hypothesis may say that two populations have the same gene frequencies, the same zygotic frequencies, or both; appropriate procedures exist for testing this hypothesis. Or, again, the hypothesis may say that gene frequencies have remained constant from one generation to another as the Hardy-Weinberg equilibrium also predicts; this hypothesis, too, is testable once gene frequencies have been determined.

A majority of mutations that are useful in population studies are recessive. Only the recessive homozygotes can be recognized by an inspection of the individual members of a population. Now, if a recessive allele has a frequency q, the frequency of individuals homozygous for this allele will be q^2. Only a fraction q of the q recessive alleles can be detected by inspection of the phenotype; the remainder, $1 - q$, are concealed. If q is small, $1 - q$ is nearly 1.00; nearly all rare recessive alleles are hidden from sight.

To see what lies beneath the outward uniformity of a population, the geneticist must obtain individuals homozygous for those alleles that are hidden from view. He inbreeds his experimental material. And, by inbreeding, he attempts to obtain an estimate of q instead of q^2—of *gene* frequency rather than of *zygotic* frequency. In one form or another, whether it is brought about by sophisticated mating procedures in *Drosophila,* by artificial diploidization of haploid plants, or by consanguinity as in man, inbreeding is the procedure depended upon by population geneticists to reveal otherwise hidden mutations.

In this chapter, we shall describe some of the standard procedures used by Drosophila geneticists in studying the genetics of populations. A description of procedures without a corresponding description of experimental data would be highly artificial; the data themselves and some of the statistical procedures used in their analysis will also be described. And, as we shall see, populations do indeed have a storehouse of concealed variability.

Brother-Sister Matings

A commonly used procedure for inbreeding in cross-fertilizing experimental organisms is the mating of brother with sisters. Figure 3-1 illustrates

TABLE 3-1

DISTRIBUTION OF MUTANTS RECOVERED AMONG THE F_2 PROGENY OF 736 PAIR MATINGS OF WILD D. MULLERI.

Visible mutations revealed in this manner represent a relatively small fraction of the total number of mutations present in the sampled population. (Spencer, 1957, Table 3.)

	NUMBER OF MUTANTS OBSERVED						TOTAL CULTURES
	0	1	2	3	4		
Number of cultures	513	189	28	6	0		736
Average number of mutants per culture						0.36	
Variance						0.36	

the entire procedure as it might be used for virtually any animal species. Females—flies, for example—taken from a natural population are placed individually into culture bottles. A number of F_1 sons and daughters of each parental culture are used as parents for a corresponding F_2 culture. If either the original female or her unknown mate was heterozygous for a rare visible (recessive) mutation v, then one half of the F_1 progeny would be heterozygous for this gene. Consequently, one fourth of the matings between F_1 brothers and sisters would involve heterozygotes, Vv, of each sex; one fourth of their progeny—one sixteenth of the total progeny of a mass culture—would then be homozygous, vv.

The investigator himself sets the limits to the effectiveness of this simple procedure; the data obtained (see Table 3-1 for one such experiment) depend upon his ability to detect visible mutations in his material. Individual investigators vary so greatly in this ability that comparisons between studies made by different persons are not reliable. Theoretically, one could test F_2 individuals singly for fertility on the assumption that some rare recessive mutations cause sterility; unfortunately, the labor required would be excessive. Furthermore, if the species were an unfamiliar one, culture conditions might be so unsatisfactory as to swamp in a mass of barren crosses the sterile cultures expected on theoretical grounds.

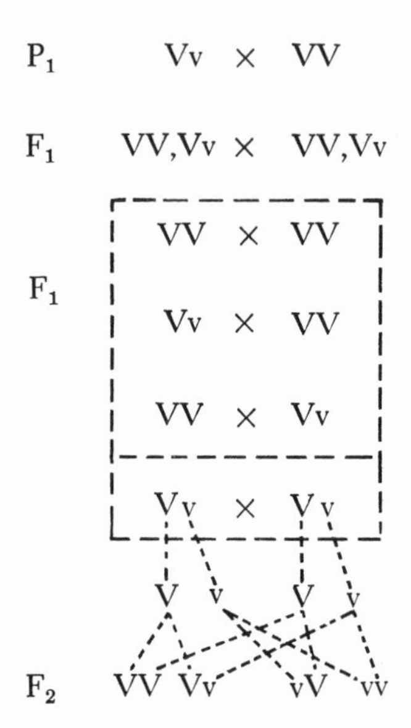

FIGURE 3-1. Breeding procedure for the extraction of visible mutations (v) from wild populations of *Drosophila*. (After Spencer, 1947.)

Modified *ClB* Procedures

In the case of many *Drosophila* species, standard procedures exist for rendering individuals homozygous for virtually all genes carried by an entire chromosome. A variety of terms are used in different laboratories to describe the process as a whole; "modified *ClB*" is a formal phrase that gives credit to the original technique of this sort, the one devised by H. J. Muller for use in his early studies of mutation in *D. melanogaster*.

Figure 3-2 represents the *ClB*-type crosses used in an early analysis

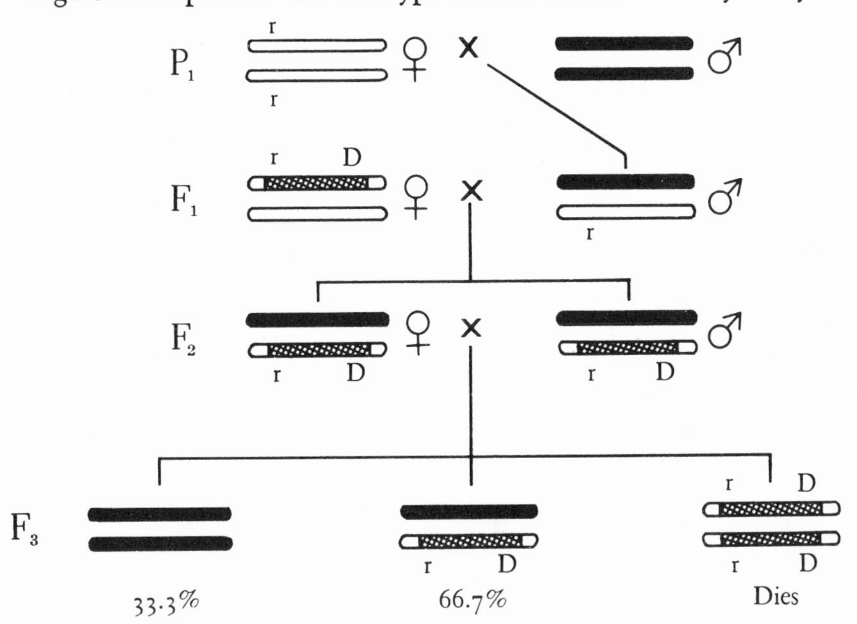

Figure 3-2. Series of crosses used to obtain flies homozygous for wildtype chromosomes in *D. pseudoobscura*. Black: chromosomes from natural population; white: chromosomes carrying recessive (*r*) or dominant (*D*) genetic marker genes; cross-hatched: inverted sections that suppress crossing over. (After Dobzhansky and Spassky, 1953.)

of *D. pseudoobscura*; except for the actual marker genes used, the procedure is nearly identical to that used in the analysis of any autosome in other *Drosophila* species. (Figure 3-3 illustrates a comparable test for chromosome two of *D. melanogaster*.)

Wildtype males taken from the population to be studied are crossed individually with virgin females homozygous for *r*, a recessive mutation. Sons, necessarily +/*r*, are mated individually—one son for each parental male—with *rD*/*r* females. Males and females showing *D* (a dominant mutation) but not *r* (and, consequently, *rD*/+) are picked from each

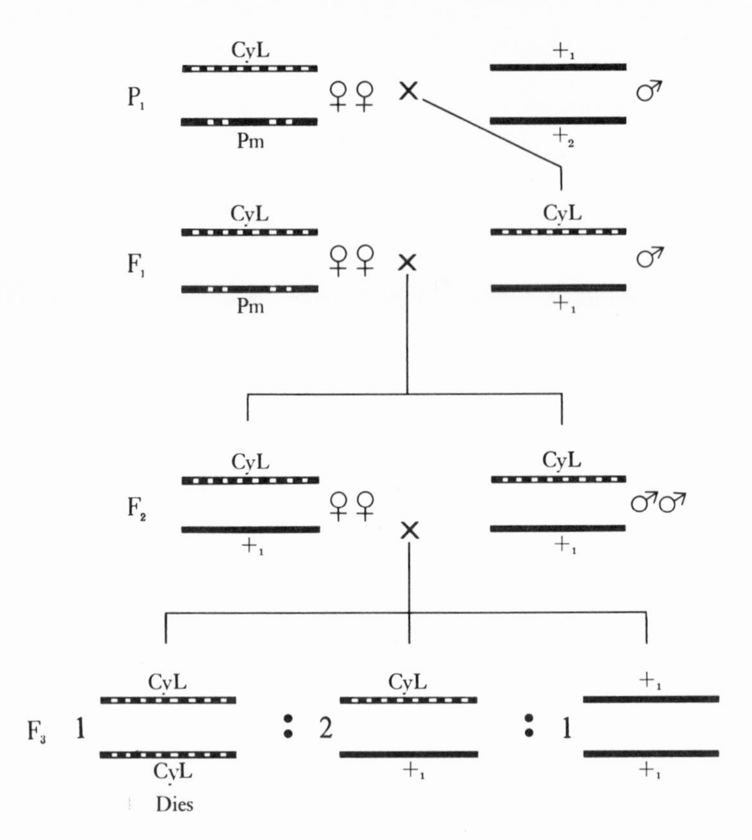

FIGURE 3-3. Series of crosses used to obtain flies homozygous for wildtype chromosomes in *D. melanogaster*. Wildtype chromosomes are identified $+_1$ and $+_2$; chromosomes carrying crossover suppressors are genetically marked with the dominant mutations *Curly* (*Cy*), *Lobe* (*L*), and *Plum* (*Pm*). (After Wallace and King, 1951.)

F_2 culture and are mated, brother with sister. One quarter of each F_3 progeny consists of *rD*/*rD* homozygotes; these individuals die shortly after hatching because the dominant genes used in these procedures are generally lethal when homozygous. Of the remaining offspring, two thirds are heterozygous for the dominant marker gene (*rD*/+) and can be recognized. The remaining one third (that is, one fourth of the original four fourths) are homozygous for the wildtype chromosome. They are homozygous, then, for the various alleles found at the different loci on this chromosome. If a recessive lethal mutation were to occupy any locus on the tested chromosome, the expected homozygous class would fail to appear in the F_3 culture; a culture of this sort would consist entirely of flies heterozygous for the dominant marker gene. A good many cultures are, in fact, of just this sort.

The classification of the flies in a culture bottle into two phenotypic classes—those heterozygous for a dominant marker *D*, and those lacking this marker—does not require particularly sharp eyesight or any other special ability on the part of the investigator. Thus, analyses of the sort described can be made by different persons and their results will be comparable. Furthermore, the analysis need not be limited to lethal genes alone—those which permit fewer than 10 percent of the expected number of homozygous flies to survive, for example. On the contrary, recessive viability mutations can be defined as mutations that alter (lower, for the most part) the expected frequency of the homozygous class. Accordingly, information on the entire spectrum of viability effects of the studied chromosomes can be obtained from data which include the frequency of homozygous individuals in each F_3 culture. Table 3-2 lists

TABLE 3-2

Sample Data Showing the Effects of Different Chromosomes and Their Heterozygous Combinations on the Observed Frequency of Wildtype Flies in Tests Made by the *CyL* Technique Illustrated in Figure 3-3.

Note that each heterozygous combination (2/3, for example) carries the chromosomes tested immediately above and below in homozygous condition (2/2 and 3/3). (Wallace, unpublished data.)

TYPE OF TEST	CHROMOSOME DESIGNATION	WILD	NO. OF FLIES TOTAL	% WILD
Homozygous	1/1	0	236	0
Heterozygous	1/2	*126*	*368*	*34.2*
Homozygous	2/2	0	274	0
Heterozygous	2/3	*118*	*361*	*32.7*
Homozygous	3/3	113	380	29.7
Heterozygous	3/4	*105*	*315*	*33.3*
Homozygous	4/4	80	236	33.9

the results for several cultures observed in this type of study; Figures 3-4 and 3-5 illustrate frequency distributions prepared from comparable data.

In passing it may be noted that the homozygous individuals in cultures which contain them can be examined for morphological peculiarities; the results of an examination of this sort will again depend upon the skill of the investigator in recognizing abnormalities. These same homozygotes can be tested for fertility; males can be mated with virgin females of laboratory stocks while females (nonvirgin) can be cultured together with their heterozygous brothers. Finally, developmental rates

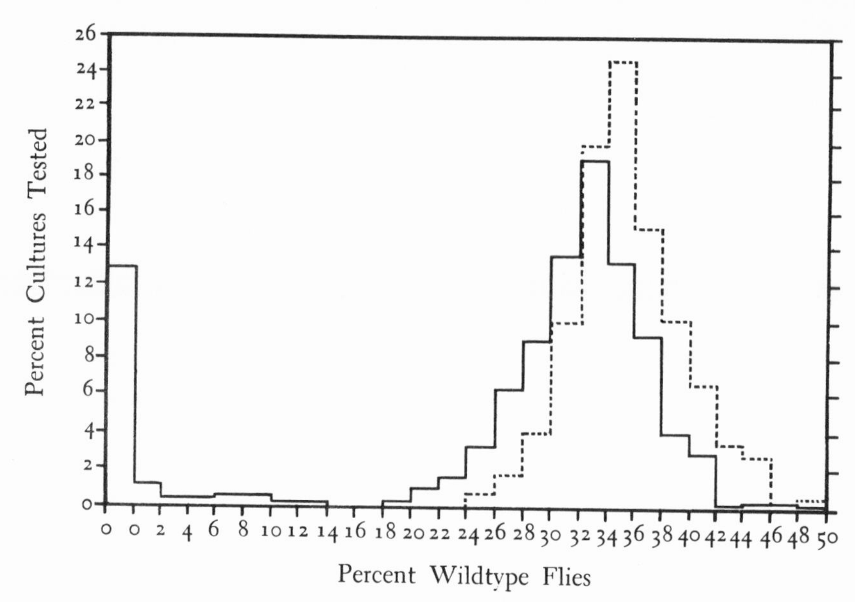

FIGURE 3-4. Percent wildtype flies obtained in cultures in which these flies were homozygous (solid line) or heterozygous (dashed line) for chromosomes encountered in wild populations of *D. pseudoobscura.* (After Dobzhansky and Queal, 1938.)

of the homozygotes can be determined—relative to the heterozygous, genetically marked class—by noting the frequency of homozygous flies in counts made on successive days.

The modified *ClB* procedure just described yields a great deal of information; the most valuable, viability as measured by means of the percent homozygous flies per culture, is expressed in terms relative to a heterozygous class. Control data measured in the same manner are needed so that meaningful statements can be made about wildtype homozygotes. Control cultures are cultures in which the wildtype flies, rather than being homozygous as in the cultures described immediately above, are heterozygous for two different chromosomes taken at random from among those being tested. A commonly used procedure is to number the F_3 cultures in which homozygotes will appear ("homozygous" cultures) consecutively as shown in Tables 3-2 and 3-3, and to set up control heterozygotes by mating *rD*/+ males carrying one wildtype chromosome with *rD*/+ females carrying the wildtype chromosome identified by the next larger number: $+_1/+_2$, $+_2/+_3$, $+_3/+_4$, and so on. Data from crosses of this sort are shown in Tables 3-2 and 3-3 interspersed among the homozygous tests; the dashed line in Figure 3-4 shows the frequency distribution of control cultures in respect to the

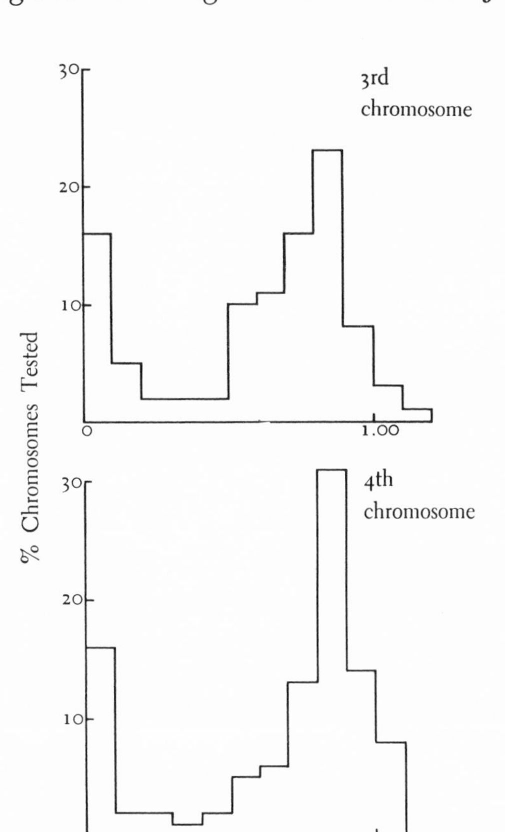

FIGURE 3-5. Distributions of viabilities of flies homozygous for individual third and fourth chromosomes obtained from wild populations of *D. persimilis*. Viability in this case (unlike that in Figure 3-4) is expressed in terms of average viability of flies heterozygous for two different wildtype chromosomes. (After Dobzhansky and Spassky, 1953.)

percent wildtype flies each contained. (Similar frequency distributions of control cultures are not shown in the two drawings of Figure 3-5; the mean frequency of wildtype flies in these controls, however, have been used as the standard of measurement for the horizontal scale.)

The following facts may be noted here to be discussed in detail below:

1. Very few control cultures lack heterozygous wildtype flies.
2. The average frequency of wildtype flies in heterozygous cultures

is generally somewhat greater than that of nonlethal, nonsemilethal (of the "quasi-normal") homozygotes.

3. Only very rarely are the wildtype heterozygotes of the control cultures morphologically abnormal or sterile.

4. Where tests have been made, wildtype flies of the control cultures emerge in greater numbers during the early counts of the F_3 cultures than do those of the homozygous cultures; heterozygotes, that is, tend to develop more rapidly than do homozygotes.

Modified *ClB*—Four-Class Tests

The procedure outlined immediately above is restricted for the most part to a study of mutations whose effects are recessive or very nearly so. The reason for this limitation can be explained by considering a culture in which the initial numbers of genetically marked heterozygotes and wildtype homozygotes are as $2X$ and $1X$. Assume, for example, that the wildtype chromosome carries a dominant gene that causes the death of a fraction Z of both homozygous and heterozygous carriers; after this gene has had its effect, the numbers of genetically marked heterozygotes and wildtype homozygotes are as $2X(1 - Z)$ and $X(1 - Z)$. The final frequency of homozygotes in this culture equals $X(1 - Z)/3X(1 - Z)$, or $1/3$. But this is precisely the ratio of the initial zygotes. Dominant viability modifiers, consequently, are not revealed by the relative frequencies of the genetically marked and wildtype flies in the homozygous cultures when the genetically marked flies carry the wildtype chromosomes being tested.

Information about the possible dominance of viability genes can be obtained by means of a simple modification of the standard procedure; Figure 3-6 shows the modification as applied to the *CyL* test commonly used in studying *D. melanogaster* populations (Figure 3-3): F_3 cultures are set up using male and female parents heterozygous for two different dominant genetic markers, D_1 and D_2. The result is a series of cultures in each of which four classes of flies are expected in (theoretically) equal numbers:

$$D_1/D_2 : D_1/+ : D_2/+ : +/+$$

The D_1/D_2 doubly marked heterozygotes appear in each culture. They do not carry the wildtype chromosome being tested. Consequently, they can be used as a standard of comparison for both the wildtype flies and for $D_1/+$ and $D_2/+$ flies which carry the wildtype chromosome in heterozygous condition. The viabilities of the three classes of flies other than the double heterozygotes, D_1/D_2, are most easily expressed as a ratio of the number of the given class to that of the D_1/D_2 class; for mathematical reasons, the number of D_1/D_2 flies is increased by one in

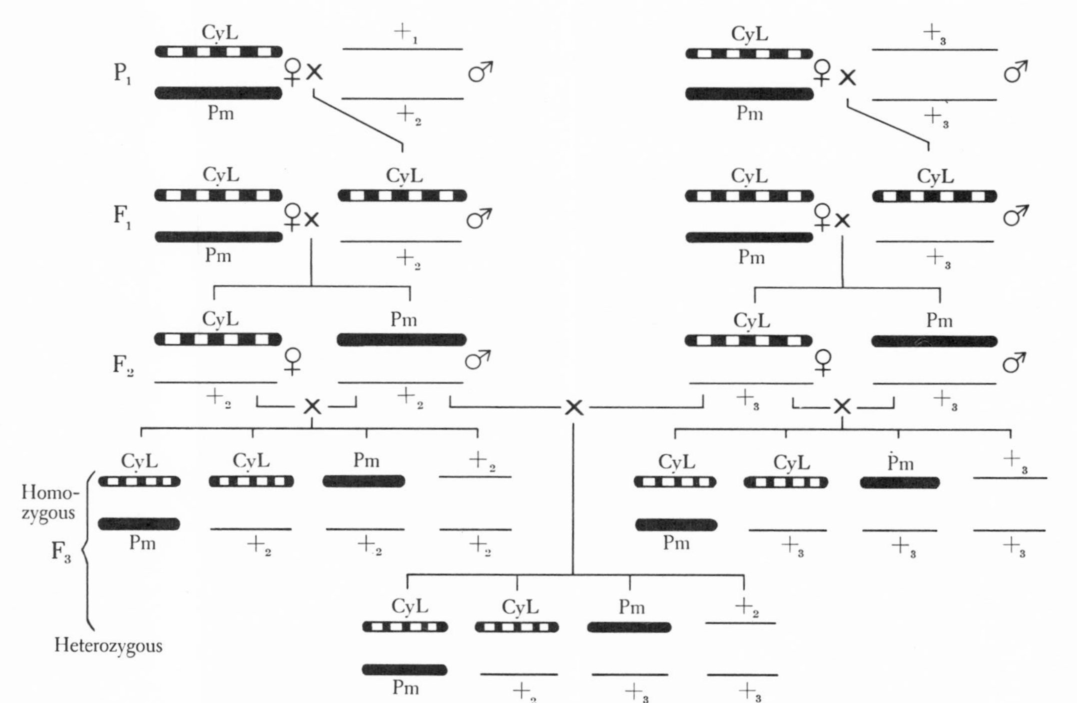

FIGURE 3-6. Series of crosses used to obtain flies homozygous and heterozygous for wildtype and genetically marked chromosomes. Note (1) that each F_3 culture contains one class of flies (*CyL/Pm*) that does not carry the wildtype chromosome, and (2) that the sources of the wildtype chromosomes in the *CyL*/+ and *Pm*/+ flies of the heterozygous F_3 cultures are known. (After Wallace, 1956.)

computing this ratio (Haldane, 1956). Figure 3-6 outlines for *D. melanogaster* a series of crosses leading to both homozygous and heterozygous, control cultures. Table 3-3 lists data obtained from a number of test cultures of this sort. In general, all the information that can be gotten from the simpler, two-class experiment can be gotten (with somewhat more effort) from the four-class technique as well. In addition, the four-class cultures tell us a great deal more about the dominance of "recessive" mutations than we could learn otherwise. [From the data listed in Table 3-4, it is obvious that the double heterozygote, *CyL/Pm,* has a consistently lower viability than those of *CyL/+*, *Pm/+*, or +/+ (controls). Merrell (1965) has suggested that flies with low viabilities cannot properly be used as standards of viability. His suggestion is based on misapprehension concerning the role of a "standard." Just as with measuring sticks, a standard is something against which to measure something else. The nature of the standard is unimportant; the length of a newborn baby, for example, is the same whether expressed in inches or in centimeters.]

TABLE 3-3

SAMPLE DATA SHOWING THE EFFECTS OF DIFFERENT CHROMOSOMES AND THEIR HETEROZYGOUS COMBINATIONS ON THE RELATIVE FREQUENCIES OF *CyL/+*, *Pm/+*, AND +/+ FLIES IN TESTS MADE BY THE *CyL-Pm* TECHNIQUE ILLUSTRATED IN FIGURE 3-6.

Note that each heterozygous wildtype combination carries chromosomes tested immediately above and below in homozygous condition. (Wallace, unpublished data.)

TYPE OF TEST (+/+)	CHROMOSOME DESIGNATION	RELATIVE FREQUENCIES (*CyL/Pm* = 1.00) *CyL/+*	*Pm/+*	+/+
Homozygous	1/1	1.15	1.34	0.95
Heterozygous	1/2	*0.93*	*1.05*	*1.08*
Homozygous	2/2	1.17	1.16	0.01
Heterozygous	2/3	*1.12*	*1.58*	*1.20*
Homozygous	3/3	1.26	2.29	0
Heterozygous	3/4	*1.11*	*0.86*	*0.93*
Homozygous	4/4	1.00	1.02	0

MODIFIED *ClB*—MULTIPLE AUTOSOMAL STUDIES

The high efficiency of all *ClB*-type procedures results from the simultaneous analysis of a great many gene loci. Instead of asking, What is the rate at which lethals occur at a given locus?, the *ClB* technique enabled Muller to ask: What is the frequency with which lethals arise

TABLE 3-4

FREQUENCIES OF "GENOME" LETHALS IN D. MELANOGASTER (SECOND AND THIRD CHROMOSOMES TESTED SIMULTANEOUSLY BY MEANS OF THE TECHNIQUE SHOWN IN FIGURE 3-7) CAPTURED AT FOUR COLLECTING SITES NEAR BOGOTÁ, COLOMBIA.

The sites were spaced linearly at 30-meter intervals.

SITE	NO. OF TESTS	NO. OF LETHALS AND SEMILETHALS	% LETHAL AND SEMILETHAL
A	30	25	83.3
B	34	29	85.3
C	19	15	78.9
D	36	26	72.2
Combined	119	95	79.8

at at least one of several hundred loci? Since mutation rates for lethals at individual loci are low (about 10^{-5}), the efficiency of the *ClB* technique is several hundred times that of single-locus studies. Approximately the same is true in respect to frequencies of lethals in populations; the frequency of lethal alleles at each locus is low, so that an analysis of entire chromosomes in which several hundred gene loci are studied simultaneously is correspondingly more efficient than an analysis of but one locus.

The limiting situation of this type of reasoning is reached when the experimental procedure enables one to analyze the whole of one haploid set of autosomes in a single series of crosses. By the use of reciprocal translocations involving autosomes marked with dominant mutations and carrying crossover suppressors, this goal has now been approached in *D. melanogaster*. Figure 3-7 shows how the so-called (*CyL-Ubx*)/(*Pm-Sb*) stock of flies can be used for the simultaneous analysis of both chromosomes 2 and 3 in this species; between them, these two autosomes account for some 98 percent of all autosomal material. Of the three autosomes, only the small chromosome 4 is uncontrolled by this technique. The results obtained by an analysis of 119 "genomes" of flies collected near Bogotá, Colombia, have been listed in Table 3-4; the frequency of lethal and semilethal genomes in four closely spaced collecting sites was roughly 80 percent. Had the two autosomes been tested separately, frequencies of lethals and semilethals would have been about 55 percent for each autosome.

SPECIAL USES OF *ClB* TECHNIQUES

It must be obvious by now that chromosomes marked with dominant mutations and carrying crossover suppressors act as delicate forceps by

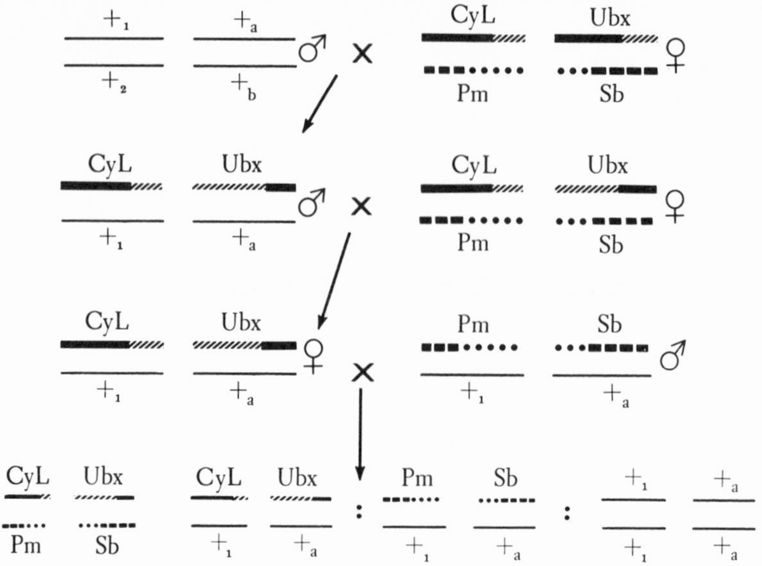

FIGURE 3-7. Use of translocated, genetically marked chromosomes containing crossover suppressors in running simultaneous tests on the viabilities of flies homozygous for wildtype second and third chromosomes in *D. melanogaster*.

which other chromosomes can be manipulated during a series of crosses. There is no need for restricting this manipulation to a study of chromosomes exactly as they are carried in wildtype flies. Indeed, in radiation genetics, the *ClB* techniques are used to determine the effect of radiation on chromosomes, usually (but not exclusively) in relation to the induction of lethal mutations. On the other hand, the same procedures can be used to analyze the array of recombination products produced by interpopulation hybrid females, females that cannot arise without the help of an experimenter (see Figure 3-8). Similar procedures have been used to study the recombination chromosomes produced by females heterozygous for two different chromosomes of the same populational origin (see Figure 3-9); from certain combinations of nonlethal chromosomes, lethals (designated "synthetic" lethals) can be recovered in higher-than-expected frequencies.

A SECOND LOOK AT WILDTYPE HOMOZYGOTES AND THEIR CONTROLS

In the above discussion of experimental procedures, several frequency distributions were illustrated; they show the proportion of chromosomes (or heterozygous combinations) tested which yield certain percentages

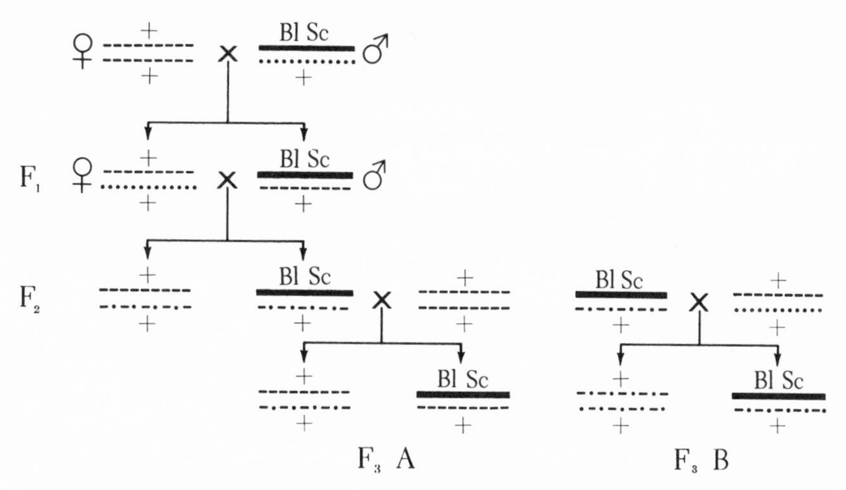

FIGURE 3-8. Series of crosses used to compare the viability effects of wildtype third chromosomes of *D. pseudoobscura* from different geographic localities with that of interlocality recombinant chromosomes. (After Brncic, 1954.)

of wildtype flies in the F_3 test cultures. In the case of homozygotes, these curves are bimodal; a great many chromosomes are lethal (a lethal chromosome may be defined as one that produces less than 10 percent of the expected frequency of wildtype flies, for example), while the remainder form a fairly symmetrical distribution with a peak somewhere near the expected value of 33⅓ percent. The corresponding distribution of viabilities of control heterozygotes consists almost entirely of the peak centered near the expected frequency of 33⅓ percent; as a rule, the peak for the control cultures falls distinctly above that of the homozygous tests. A few—very few, ordinarily—heterozygous tests yield no or very few wildtype flies.

For the moment, we can refer to chromosomes tested in homozygous condition as "drastic" and "quasi-normal." By "quasi-normal" we mean a chromosome that permits more than 50 percent of its carriers to survive to the adult stage under the usual conditions of these standard tests (half-pint culture bottles, somewhat overcrowded culture conditions); "drastics," on the other hand, are chromosomes that permit fewer than 50 percent of their homozygous carriers to survive; the majority of these will be lethal by our earlier standard. Obviously, in a series of random combinations of two different chromosomes, such as the series of control heterozygotes, some will include two quasi-normals (N_1/N_2), others will carry a quasi-normal and a drastic (N/D), while the remainder will carry two drastics (D_1/D_2). It is in the latter combinations, almost without exception, that the lethal and near-lethal control com-

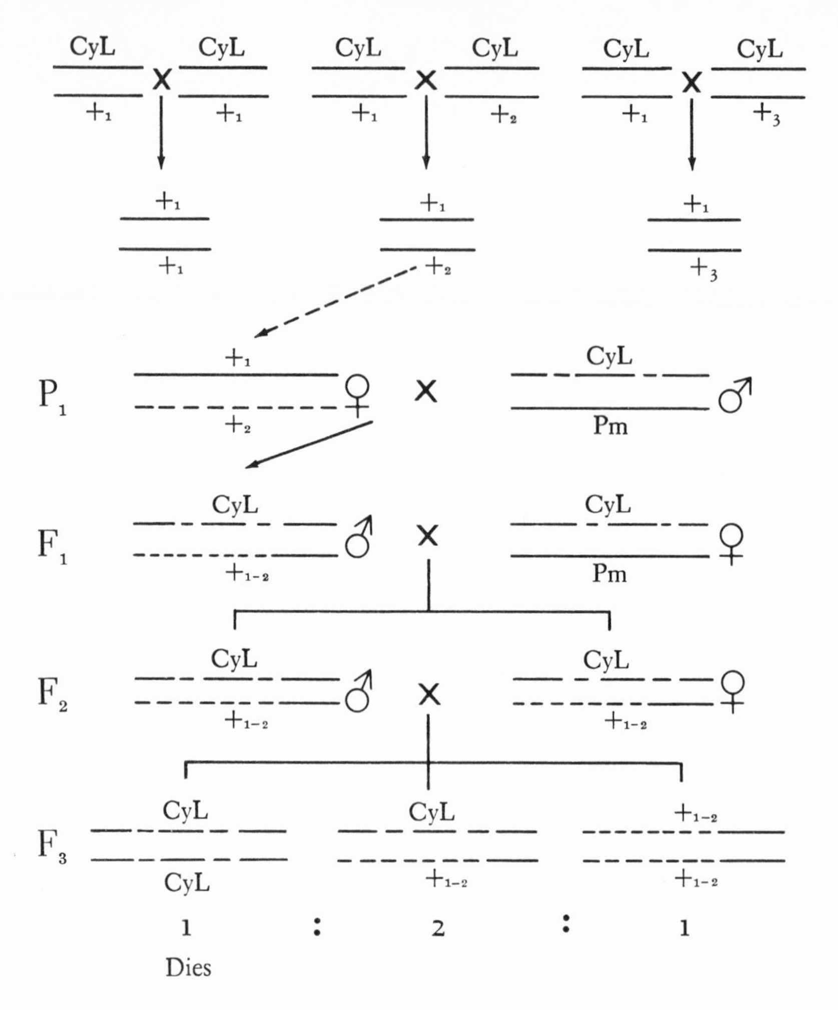

FIGURE 3-9. Series of crosses suitable for studying the array of viability effects of recombinant chromosomes generated from known pairs of wildtype chromosomes in *D. melanogaster.* (After Wallace et al., 1953.)

binations fall. The remainder of the D_1/D_2 combinations fall, for present purposes, with the bulk of all control tests which are centered near the expected 33⅓ percent wildtype flies.

That individuals carrying two lethal chromosomes need not die can bear repeating. There are many gene loci on one chromosome; a lethal gene at any one of these is sufficient to cause the death of homozygous individuals. An individual that carries two different lethal chromosomes, however, need not die, because the two lethals need not be (and, in fact, most often are not) at the same locus. In this case, each lethal allele is "covered" by the nonlethal allele of the homologous chromosome. The few control combinations that are in fact lethal are, for the most part, combinations of lethal chromosomes whose lethal alleles occupy the same locus. Indeed, if this is not the case, we shall nevertheless treat such combinations as if the lethal genes involved do indeed affect the same locus and hence are *allelic*.

The probability of allelism of lethal chromosomes serves as a means for estimating the number of loci on a chromosome at which lethal mutations can occur. For this purpose, the lethals must be independent in origin; either they must be obtained from widely separated geographic localities or they must be of independent mutational origin.

The number of potentially lethal-bearing loci on a chromosome and the probability of allelism are related in a simple manner. If there were but one locus in a chromosome at which lethal mutations could arise, all combinations of two lethals would prove to be lethal. If there were two loci, one half of all possible combinations would be lethal. If there were three, then one third of all possible combinations would prove to be lethal. And since approximately one combination in three hundred is lethal in the case of chromosome 3 of *D. pseudoobscura*, it follows that there are about three hundred loci at which lethals can occur (Dobzhansky and Wright, 1941; Wright et al., 1942). Similarly, since about one combination in four or five hundred is lethal in the case of chromosome 2, a large V-shaped autosome, of *D. melanogaster*, it seems that there are some four or five hundred loci on it at which lethal mutations can occur (Ives, 1945; Wallace, 1950). The discussion given here applies only to combinations of lethal chromosomes that carry but one lethal mutation each; chromosomes with two or more lethal genes have correspondingly greater probabilities of proving to be allelic to other lethal-bearing chromosomes. The correction that adjusts for this complication has been given by Dobzhansky and Wright (1941) and, modified from them, by Wallace (1950). The corrected estimate of n, the number of loci on a chromosome at which lethals can occur, is

$$n = \frac{1}{I}\left(\frac{m}{Q}\right)^2$$

where I is the observed frequency of allelism of lethal chromosomes in random combination, Q the frequency of lethal chromosomes in the population of chromosomes from which those to be tested were obtained, and m the frequency of lethal genes in the original population, taking into account that some lethal chromosomes may carry two or more lethal genes. The Poisson distribution, which will be discussed later in this chapter, offers a means for estimating m from Q.

Statistical Terminology: Mean and Variance

A great many of the words used in population genetics have been borrowed from statistics. Instruction in statistics is not an aim of this book; a review of the meanings of some of the more common terms, however, and a discussion of several distributions as well as the meaning of "significance" will not be entirely out of place.

The term "average" or "mean" is used repeatedly in discussions of populations. It is calculated as the sum of all individual values divided by the number of individuals. Alternatively, it is the sum of various values times the relative frequencies with which they occur. If we let letters *a*, *b*, and *c* stand for three values, the preceding statements can be represented as follows:

$$\bar{x} = \frac{a + a + a + b + c + c + c + c + c + c}{10}$$

$$= \frac{3a + b + 6c}{10}$$

$$= 0.3a + 0.1b + 0.6c$$

"Variance" is a term that refers to a commonly used measure of the dispersion of observations (or of the values of individuals in respect to a certain characteristic) about their mean; specifically, variance is the mean of the squared deviations of individual values from the mean. The greater the dispersion about the mean, the greater the variance; the more tightly grouped the values, the smaller the variance.

The variances of independent variables are additive. Suppose, for example, that two persons try to put one dozen lead BB shot into each freight car of a toy train as it speeds around a circular track. Each person will make mistakes; some pellets will miss entirely and fall on the floor while others will go into the wrong cars. If the shot given to the two persons differ (copper-plated versus nickel-plated), a mean and variance can be computed for the number of shot per car for each person. These means and variances may differ. One person may have a higher mean than the other because he misses less frequently; one person may be more uniform in distributing his shot among the different cars than the other. An overall mean based on the sum of the two types of shot can be computed; this mean will be the sum of the two individual means. Deviations from the grand mean can be computed, squared, and an average (the variance) can be obtained. This variance will be the sum of the variances of the two individuals taken separately if there is no correlation between the two.

Variances of means are smaller than variances of individual observations. During a single day, a lecture room may be occupied by several groups of students. In each successive class there are likely to be short students and tall students. Each class has a certain mean height determined by the distribution of the individual class members. The different classes will have different means, but these means will be much more like one another than the heights of the individual students of each class. The variance of the mean equals the variance of individual values divided by the number of individuals used in computing each mean.

Some Common Statistical Distributions

One of the most familiar distributions is the "normal" distribution, the bell-shaped distribution in which individual observations tend to cluster about some mean value while observations falling at increasing distances from the mean occur with progressively lower frequencies.

The area under the normal curve is related in a simple manner to distances from the mean if distance is measured in terms of standard deviations (a standard deviation is the square root of the variance). About 95 percent of the total area lies between points two standard deviations below and two standard deviations above the mean, approximately 99 percent between points three standard deviations below and above the mean. Thus, we can use the known mean and standard deviation of a given group in deciding whether a given observation is likely to belong to the group. For example, if the mean age of graduating college seniors is 21 years and the standard deviation is 6 months, then it is unlikely that a middle-aged man seen on the campus between classes is a graduating senior. From his age alone, it may be difficult to decide, for example, whether he is a professor or one of the maintenance staff; other characteristics must be known before this decision can be made.

A second distribution, one familiar to all genetics students, is the binomial distribution. This distribution describes the probability of observing given numbers of certain items in a sample of specified size when the relative frequencies of the various items are known. The distribution is employed in solving the problems given on innumerable genetics examinations, lab quizzes, and practice problems: What is the probability of obtaining exactly three boys and two girls in a family of five children? What is the probability that there will be no blue-eyed children among three when the parents are known to be heterozygous for the appropriate allele?

In courses of general genetics, the binomial expansion is used to calculate precise probabilities for small samples of defined composition. In population genetics we are more likely to be interested in larger samples such as in the subdivision of a large population into smaller ones still of considerable size. Or, in reference to the *ClB*-like techniques described earlier, we may be interested in the variation expected in the frequency of wildtype flies among cultures of several hundred flies each. In these cases, we take advantage of the approach of the binomial to the normal distribution when the sample size becomes large. The mean frequency and the variance are p and pq/n, where the probability that a given event will occur in a single try is p and n is the sample size.

The Poisson distribution is the third and last distribution to be discussed here. It is probably the least familiar but potentially most useful of the better known statistical distributions. Persons with algebraic training can demonstrate to their satisfaction that the Poisson distribution can be derived from the binomial distribution. The condition that leads from the one to the other is that the probability p of an event become very small (approach zero) but do it in such a manner that the number of events expected in a sample of size n be a small, but finite, number (m). In this case, the probabilities that the event will occur zero, one, two, three, or more times are

$$e^{-m}, \quad me^{-m}, \quad \left(\frac{m^2}{2}\right)e^{-m}, \quad \left(\frac{m^3}{6}\right)e^{-m}, \ldots$$

where m is the mean number of events expected ($= np$), e is the base of natural logarithms ($= 2.7183 \ldots$), and the denominators are factorials 0! ($=1$), 1!, 2!, 3!, ... The sum of these terms equals 1, as frequency distributions must if they are to represent all possible outcomes, no more and no less. An important characteristic of the Poisson distribution is that its variance equals its mean (see Table 3-1).

The Poisson distribution is useful in solving a wide variety of problems. Haldane used it during World War II to demonstrate that, despite the apparent patchiness of bomb hits in London, the proportions of blocks hit zero times, once, twice, three, four, and more times were those expected on the basis of the Poisson distribution; consequently, there was no reason to believe that the Germans were attempting to bomb particular blocks more intensively than others.

For some types of observations, only two classes can be discerned such as survivors and nonsurvivors for irradiated bacteria, or lethal and nonlethal chromosomes. A bacterium that has survived exposure to radiation has suffered no "hits" where hits are often Poisson events leading to death; those that have been killed have been hit at least once but possibly several times. A nonlethal chromosome is one with no lethal genes; a lethal chromosome has at least one lethal gene—possibly more. In each of these examples, the mean number of "hits" (a lethal gene is a "hit" in this case) per bacterium or per chromosome can be calculated from the frequency of the zero class (survivors or nonlethals). If the frequency of the zero class equals P, then

$$P = e^{-m}$$

or

$$m = -\ln P$$

where "ln" means "natural logarithm" or "$\log_e$." At times it is particularly useful to know when the mean number of Poisson events is 1; since e^{-m} equals approximately 37 percent when m equals 1, this value for the zero class is an especially important one.

Tests of Significance

We are not concerned here with the calculations upon which tests of significance are based; these calculations are those dictated by the test—chi-square, t-test, or F-test—actually employed. We are concerned instead with the underlying reasons for the calculations: What are we testing? What is our hypothesis? What does the probability obtained really mean?

In performing any test, we assume that our observations do *not* differ from those expected on some theoretical basis. It is impossible to ask whether our observations *differ* from expectations because there are an infinite number of ways in which the two may differ. Agreement, on the contrary, is *unique*. Thus, the probability values listed in the statistical tables are probabilities that observations would differ as much or more from the expected values by chance alone were the observations actually drawn from a population which was precisely that expected. If this probability is small, we usually decide that our observations do not in fact represent a sample drawn from our theoretical population. Thus, if only one undergraduate in 1000 is forty years old or older, we may decide (perhaps erroneously) that our middle-aged friend in the previous example is *not* an undergraduate. Our decision is based upon the extreme rarity of middle-aged undergraduates.

A Third Look at Wildtype Homozygotes and Their Controls

The bimodal frequency distributions typical of chromosomes of natural populations when tested in homozygous condition (Figure 3-4, for example) are difficult to describe; the difficulty is exaggerated by the need to describe the frequency distribution of homozygous tests in relation to a second bell-shaped distribution that represents the control tests. Two techniques have been devised for making this description. The first attempts to assort chromosomes into pigeonholes so that, following their classification, one can talk of the frequencies of lethals, semilethals, subvitals, normals, and supervitals in the sample tested (Wallace and Madden, 1953; Dobzhansky and Spassky, 1953). The alternative technique, utilizing the ratio of mean frequencies of wildtype flies in the homozygous and heterozygous cultures, specifies the proportion of lethals alone that would be needed to give a corresponding ratio (Greenberg and Crow, 1960). Lethals and semilethals can, of course, be omitted from the data on homozygous cultures; the calculations then reveal what frequency of lethals would be required to lower a ratio from some normal value to that actually observed. Hence the term "lethal equivalents."

The task of resolving the tests of homozygotes into definite pigeonholes is not difficult for lethal and semilethal chromosomes; these differ so much from the remaining quasi-normal chromosomes that their classification is an easy matter. The remaining portion of the distribution curve, that in which more than 50 percent of the expected number of wildtype homozygotes appear as adult flies, must be subdivided so that meaningful estimates of the frequency of chromosomes carrying detrimental gene mutations (subvitals) and those free of such genes (normals) are obtained. Worse, the estimates must be made from one bell-shaped distribution under circumstances where the standard of comparison is another bell-shaped distribution of frequencies of wildtype flies seen in control cultures.

The task is simplified by establishing certain rules. First, "normal" must be defined as a *range*, not as a *point*; otherwise the proportion of normal chromosomes is zero. Second, of those things which contribute to the bell-shaped distribution of homozygous and heterozygous frequencies, only those which are related to genetic differences between chromosomes and chromosome combinations interest us. The largest contribution to the variance of observed frequencies is that of the binomial variance related to the number of flies counted per test (F_3) culture. Another large contribution to the variance of homozygous cultures is the variation observed when the same chromosome is tested in replicate cultures—the environmental variance. The genetic variance, that in which we are interested, is obtained by subtracting the binomial variance and the environmental variance from the total variance observed; this can be done because variances, as we mentioned earlier, are additive when they are due to independent causes.

The range of viabilities that we shall consider normal could be defined in any convenient way. The definition that has been used in the past (see Figure 3-10) extends from two standard deviations (calculated on the basis of genetic variance) below the mean of the control cultures to two standard deviations above; this definition claims that virtually all (95 percent) of the control heterozygotes are normal. By using a table of areas under the normal curve, one can easily determine the area of the distribution curve (using genetic variance, again) of homozygous frequencies that lies below the lower limit of the normal range (this area is the proportion of subvitals among quasi-normals), that which lies between the lower and upper limits of the normal range (this area is the proportion of normals among quasi-normals), and that which

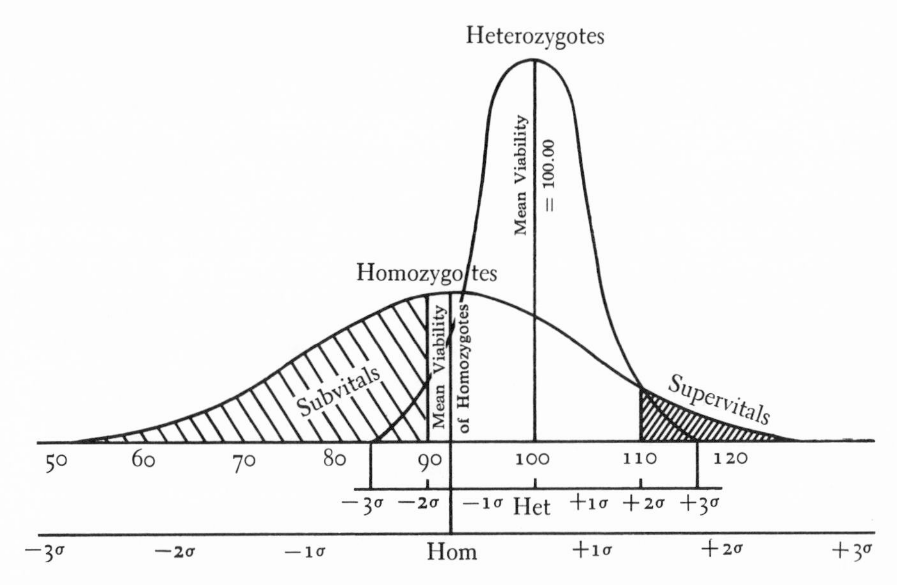

FIGURE 3-10. Calculation of the proportions of subvital, normal, and supervital chromosomes (tested in homozygous combinations). Normal viability is defined as that falling within 2 standard deviations of the average frequency of flies heterozygous for two different wildtype of chromosomes. (After Dobzhansky and Spassky, 1954.)

lies above the upper limit of the normal range (this area, usually very small, represents the proportion of supervitals among homozygotes). In passing, it should be noted that the removal of extraneous sources of variations is an essential feature of this procedure. If sampling variance is allowed to remain, for example, the estimated proportion of subvitals present in the population will prove to be dependent upon the number of flies counted per culture; this is absurd, because subvitality is a characteristic of chromosomes, not of experimental procedures.

The calculation of *lethal equivalents* from the frequency distributions of homozygous and heterozygous control cultures utilizes average frequencies and the Poisson distribution. The procedure is most easily explained by an example. Suppose that of a sample of chromosomes, 20 percent were lethal when homozygous (0 percent wildtype flies per test culture) while the remaining 80 percent gave the same frequency of wildtype flies as the average of the control cultures (100 percent wildtype flies relative to the controls). The average frequency of wildtype flies in all cultures would equal (0.20 × 0 percent) plus (0.80 × 100 percent), or 0.80. The average frequency is obviously identical to the proportion of nonlethals. From the Poisson distribution we learn that the frequency of lethal genes, m, in this sample equals $-\ln(0.80)$, because the nonlethals are chromosomes with no lethal genes. From an appropriate table of logarithms we find that m equals 0.223. The frequency of lethal chromosomes is smaller than 0.223 because some lethal chromosomes really carry two or more lethal genes.

Suppose now that we study a second sample of chromosomes in which all chromosomes tested prove to be identical and each yields 80 percent as many wildtype flies per homozygous culture as does the average control culture. By the pigeonhole analysis we would say that this sample consists entirely of subvitals, but the degree of subvitality would be left unspecified. By the lethal equivalent procedure, we point out that a frequency of wildtype flies only 80 percent that of the control is that which would be obtained if lethals were the only deleterious mutation and their frequency m were 0.223. Thus, we can say that there is an average of 0.223 lethal equivalents per chromosome in the second sample. The deleterious effects of nonlethal mutations are looked upon in this case as fractions of the effects produced by lethal genes. When viewed in this manner, two semilethals, five one-fifth lethal, or one hundred one-hundredth lethals are equivalent to one lethal—that is, constitute a "lethal equivalent."

These two analytical procedures represent attempts at reducing the information contained in a complex table or graph to a set of figures amenable for mathematical manipulation. Each procedure discards some of the original information. One tells us the proportion of subvitals but does not tell us the degree of subvitality; the second tells us the average degree of subvitality (in terms of lethal equivalents) but loses sight of the original distribution pattern. Some of this lost information can be regained by dividing the lethal equivalents of quasi-normal chromosomes by the frequency of subvitals; the result (Table 3-5) is the average degree of subvitality per subvital chromosome. From the table we can see that this value is quite stable; it varies from 0.19 to 0.33, depending upon the chromosome and species studied. The average effect of subvital chromosomes appears to be roughly 0.25—one-quarter lethal.

Concealed Variability

At the beginning of this chapter we pointed out that the dominance of wildtype genes prevents us from seeing the recessive mutations which are present in a population of diploid organisms. Virtually the entire chapter was then devoted to schemes—biological and statistical—for revealing and measuring this store of concealed recessives. Table 3-6, a summary of the results of a number of analyses of various *Drosophila* species, reveals that this store is indeed impressive. As an exercise in

TABLE 3-5

ESTIMATION OF THE AVERAGE DEGREE OF SUBVITALITY OF SUBVITAL CHROMOSOMES MEASURED IN LETHAL EQUIVALENTS.

Lethal equivalents for quasi-normal chromosomes have been computed according to the method of Crow and Temin (1964), the frequencies of subvitals by the method of Wallace and Madden (1953).

	PSEUDOOBSCURA			PERSIMILIS			PROSALTANS	
Average viabiltiy of quasi-normals	0.750	0.771	0.856	0.879	0.834	0.779	0.917	0.966
Lethal equivalents	0.29	0.26	0.16	0.13	0.18	0.25	0.09	0.04
Frequency subvitals	0.94	0.78	0.70	0.67	0.80	0.98	0.33	0.15
Lethal equivalents ÷ subvitals	0.31	0.33	0.23	0.19	0.23	0.26	0.27	0.27

TABLE 3-6

CONCEALED GENETIC VARIABILITY OF THREE DROSOPHILA SPECIES: D. PROSALTANS, D. PSEUDOOBSCURA, AND D. PERSIMILIS.

(Dobzhansky and Spassky, 1953; Sankaranarayanan, 1965.)

	PROSALTANS		PSEUDOOBSCURA			PERSIMILIS		
	2nd	*3rd*	*2nd*	*3rd*	*4th*	*2nd*	*3rd*	*4th*
Lethals and semilethals (%)	33	10	33	25	26	26	23	28
Subvitals (%)	33	15	94	78	70	67	80	98
Supervitals (%)	0	3	0	0	0	0	3	0
Sterility (%)	20	11	19	24	16	32	30	27

numerology, we can ask: What proportion of all individuals of the species *D. pseudoobscura* is free of lethal, semilethal, and subvital autosomes? The answer is obtained by squaring the frequency of nonlethal (and nonsemilethal) second, third, and fourth chromosomes listed in Table 3-6, as well as the frequency of nonsubvitals for the same chromosomes, and then multiplying the six squares. (Squaring is necessary because there are two of each chromosome in a diploid individual.) The answer obtained is approximately 0.0000004, four in ten million. We are not seriously wrong when we say that no fly—and no human being, either—is free of deleterious mutations. As a *practical* matter, that is, no diploid individual exists which is homozygous for normal alleles at every gene locus. This was a startling calculation in the 1920s and 1930s when the majority of biologists were impressed by the superficial similarity of wildtype individuals. More importantly, however, as a *theoretical* matter, we shall have occasion to ask later whether *any* individual of a cross-fertilizing species can be completely homozygous and, simultaneously, completely normal. That is, we shall ask whether normal homozygotes are not impossible in theory as well as extremely rare in practice.

4

THE HARDY-WEINBERG EQUILIBRIUM

It was once said that a hen is an egg's way of producing another egg. To people who think in terms of individuals and of the importance of individuals, this may seem to be a shocking inversion of obvious facts. Nevertheless, in understanding populations it is often more convenient to think of a population as an assemblage of genes existing through time than to regard it as a collection of individuals reproducing generation after generation. An assemblage of genes can be described in terms of the frequencies of various alleles (gene frequencies); collections of diploid individuals must be described in terms of frequencies of various combinations of alleles (zygotic frequencies). The latter is considerably more complicated than the former. It is an easy matter, for example, to enumerate the 52 cards of an ordinary deck of cards; it is an entirely different matter to write down the 1326 possible pairs of these cards.

Gametic and zygotic frequencies are evidently not unrelated. In the following pages we shall see how each is computed, learn how the two are related, and spell out explicitly some of the assumptions upon which the specified relationship depends. In doing this, we shall be reconstructing arguments and arriving at conclusions reached by Pearson (1904), Hardy (1908), Weinberg (1908), and Chetverikov (1926). It seems strange that a phenomenon which is really a corollary of Mendel's law of segregation was not described until eight years or more after the rediscovery of Mendel's paper. It is stranger still that to a large extent the significance of the Hardy-Weinberg equilibrium was not generally appreciated until it was "popularized" in *Genetics and the Origin of Species* (Dobzhansky, 1937).

The Calculation of Gene Frequencies

In Chapter 3, frequencies of various kinds of chromosomes were discussed. These frequencies were arrived at by testing a series of chromosomes (one from each of a large number of individuals) and, having classified them according to some scheme (lethal and quasi-normal, for example), by calculating the proportion of chromosomes found in each class (see, for example, Table 3-6).

Gene frequencies are computed in a similar way. In this case, however, they will be based upon the sampled individuals themselves. We shall assume that both homozygous and heterozygous individuals can be recognized.

Let the frequencies (zygotic frequencies, that is) of AA, Aa, and aa individuals be X, Y, and Z for both males and females. The frequency of A (gene frequency) is *defined* as $X + \frac{1}{2}Y$, and the frequency of a is *defined* as $\frac{1}{2}Y + Z$. We assume that A and a are the only alleles, so that the sum of X, Y, and Z is 1.00; consequently, $(X + \frac{1}{2}Y) + (\frac{1}{2}Y + Z)$ is also equal to 1.00. Those who are uncomfortable with defined frequencies may take comfort in the thought that only one of the two gametes that gave rise to each Aa individual carried gene A; the other carried gene a. Or, thinking of the gametes which an individual will produce, Aa individuals will produce gametes one half of which will carry A while the other half will carry a. The justification for the definition is unimportant. That the gene frequencies are *as defined,* on the contrary, is tremendously important.

The Derivation of the Hardy-Weinberg Equilibrium

Assume that the AA, Aa, and aa males and females of the preceding paragraph mate at random. The frequencies of the different types of matings as well as of the various types of proportions of offspring produced by each can be represented by a diagram such as that shown in Figure 4-1. The adjoining margins of the square represent the two sexes and each is divided into sections X, Y, and Z (the total length of a side equals 1.00) to represent the three genotypes AA, Aa, and aa. The lines that mark the sections intersect to form nine compartments; the area

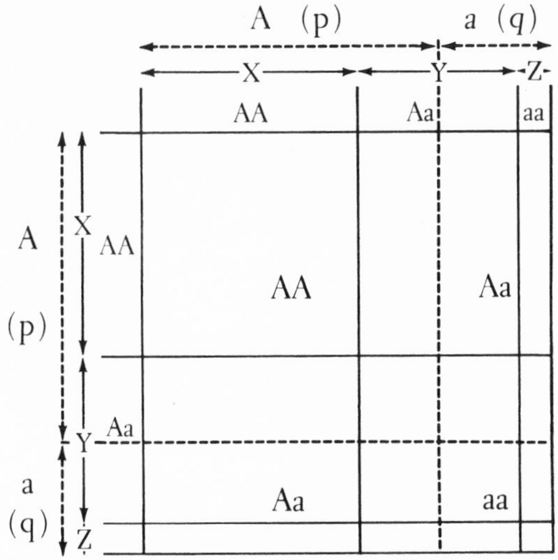

FIGURE 4-1. Hardy-Weinberg equilibrium. A geometric "proof" that whatever the initial proportions of homozygotes and heterozygotes (provided that they are the same in the two sexes), the proportions in the progeny can be expressed as p^2 *AA* : $2pq$ *Aa* : q^2 *aa*. (After Wallace and Srb, 1964.)

of each represents the probability of one type of mating while the sum of all nine probabilities equals 1.00, the total area of the square. Each of the nine compartments has been subdivided according to the proportions of offspring of various genotypes produced; some compartments are left intact, others are divided in half, and one ($Aa \times Aa$) is divided into quarters. The subdivisions as a whole, then, represent the proportions of offspring of various genotypes in the next generation; they too, add up to 1.00

It is important to recall that the values X, Y, and Z of the preceding paragraph had no specified relation to one another except that their sum should be 1.00. Consequently, we find that the offspring produced have no simple proportions when these are expressed in terms of X, Y, and Z. The frequency of *AA* individuals among offspring equals, for example, $X^2 + XY + \frac{1}{4}Y^2$ (Figure 4-1). The relative proportions of offspring of the three genotypes do have a simple relationship in terms of gene frequencies. Recalling that the frequency of *A* (equals p) has been defined as $X + \frac{1}{2}Y$, while that of *a* (equals q) has been defined as $\frac{1}{2}Y + Z$, we can see that the subdivisions shown in Figure 4-1 can be combined to give

$$p^2\ AA : 2pq\ Aa : q^2\ aa$$

The relative frequencies of *AA*, *Aa*, and *aa* individuals among offspring (p^2, $2pq$, and q^2) need not equal the original frequencies of these genotypes (X, Y, and Z) among parents. The new frequencies can, however, be regarded as representing X_1, Y_1, and Z_1, because the X's, Y's, and Z's are arbitrary. It follows, though, that the frequency of $A = X_1 + \frac{1}{2}Y_1 = p^2 + pq = p(p + q) = p$, while that of $a = \frac{1}{2}Y + Z = pq + q^2 = q(p + q) = q$. The frequencies of the two alleles, *A* and *a*, have not changed during the passage from parents to offspring. Nor will they change in succeeding generations. Consequently, a randomly mating population of the sort described can be represented in the following way:

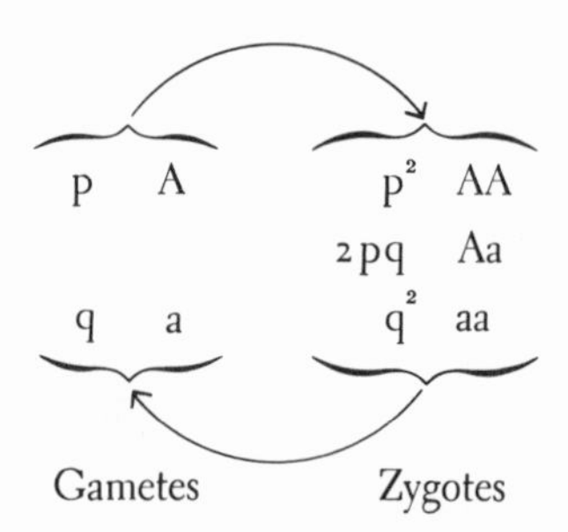

Undisturbed, the cycle will persist indefinitely. Each generation produces gametes that carry the alleles *A* and *a* in proportions p and q; these in turn unite at random to give rise to zygotes *AA*, *Aa*, and *aa* in the proportions p^2, $2pq$, and q^2.

The implications of the Hardy-Weinberg equilibrium are less obvious to persons who look upon populations as collections of individuals (as many biologists are inclined to do) than to those who look upon individuals merely as temporary carriers of genes. First, those who think in terms of individuals think as well in terms of *dominance* and *penetrance,* terms that can be misleading if applied erroneously to genes and their fate in populations. Second, persons familiar with individuals know, too, that many, if not the majority, of these never survive. This sort of knowledge does not prompt one to think in terms of stable equilibria; on the contrary, it emphasizes the possibilities for change within a population.

Nevertheless, if we concentrate on gene frequencies and if we carefully avoid any statement which implies a differential gain or loss of *genes,* it is immediately obvious that gene frequencies must remain constant from generation to generation. If the relative frequencies of two alleles, *A* and *a*, are p and q, then the probabilities that pairs of genes (diploid individuals) will consist of *AA*, *Aa*, and *aa* are given by the binomial expansion as p^2, $2pq$, and q^2. If I have a total of 100 pennies and dimes in my pocket, and if I place these, two coins at a time, on the desk in front of me, I have neither more nor fewer coins of each kind than I had when they were in my pocket. The frequencies of the two types of coins change only if I state that in transferring them from my pocket to the desk, or from the desk back to my pocket, I lose some and that I lose one type more often than I do the other.

Data that very nearly correspond to the diagram shown in Figure 4-1 are listed in Table 4-1. These data consist of the frequencies of *M*, *N*,

TABLE 4-1

PROPORTIONS OF *M*, *MN*, AND *N* CHILDREN BORN TO PARENTS WHO, IN TURN, HAVE BEEN CLASSIFIED ACCORDING TO THESE BLOOD TYPES.

FAMILIES		CHILDREN			
TYPE	NO.	*M*	*MN*	*N*	TOTAL
$M \times M$	119	272	1	0	273
$M \times N$	142	1	315	0	316
$N \times N$	33	0	0	72	72
$MN \times M$	341	408	387	1	796
$MN \times N$	250	3	334	300	637
$MN \times MN$	275	163	317	160	640
Total	1160	847	1354	533	2734

and *MN* blood types among the children of parents whose blood types are also known. A total of 1160 families or 2320 parents were tested; 2734 children were classified.

The *MN* blood group system is based on a single pair of alleles: $L^M L^M$ individuals have blood type *M*; $L^N L^N$, *N*; and $L^M L^N$ heterozygotes, *MN*. Of the 2734 children tested, six cannot be explained on this basis; these children may represent cases of erroneous blood typing, mutations, or of extramarital parentage.

An examination of the parents shows that among them the frequencies of *M*, *N*, and *MN* persons were 0.311, 0.197, and 0.492, in that order. The frequencies of L^M $(= p)$ and L^N $(= q)$ are

$$0.311 + 0.246 = 0.557 = p$$

$$0.197 + 0.246 = 0.443 = q$$

The frequencies of *M*, *N*, and *MN* persons expected under the Hardy-Weinberg equilibrium, then, are

$$p^2 = 0.310 = \text{frequency of } M$$

$$q^2 = 0.196 = \text{frequency of } N$$

$$2pq = 0.494 = \text{frequency of } MN$$

Of the 2734 children examined, 0.310 were *M*, 0.196 were *N*, and 0.495 were *MN*. These frequencies are very nearly identical to the corresponding parental frequencies. The frequencies of L^M and L^N among children were 0.558 and 0.442; again, these are nearly identical with the parental frequencies. The expected frequencies of *M*, *N*, and *MN* children are 0.311, 0.195, and 0.493.

Finally, using procedures comparable to the calculation of zygotic from gene frequencies, we can ask whether the observed frequencies of various family combinations are as we would expect on the basis of random mating. With the exception that the reciprocal combinations of unlike parents have been combined in the table, these frequencies represent the nine main divisions of the square shown in Figure 4-1. The observed and expected frequencies of various family combinations are listed below:

	$M \times M$	$M \times N$	$N \times N$	$MN \times M$	$MN \times N$	$MN \times MN$
Observed, %	10.3	12.2	2.8	29.4	21.6	23.7
Expected, %	9.7	12.3	3.9	30.6	19.4	24.2

Once more, we see that the agreement between observed and expected frequencies is very close.

Corresponding Gene and Zygotic Frequencies

The Hardy-Weinberg equilibrium enables us to pack a great deal of information into a single number. If we know, for example, that only two alleles are present in a randomly mating population, and we are told that the frequency of one of these (say, *A*) is 20 percent, we learn from that single item that

1. The frequency of the alternative allele (*a*) is 80 percent.
2. The expected frequency of *AA* individuals is 4 percent.
3. The expected frequency of *Aa* individuals is 32 percent.
4. The expected frequency of *aa* individuals is 64 percent.

Thus, knowing one gene frequency, we know the relative frequencies of the two alleles as well as the relative frequencies of the three types of zygotes.

The relative proportions of individuals of the three genotypes have been illustrated diagrammatically in Figure 4-2 for various values of p. It is important to note that the frequencies of the two homozygotes form exponential curves. Squares of numbers less than 1.00 are always smaller than the numbers themselves; squares of small fractions are, however, very small. Consequently, if $p = 0.10$, eighteen individuals of every hundred are expected to be heterozygotes, but only one individual is expected to be *AA*.

From Figure 4-2, it is clear that the frequency of heterozygotes cannot exceed 50 percent; this value is attained when both p and q are 0.50. This, of course, is the value obtained by mating F_1 hybrids in rearing an F_2; as Pearson (1904) pointed out, it is the value one would get indefinitely by the random mating of individuals of the F_2, F_3, and later generations.

Evolution and the Hardy-Weinberg Equilibrium

It is unfortunate that Darwin was unaware of Mendel and his work. According to Crew (1966), Mendel was familiar with some of Darwin's

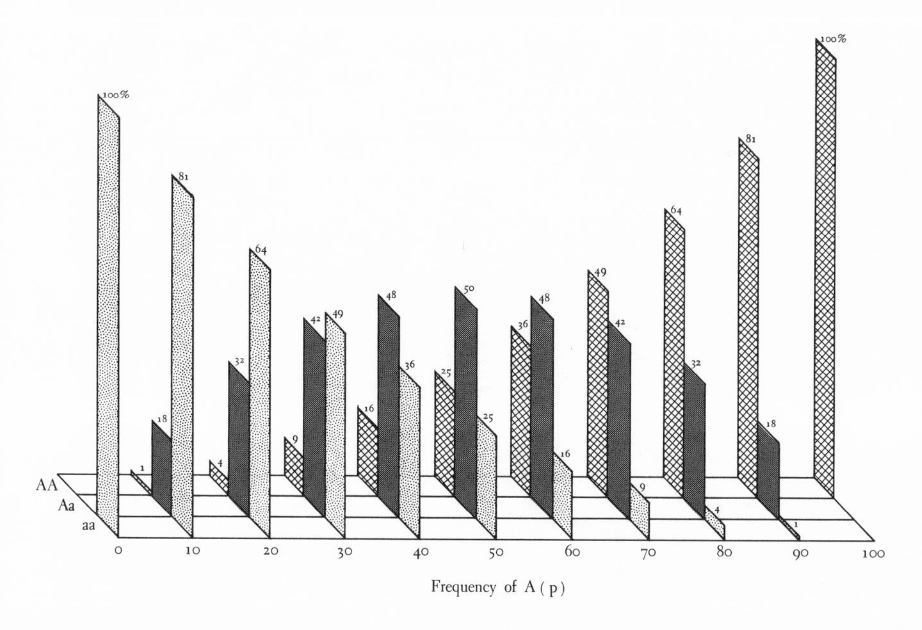

Figure 4-2. Hardy-Weinberg proportions of homozygotes and heterozygotes for various gene frequencies.

publications; at least three of Darwin's books were included in Mendel's personal library. During much of Darwin's scientific life he was concerned with the mechanism of inheritance. Between 1862 and 1879, Darwin published three books on hybridization, fertilization, and variation in plants. It is one of the great ironies of evolutionary biology that Mendel's paper was available but unused for much of this period.

Darwin's interest in the mechanism of inheritance arose directly from his theory of evolution by natural selection. For natural selection to operate, a number of conditions must be met. First, there must be variation upon which selection might act. Second, this variation must be heritable. Third, survival of individuals with differing characteristics must be differential rather than random. These three conditions will bring about evolutionary changes in a population but only while variation exists. Darwin was well aware of this last point; his interest in heredity was an outcome of his search for the source of and the means for preserving heritable variability.

An illustration of the nature of variation in wing length in *D. melanogaster* and the demonstration that this variation is indeed heritable are shown in Figure 4-3. The regression line that summarizes the relation between the wing lengths of offspring and those of their parents runs from lower left to upper right in the chart (positive slope): short-

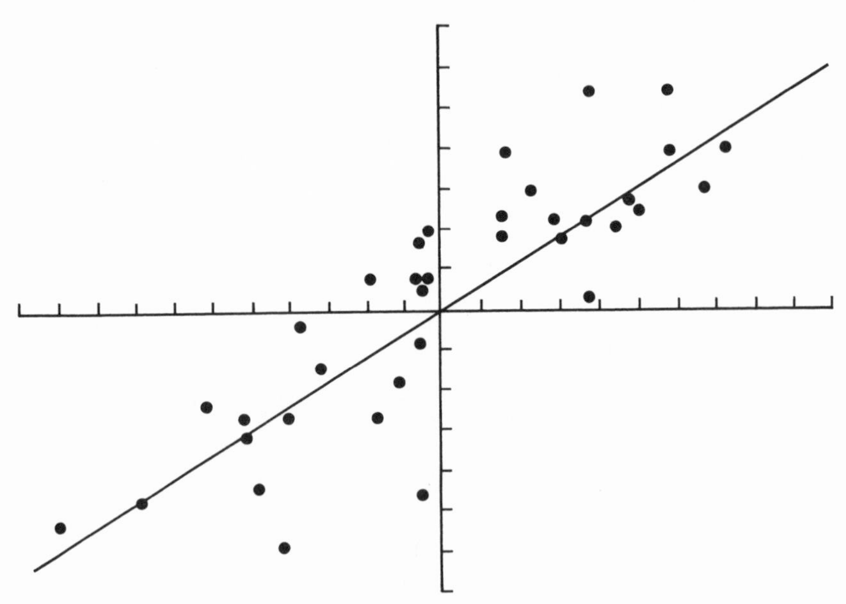

FIGURE 4-3. Regression of offspring (vertical axis) on parents (horizontal axis) for wing length in *D. melanogaster*. Axes are marked in 0.01-mm intervals; the origin represents the mean of both parents and offspring. (After Falconer, 1960, 171.)

winged parents tend, on the average, to have short-winged offspring, whereas long-winged parents have long-winged offspring.

In the absence of information on the particulate nature of heredity, Darwin accepted the notion of blending inheritance generally held by his contemporaries. "Blending inheritance" in this sense means a blending of genetic material such that the gametes of children of dissimilar parents uniformly contain average or intermediate hereditary material rather than being divided into discrete classes in which the original materials have segregated without contamination. If genetic material did blend, as Darwin believed, then the mere process of reproduction, the creation of one generation by the preceding one, would reduce the existing variation by one half. This process is illustrated in Figure 4-4. A hypothetical population is shown as consisting of individuals whose value in some quantitative sense is said to be 2, 4, and 6; one third of each sex has each value. The mean value of these individuals is 4; the variance (calculated in the figure) is $8/3$. Now, if matings occur at random and if the offspring of each mating have values equal to the means of their parents, the mean of the second generation is still 4 but the variance is only $4/3$. The variance of the offspring ($4/3$) is only one half that of their parents ($8/3$).

A loss of one half of the genetic (= heritable) variance of a population each generation is a tremendous loss. As Darwin was well aware,

Females \ Males	2	4	6
2	2	3	4
4	3	4	5
6	4	5	6

Parents

Mean: $2(1/3) + 4(1/3) + 6(1/3) = 12/3$ or 4

Variance: $(-2)^2(1/3) + (0)^2(1/3) + (2)^2(1/3) = 8/3$

Offspring

Mean: $2(1/9) + 3(2/9) + 4(3/9) + 5(2/9) + 6(1/9) = 36/9$ or 4

Variance: $(-2)^2(1/9) + (-1)^2(2/9) + (0)^2(3/9) + (1)^2(2/9) + (2)^2(1/9) = 4/3$

FIGURE 4-4. Halving of variability under blending inheritance (arbitrary units).

it is a loss that would be quite fatal to the theory of natural selection. If variation were actually to be lost at this rate, that which exists at any one moment could be represented as in Figure 4-5; one half of the existing variation must be completely new, one half of the remainder (1/4) would be only one generation old, 1/8 would be two generations old, 1/16 three generations, and so forth. Less than 1/1000 of the existing variation would be nine generations old. In the absence of a mechanism for storing variation, evolution would necessarily proceed on the basis of that which had just arisen within the past generation or two. The problem was so formidable and the solution to it so necessary but so elusive that Darwin once wrote that it seemed as if fertilization were a mixing rather than a true fusion of individuals (Fisher, 1958, p. 1). It

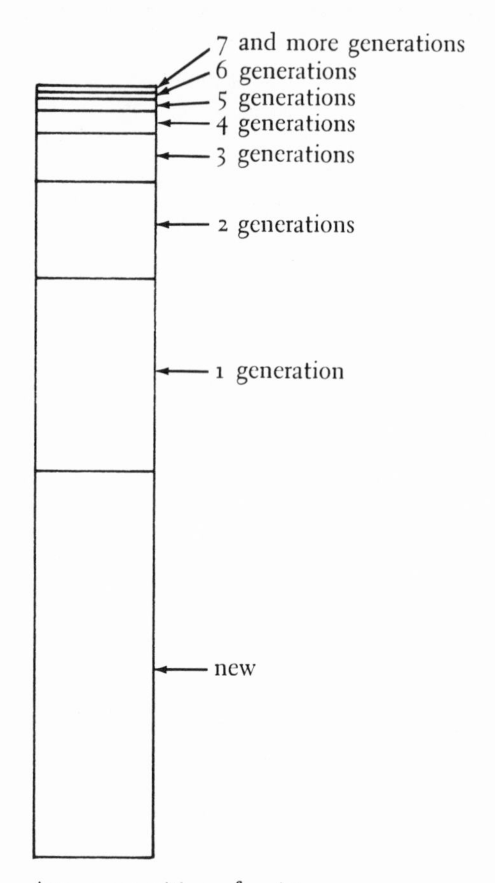

FIGURE 4-5. Age composition of existing genetic variation in a population under blending inheritance; less than 2 percent of all variation would be more than five generations old.

is indeed unfortunate that he did not happen to see the report of Mendel's work.

The Hardy-Weinberg equilibrium, of course, removes the apparent dilemma described above. Hereditary variation is not lost. More exactly, variation is not lost from a population as a consequence of reproducing. As we have seen, the relative frequencies of *AA*, *Aa*, and *aa* individuals are p^2, $2pq$, and q^2 if the frequencies of *A* and *a* are p and q. These frequencies remain constant generation after generation. Consequently, if we let these genotypes have phenotypic values X_{AA}, X_{Aa}, and X_{aa}, then the mean value, $\overline{X}$, equals $p^2X_{AA} + 2pqX_{Aa} + q^2X_{aa}$ (page 46). This value can be computed readily because the various *X*'s and their frequencies are known. As long as the frequencies of *A* and *a* remain unchanged, the value of $\overline{X}$ also remains unchanged.

Variance has been defined (page 46; see Figure 4-4) as the mean squared deviation from the mean. The deviations of the different genotypes in our hypothetical example are $X_{AA} - \overline{X}$, $X_{Aa} - \overline{X}$, and $X_{aa} - \overline{X}$. The variance of the population equals

$$p^2(X_{AA} - \overline{X})^2 + 2pq(X_{Aa} - \overline{X})^2 + q^2(X_{aa} - \overline{X})^2$$

This variance is constant because p, q, X_{AA}, X_{Aa}, X_{aa}, and $\overline{X}$ are all constants. Consequently, under Mendelian inheritance genetic variation in a population does not disappear; it persists unchanged from generation to generation. In reality, this is true only for an infinitely large population; we shall see later that variation is slowly lost in populations of finite size. Nevertheless, the rate of loss is a mere trickle compared to the loss of one half of all existing variation each generation under blending inheritance.

Uses of the Hardy-Weinberg Equilibrium

The great theoretical importance of the Hardy-Weinberg equilibrium is that described above: It shows that genetic variation does not disappear in the passage from generation to generation. On a more practical level, it is useful in a number of respects. First, it provides a basis for analyzing data obtained by sampling populations. Do the observed zygotic frequencies agree with the observed gametic frequencies? Because a number of assumptions are implicit in the calculation of zygotic frequencies by the binomial expansion, agreement with Hardy-Weinberg expectations implies that these assumptions have been met. A poor fit of observed and expected zygotic frequencies, on the other hand, suggests that one or more of these assumptions is not satisfied within the particular population sampled. Second, in calculating gene and zygotic frequencies, information is obtained that permits us to compare samples taken from the same population at different times or from populations inhabiting different geographic regions; significant differences between such samples reveal microevolutionary changes occurring in either time or space.

Testing the Goodness of Fit. The Hardy-Weinberg equilibrium was illustrated by means of data on the *MN* blood group system in human populations (Table 4-1). In the earlier discussion of these data, it was enough for our purposes to point out that observations and expectations were in excellent agreement. Now we want to see on just what basis such a statement is made. To illustrate the procedure, data on parents only will be considered.

Of 2320 parents, 721 (31.1 percent) were type *M*, 1141 (49.2 percent) type *MN*, and 458 (19.7 percent) type *N*. The frequencies of the genes L^M and L^N (to three decimal places) are 0.557 ($= p$) and 0.443 ($= q$). On the basis of the Hardy-Weinberg equilibrium, we expect to see the following numbers of *M*, *MN*, and *N* parents: 719.2 ($= p^2 \times 2320$), 1146.1 ($= 2pq \times 2320$), and 454.7 ($= q^2 \times 2320$). The chi-square test for the goodness of fit of these observations is made as follows:

GENOTYPE	*M*	*MN*	*N*	TOTAL
Observed	721	1141	458	2320
Expected (E)	719.2	1146.1	454.7	2320
Difference (d)	1.8	−5.1	3.3	0
d^2	3.24	26.01	10.89	
d^2/E	0.005	0.023	0.024	

The sum of d^2/E equals chi-square (1 degree of freedom); the probability of getting deviations as large or larger than those observed by chance alone is approximately 90 percent. It appears, then, that the frequencies of zygotes observed do in fact agree with expectations based on the Hardy-Weinberg equilibrium.

To those who are accustomed to testing observations against expected ratios, it may appear that the number of degrees of freedom in the above calculation is too small. The reason for the single degree of freedom is that there is no theoretical expectation for gene frequencies in a population; p (and, consequently, q) is estimated from the sample data. In calculating a chi-square, one degree of freedom is lost for every value of this sort (parameter) which is estimated from the data themselves.

The use of the Hardy-Weinberg equation appears to some persons as an exercise in circular reasoning: Zygotic frequencies are observed; these are used to calculate gene frequencies. Is it possible for observed and expected frequencies to differ in this case? The following data taken from observations by Dobzhansky and Pavlovsky (1955) show that the procedure is not entirely circular. The observations concern two gene arrangements (A and D) of one autosome of *D. tropicalis*; individuals can be homozygous for one or the other of the two arrangements (A/A or D/D) or heterozygous for the two (A/D). The observations and calculations based on them are listed below:

GENOTYPE	A/A	A/D	D/D	TOTAL
Observed number	3	134	3	140
Frequency	0.02	0.96	0.02	1.00
Frequency of $A = p = 0.50$				
Frequency of $D = q = 0.50$				
Frequency expected	0.25	0.50	0.25	
Expected number	35	70	35	140
Difference	—32	64	—32	

Chi-square, the sum of d^2/E, in this example equals nearly 117; again, it is a chi-square with only 1 degree of freedom. The probability of getting deviations as large or larger than these were the three genotypes actually occurring in Hardy-Weinberg proportions is virtually zero.

From the above calculations it appears that the observed and expected distributions of zygotic frequencies differ. Ostensibly, it appears that there is a shortage of the two classes of homozygotes and an excess of heterozygotes. It appears, in fact, that within this population of *D. tropicalis* the A and D gene arrangements form a balanced lethal system. Other tests have revealed that this inference is very nearly correct.

Geographic Variation in Gene Frequencies. Tests of many populations of the same species may show, as in the case of the *MN* blood group system examined previously, that the gene and zygotic frequencies fit the Hardy-Weinberg equilibrium. Nevertheless, it may also be apparent from the data that the different populations differ from one another. Differences in the frequencies of different gene arrangements in *D. pseudoobscura* can be used as an illustration. In Table 4-2 are listed the frequencies of certain gene arrangements observed in a number of localities; some of these places are as far from one another as Texas is from California, whereas others are separated by only one or two kilometers. In the case of distant populations, differences may be qualitative; gene arrangements present in one locality may be missing entirely in another. In less remote localities (see Figure 4-6), differences for the most part involve the relative frequencies, not presence and absence, of various gene arrangements.

Temporal Changes in Gene Frequencies. Repeated samples of individuals inhabiting a certain locality frequently reveal that the genetic composition of the population does not remain constant. At times, the variation is cyclic; the Piñon Flats population of *D. pseudoobscura* goes

TABLE 4-2

FREQUENCIES (PERCENT) OF CHROMOSOMES CARRYING DIFFERENT GENE ARRANGEMENTS IN NATURAL POPULATIONS OF D. PSEUDOOBSCURA SEPARATED BY VARIOUS DISTANCES.

The approximate number (n) of chromosomes examined per locality is also listed.

Localities separated by hundreds or thousands of miles: (n = 1000 or more)

LOCALITY	GENE ARRANGEMENT				
	ST	*AR*	*CH*	*PP*	OTHERS
Austin, Texas	0	18	0	76	6
Death Valley, Calif.	38	40	19	0	3
San Jacinto Mt., Calif.	41	26	29	0	4

Localities separated by tens of miles (within San Jacinto Mountain) (n = 3000 or more)

LOCALITY	GENE ARRANGEMENT				
	ST	*AR*	*CH*	*PP*	OTHERS
Keen Camp	30	26	40	0	4
Piñon Flats	41	26	28	0	5
Andreas Canyon	58	24	15	0	3

Localities separated at most by 2 miles (within Keen Camp) (n = 350 or more)

LOCALITY	*ST* ONLY	
	1939	1940
Keen A	32	35
Keen B	32	31
Keen C	23	30
Keen D	25	29
Keen E	27	32

through a well-defined cycle of gene-frequency changes each year (Table 4-3; see, too, Dobzhansky, 1943). Other populations of the same species may show relatively long-term changes of frequencies which are still poorly understood (Table 4-3; Dobzhansky, 1952a). One of the most

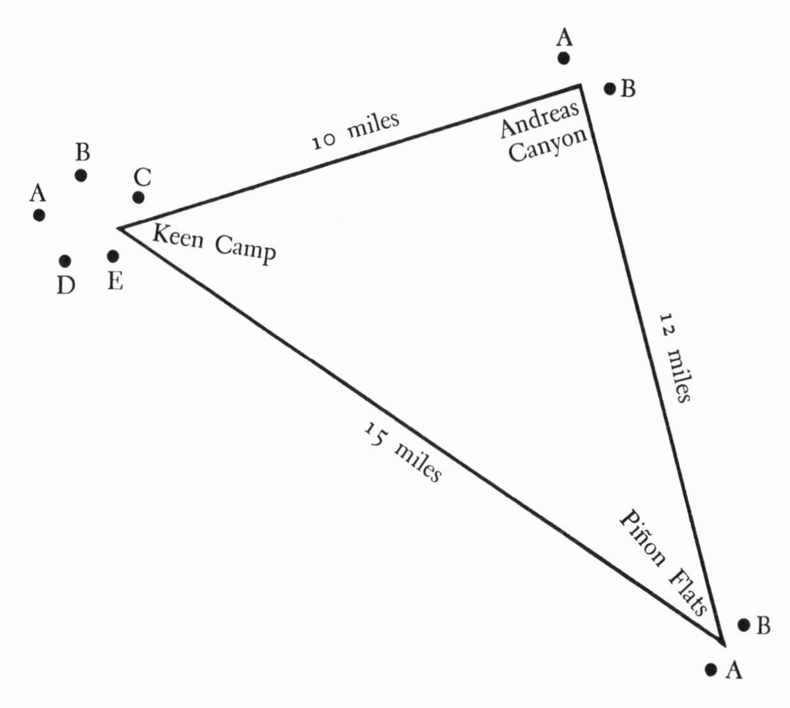

FIGURE 4-6. Collecting stations and localities on Mt. San Jacinto, California. (After Wright et al., 1942.)

spectacular changes has been the increase in frequency of the Pikes Peak (*PP*) gene arrangement of *D. pseudoobscura* in the Pacific coast populations of this species. Prior to 1942, for example, only four chromosomes of this arrangement had been seen among some 20,000 chromosomes examined from the state of California. The Pikes Peak gene arrangement was first observed in samples taken on Mt. San Jacinto in Southern California by Epling and his colleagues (Epling et al., 1953); by that time its frequency in that locality was approximately 10 percent. It is now known to occur at frequencies approaching 10 percent over much of the Pacific coast from British Columbia to the Mexican border.

Pitfalls in the Use of the Hardy-Weinberg Equilibrium. For many early geneticists, the important feature of Mendel's work was the predictions one could make regarding ratios of various offspring following certain crosses. The *Laws* of inheritance rather than its physical basis were given top billing. In a very real sense, population genetics is still in this stage. To many people, the demonstration that alternative alleles are distributed among members of a population as the Hardy-Weinberg equilibrium predicts is exciting in itself. Perhaps this is the most serious pitfall of

TABLE 4-3

TEMPORAL CHANGES IN THE FREQUENCIES (PERCENT) OF DIFFERENT GENE ARRANGEMENTS IN TWO POPULATIONS OF D. PSEUDOOBSCURA. (Dobzhansky, 1947.)

Cyclic seasonal changes at Piñon Flats, California

MONTH	*ST*	*AR*	*CH*	*TL*	NO.
March	52.1	18.2	23.1	6.5	1054
April	40.3	27.7	28.2	3.7	801
May	33.6	29.0	31.3	6.1	642
June	27.9	27.6	39.4	5.1	1130
July	41.9	21.8	30.6	5.6	124
August	42.4	28.4	25.8	3.4	264
September	47.6	22.5	25.7	4.3	374
October–December	49.6	26.3	19.8	4.3	464

Directional changes of prolonged duration at Keen Camp, California

YEAR	*ST*	*AR*	*CH*	*TL*	NO.
1939	27.8	30.4	38.3	3.5	1986
1940	31.5	22.6	41.9	4.0	2382
1941	34.7	24.1	37.1	4.1	764
1942	36.0	16.4	40.3	7.2	414
⋮					
1945	41.0	22.2	29.2	7.6	288
1946	50.0	15.3	28.1	6.7	800

all in the use of these calculations! The Hardy-Weinberg equilibrium should be tested in each instance, not because it represents the end of a study but because it is by such tests that discrepancies are found. Deviations from the Hardy-Weinberg expectations offer an opportunity to examine a population for migration, nonrandom mating, selection, and all other factors which tend to upset the theoretical expectations.

Attempting to Get More Information Than the Sample Contains. A sample of individuals taken from a population consists of a certain number of each of a number of genotypes. From these, gene frequencies can be computed. These, in turn, can be used to determine whether the observed frequencies of association are those expected on the basis of random association. The comparison of observed and expected distributions is a legitimate comparison. Meaningful reference to specific deviations of individual genotypes requires additional information. We saw (page 67) that the frequencies of *A/A*, *A/D*, and *D/D* individuals in one population of *D. tropicalis* differed markedly from the Hardy-Weinberg expectation; because nearly all zygotes sampled were heterozygotes (*A/D*), it would appear that homozygotes were virtually lethal in this population. To draw this conclusion, however, we must make an implicit assumption that gene frequencies are constant within the sampled population. If this assumption is not justified, it is impossible to reconstruct the original gene frequencies from those observed in the sample. Consequently, if we are unable to assume that gene frequencies have remained constant in this population of *D. tropicalis*, it is impossible to talk of the relative viabilities of the genotypes on the basis of the one sample; there are an infinite variety of starting zygotic frequencies which could, as the result of differential mortality, lead to the observed 3 : 134 : 3 ratio of *A/A*, *A/D*, and *D/D* individuals.

The Hardy-Weinberg Equilibrium Is Insensitive. Very often persons try to use the Hardy-Weinberg equation to detect phenomena that are quite beyond its resolving power. As an example, we can cite Ford (1964, p. 111), who mentions that the misclassification of the *medionigra* (heterozygote) form of *Panaxia dominula* as the typical form *dominula* (homozygote) may occur but probably does not occur often because the frequencies of the three forms as classified fit the Hardy-Weinberg distribution.

Because misclassification would alter the calculated gene frequencies, the agreement between observed and expected frequencies offers very little evidence concerning the frequency with which heterozygotes might be misclassified. This point can be illustrated as shown below where 50 percent of *Aa* individuals are misclassified as *AA*:

GENOTYPE	*AA*	*Aa*	*aa*
True frequency	0.81	0.18	0.01
Apparent frequency	0.90	0.09	0.01
Apparent gene frequencies: *A*, 0.945; *a*, 0.055			
"Expected" frequencies	0.893	0.104	0.003

An extremely large sample of individuals would be required to show that the 0.90–0.09–0.01 distribution erroneously tabulated does in fact differ from the Hardy-Weinberg distribution expected on the apparent gene frequencies. Ford is probably right when he says that extremely few individuals are misclassified; the Hardy-Weinberg equation, however, offers little to support this contention.

Pooling Data. A number of collections are frequently lumped into a single large test of the Hardy-Weinberg distribution (for an exception to this statement, however, see Dobzhansky and Levene, 1948). We shall return to this point in greater detail when discussing inbreeding in a later chapter; at the moment, it will suffice to say that data pooled in this way are systematically biased. If the gene frequencies in the various samples differ, the pooled data possess fewer than the "expected" number of heterozygous individuals.

An extreme example can be cited to support the above claim. Suppose a collector unwittingly samples populations in neighboring localities which are, in fact, effectively isolated from one another by distance. Suppose, too, that the frequency of *A* in one population is 100 percent while its frequency in the other is 0 percent (100 percent *a*). If the two samples were of equal size, the collector will calculate gene frequencies as $p = q = 0.50$, from which he will conclude that the frequency of heterozygotes, *Aa,* should be 50 percent. Actually, there will be no heterozygotes in the samples because there were none in either of the two populations. Lumping data from what are partially isolated populations or from generations for a population in which gene frequencies are changing always leads to "expected" frequencies of heterozygotes which are greater than those which actually exist.

Both Ford (1964, p. 131) and Dobzhansky (1951b, p. 131) refer to data obtained by Cunha (1949) on a polymorphic species of fruit flies, *D. polymorpha.* In collections totaling 8070 individuals (State of Paraná, Brazil), there were 3969 *EE,* 3174 *Ee,* and 927 *ee* flies. The frequencies of the alleles *E* and *e* were, consequently, 0.69 and 0.31. The expected numbers of the three genotypic classes are 3841, 3454, and 775; there is a clear excess of homozygotes of both sorts and a deficiency of heterozygotes. Heed (1963) has described a third allele at the *e*-locus of this species which, if present but not recognized in the Brazilian populations sampled by Cunha, would have led to a spurious deficiency of heterozygotes. On the other hand, the original data reveal that the frequencies of *E* and *e* vary from season to season and from place to place. Furthermore, we have seen above in discussing dispersion that local populations of *Drosophila* may be extremely restricted. It seems possible, then, that at least part of the apparent deficiency might have come from pooling samples from many small populations; the large numbers of Brazilian flies tested suggest that many small samples were pooled in the original study.

The Hardy-Weinberg Equilibrium and Multiple Alleles

The Hardy-Weinberg calculations are not restricted to loci occupied by two alleles only; they can be extended to cover cases where three or more alleles exist. The many different gene arrangements that are known for the chromosomes of numerous *Drosophila* species form what are effectively multiple allelic series.

The gene frequency of a particular allele in the case of multiple alleles is calculated as the sum of the frequency of the homozygous class plus one half that of every heterozygous class carrying the allele in question. The sum of the gene frequencies—p, q, r, s, . . —equals 1.00. If these are the frequencies of the alleles a_1, a_2, a_3, a_4, . . ., the expected zygotic frequencies obtained by the binomial expansion are:

$$\begin{array}{ccccccccccccccc} a_1a_1 & a_1a_2 & a_1a_3 & a_1a_4 & \cdots & a_2a_2 & a_2a_3 & a_2a_4 & \cdots & a_3a_3 & a_3a_4 & \cdots & a_4a_4 \\ p^2 & 2pq & 2pr & 2ps & \cdots & q^2 & 2qr & 2qs & \cdots & r^2 & 2rs & \cdots & s^2 \end{array}$$

An example based on the gene arrangements of chromosome 3 of *D. pseudoobscura* is given in Table 4-4.

TABLE 4-4

COMPARISON OF OBSERVED AND EXPECTED FREQUENCIES OF INDIVIDUALS CARRYING HOMOZYGOUS OR HETEROZYGOUS COMBINATIONS OF FIVE DIFFERENT GENE ARRANGEMENTS OF CHROMOSOME 3 OF D. PSEUDOOBSCURA.

The samples were taken at Aldrich Farm (near Austin), Texas, on April 1939 and March 1940.

	APRIL 1939		MARCH 1940	
GENOTYPE	OBSERVED	EXPECTED	OBSERVED	EXPECTED
PP/PP	79	85	70	75
AR/AR	3	6	8	13
TL/TL	0	0	2	1
OL/OL	0	0	0	0
EP/EP	0	0	0	0
PP/AR	53	45	73	62
PP/TL	14	12	15	16
PP/EP	5	4	0	0
PP/OL	1	1	7	6
AR/TL	2	3	5	6
AR/OL	0	0	3	3
AR/EP	0	1	0	0
TL/OL	0	0	0	1
TL/EP	0	0	0	0
OL/EP	0	0	0	0
	157	157	183	183

The Hardy-Weinberg Equilibrium in the Case of Two Loci

If two alleles exist at each of two different gene loci, the expected frequencies of the nine possible genotypes are obtained by multiplying the two individual binomial expansions. Thus, if the expected frequencies of *AA*, *Aa*, and *aa* individuals are p^2, $2pq$, and q^2 and those of *BB*, *Bb*, and *bb* individuals are r^2, $2rs$, and s^2, then the expected frequencies of the nine genotypes are:

AABB	p^2r^2
AABb	$2p^2rs$
AAbb	p^2s^2
AaBB	$2pqr^2$
AaBb	$4pqrs$
Aabb	$2pqs^2$
aaBB	q^2r^2
aaBb	$2q^2rs$
aabb	q^2s^2

Lewontin and White (1960) have published data on the frequencies of chromosomal rearrangements in the Australian grasshopper, *Moraba scurra*; two different chromosomes were analyzed. A portion of their data is given in Table 4-5 to illustrate zygotic frequencies in the case of two independent "loci."

Linkage and "Independence." Unless linkage is absolute, linked genes tend to behave within populations as if they were "independent." They

TABLE 4-5

Numbers of Individuals in One Local Population of Moraba scurra Observed Carrying Various Combinations of Standard and Inverted Chromosomes of Two Different Chromosome Pairs (**CD** and **EF**).

ST, standard arrangement of either chromosome; *BL*, inverted sequence of the **CD** chromosome; *TD*, inverted sequence of the **EF** chromosome. (Lewontin and White, 1960.)

	chromosome **CD**			
chromosome **EF**	*ST/ST*	*ST/BL*	*BL/BL*	total
ST/ST	7	100	324	431
ST/TD	3	22	118	143
TD/TD	0	4	6	10
Total	10	126	448	584

are independent in the sense that, in the absence of selection, the zygotic frequencies for various combinations of alleles at the two loci are given by the product of the frequencies of the two separate loci.

Suppose that at each of two loci on the same chromosome there exist two alleles: *A* and *a*, *B* and *b*. The frequencies of these alleles are p and q ($p + q = 1.00$) and r and s ($r + s = 1.00$). These alleles will be associated with each other at random if chromosomes bearing *AB*, *Ab*, *aB*, and *ab* have frequencies *pr*, *ps*, *qr*, and *qs*, respectively. We will now show that a population *at equilibrium* in the absence of selection does in fact contain these four possible types of chromosomes with precisely the frequencies listed above.

Within a randomly mating population, chromosomes carrying the gene combinations *AB*, *Ab*, *aB*, and *ab* occur with frequencies f_1, f_2, f_3, and f_4 (not necessarily equal to the *pr*, *ps*, *qr*, and *qs* of the preceding paragraph). These chromosomes exist as a variety of combinations:

AB/AB, AB/Ab, AB/aB, AB/ab, Ab/Ab, Ab/aB, Ab/ab, aB/aB, aB/ab,

and *ab/ab*. The types of gametes produced by eight of these ten genotypic classes are not affected by recombination; individuals homozygous for alleles at one or both loci produce the same gametes whether recombination does or does not occur. In the case of the double heterozygotes, recombination chromosomes differ from the nonrecombinants. If the total recombination frequency equals x, the gametes produced by the double heterozygotes will be:

GENOTYPE	NONCROSSOVER		CROSSOVER	
AB/ab	*AB*	*ab*	*Ab*	*aB*
	0.5–0.5x	0.5–0.5x	0.5x	0.5x
Ab/aB	*Ab*	*aB*	*AB*	*ab*
	0.5–0.5x	0.5–0.5x	0.5x	0.5x

Now, the frequency of a given crossover chromosome newly produced within the population is given by the product of its frequency among the gametes of the appropriate double hybrid and the frequency of the hybrid in the population. Thus, as the result of recombination in *AB/ab* individuals, new *Ab* chromosomes arise with a frequency $(0.5x)(f_1f_4)$. However, as the result of recombination within *Ab/aB* individuals, *Ab* chromosomes are lost with a frequency equal to $(0.5x)(f_2f_3)$. At equilibrium, of course, the frequencies of the various types of chromosomes are constant and, consequently, gains must equal losses. Therefore, at equilibrium, $(0.5x)(f_1f_4)$ equals $(0.5x)(f_2f_3)$, or f_1f_4 equals f_2f_3. The identical conclusion would have been reached no matter which of the four chromosomal types had been chosen as an example.

From the definitions of f_1, f_2, f_3, and f_4 and p, q, r, and s, we can see that the following relationships must be so:

$$\begin{aligned} f_1 + f_2 &= p & \quad\text{or}\quad & f_1 = p - f_2 \\ f_3 + f_4 &= q & & f_3 = q - f_4 \\ f_1 + f_3 &= r & & f_3 = r - f_1 \\ f_2 + f_4 &= s & & f_4 = s - f_2 \end{aligned}$$

The above tabulation contains expressions for f_3 in terms of f_1 and f_4. A further expression, relating f_3 to f_2, can be obtained as follows:

$$\begin{aligned} f_1 + f_2 &= p \\ f_1 + f_3 &= r \end{aligned}$$

By subtraction, $f_2 - f_3 = p - r$, or $f_3 = r - \mathrm{p} + f_2$. By substitution into the equation $f_1 f_4 = f_2 f_3$, we obtain

$$\begin{aligned} (p - f_2)(s - f_2) &= f_2(r - p + f_2) \\ ps - f_2 s - f_2 p + f_2^2 &= f_2 r - f_2 p + f_2^2 \\ ps - f_2 s &= f_2 r \\ ps = f_2(r + s) &= f_2 \end{aligned}$$

That is, the frequency of chromosomes of type *Ab* within the population at equilibrium equals the product of the frequency of *A* and the frequency of *b*. Consequently, *A* and *b* are associated at random in respect to one another despite their linkage. The same type of calculation could be made for the other three types of chromosomes with corresponding results:

$$\begin{aligned} f_1 &= pr \\ f_2 &= ps \\ f_3 &= qr \\ f_4 &= qs \end{aligned}$$

Alleles, either linked or unlinked, are associated at random within populations of cross-fertilizing individuals unless (1) linkage is absolute ($x = 0$ in the above calculations) or (2) some event such as differential survival of certain gene combinations prevents the attainment of linkage equilibrium within the population.

Assumptions Underlying the Hardy-Weinberg Equilibrium

The Hardy-Weinberg equilibrium is based on a number of assumptions. The reason for checking population samples against expectations based on the Hardy-Weinberg equilibrium is, as we mentioned earlier, to see whether in fact these assumptions hold. In the following paragraphs, the assumptions will be discussed in some detail.

Random Mating. The use of a square to illustrate the matings of males and females of different genotypes (Figure 4-1) implies that these matings take place at random—that the frequencies of different sorts of matings are solely dependent upon the relative frequencies of the different types of individuals. Although matings in many instances are seemingly random (see Table 4-1), this is not always the case. Marriages between persons are generally random in respect to hidden genetic traits such as blood group systems; they are usually far from random in respect to genetically determined visible characteristics. With the increased use of marriage counseling, carriers of hidden, grossly deleterious, recessive genes will probabably avoid marrying one another.

Nonrandom mating patterns do not alter gene frequencies except through some associated selection process. They do affect the relative frequencies of different genotypes, however. Self-fertilization and matings between close relatives increase the frequency of homozygous individuals at the expense of heterozygotes. Matings between dissimilar individuals, on the contrary, tend to exaggerate frequencies of heterozygotes.

Migration. The Hardy-Weinberg equilibrium applies to closed populations in which all individuals of one generation have descended from parents who were members of the same population. If two populations differ in gene and zygotic frequencies and if individuals from one migrate periodically to the other, then the zygotic frequencies in the second do not fit the Hardy-Weinberg expectation (except fortuitously perhaps), nor do the gene frequencies of this population remain constant. Alternatively, if emigrants from a population do not represent a random sample of the various genotypes, gene frequencies among the remaining individuals will differ from the initial ones.

Selection. The Hardy-Weinberg equilibrium assumes that individuals of different genotypes make equal contributions to the next generation; only under such an assumption can the subdivisions shown in Figure 4-1 be said to add up to 1.00. If, for example, the proportions of surviving individuals differ from genotype to genotype, then zygotic frequencies need not fit the Hardy-Weinberg expectations, nor need gene frequencies remain constant from generation to generation.

Mutation. In computing zygotic frequencies from gametic ones and, further, in defining gene frequencies, no provision was made for the change of one allele into another. Mutational changes of this sort, unless they occur equally frequently in opposite directions, lead to changes in gene frequencies.

Sampling Error. The Hardy-Weinberg calculations treat gene frequencies as constants with no error variance; populations, that is, are assumed to be infinite in size. On the contrary, all populations are finite in size;

their members are limited in number. Consequently, zygotic frequencies (and gene frequencies to which they give rise) are not constant but fluctuate from generation to generation. Furthermore, since the Hardy-Weinberg equilibrium is not a *stable* equilibrium that tends to return to its original value if displaced, the sampling fluctuations can continue until one allele or another has been lost from the population.

A Final Caveat

At this point it is worth repeating the definition of gene frequency for the simple case where two alleles, *A* and *a*, exist in a population. The frequency of *A* is defined as the frequency of *AA* individuals plus one half that of *Aa* heterozygotes; correspondingly, the frequency of *a* equals one half the frequency of *Aa* heterozygotes plus the frequency of *aa* individuals. Calculated in this way the sum of p and q, the frequencies of *A* and *a*, will equal 1.00. Furthermore, all *A*'s and *a*'s have taken part in the determination of their respective frequencies.

If only two phenotypic classes can be recognized (where, for example, *AA* and *Aa* are indistinguishable), the frequency of the gene *a* can be *estimated* by taking the square root of the frequency of *aa* individuals. Such an estimate is useful in a number of instances. However, one cannot test goodness of fit in this case because the expected classes are mathematically identical to those observed. Nor is there any advantage in comparing different populations by comparing gene frequencies obtained in this manner; the original observations serve as a better basis for comparison.

To repeat, if in a sample of individuals taken from a population, the frequencies of *AA*, *Aa*, and *aa* individuals are X, Y, and Z, then the frequencies of the alleles *A* and *a* equal $X + \frac{1}{2}Y$ and $\frac{1}{2}Y + Z$.

5

MIGRATION

Outside the laboratory, local populations are probably never closed. Individuals are always entering and leaving. Only in the large sense of a species inhabiting a remote area—such as the progenitors of Darwin's finches after their arrival on the Galapagos Islands—can we neglect immigrant individuals. But in such large systems the species still subdivides itself into local populations from which emigrants depart and to which immigrants come from elsewhere. It is the continual movement of migrant individuals that genetically unites the many local populations of a species; these individuals are the links which cause all members of a species to share ultimately a common gene pool. In the absence of migrants, carrying as it were their genetic wares and offering these to their more sedentary cousins, even cross-fertilizing species would become differentiated into independent populations, as are self-fertilizing plants and clonal or asexual organisms.

In the present chapter we want to examine some of the algebra of migration together with appropriate illustrative examples. Then, since migration and dispersal are closely intertwined, we shall turn once more to a discussion of dispersion. Finally, although it raises questions about selection that we are not yet prepared to answer, we shall discuss the fate of immigrant genes in populations.

The Algebra of Migration

Migration, like many aspects of population genetics, can be treated in an entirely formal manner without reference to a particular organism.

The procedure has been illustrated in Figure 5-1. Two populations are shown at a time prior to the onset of migration; then one of these, together with its new immigrants, is shown at a somewhat later time.

Suppose that the frequencies of the alleles A and a in population 1 are p_1 and q_1, while the corresponding frequencies in population 2 are p_2 and q_2. In each population, the sum of p and q is 1.00. Suppose, too, that as a result of the migration of individuals from one population to another, population 1 consists of m immigrant individuals and $1 - m$ natives. What is the frequency of A in this population now that it has gained new members?

An average is obtained by summing the products of the values to be averaged times the frequency with which each occurs (page 46). The final or average frequency of $A(p_f)$ in population 1 following migration is

$$mp_2 + (1 - m)p_1$$

or

$$p_f = p_1 + m(p_2 - p_1)$$

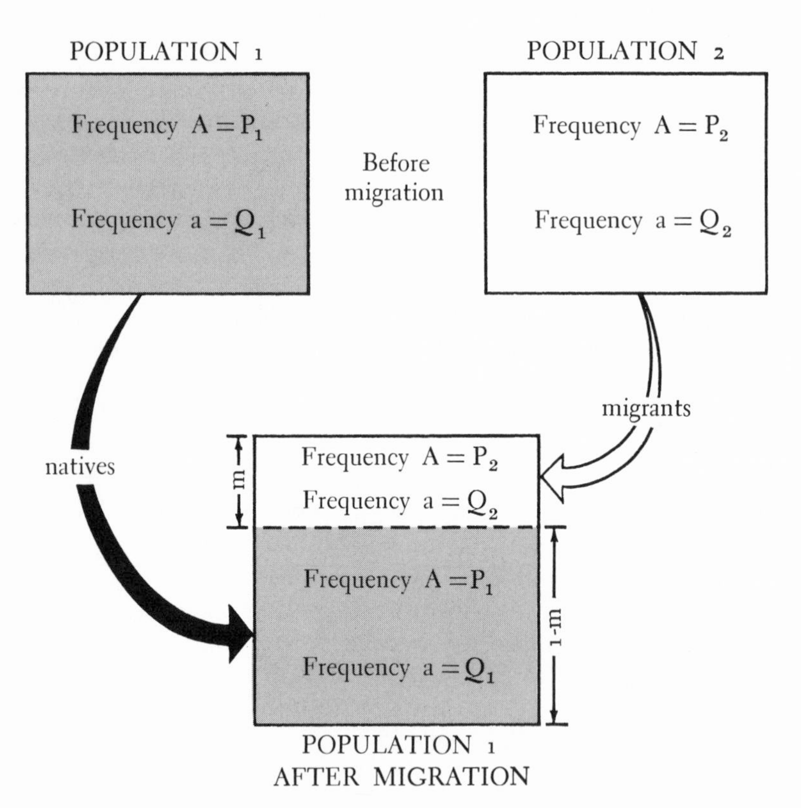

FIGURE 5-1. Diagrammatic representation of migration and its effect on gene frequencies. (After Wallace, 1964.)

If the *change* in gene frequency is represented as the difference obtained by subtracting the old frequency from the new (so that the difference will be negative if the new frequency is smaller), then the change in the frequency of allele *A*, Δp, equals

$$m(p_2 - p_1)$$

From the expression for Δp—$m(p_2 - p_1)$—it is obvious that the effect of migration on the gene frequency within a population depends upon (1) the proportion of immigrants and (2) the difference in gene frequencies in the two populations. If either of these drops to zero, the gene frequency within a population remains unaltered. If neither is zero, the frequency of *A* in a population must be altered by migration. If both m and $(p_2 - p_1)$ are small, though, the product of the two can be very small indeed.

The equation arrived at immediately above,

$$p_f = p_1 + m(p_2 - p_1)$$

can be used to solve for any one item (p_1, p_2, p_f, or m) if values for the other three are known. In the form it is written, we could easily solve for (or predict) the final frequency of the allele $A(p_f)$ if we knew p_1 and p_2 as well as the proportion of migrants (m) that make up population 1.

A more general problem, however, is that in which the proportion of migrants is unknown but the three gene frequencies—p_1, p_2, and p_f—are either known or can be estimated. In this case, the equation can be rewritten as

$$m = \frac{p_f - p_1}{p_2 - p_1}$$

It is in this form that we shall use the equation to analyze an example of migration in man.

Racial Intermixture

An excellent analysis of "migration," in the form of racial intermixture involving American Negroes and whites, has been made by Glass and Li (1953). A somewhat simplified account of this study will be used to illustrate the algebraic analysis given above. The Negro population will be designated "population 1," into which migrants move from "population 2," the white population of the United States. The American Negro population, then, corresponds to population 1 after migration has occurred. We do not have information on gene frequencies in Negro populations in Africa during the eighteenth and early nineteenth centuries;

in place of these unknown frequencies, averages of gene frequencies found in appropriate tribes at the present time will be used.

A large number of genes can be used, at least in theory, for studying the intermixture of the white and Negro races in the United States. In practice, some are unsuitable. The *MN* blood group system, which served to illustrate the Hardy-Weinberg equilibrium (Table 4-1), cannot be used because the gene frequencies in the two races are too nearly alike. The frequencies of still other genes, genes that will be considered shortly, are subject to modification by the differential survival of various zygotic types; these survival patterns are not the same in the United States as they are in Africa. Hence, such genes cannot be used.

A number of genes which are suitable for a study of racial intermixture in the United States are listed in Table 5-1. Four of the alleles listed

TABLE 5-1

ESTIMATION OF THE PROPORTION OF MIGRATION THAT HAS OCCURRED FROM THE WHITE TO THE NEGRO POPULATION IN THE UNITED STATES.

The genes upon which the estimates are based are identified under "allele" and "system." p_1, p_f, and p_2 are the frequencies of the alleles in African Negroes (estimated), American Negroes, and American whites. m, the total proportion of migration; m', the proportion of migrants per generation.

ALLELE	SYSTEM	p_1	p_f	p_2	m	m'
R^0	Rh	0.62	0.45	0.03	0.288	0.033
R^1	Rh	0.06	0.15	0.42	0.250	0.028
t	PTC	0.18	0.30	0.55	0.324	0.038
r	Rh	0.22	0.27	0.38	0.313	0.037
I^B	ABO	0.17	0.13	0.08	0.444	0.057
I^A	ABO	0.15	0.18	0.25	0.300	0.035
R^2	Rh	0.06	0.10	0.15	0.444	0.057

(R^0, R^1, R^2, and r) belong to the Rh blood group system. Two others (I^A and I^B) are members of the ABO blood group system. The remaining allele (t) is the recessive allele of the pair involved in the ability to taste phenylthiocarbamide (PTC); *tt* individuals are nontasters.

The three gene frequencies—p_1, p_2, and p_f—needed to estimate the proportion of immigrants in a population are given in Table 5-1. Frequencies (an average of three studies made on different tribes) that represent the Negro population before migration are listed under p_1; frequencies observed in American Negroes (that is, the Negro population plus immigrants) are listed under p_f; frequencies representative of the white population of the United States are listed under p_2. The substitution of these values into the equation given earlier leads to an estimate of *m,* the fraction of immigrants. It is obvious from the values listed in Table 5-1 that this fraction is a substantial one; estimates range from 25 to 44 percent. In estimating *m,* a fraction is used in which $p_2 - p_1$ is the denominator; consequently, the greater this difference, the more reliable is the estimate of *m.* Since the first two or three entries in the table are the most reliable, it seems that 0.30 is a reasonable estimate of *m.*

A particularly useful description of migration is its rate per generation. In the studies of the Negro population, we have based our analysis on changes in gene frequencies that have been occurring throughout the eighteenth, nineteenth, and twentieth centuries. To reduce the calculated value of *m* to a smaller "per generation" value, we need an estimate of the number of generations during which migration has been occurring, together with an arithmetic technique for manipulating our earlier equation. A generation for human beings requires about 25 to 30 years, so ten generations is a reasonable estimate of the time during which racial intermixture has been taking place in the United States.

To manipulate our equation, we can note that although a certain fraction of migrants may enter a population during each generation, the total frequency of *m* is not permitted to exceed 100 percent. In effect, new migrants replace old migrants as the proportion of migrants increases. Under such conditions, the frequency of natives (nonmigrants) declines exponentially. Thus, using gene R^0 as an example, the frequency of natives after ten generations has been found to be 0.712. This frequency equals x^{10}, where x is the proportion of natives after a single generation of migration. Solving for x, we find that it equals 0.967; the average proportion of migrants each generation, m', equals 0.033. Estimates of the proportion of migrants entering the Negro population each generation are listed in Table 5-1 under m'. Considering the uncertainties in the estimates of both p_1 and the number of elapsed generations, it seems reasonable to conclude that some 3 to 4 percent of the genes in the American Negro population come each generation from that segment of the population known as the white population.

Migration and Dispersal

For most organisms, dispersal is an essential part of migration. Man through legal or social procedures has the capacity to erect mating barriers between groups that are otherwise intermingled at work and at play. For this reason we can speak of intermixture between groups of persons as migration even though very little actual travel may be involved. Similarly, the religious and social castes in India (to the extent that they still exist) represent isolated populations between which migration (intermixture) occurs at low frequencies. Ordinarily, however, the members of a species that inhabit a given locality interbreed with one another; migrants are individuals that have arrived from outside the locality. The arrival of immigrants depends, in turn, upon their ability to disperse from their original homeland.

In Chapter 3, lethal chromosomes were shown to make up a sizable fraction of all chromosomes of *Drosophila* populations. Random combinations of lethal chromosomes are, for the most part, not lethal to the flies carrying them; in such combinations the lethal genes of the two chromosomes must not occupy the same locus. On the other hand, a few combinations of two lethals of seemingly different origins do kill their carriers; in this case, the lethal genes presumably occupy the same locus. If the tested lethals are from widely separated localities—hundreds of miles apart, for example—the frequency of allelism is taken as a measure of the number of loci at which lethal mutations can occur. In this type of study, the lethal chromosomes involved are regarded as being independent in origin. Consequently, if some carry identical lethals, the number of loci at which lethals can occur must be limited.

When lethal chromosomes are obtained from geographically restricted areas, the frequency of allelism is greater than that observed when collection sites are more distant. The probability of allelism due to the finite number of loci at which lethals can occur is present, of course, in all tests. To this low, constant frequency of allelism, however, has been added another source of allelism—identity by common descent. In restricted localities, two lethal chromosomes may carry identical lethals because each is a direct descendant by replication of a single ancestral lethal gene. In this case, the two lethal genes (and the chromosomes bearing them) must be allelic to one another because they are identical. As a consequence, the frequency of allelism is higher among lethal chromosomes taken from a single locality than it is among those collected at remote distances.

The dependence of the frequency of allelism of lethals on distance was reported by Dobzhansky and Wright (1941) and Wright et al. (1942). The lethals used in these studies were obtained from flies (*D. pseudoobscura*) captured at various places on Mt. San Jacinto in Southern California (Figure 4-6). Collections were made at three localities—Andreas Canyon, Keen Camp, and Piñon Flats—separated from one another by ten to twenty kilometers. At each locality there were two or more collecting stations—Andreas A and B, Keen A, B, C, D, and E, and Piñon A and B—separated by distances of one to two kilometers. Each collecting station consisted of a series of small baited traps placed in a circle roughly one hundred meters in diameter.

The results of the tests of lethals from Mt. San Jacinto are presented in Table 5-2. Their interpretation is quite clear: The smaller the area

TABLE 5-2

DEPENDENCE OF THE PROBABILITY OF ALLELISM OF LETHALS UPON DISTANCE AND TIME.

(After Wright et al., 1942.)

	TOTAL TESTS	ALLELIC COMBINATIONS	FREQUENCY OF ALLELES
Within station	2068	44	0.0213
Between stations, within locality	2284	20	0.0088
Between localities	706	4	0.0057
Between regions	6294	26	0.0041
Within station (simultaneous)	594	15	0.0253
Within station (different times)	1474	29	0.0197
Between stations, within locality (simultaneous)	691	9	0.0130
Between stations, within locality (different times)	1593	11	0.0069

from which lethal chromosomes are obtained, the higher the probability (frequency) of allelism. The second part of the table shows that *time* can be substituted for *distance* in the preceding statement: The more nearly simultaneously two lethals are taken from a restricted locality, the more likely they are of being allelic. In this case, however, differences observed in the data are not statistically significant; they merely agree with expectation.

The data on the dispersion of flies (Chapter 2) allow us to speak more definitely about distances and the allelism of lethals, especially that portion of the frequency of allelism caused by common descent. We saw that the dispersal of flies from a point was such that the logarithm of the number of flies recaptured at various distances decreased more or less linearly with the square root of distance. At least in a gross sense, this dispersal pattern was true of the lifetime dispersal (Table 2-5; Figure 2-2). Because lethals are carried from place to place by dispersing flies, it follows that the logarithm of the frequency of allelism (common descent, only) should also decrease more or less linearly with the square root of distance.

An experiment to test the suspected relationship between distance and allelism of lethals has been made using *D. melanogaster* (Wallace et al., 1966; Wallace, 1966b); the results agree well with theoretical expectations. Through the use of the (*CyL-Ubx*)/(*Pm-Sb*) technique (see page 41 and Table 3-4), a fair number of lethals (95) were recovered from relatively few tests (119); the frequency of lethal "genomes" (second and third chromosomes tested simultaneously) was approximately 80 percent. The flies used in these experiments were collected near Bogotá, Colombia, at four trapping sites that were spaced linearly at 30-meter intervals.

The lethals obtained at these four sites were intercrossed both within and between collecting sites. Not all possible tests were made; those which were made are summarized in Table 5-3. Ostensibly, the probability of allelism declined with increasing distance. The numbers are not large enough to give significant results with a chi-square test. The slope of the regression of logarithm of these frequencies on the square

TABLE 5-3

OBSERVED FREQUENCIES OF ALLELISM BETWEEN LETHAL "GENOMES" IN D. MELANOGASTER.

The crosses have been grouped according to the distance between the collecting sites (see Table 3-4) from which the lethals were obtained.

DISTANCE	TYPE OF CROSS	NO. TESTS	ALLELIC COMBINATIONS	FREQUENCY
0 meters	$A \times A$	132	11	0.083
	$B \times B$	232	9	0.039
	$C \times C$	72	2	0.028
	$D \times D$	193	7	0.036
Total		629	29	av. 0.0461
30 meters	$A \times B$	307	15	0.049
	$B \times C$	220	5	0.023
	$C \times D$	186	6	0.032
Total		713	26	av. 0.0365
60 meters	$A \times C$	134	5	0.037
	$B \times D$	452	14	0.031
Total		586	19	av. 0.0324
90 meters	$A \times D$	327	9	0.028
Total		327	9	av. 0.0275
Grand total		2255	83	av. 0.0368

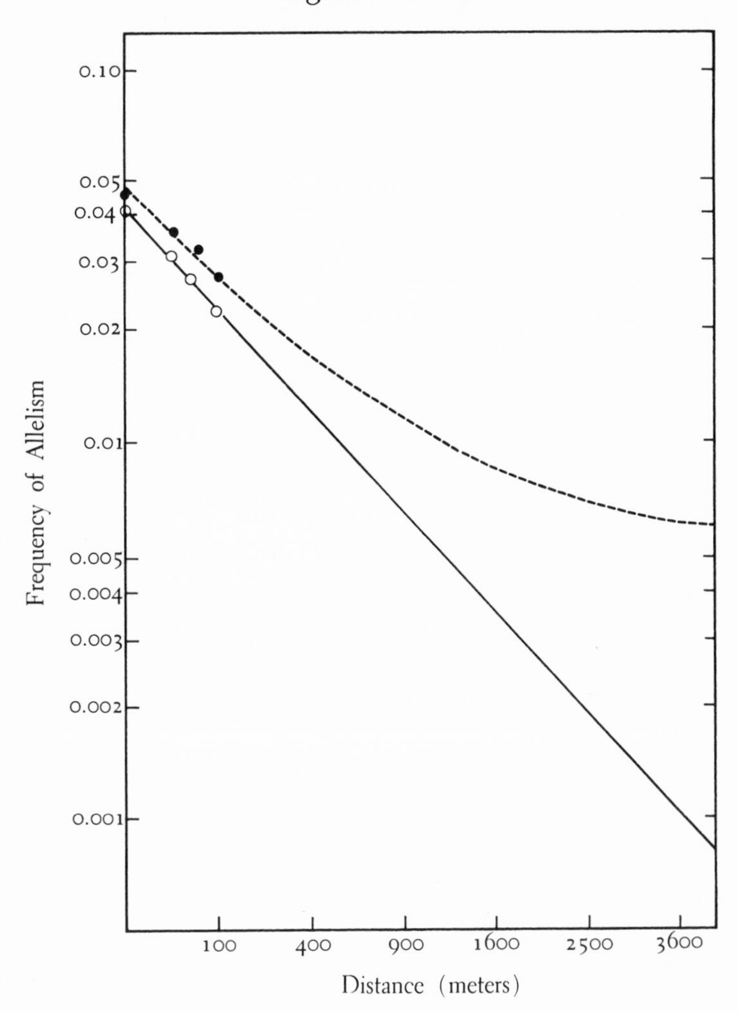

FIGURE 5-2. Probability that two lethal chromosomes are allelic represented as a function of distance between the sites at which they were obtained. Solid line: allelism attributable to inbreeding; dashed line: total allelism. Data obtained from *D. melanogaster* collected near Bogotá, Colombia. (After Wallace, 1966b.)

root of distance (Table 5-4 and Figure 5-2) does differ significantly from zero (zero slope would mean that no relation exists between allelism and distance). The observed decrease in the frequency of allelism with increasing distance, even in the case of distances as small as 90 meters, is probably meaningful. From these data it appears that lethals do spread from locality to locality much as flies themselves.

TABLE 5-4

CALCULATIONS LEADING TO AN ESTIMATION OF THE REGRESSION OF THE LOGARITHM OF THE FREQUENCY OF ALLELISM (Y) ON THE SQUARE ROOT OF DISTANCE (X) BETWEEN SITES WHERE LETHALS WERE COLLECTED (SEE TABLE 5-3).

X	Y
0	2.614
5.48	2.498
7.75	2.438
9.49	2.352

$\overline{X}$ = 5.68 (average distance: 32.26 meters)
$\overline{Y}$ = 2.4755 (average frequency: 0.0299)
b = −0.0262
s_b = 0.0035
t = 7.49 (2 degrees of freedom)
p = 0.01 − 0.02
Y intercept ($X = 0$) = 2.6244
(average frequency at 0 meters: 0.0421)

DIFFUSION OF *orange* IN POPULATIONS OF D. PSEUDOOBSCURA

The analysis of the allelism of lethals is but one technique by which the diffusion of genes through populations can be studied. The artificial introduction of a mutant gene and the analysis of its spread represent an alternative procedure. Dobzhansky and Wright (1947) have carried out an enormous experiment of this sort on *D. pseudoobscura* at Mather, California.

The experiments performed at Mather involved a study of the dispersal of flies as well as an analysis of the diffusion of the mutant gene, *orange* (eye color), within the local population. An analysis of the lifetime dispersion of some 3800 orange-eyed flies yields results very much like those presented in Table 2-5 for the lifetime dispersal data obtained in Southern California. The logarithm of the total number of flies recaptured daily over the 6-day period decreases linearly with the square root of distance. The slope of the regression equals —0.109; the comparable slope for the earlier data was —0.102 (Table 2-5). The error of the slope for the Mather data is small (0.007); consequently, the slope of the regression is highly significant.

Following the analysis of the dispersal of released flies over a period of six days, approximately 25,000 additional orange-eyed flies were released at the center of the trapping field over the following twenty days

(July 23 to August 11, 1945). The release was prolonged so that the area immediately surrounding the point of release would not be grossly overcrowded. (Despite this precaution, the area was probably overcrowded.) Almost immediately (August 10 to 16), trapping of the released flies at 250-meter intervals, up to 1500 meters from the point of release, was begun. The results of these collections are presented in the second column of Table 5-5; the bulk of the survivors are found at the

TABLE 5-5

RECAPTURE DATA FOR *orange* D. PSEUDOOBSCURA AT VARIOUS DISTANCES FROM POINT OF RELEASE IN MATHER, CALIFORNIA, TOGETHER WITH DATA ON THE FREQUENCY OF THE *orange* ALLELE AT THE SAME RECAPTURE SITES.

"Days" represents days following release. (Dobzhansky and Wright, 1947.)

	RECAPTURED FLIES (PER DAY)		FREQUENCY OF *orange*, %	
DISTANCE	0–24 DAYS	11–44 DAYS	11–44 DAYS	300 DAYS
0	168.5	34.3	9.3 (130)	3.0 (646)
250	41.5	—	—	—
500	6.7	4.7	4.6 (88)	1.8 (1444)
750	2.3	—	—	—
1000	1.5	0.6	2.9 (36)	0.5 (646)
1250	2.0	—	—	—
1500	1.0	—	—	—

point of release. One or two orange-eyed flies were captured per day, however, even as far from the point of release as 1500 meters.

The recapture of orange-eyed flies was continued from August 22 through September 5, a period covering 11 to 44 days from the prolonged period of release. During this time both survivors and newly hatched orange-eyed flies were found among the mutants recovered. The number of orange-eyed flies captured per day during this second period of trapping are listed in column 3 of Table 5-5. A further test was made: Chromosomes of the phenotypically wildtype males were tested by mating these males individually with virgin *orange* females; *orange* genes were found among the tested chromosomes (Table 5-5, column 4). Gene frequencies for the mutant *orange* decreases with increasing distance from the point of release at a rate smaller than that of orange-eyed flies; this is expected because frequencies of the latter are determined by the square of the former.

Finally, nearly one year following the release of the mutant flies, several thousand flies were captured in the experimental field and examined; no orange-eyed flies were observed. Again, chromosomes were tested by mating phenotypically wildtype males to virgin *orange* females. As shown in the last column of Table 5-5, *orange* alleles were still present in the population and still most common at the point of release. The *orange* mutation did not spread readily among the flies of the Mather region. Despite the release of nearly 30,000 orange-eyed flies at Mather, the frequency of the *orange* gene was only 3 percent one thousand meters from the point of release approximately one month following release, and only 0.5 percent the following year. (Before releasing the orange-eyed flies, an analysis of wildtype chromosomes revealed that the "control" frequency of the *orange* gene in this population was about 0.2 percent.)

THE FATE OF MIGRANT GENES

In calculating the intermixture of Negro and white races in the United States, we tacitly assumed that the different alleles were not subject to appreciable selection through the differential survival or fertility of their carriers. This assumption applied to alleles both within the race from which they came and within that to which they migrated. The assumption is reasonable since the alleles used in the analysis are normally present in both races, white and Negro, even though their frequencies differ.

The data on the introduction of the mutant gene, *orange*, into the Mather population of *D. pseudoobscura* differ from those on racial intermixture in man. At least they seem to do so. In describing racial intermixture, a model involving migrant individuals was used to gain information about changing frequencies of genes. The changes in gene frequencies were caused by the movement of migrant genes carried by individuals. In the case of the *orange* allele in Mather, the frequency did not remain constant through time. The orange-eyed flies that were released died, of course, as flies do. Before dying, however, these flies had been seen mating with wildtype natives; consequently, it is known that the mutant gene was introduced into the breeding population. Nevertheless, the frequency of *orange* dropped considerably in the year between release and final test.

A similar elimination of a mutant gene would have been observed in the study of Negro-white intermixture had the study been based on the allele responsible for sickle cell anemia. This allele, Hb^S, is common in populations of African Negroes inhabiting malarial regions of that continent; it is rare in all other populations. Its rareness in other populations results from the severe (usually fatal) anemia suffered by homozygous individuals ($Hb^S Hb^S$). Despite its severe effect on homozygotes, the gene is common in malarial regions, seemingly because its heterozygous carriers tolerate malaria better than do individuals homozygous for the normal hemoglobin allele, Hb^A. Malaria is not a serious disease in the United States, so the Hb^S allele carried by many of the original Negroes would have decreased gradually in frequency because of the early death of the Hb^SHb^S homozygotes. Consequently, an estimate of racial intermixture through a study of the present frequencies of the Hb^S allele would have differed considerably from those listed in Table 5-1. The frequency of migrant genes in the American Negro population would have been grossly overestimated because the low frequency of Hb^S would have been ascribed, in the absence of other information, entirely to dilution by alleles entering from the white population. Obviously, if gene frequencies are to be used in gaining information about the behavior of individuals, the genes must be neutral or very nearly so in their effects on their carriers.

What we have said immediately above regarding the near uselessness of the sickling gene for studying racial intermixture over prolonged periods of time applies to seemingly "neutral" alleles over much greater lengths of time. A suggestion was once made (Bernstein, 1925) that mankind consisted originally of three pure races—type *A*, type *B*, and type *O*. The gene frequency of the appropriate allele—I^A, I^B, and I^O—in each race was 100 percent. Populations of men living today were said to represent mixtures of these original races. No support for this suggestion has been found. It takes very little effort, in fact, to adduce contradictory evidence. The three original "pure races," if we are to be consistent regarding their purity, must have been homozygous for one or the other of various alleles now known to exist at many other gene loci. Again, to be consistent, each segment of mankind that is said to be a mixture of known proportions of the three races must have frequencies of alleles at the other loci compatible with the proportions of alleles of the ABO blood group system. The distribution of gene frequencies at the other gene loci cannot be made to fit the notion that mankind was composed initially of a small number of pure races. Genes such as those of the ABO and Rh blood group systems are suitable for the analysis of migrations over periods of one or two centuries; even these genes, however, are influenced too much by natural selection to be useful for reconstructing events extending over hundreds or thousands of centuries.

We have already seen how the recessive gene, *orange*, was nearly eliminated from the Mather population of *D. pseudoobscura* following its introduction there in enormous numbers. One year after the release of nearly 30,000 individuals, less than 3 percent of the alleles taken at the point of release was *orange*. A similar course of events accompanied the release of many (36,000) *D. melanogaster* heterozygous for the recessive gene *ebony* in Great Britain (Gordon, 1935); four months after release, the frequency of *ebony* had fallen from 50 to 11 percent. Finally, data shown in Figure 2-3 show how an introduced gene arrangement was eliminated from a population of *D. funebris.*

A number of experiments have been made in which mutant genes have been introduced into laboratory populations of *D. melanogaster.* Carson (1961) introduced single females heterozygous for the recessive mutations *sepia, rough,* and *spineless* (together with the corresponding wildtype alleles) into populations that were otherwise homozygous wildtype or homozygous *sepia, rough,* and *spineless.* Very rapid changes in gene frequencies occurred following both types of "contamination" (Figures 5-3 and 5-4). In the case of each mutation, final frequencies in both

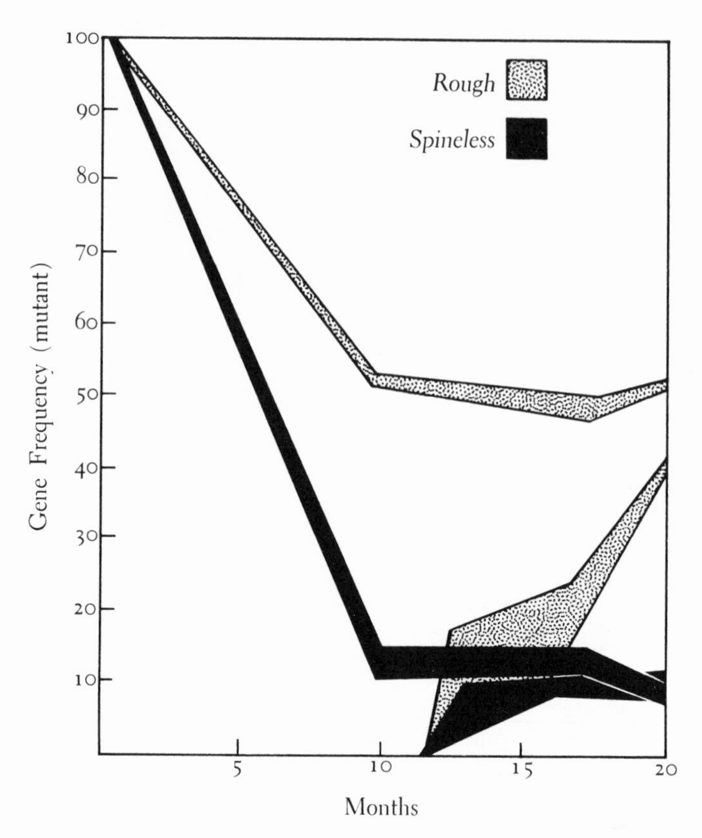

FIGURE 5-3. Changes in the frequencies of *rough* and *spineless* in laboratory populations of *D. melanogaster* following the introduction of (1) single mutant chromosomes into otherwise homozygous wildtype populations, and (2) single wildtype chromosomes into homozygous mutant populations. (After Carson, 1961.)

types of introductions (single mutant into pure wild; single wild into pure mutant) tended to be the same. The single migrant wildtype allele displaced the already present mutants of the *sepia-rough-spineless* populations until the mutant frequency had been lowered to an apparent equilibrium point.

A *sepia* allele recovered from wildtype *D. melanogaster* captured in North Carolina was introduced into populations of various wildtype strains; its fate was clearly dependent upon the wildtype strain (Wallace, 1966c). The wild strains of this study were originally from Bogotá, Barcelona, California, and North Carolina. The frequencies of *sepia* homozygotes and of the *sepia* allele (gene frequency) after one year in these populations are listed in Table 5-6. The data make it quite clear that the final frequency of *sepia* in each instance depended upon the source of the wild flies. Surprisingly, *sepia* was very nearly eliminated from the populations whose wildtype flies were from North Carolina; these flies

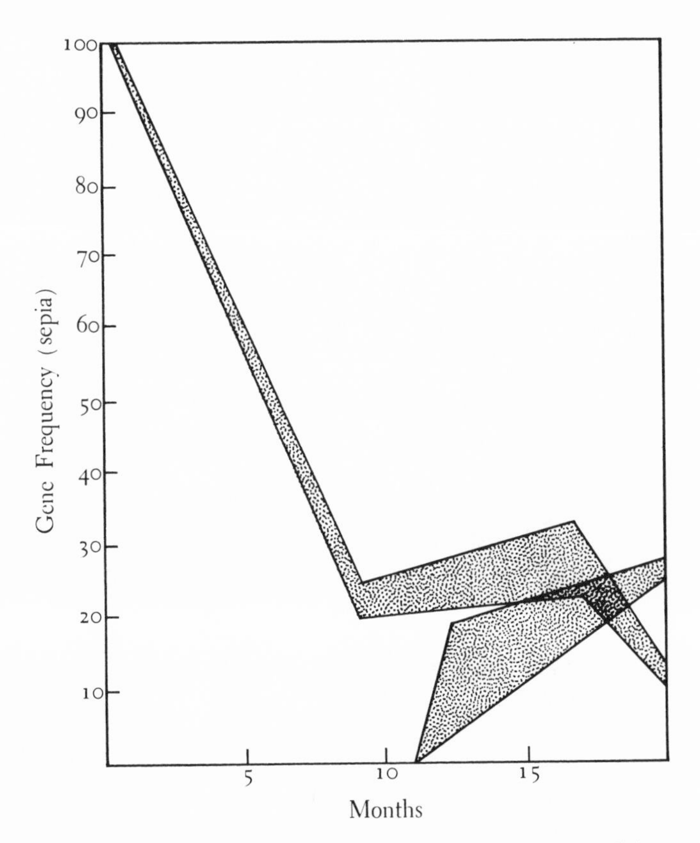

FIGURE 5-4. Changes in the frequency of *sepia* in laboratory populations of *D. melanogaster* (see the legend for Figure 5-3).

TABLE 5-6

OBSERVED FREQUENCY OF *sepia* HOMOZYGOTES AND THE CALCULATED FREQUENCY OF THE *sepia* ALLELE IN FOUR "POPULATIONS" INVOLVING WILDTYPE FLIES OF DIFFERENT GEOGRAPHIC ORIGINS.

Each "population" consisted of thirty cultures (three sets of ten each) maintained by the mass transfer of adults each generation. The observations reported here were made one year after the cultures were started. (Wallace, 1966c.)

SOURCE OF WILDTYPE	FREQUENCY OF *sepia* HOMOZYGOTES OBSERVED, %	NUMBER OF INDIVIDUALS EXAMINED	CALCULATED FREQUENCY OF *sepia* GENES, %
North Carolina	0	7731	0.4
Bogotá	16.5	8766	37.5
Barcelona	2.8	8925	20.2
California	38.5	8380	62.2

had been found originally to be carrying the *sepia* allele used in these studies.

INTROGRESSION

"Migration" is not limited in its effects to the transfer of genes from one local population to another; it can also lead to the introduction of genes from one species to another. A species, as we use the term, is a group of individuals that for one or more reasons does not regularly exchange genes with any other similar group. Excluded, of course, from the reasons for such reproductive isolation is simple distance—spatial isolation.

The statement that different species do not exchange genes is, in most cases, a true description of events in nature. Nevertheless, because living organisms do evolve, because isolating mechanisms evolve, and because, in most instances, evolution is a gradual process, exceptions have been found. Such exceptions, together with methods by which they can be studied, have been described by Anderson (1949). Because the interesting organisms are often not suitable for genetic analysis in the laboratory, the genetic variants that reveal introgression of genes from one species into another are those with obvious phenotypic effects; the most valuable ones, as in the case of migration in man, are those for which the two species differ markedly.

Our interest in introgression at this time is limited to it as an example of migration, of genic intermixture. An excellent example involving two species of towhee in Mexico, *Piplio erythrophthalmus* and *P. ocai,* has been reported by Sibley (1954). Members of these two species differ in plumage color in six areas of the body: pileum, back- and wingspots, back, throat, flank, and tailspots. For each of these six areas, the color usually found in *ocai* was scored 0, while that characteristic of *erythrophthalmus* was scored 4. Colors roughly intermediate to the two species were scored 2, while those displaced in one direction or the other from the midvalue were scored 1 or 3, depending upon the direction of the displacement. In this way, a scale was constructed extending from 0 to 24 (6 areas × 4, the maximum value for each area); individuals of *ocai* tend to fall near 0 on this scale, while those of *erythrophthalmus* tend to fall near 24. Individuals carrying mixtures of genes from the two species should tend to fall between the two extremes.

One area in southwestern Mexico studied by Sibley is shown in Figure 5-5; a series of histograms representing the samples of towhees taken at seven sites within this area is shown in Figure 5-6. The histograms show that at a number of localities the bird populations have intermediate indices and, in addition, form highly variable populations. The genes of the two species are intermixed in these localities. Outside the narrow zone in which the two species meet, the "hybrid indices" take on values near 0 and 24, those of *ocai* and *erythrophthalmus*, respectively.

The Movement of Individuals versus the Migration of Genes

Our treatment of migration has dealt to a large extent with the movement of genes from one group to another. This emphasis merely reflects our interest in the genetics of populations rather than in charting the movements of individuals. The algebraic treatment of the problem began, to be sure, with the stipulation that a fraction m of one population consisted of migrants from another. However, the genes carried by these individuals were then treated in an identical manner as those of the natives. The attitude we have taken toward genes and individuals can be illustrated by that which we would take regarding the tremendous annual "migration" of students from many geographic localities to Ithaca, New York, each fall together with their departure at the close of the academic year. To a sociologist this "migration" may have tremendously important consequences for the economy of the Ithaca area; to a population geneticist this ebb and flow of students has remarkably little interest, because it scarcely affects the local gene pool.

It is difficult, if not impossible, to predict in advance the fate that awaits migrant genes in a new population. The data on orange-eyed *D. pseudoobscura* and ebony-bodied *D. melanogaster* suggest that introduced genes fare rather poorly. The abnormal aspects of these particular mutants cannot be cited in an effort to explain their fate for that of equally abnormal mutants *sepia, rough,* and *spineless* was quite different.

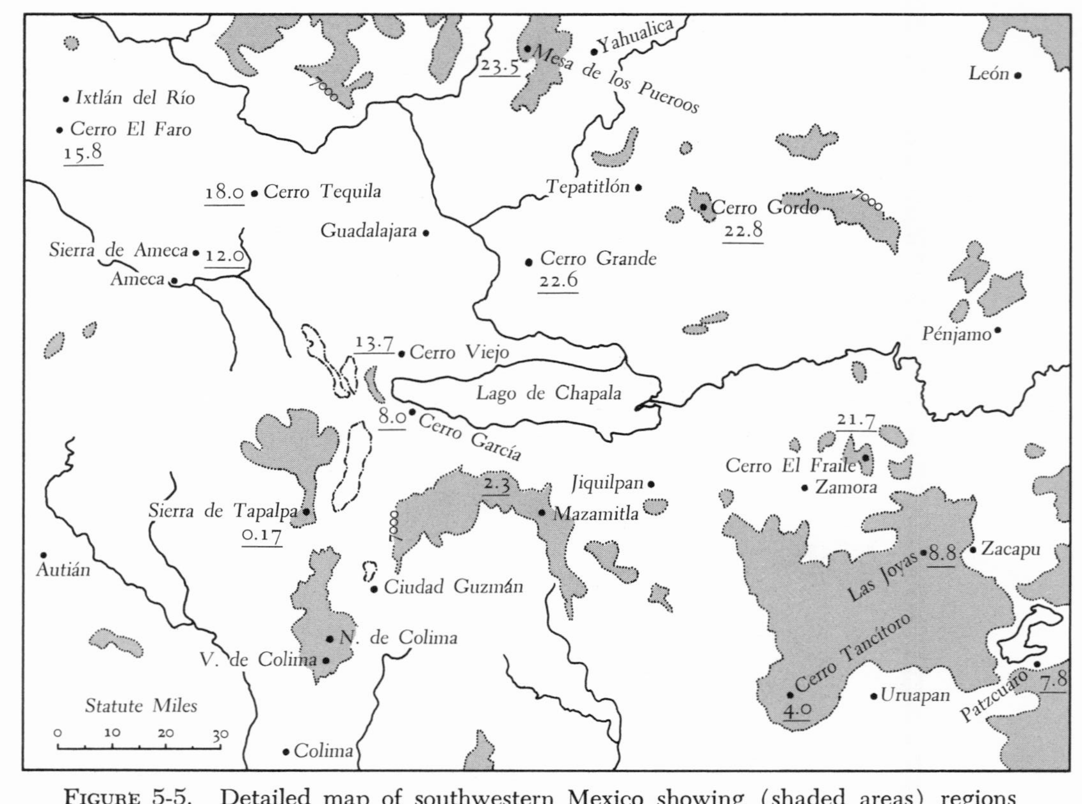

FIGURE 5-5. Detailed map of southwestern Mexico showing (shaded areas) regions favorable for towhees. Numbers are hybrid indices. (After Sibley, 1954.)

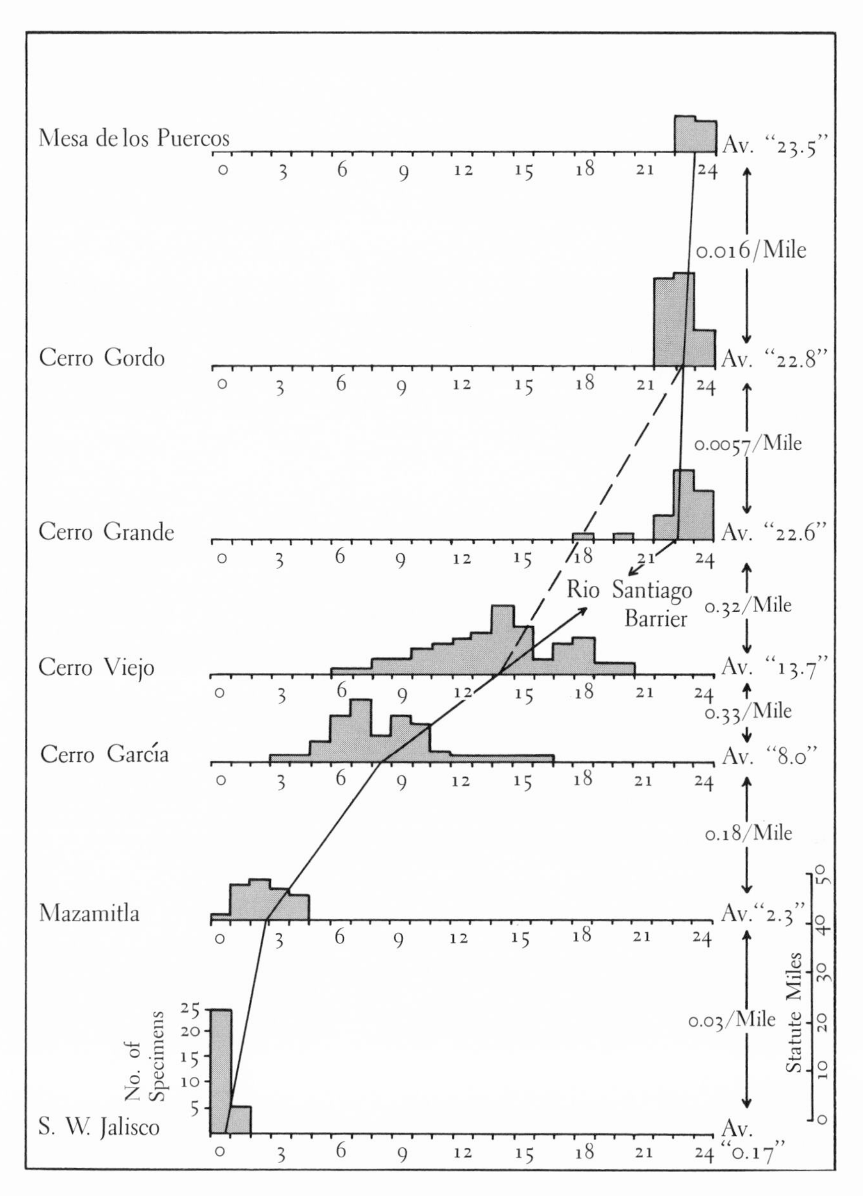

Figure 5-6. Histograms of hybrid indices calculated for some hybrid populations of towhees in Mexico. (After Sibley, 1954.)

The gene arrangement studied by Dubinin and Tiniakov (1946; see Figure 2-3) was a naturally occurring one, but it failed to establish itself in the host population.

In closing this chapter it may be worthwhile to refer once more to the results of Carson (1961) and to the implication which these results may have for local populations of the sort illustrated in Figure 2-7. Local populations, according to the figure and the arguments on which it is based, are much more restricted and inbred than we would normally suspect. Nevertheless, because of the dispersal pattern described in Chapter 2 for *Drosophila,* migrant individuals, although low in number, are scattered over large distances throughout the entire landscape. The local populations may resemble rather closely Carson's laboratory populations into which he introduced single females carrying "migrant" chromosomes. Very few immigrant individuals may enter a given local population, but the genes these individuals do introduce may at times increase in frequency with explosive speed.

6

CHANCE AND INBREEDING

As we saw earlier, the Hardy-Weinberg equilibrium is based upon a number of assumptions. Among these is the absence of chance fluctuations in gene frequencies; this assumption would be met if populations contained infinite numbers of individuals. Real populations do not. And so, the purpose of the present chapter is to examine the consequences of finite size on the genetics of populations.

No matter what plan is followed in deciding the topics to be treated in this chapter, the result will consist of an interwoven account of (apparently) loosely related subjects. Since that is the case, I propose to tighten up the discussion by directing it specifically at the following statement by R. A. Fisher (1958, pp. 9 and 10): "In a population breeding at random in which two alternate alleles . . . exist in the ratio p to q, the three genotypes will occur in the ratio $p^2 : 2pq : q^2$, and these assure that their characteristics will be represented in fixed proportions of the population . . . provided that the ratio $p : q$ remains unchanged. This ratio will indeed be liable to slight changes; first by chance survival . . .; and secondly by selective survival . . . The effect of chance survival is easily susceptible to calculation, and it appears . . . that in a population of n individuals breeding at random the variance will be halved by this cause acting alone in 1.4 n generations . . . [It] will be seen that this cause of diminution of hereditary variance is exceedingly minute, when compared to the rate of halving in one or two generations by blending inheritance."

As a central problem for this chapter, I suggest that we gain an understanding of these few sentences. In particular, I suggest that we attempt to discover why Fisher says that the variance will be halved in 1.4 n generations; if we understand this one point, we will have a good grasp of the role chance plays in modifying gametic and zygotic frequencies within populations.

Equilibria: Stable and Otherwise

We have referred repeatedly to the Hardy-Weinberg equilibrium in discussing the interrelations of gene and zygotic frequencies within populations of cross-fertilizing diploid individuals. If the frequencies of A and a are p and q, then AA, Aa, and aa individuals are expected to occur with frequencies p^2, $2pq$, and q^2. Furthermore, given that individuals have the frequencies indicated, then the frequencies of A and a are p and q. This system endures in the state described until it is disturbed. Should the frequencies of A and a be altered by some accident (the accidental death of some aa individuals, for example) to p_1 and q_1, then the zygotic frequencies will be p_1^2, $2p_1q_1$, and q_1^2. Within a population of this composition, the frequencies of A and a are p_1 and q_1. Once more the system endures until disturbed.

There is nothing in the Hardy-Weinberg equilibrium that includes a tendency to restore original conditions. The equilibrium is not a stable one. Instead, it is a pattern of association of alleles, an association that depends upon whatever frequencies the alleles possess at the moment; the association has no effect upon these frequencies. A stable equilibrium can be represented as a smooth saucer containing a marble; no matter where the marble is placed within the saucer, it returns sooner or later to the lowest point—the point of equilibrium. A comparable analogy for the Hardy-Weinberg equilibrium is a marble resting on a level table top; place the marble where you will upon the table top and it will remain in place. Move it and it remains in the new position. There is no tendency to return to the earlier position.

A third type of equilibrium, one that will be encountered in a later chapter (page 273), is the unstable equilibrium. A population at an unstable equilibrium corresponds to a pencil balanced on end or to a marble resting on the very top of an inverted saucer. Although an equilibrium point corresponding to these examples can be demonstrated mathematically, displacement from that point leads to a series of subsequent changes—the pencil falls, the marble rolls off the saucer, or the gene frequencies take on new values.

Wahlund's Principle

The Hardy-Weinberg equilibrium is not a stable equilibrium. Gene frequencies in populations *tend* to remain at whatever value they have at the moment; having changed, on the other hand, they have no tendency to return to earlier values. Gene frequencies have no memories under the Hardy-Weinberg equilibrium.

Local subdivisions of a large population, according to the above scheme, should most likely differ from one another in gene frequencies. This is so even if all subdivisions came originally from one source and, consequently, started with identical gene frequencies. Using the table top and marble analogy once more, a series of local populations can be represented as a series of marbles on the table top. Each can be displaced without returning to its original spot. There are many spots to which marbles can be displaced; consequently, after some time they are most likely to be found scattered about at different spots on the table top.

The problem we pose at the moment concerns the gene and zygotic frequencies in these separate populations and how these compare with those which would be found if all individuals were members of one large, freely interbreeding population. We touched on this problem earlier (page 72) when we mentioned the pitfalls that threaten the person who pools data when comparing observed and expected zygotic frequencies. Even now, however, we shall not give a comprehensive treatment of Wahlund's principle; Li (1948, Chap. 19) gives a thorough discussion of the subdivision of a population and its consequences.

Suppose that within a given geographical region a species exists in the form of n local populations of equal size. The frequencies of alleles A and a within these populations have values $p_1 : q_1$; $p_2 : q_2$; . . .; $p_n : q_n$, where $p_i + q_i$ equals 1.00 (Table 6-1). The zygotic frequencies of AA, Aa, and aa in the populations will be $p_1^2 : 2p_1q_1 : q_1^2$, $p_2^2 : 2p_2q_2 : q_2^2$, . . . and $p_n^2 : 2p_n q_n : q_n^2$. The zygotic distributions expected under the Hardy-Weinberg equilibrium are expected *within* local populations; the distribution within each population is determined by the gene frequencies of that population.

If all individuals of this particular region were members of a single population, the frequencies of p and q would be averages of the individual values of the different populations. The average frequency of A, $\bar{p}$, would equal the sum of the individual p's divided by n, Sp_i/n. Similarly, $\bar{q}$ would equal Sq_i/n. The zygotic frequencies expected in the consolidated population would be $(\bar{p})^2 : 2(\bar{p})(\bar{q}) : (\bar{q})^2$.

By ignoring the subdivided state of the population and by basing computations on all individuals in the entire area, relative frequencies of AA, Aa, and aa individuals among all those inhabiting the geographic region can be calculated. These are given by the averages of the n individual frequencies:

$$\overline{p^2} = Sp_i^2/n$$
$$\overline{2pq} = S2p_iq_i/n$$
$$\overline{q^2} = Sq_i^2/n$$

TABLE 6-1

CALCULATIONS SHOWING THE RELATION BETWEEN OBSERVED AND EXPECTED FREQUENCIES OF HOMOZYGOTES AND HETEROZYGOTES WHEN DATA FROM SEVERAL POPULATIONS ARE POOLED.

POPULATION	GENE FREQUENCY		ZYGOTIC FREQUENCY		
	A	a	AA	Aa	aa
1	p_1	q_1	p_1^2	$2p_1q_1$	q_1^2
2	p_2	q_2	p_2^2	$2p_2q_2$	q_2^2
3	p_3	q_3	p_3^2	$2p_3q_3$	q_3^2
.	.	.	.	.	.
.	.	.	.	.	.
.	.	.	.	.	.
i	p_i	q_i	p_i^2	$2p_iq_i$	q_i^2
.	.	.	.	.	.
.	.	.	.	.	.
.	.	.	.	.	.
n	p_n	q_n	p_n^2	$2p_nq_n$	q_n^2
Average	$\bar{p}$ or $\Sigma p_i/n$	$\bar{q}$ or $\Sigma q_i/n$	$\overline{(p_i^2)}$ or $\Sigma p_i^2/n$	$\overline{(2p_iq_i)}$ or $\Sigma 2p_iq_i/n$	$\overline{(q_i^2)}$ or $\Sigma q_i^2/n$

$s_p^2 = \Sigma p_i^2/n - \bar{p}^2 = s_q^2$

$\overline{(p_i^2)} = \bar{p}^2 + s_p^2$; similarly,

$\overline{(q_i^2)} = \bar{q}^2 + s_p^2$. Since $\bar{p}^2 + 2\bar{p}\bar{q} + \bar{q}^2 = 1$,

$\overline{(2p_iq_i)} = 2\bar{p}\bar{q} - 2s_p^2$

The zygotic frequencies expected if all populations were consolidated into one freely interbreeding population and those that exist as averages obtained by adding through the separate populations differ; $\bar{p}^2$, for example, is not the same as $\overline{p^2}$. The two sets of frequencies are related to one another in a simple manner through the variance of $p(s_p^2)$:

$$\overline{p^2} = Sp_i^2/n$$
$$\bar{p}^2 = Sp_i^2/n - s_p^2$$
$$\bar{p}^2 = \overline{p^2} - s_p^2$$

The expectation based on the average value of p $(\bar{p})$ is smaller than the average frequency of homozygous AA obtained by averaging p^2's. The subdivided population contains more homozygous AA individuals than one expects on the basis of an average frequency of A obtained by averaging through (pooling) its isolated subdivisions.

The designations p and q given to the frequencies of A and a are arbitrary, so it is obvious that the calculations given immediately above apply to aa individuals as well:

$$\bar{q}^2 = \overline{q^2} - s_p^2$$

(Since $q = 1 - p$, s_q^2 must equal s_p^2.)

Finally, since

$$\bar{p}^2 + 2\bar{p}\bar{q} + \bar{q}^2 = 1.00$$

and

$$\overline{p^2} + \overline{2pq} + \overline{q^2} = 1.00$$

then

$$2\bar{p}\bar{q} = \overline{2pq} + 2s_p^2$$

The frequency of heterozygous individuals expected on the basis of average gene frequencies is greater than the average of actual frequencies by twice the variance of gene frequency. This is the basis for our earlier claim (page 72) that pooled data are always biased toward an apparent shortage of heterozygous individuals.

Wahlund's principle has an important consequence for human populations. Whenever travel between formerly isolated communities is increased so that marriages occur between members of once separate communities, the frequency of homozygous individuals declines while that of heterozygotes increases. Modernization of communication networks, through their effects on marriage patterns, reduces the number of human beings that are genetically defective because of homozygosis for mutant alleles.

This section can be summarized as follows: If a "population" of individuals consists in reality of a number of isolated subpopulations in which the frequencies of p and q differ from one subpopulation to another, the actual frequency of heterozygotes in the entire area is smaller than that predicted on the basis of average gene frequencies, $\bar{p}$ and $\bar{q}$, by twice the variance of gene frequency. The actual frequencies of the two homozygotes are correspondingly increased over expectation—each by the variance of gene frequency.

Chance: A "Genetic Drift" Machine

In the two previous sections, we have learned (1) that gene frequencies under the conditions of the Hardy-Weinberg equilibrium, have no tendency to return to a former frequency if displaced; and (2) that the zygotic frequencies in a series of isolated subpopulations, when averaged, are not related as the Hardy-Weinberg equilibrium predicts but deviate from expectation in a simple manner.

In the present section we shall learn something about one of the processes by which populations come to have different gene frequencies—chance fluctuations in frequency or, as it has been called, *random genetic drift.* Inasmuch as the continued divergence of populations by chance—the scattering of marbles over the level surface of the table top in our earlier analogy—involves the superposition of one probability on another in successive generations, the theoretical aspects of this divergence can become quite formidable.

The operation of chance in determining gene frequencies in population will be illustrated empirically by the use of red and white beans, and then—in the following section—by a description of an actual experiment. The empirical model involves samples of ten beans, some of which are red and some white. Excluding the terminal frequencies of 0 and 100 percent red, there are only nine possible frequencies of red beans in a

TABLE 6-2

RESULTS OF SEVEN SAMPLE GAMES PLAYED ON THE RANDOM DRIFT MACHINE AS DESCRIBED IN THE TEXT.

	GAMES						
DRAWS	1	2	3	4	5	6	7
0	5	5	5	5	5	5	5
1	4	4	4	8	6	6	5
2	7	3	6	8	6	6	4
3	5	4	9	8	7	5	3
4	7	1	9	7	6	3	4
5	6	1	9	7	6	3	4
6	7	0	9	9	7	2	3
7	6	—	7	10	6	3	3
8	8	—	7	—	6	4	2
9	9	—	4	—	3	6	3
10	9	—	5	—	4	7	1
11	10	—	5	—	4	7	2
12	—	—	6	—	0	6	0
13	—	—	7	—	—	7	—
14	—	—	10	—	—	7	—
15	—	—	—	—	—	7	—
16	—	—	—	—	—	8	—
17	—	—	—	—	—	9	—
18	—	—	—	—	—	10	—
19	—	—	—	—	—	—	—

sample of ten: 10, 20, 30, 40, 50, 60, 70, 80, and 90 percent. In preparation for the experiment, then, nine rather large containers are set up in which some 250 to 300 red and white beans are placed: 10 percent red in the first container, 20 percent in the second, and so forth.

The nine containers are merely a convenience; they make the experiment go faster. The experiment consists of games. Each game is started by drawing 10 beans at random from the container holding 50 percent red and 50 percent white beans (see Table 6-2). Among the beans of the first draw, there may be only 4 (40 percent) red ones, as in the first game of Table 6-2. The 10 beans are returned to the 50 : 50 container from which they were drawn, and a second drawing of ten beans is made —this time from the container with 40 percent red beans, because the result of the first draw was 40 percent. In the first game of Table 6-2, the second drawing contained 7 (70 percent) red beans. These beans are returned to the 40 : 60 container from which they were drawn, and the next drawing is made from the container holding 70 percent red beans. The steps of this game have their counterparts in biological populations: A drawing of ten beans represents a population of five diploid individuals (our mechanical analogy neglects the need for separate sexes); for these individuals there are calculable gene frequencies (frequencies of red and white beans). The ten individuals produce many gametes (the large container holding the given frequency of red and white beans) which, in turn, give rise to five new individuals (the new sample of ten

TABLE 6-3

Results of 100 Games From the Random Drift Machine.

Entries are the number of games in which various numbers of red were found remaining after the stated number of draws.

	NO. OF REDS										
DRAWS	0	1	2	3	4	5	6	7	8	9	10
0	—	—	—	—	—	100	—	—	—	—	—
1	0	2	4	11	17	24	24	11	4	2	1
3	3	1	9	11	13	17	11	14	7	9	5
5	9	7	6	8	11	6	10	11	10	12	10
10	28	4	4	2	5	5	3	4	4	8	33
15	36	2	1	2	3	2	4	2	2	0	46
20	37	0	2	3	0	3	0	3	1	2	49
25	42	0	0	0	1	0	3	0	1	0	53
30	42	0	0	0	1	0	1	0	0	1	55
35	43	0	0	0	0	0	0	0	0	0	57

beans) with a new gene frequency. A "game," then, represents a breeding population; a "drawing" represents a single generation of individuals; the container with the same frequency of red beans as a given drawing represents the gametes produced by that generation. From the data listed in Table 6-2, one sees that the frequency of red beans in successive drawings fluctuates from drawing to drawing and eventually reaches 0 or 100 percent. When a game reaches either of these terminal frequencies, drawings cease, because every subsequent drawing would contain the same terminal frequency. All games begin by drawing from the 50 : 50 container, so fixation at 0 or 100 percent should be equally probable; it appears from Tables 6-2 and 6-3 that this is so for the games reported here.

The first drawing consists of ten beans taken from a container holding 50 percent red and 50 percent white beans. The binomial distribution allows us to calculate the proportions of all drawings from this container that should yield 0, 1, 2, . . ., 10 red beans:

NO. OF RED BEANS	TERM OF BINOMIAL EXPANSION	NUMERICAL EQUIVALENT
0	$(\frac{1}{2})^{10}$	0.1
1	$10(\frac{1}{2})^{9}(\frac{1}{2})$	1.0
2	$45(\frac{1}{2})^{8}(\frac{1}{2})^{2}$	4.4
3	$120(\frac{1}{2})^{7}(\frac{1}{2})^{3}$	11.7
4	$210(\frac{1}{2})^{6}(\frac{1}{2})^{4}$	20.5
5	$252(\frac{1}{2})^{5}(\frac{1}{2})^{5}$	24.6
6	$210(\frac{1}{2})^{4}(\frac{1}{2})^{6}$	20.5
7	$120(\frac{1}{2})^{3}(\frac{1}{2})^{7}$	11.7
8	$45(\frac{1}{2})^{2}(\frac{1}{2})^{8}$	4.4
9	$10(\frac{1}{2})\ (\frac{1}{2})^{9}$	1.0
10	$(\frac{1}{2})^{10}$	0.1

A comparison of these expected numbers with those listed in Table 6-3 reveals that the agreement between the two is quite good (see, too, Figure 6-1).

Many biology students do not have sufficient mathematical training to enable them to calculate the expected distributions for the second and subsequent drawings. Intuitively, however, one can see a number of features about these further drawings:

1. A game which has reached 0 percent red or 100 percent red must remain at that frequency; therefore, the numbers in these columns cannot decrease.

2. Any population, but especially those with very low or very high frequencies of red beans, can become fixed at any time at 0 or 100 percent red; consequently, these columns accumulate new populations con-

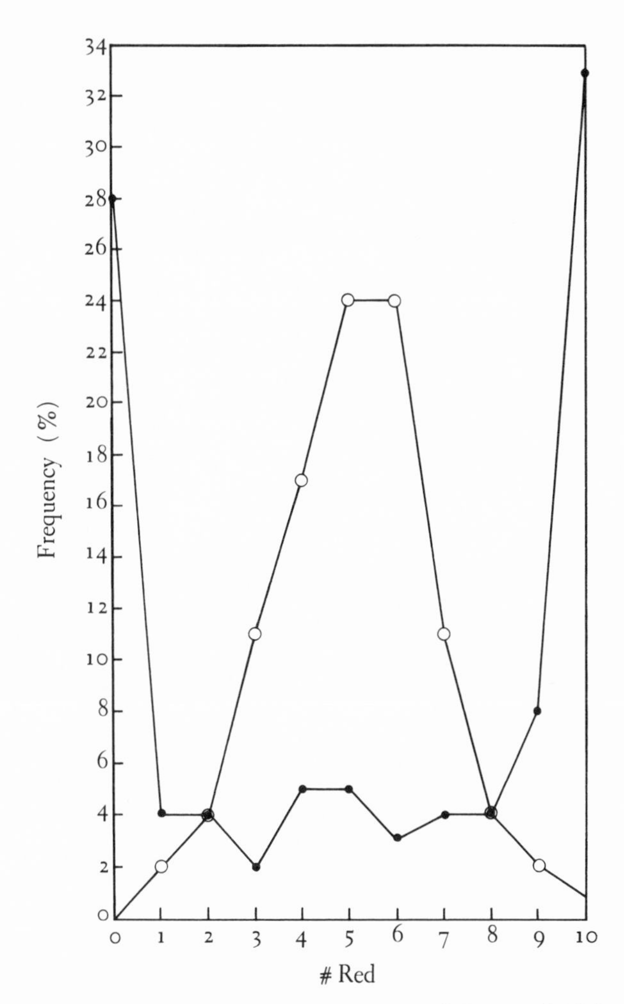

FIGURE 6-1. Results of 100 games consisting of draws of 10 beans from containers with 10, 20, . . ., 90 percent red beans. The first draw was always from the container with 50 percent red beans; later draws were from the containers whose contents equaled the results of the previous draw. The bell-shaped curve shows the results of the first draw; the U-shaped curve shows those of the tenth drawing.

tinually at the expense of the remaining—unfixed, or segregating—populations.

3. As populations take on a variety of intermediate values and especially as more and more of them become fixed at 0 and 100 percent red, the initial peak at 50 percent red disappears; the entire distribution becomes U-shaped (Figure 6-1).

4. Because 50 percent red was the starting point for the games described, the distribution tends to be symmetrical; a change in the starting conditions (to 30 percent red : 70 percent white, for example) would have led to a corresponding change in the relative frequencies of populations fixed at 0 and 100 percent red. We can be very precise about this point: The proportion of games that are expected to terminate at 100 percent red beans is precisely equal to the proportion of red beans in the starting container. This expectation applies to games in which there is no bias in the selection of beans in drawing the samples from the containers.

Chance: An Experiment with Drosophila

The outcome of experiments done on the genetic drift machine can be predicted by theoretical models based on complex probability calculations and can be verified as well by experiments with living organisms. One such experiment (Buri, 1956) has been analyzed in detail by Falconer (1961). Kerr and Wright have published the results of three experiments in successive issues of EVOLUTION (1954); the first of these experiments will be described here.

Kerr and Wright studied ninety-six lines (a "line" corresponds to a "game" of the previous section or to a "population" in the sense we use this term) each of which contained at the outset two alleles, *forked* (f) (a recessive, sex-linked bristle mutation) and nonforked ($+^f$), in equal frequencies. The zygotic constitution of the initial generation in each case was as follows: $1f/f : 2f/+^f : 1+^f/+^f : 2f/Y : 2+^f/Y$. The initial parents were eight in number; in each succeeding generation, eight flies were chosen at random from the progeny to be parents for the succeeding generation. As in the case of the red and white beans, the frequency of *forked* and its wildtype allele fluctuated from generation to generation. If all eight flies chosen at a given time were *forked*, the population was fixed at 100 percent f (0 percent $+^f$). If on the other hand, all eight were wildtype in appearance and if there were no *forked* flies among their progeny, the population was fixed at 0 percent f (100 percent $+^f$). Gene frequencies for the unfixed populations were not determined, because *forked* is recessive and, consequently, females heterozygous for *forked* cannot be distinguished phenotypically.

The results of the experiment with *forked* are listed in Table 6-4 and are shown graphically in Figure 6-2. The proportions of all lines that had become fixed at 100 percent f or 0 percent f at various generations, as well as the proportion of segregating populations, are listed. As far as one can tell from the available data, the populations tend to become fixed for the two alleles equally frequently; there was, however, an ap-

TABLE 6-4

GENETIC DRIFT IN SMALL POPULATIONS OF D. MELANOGASTER.

For each generation are listed the number of populations that contained only wildtype or mutant (*fork*) alleles or that were still segregating. (Kerr and Wright, 1954.)

GENERATION	WILD	UNFIXED	FORKED
0	0	96	0
1	1	94	1
2	1	92	3
3	2	87	7
4	7	79	10
5	10	70	16
6	11	66	19
7	16	59	21
8	17	56	23
9	20	52	24
10	24	47	25
11	29	39	28
12	31	37	28
13	34	34	28
14	37	30	29
15	38	29	29
16	41	26	29

parent tendency following generation 11 for more populations to lose *forked* than to lose its wildtype allele. With the fixation of populations occurring steadily, a correspondingly steady loss of segregating populations occurred throughout the course of the experiment. It is interesting to note that at generation 15, 67 of 96, or about 70 percent, of the *Drosophila* populations had become fixed, whereas in the earlier study on red and white beans, 82 percent had become fixed at the same "generation." The slight difference can easily be ascribed to the differences in "population" size: A sample of four males and four females contains *twelve* representatives of a sex-linked gene; the beans were sampled *ten* at a time. The smaller the sample size, the greater are the fluctuations in gene frequency, and the greater the chance that an allele will be lost or fixed at 100 percent.

THE SUBDIVISION OF A LARGE POPULATION

In this section we shall continue to be interested in the frequencies of a gene in different populations but now we shall pay more attention to

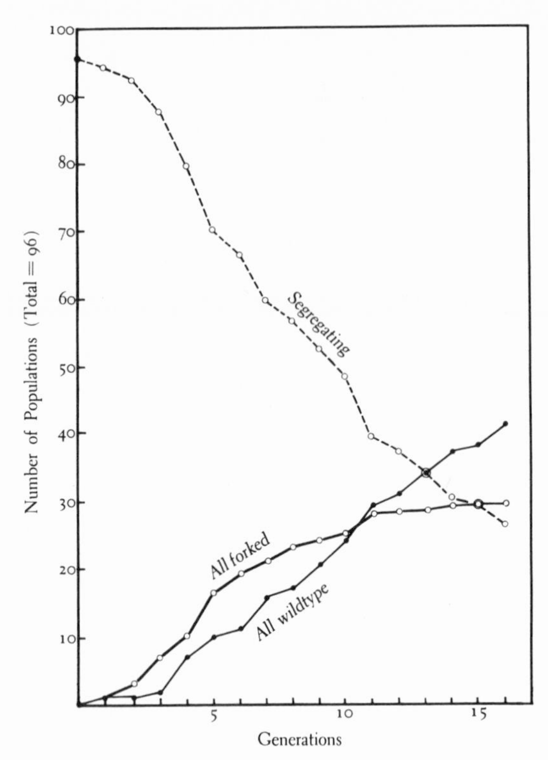

FIGURE 6-2. Decrease with time in the proportion of segregating populations and the increase in those containing only *forked* or wildtype flies in an experiment on small populations of *D. melanogaster*. (After Kerr and Wright, 1954.)

the variation in gene frequency from population to population—to the variance, s_p^2. In the two preceding sections we showed that populations become dissimilar in isolation; some become fixed for one allele, others for the alternative allele, while the remainder (ever decreasing in number) take up all possible intermediate frequencies. Without deriving the theoretical values, we will attempt nevertheless to discuss the variance of gene frequency between populations as this variance changes from generation to generation.

Suppose at the outset that we have an enormous population in which the frequency of A is p and that of a is q, and that from this population a number of smaller ones are set up by drawing off samples of individuals to act as parents. What will be the variance of p among these different samples? We have already done this experiment using the random drift machine! One hundred populations of five individuals (ten "genes") each were drawn from the enormous starting population (the container holding 50 percent red and 50 percent white beans); the composition of these samples are listed in Table 6-3 and have been illustrated in Figure 6-1. This is the one sample most biology students can analyze in terms of expected frequencies of red beans; these expectations are given by the binomial expansion and have been listed on page 106 (see Figure 6-1). Had the gene frequencies not been specified as 50 : 50 but as $p : q$, the binomial expansion suitable for calculating expectations would have been $(p + q)^{2N}$, where $2N$ equals the number of genes (N represents the number of diploid individuals). The variance of p equals $pq/2N$. Consequently, a series of small populations of N individuals each, all of which are drawn from a large population in which the frequencies of A and a are p and q, will have an average frequency of A equal to p but will have frequencies of A distributed about p in such a manner that the variance of p (s_p^2) will equal $pq/2N$.

If the small populations are allowed to perpetuate themselves (as small populations of size N), the gene frequencies will become more and more dissimilar until—as we saw both with the beans and the forked-bristled flies—a number have only allele A while the remainder have only allele a. At that time, what is the between-population variance of p about the mean value?

The initial frequency of A equaled p, so a proportion p of all fixed populations will have only the A allele. The proportion of populations that lose A completely (fixation of allele a) will be q. The variance of p can be calculated readily as follows:

PROPORTION OF POPULATIONS	FREQUENCY OF A WITHIN POPULATION	$p - \bar{p}$	$(p - \bar{p})^2$
p	1.00	$1 - p$	$(1 - p)^2$
q	0	$-p$	p^2

$$\bar{p} = p \qquad s_p^2 = p(1 - p)^2 + qp^2 = pq$$

From these calculations we see that when all populations are fixed at 0 or 100 percent—a situation in which the average deviation from the mean is as large as it possibly can be—the variance of the frequency of p in the different populations equals pq.

It is not possible to derive here the mathematical expression that describes the gradual increase in variance from zero (the original population), to $pq/2N$ (first generation), to pq (after all populations have reached fixation). The equation that describes this increase is

$$s_p^2 = pq\left[1 - \left(1 - \frac{1}{2N}\right)^t\right]$$

For the three times mentioned in the preceding sentence, this equation gives the correct answers:

When t equals 0, $s_p{}^2 = 0$

When t equals 1, $s_p{}^2 = \frac{pq}{2N}$

When t approaches infinity, $s_p{}^2$ *approaches* pq

Furthermore, if N is large and t is small, $(1 - 1/2N)^t$ is approximately $1 - t/2N$. In early generations (1, 2, 3, . . .), consequently, the variance of p would be approximately proportional to time: $pq/2N$, $2pq/2N$, $3pq/2N$, . . .

To summarize, in this section we have gained some insight into the pattern by which gene frequencies of finite populations become increasingly different under isolation. Basically, everything that we have said follows from the "lack of memory" in a population under the Hardy-Weinberg equilibrium.

Inbreeding and Homozygosis

Falconer (1960, p. 60) has pointed out that some inbreeding must occur in any population. By inbreeding is meant the bringing together at fertilization of two alleles that are identical by descent from some specified earlier generation. That inbreeding must occur can be seen from the enormous number of ancestors an individual has, at least in theory: One generation back, two parents; two generations back, four grandparents; three generations back, eight great-grandparents; . . .; x generations back, 2^x great-great . . . great-grandparents. Suppose an isolated population exists that consists each generation of 500 adult individuals; a relatively small laboratory population of *Drosophila* might be maintained at this size. The number of individuals in this population is not large enough to encompass the thousand or more (2^{10}) ancestors to which any individual's pedigree can be traced in the tenth preceding generation. If the number of individuals in the population is less than the theoretical number of ancestors, then some of these individuals must have occupied several places on the family tree. And, consequently, there is a certain probability that the two alleles carried by an individual at a certain locus are the same, not because they are the same mutant form of the gene, but because they are identical through replication—one replicate having arrived at the individual by one line of descent and the other replicate by a different line.

Having defined inbreeding as the bringing together in one individual of two alleles that are identical through descent over a specified number of generations, we can assign a value, F, to the probability that two alleles are indeed identical because of inbreeding. The role F plays in the zygotic composition of a population is shown in Figure 6-3. The

	1-F	F
AA	p^2	p
Aa	2pq	—
aa	q^2	q

FIGURE 6-3. Subdivision of a population into two segments by inbreeding: (1) the segment within which the Hardy-Weinberg equilibrium applies, and (2) that which contains only homozygous individuals.

frequencies of the alleles A and a are given as p and q. A segment, F, of the population consists of individuals in which the two alleles are identical by descent; of necessity, these individuals are homozygous and the relative frequencies of AA and aa are equal to p and q, the gene frequencies. The remainder of the population, $1 - F$, contains zygotes of the three genotypes in proportions given by the Hardy-Weinberg equilibrium.

The frequency of AA individuals in the population shown in Figure 6-3 equals $(1 - F)p^2 + Fp$. This, in turn, equals $p^2 + F(p - p^2)$, or $p^2 + Fpq$. Similarly, the frequency of the other homozygote, aa, equals $(1 - F)q^2 + Fq$, or, after some minor algebraic manipulations, $q^2 + Fpq$. The frequency of heterozygotes equals $(1 - F)(2pq)$, or $2pq - 2Fpq$.

The above expression for the proportions of various genotypes in a partially inbred population are identical in form to those obtained for a subdivided population under Wahlund's principle. The following tabulation makes this identity clear:

GENOTYPE	WAHLUND'S PRINCIPLE	INBREEDING
AA	$p^2 + s_p{}^2$	$p^2 + Fpq$
Aa	$2pq - 2s_p{}^2$	$2pq - 2Fpq$
aa	$q^2 + s_p{}^2$	$q^2 + Fpq$

The identity in algebraic form enables us to think of inbreeding as a form of subdivision of a large population into smaller ones. In addition, because we have an expression relating the variance of gene frequency to population size, we have a means for expressing inbreeding in terms of population size, too. In the preceding section we saw that

$$s_p{}^2 = pq\left[1 - \left(1 - \frac{1}{2N}\right)^t\right]$$

while, in the list immediately above, we see that

$$s_p{}^2 = Fpq$$

Consequently,

$$F = 1 - \left(1 - \frac{1}{2N}\right)^t$$

It now becomes clear why, in defining inbreeding, it was necessary to state the number of generations involved when referring to common descent. Presumably, all genes carried by all organisms have descended by replication from an original "gene"; in this sense, all life is "inbred." The equation bears this out because, if t is very large, F is very nearly 1.00. At the other extreme, there can be no inbreeding without an interval of at least one generation. Again, the equation agrees, because F equals zero when t equals zero.

We have now managed to relate deviations from the Hardy-Weinberg proportions in subdivided populations to variance, variance to population size, inbreeding to variance, and inbreeding to population size. The only task remaining is that which we initially set for ourselves—checking Fisher's statement about the rate at which genetic variance is lost from a population. Before doing this, however, we might rewrite $(1 - 1/2N)^t$ in a more convenient form. For a large N, $(1 - 1/2N)^t$ equals approximately $e^{-t/2N}$. The term e is the base of the natural logarithms and has an approximate value of 2.7183. It is the limit approached by $(1 + 1/x)^x$ as x becomes large; e^{-1} is the limit approached by $(1 - 1/x)^x$ as x becomes large. $(1 - 1/2N)^t$ can be rewritten, then, $(1 - 1/2N)^{2N(t/2N)}$, or $(e^{-1})^{t/2N}$, or simply $e^{-t/2N}$, when N is large.

Population Size and Loss of Genetic Variance

Our final task is to determine why the genetic variance of a population will be reduced by one half in 1.4 n generations, where n equals the number of individuals in the population. This task involves two questions. Why is genetic variance lost at all? Why is it lost at the rate mentioned by Fisher—one half in the 1.4 *n* generations?

An answer to the first question has been the implicit goal of this whole chapter. A population that consists entirely of genetically identical, homozygous individuals possesses no genetic variance. Whatever variation exists between the individual members of such a population would be entirely environmental variance. Now, we have seen that for a population of any finite size, there is a certain amount of inbreeding. The inbreeding leads to a loss of heterozygosity by increasing the proportion of homozygotes. The proportion of heterozygotes does indeed decrease as the number of generations increases. The outcome of this process has been illustrated in Figures 6-1 and 6-2.

The final question concerns the time required to reduce genetic variance by one half. Re-examine F, the coefficient of inbreeding, once more. It is the probability that two alleles at a locus are identical by descent from a pre-existing gene. In the previous section (Figure 6-3), the application of F was restricted to a single locus and F was used to subdivide a population into inbred and noninbred segments. Now, we can change our view of F. Look at all loci in a single individual; of these a proportion F will be homozygous because of inbreeding, while $1 - F$ will not be. Homozygous loci are unable to contribute to genetic variance within populations; such variation resides at segregating loci. F, as we saw above, equals

$$1 - (1 - 1/2N)^t \qquad \text{or} \qquad 1 - e^{-t/2N}$$

As t grows larger, F also grows larger and eventually approaches 1.00. The expression that measures the increase in homozygosis as F increases is $e^{-t/2N}$; by the same token, it measures the decrease in the proportion of heterozygous loci.

Suppose initially that the proportion of heterozygous loci in an individual's genome were H_0. Further, suppose that after t generations the proportion of heterozygous loci, H_t, equals $\frac{1}{2}H_0$. If this is so, we can calculate

$$H_t = H_0 e^{-t/2N}$$

$$\tfrac{1}{2}\mathrm{H}_0 = H_0 e^{-t/2N}$$

$$e^{-t/2N} = 0.5$$

$$\frac{t}{2N} = -\log_e 0.5$$

Since $-\log_e 0.5$ equals 0.7,

$$t = 1.4N$$

The time required to reduce the genetic variance of a population by one half equals 1.4 times the number of individuals in the population.

A Short Cut

In an effort to emphasize the interrelations between inbreeding and the subdivision of populations, we have taken a long and devious path in justifying Fisher's statement about the loss of genetic variation. This was intentional; had we made our point in a simpler manner, we would still have been compelled to present the material we have covered immediately above. The following argument, however, would have sufficed to make our original point.

Assume that a population of plants contains n individuals each of which produces pollen and egg cells. Within this population, each allele (for example, at the a-locus) is given a unique designation so that plant 1 carries alleles a_1 and a_2, plant 2 carries a_3 and a_4, and so forth. Assume, too, that pollen is thoroughly mixed so that fertilization occurs randomly throughout the population. It is clear, then, that a pollen grain carrying the allele a_2, for example, has one chance in n of fertilizing an egg borne by plant 1 and, furthermore, given that it has landed on plant 1, it has one chance in two of fertilizing an egg bearing an allele (a_2) identical to itself. Thus, the probability that an individual carries two alleles that are *identical by descent through a single generation* equals $1/2n$; this probability has been called F.

If the probability that two alleles carried by an individual are identical by descent through a single generation equals $1/2n$, then the probability that they are not identical equals $1 - 1/2n$. The latter probability decreases exponentially as the number of generations contained by the interval of descent increases; the probability of not being identical by descent after two generations equals $(1 - 1/2n)^2$, or, after t generations equals $(1 - 1/2n)^t$. This expression is precisely that arrived at in the section on inbreeding and homozygosis and, of course, can be used to calculate the relation between population size and loss of genetic variation.

Conclusion

The material in this chapter has been discussed in some detail, although no attempt was made to derive profound mathematical relationships. We have been led to a discussion of a variety of topics. We have seen that these seemingly different topics are in fact quite intimately related. The attempt to understand just what lay beneath Fisher's deceptively simple statement, I believe, has been tremendously rewarding. Do not think, however, that you fully understand this statement. In reviewing *The Theory of Inbreeding* (Fisher, 1949), Lewontin (1965) confessed that each year he lectures on inbreeding, and each year he realizes that he does not yet completely understand it. The interrelations between population size, the subdivision of populations, chance fluctuations of gene frequencies, interpopulation variance of gene frequencies, inbreeding, and the genetic composition of single populations are subtle indeed.

7

MATING—RANDOM OR NONRANDOM?

In calculating the proportions of individuals of various genotypes expected under the Hardy-Weinberg equilibrium, random mating between individuals of the different genotypes was taken for granted; the assumption of random mating is implicit, for example, in the use of the modified "checkerboard" (Figure 4-1) as a means for representing probabilities. In the following few pages we shall examine this assumption. The results will disappoint those who like convenient generalizations. Some matings occur at random; others do not. When matings do not occur at random, the deviation from randomness need not be consistent. Men in northern Russia, I understand, prefer brunettes, whereas the acclaimed beauties of Mediterranean countries are often blondes. Fruit flies, too, at times prefer novelties.

Much of the theoretical interest in, and the bulk of available experimental data on, mating behavior are concerned with the reproductive isolation that accompanies speciation rather than with the validity of the Hardy-Weinberg equilibrium. Nevertheless, we have at this point an excuse for discussing mating behavior and the bearing it has on the genetic composition of populations. We shall take advantage of this opportunity; we shall discuss, at least in a preliminary fashion, mating behavior, the effect of assortative (nonrandom) mating on gene and zygotic frequencies in populations, and the difficulty in separating such effects from the complications of natural selection. Enough experimental data will be presented to show the kinds of observations that yield information on mating behavior and to give some notion of the extent to which matings may deviate from randomness.

The Problems of Mating: An Introduction

As individual human beings our routine experiences are so unlike those of other organisms that we often fail to see that they, too, have problems to solve or that they have procedures for coping with these problems. Most persons living in Western societies have a great deal of leisure time; most organisms (especially small ones) do not. Most persons do not work directly with food or food products because these are "supplied" under a system of exchange; most other animals get their food by their own efforts. Human beings have a highly developed nervous system and a means of communication unrivaled in complexity; other organisms rely on sounds and smells that, by comparison, provide effective communication over very small distances indeed. On the other hand, while men worry about future events that may be triggered by incidents occurring on the opposite side of the earth, simpler organisms are concerned with only immediate problems arising in their immediate neighborhood.

The main problems in the life of most animals are getting food and water and reproducing. Reproduction involves the finding of a suitable mate and the avoidance of unsuitable ones. A suitable mate in this sense is a member of one's own species who is of the opposite sex; an unsuitable mate is anything else. The bulk of all natural sounds—insect, bird, amphibian, or mammalian—that we hear is related to the problem of getting potential mates together or keeping them together.

Even when the intimate confrontation of a pair of individuals has occurred, mating is deferred until after a courtship ritual has been performed. Again, courtship serves to assure that the mating individuals will be members of the same species; otherwise, they might fail to leave surviving or fertile offspring. Males and females of several *Drosophila* species may descend upon the same fermenting fruit; courtships between males and females of these different species must be broken off with a minimum of wasted time while those between members of the same species should terminate successfully in matings. For an insect species whose mortality rate is ten percent per day, and for which only one or two hours each day are suitable for moving about, the saving of a minute or two may be important. Flies have very little time to spare.

In many respects the courtship of lower organisms is as complex *and as stereotyped* as the procedure for opening a combination lock. Courtship begins, let us say, with action *A* on the part of the male, the female responds with *B*, whereupon the male is stimulated to do *C*, the response to which is *D*, and so it goes until mating occurs. The entire process is very much like opening a bank vault: The dial on the combination lock is turned counterclockwise to 6—tumblers fall—right to 9—more tumblers fall—one complete turn to the left and stop at 1—still more tumblers fall— · · · —until the vault door swings open freely. And just as the knowledge of the proper digits but not of their order would do a potential thief no good, so the right actions but the wrong sequence in the courtship of many animals would be futile. In fact, different species of ducks do use largely the same signs during courtship but each species has its own sequence in which signals and responses are given (Lorenz, 1958).

How then, can one assume that matings will ever occur at random among males and females in a population? Sometimes they do not. On the other hand, for a great many genetic traits matings are either random or so nearly so that deviations have not been detected. Presumably, the answer to the opening question is simple; not all genetically determined traits are intimately involved in those aspects of the phenotype that play a part in mating procedures. Many mutations are so involved, however. Mutations affecting eye size or eye color in *Drosophila*, presumably by affecting visual acuity but possibly through more general somatic effects, alter mating success. [Reed and Reed (1950) have described such an effect for the *white* mutation in *D. melanogaster;* white-eyed males were only 75 percent as adept at obtaining mates as their red-eyed sibs.] An excellent example with an unknown explanation is that of the *yellow* body-color mutation in *Drosophila;* this mutation occurs in many species and in each of these normal females tend to reject *yellow* males. *Yellow* females, on the other hand, readily accept both *yellow* and normal males (see Tan, 1946).

The Possible Consequences of Nonrandom Mating

Matings can deviate in two obvious ways from randomness—mating individuals may be more similar (positive assortative mating) on the average than are corresponding pairs chosen by some random process, or they may be less similar (negative assortative mating). If the similarity in question has a reasonably simple genetic basis, positive assortative mating promotes homozygosity while negative assortative mating (at least to the extent that it involves matings of dissimilar homozygotes) promotes heterozygosity.

Assortative mating (positive or negative) alone has no effect on gene frequencies, only on the association of alleles in the formation of zygotes. Assortative mating may, however, be important in connection with natural selection. If we postulate that *AA* individuals of either sex prefer mates of type *aa,* and if the frequencies of these two genotypes are not identical, it may be necessary to admit as well that some individuals tend to remain without mates. Negative assortative mating, in other words, may imply differential fertility—a form of natural selection.

Similar situations involving natural selection arise in the case of positive assortative mating. The mating of similar homozygotes tends to increase their frequency at the expense of heterozygotes, much as does inbreeding. However, it frequently occurs that the fitness (in terms of survival or reproduction) of one homozygote is abnormally low. Positive assortative mating, then, tends to exaggerate the frequency with which these homozygotes are formed and exposed to the action of selection.

Random and Assortative Mating in Man

In respect to his biological nature, marriages in man are or are not random largely according to whether the phenotypic traits involved are obvious to all or are cryptic. (Stated in more precise terms, marriages are not contracted at random. In respect to certain genetically determined traits, though, they appear to occur in a random fashion.) At one extreme are external characteristics associated with racial designations; matings in respect to these characteristics deviate markedly from randomness. In these respects, marriages are almost completely assortative.

In contrast to externally visible "racial characteristics," genes whose effects can be revealed only by laboratory tests have no discernible influence on marriage patterns. Parental combinations in respect to the *MN* blood group system are listed in Table 4-1; the numbers of marriage combinations observed, together with their expected numbers, have been listed once more in Table 7-1. The observed deviations are no larger than those expected by chance alone.

TABLE 7-1

Comparison of the Observed and Expected Number of Marriages Between Couples of Various *M* or *N* Blood Group Phenotypes and of Various Haptoglobin Phenotypes.

There is no evidence that marriages are contracted other than randomly in respect to these phenotypes.

MATING	OBSERVED	EXPECTED	MATING	OBSERVED	EXPECTED
$M \times M$	119	111.5	1–1 × 1–1	6	4.4
$M \times N$	142	141.0	1–1 × 1–2	21	28.8
$N \times N$	33	44.6	1–1 × 2–2	28	25.0
$MN \times M$	341	355.3	1–2 × 1–2	48	47.4
$MN \times N$	250	224.6	1–2 × 2–2	91	82.4
$MN \times MN$	275	283.1	2–2 × 2–2	30	35.8
Total	1160	1160.1		224	223.8

Another cryptic trait for which familial data including parental genotypes are available is that concerned with the electrophoretic migration of haptoglobins. Two forms of haptoglobin (one of the blood serum proteins) are recognized in this study; each form is the product of one of the two allelic forms of the responsible gene. Thus individuals identified as 1–1, 1–2, and 2–2 are genotypically Hp^1Hp^1, Hp^1Hp^2, Hp^2Hp^2, respectively, where Hp^1 and Hp^2 are codominant alleles at the *Hp* locus.

The determination of the haptoglobin composition of blood serum requires a study of the migration through a starch matrix of these protein molecules in an electrical field. The data listed in Table 7-1 suggest that marriages are contracted without the aid of high-voltage equipment; each of the parental combinations is extremely close in number to that predicted from gene frequencies under random mating. Between the randomness of matings in respect to cryptic phenotypic traits and the strong correlation between mates in respect to phenotypic traits associated with racial designations lies a series of traits for which marriage partners show intermediate correlations. Stature is an obvious example. Pearson and Lee (1903) studied the correlation between husband and wife in respect to stature (height), span (distance from finger tip to finger tip with arms outspread), and length of forearm (tip of center finger to point of elbow). The heights of husbands and wives are, as we might suspect, correlated; the correlation coefficient is 0.28. Span and length of forearm are, of course, correlated with height; consequently, it is not surprising to learn that in respect to these dimensions the correlation of husbands and wives is 0.20. These data suggest that potential marriage partners "size up" one another by height rather than by span or length of forearm. Intuitively, this seems to be a reasonable notion.

Random and Nonrandom Mating in Insects

In insects, as in man, matings in respect to certain traits are random (or so nearly so that the available data do not compel us to believe otherwise) or assortative. In the latter case, examples of both positive and negative assortative matings are known.

Parsons (1965) has described an example of assortative matings in *D. melanogaster* which nearly duplicates that for stature in man. Males and virgin females grown under two levels of larval crowding were collected. After a period of aging, individuals of the two sexes were placed together so that matings could occur. Mating pairs were drawn out of the container, pair by pair, by means of an aspirator; each pair was stored separately. When approximately one half the flies had mated, the experiment was stopped.

The two members of each mated pair were examined under a microscope and the number of sternopleural bristles on each was recorded.

Unmated flies that remained at the end of the experiment were arbitrarily paired by the observer and the sternopleural bristle numbers of these artificial pairs were also recorded. The results are listed in Table 7-2.

TABLE 7-2

CORRELATION BETWEEN NUMBERS OF STERNOPLURAL BRISTLES ON MALE AND FEMALE MEMBERS OF MATING PAIRS OF D. MELANOGASTER.

(Parsons, 1965.)

	n	♀♀ ($\bar{x}$)	♂♂ ($\bar{x}$)	CORRELATION COEFFICIENT	p	p (DIFFERENCE)
Low larval competition						
Mated	212	23.9	22.9	0.21	<0.01	
Unmated	247	23.9	22.9	−0.04	>0.50	<0.01
High larval competition						
Mated	568	20.2	19.8	0.11	<0.01	
Unmated	610	20.1	19.7	−0.05	>0.20	<0.01
Mixed (high and low)						
Mated	172	22.0	21.2	0.20	<0.01	
Unmated	239	21.6	21.0	−0.01	>0.90	<0.05

As mentioned above, the experiment involved flies raised at two different larval densities; high density means partial starvation of larvae, small adults, and a somewhat smaller average number of sternopleural bristles. The difference in bristle number for the flies raised under the two sets of conditions can be seen in Table 7-2; the partially starved flies have about twenty bristles while the others have three or four more. From the data listed in Table 7-2, it is quite clear that the mean number of bristles on mated and unmated flies are the same. The slight difference observed in "mixed" cultures is reasonable; the larger, nonstarved flies mated somewhat more rapidly than their small, underfed associates. Quite apart from mean bristle numbers of mated and unmated flies, however, is the correlation that exists between the numbers of bristles on the two members of each pair. The correlations, like that for stature in man, are not perfect but that they should exist at all is interesting. The arbitrarily paired males and females of the unmated flies show no corresponding correlation. Just what aspect of mating behavior brings about the observed correlation is unclear; it is possible that flies of approximately the same size mate more readily than do flies of different sizes.

A case of positive assortative mating in respect to a cryptic genetic trait has been described by Wallace (1948). *D. pseudoobscura* males of a given genotype (those carrying the Standard, *ST*, gene arrangement of the X chromosome) were found to mate preferentially with females homozygous for the same gene arrangement (*ST/ST*) rather than with those homozygous or heterozygous for an alternative gene arrangement (*SR*, "sex-ratio," so named because males carrying this type of X chromosome produce only daughters; see Sturtevant and Dobzhansky, 1936b). In this case, males (*SR* or *ST*) were placed ten at a time with twenty females (ten of each of two of the three possible genotypes—*ST/ST*, *ST/SR*, and *SR/SR*; all three combinations of two genotypes were tested with each type of male) and allowed to remain with them for two hours. The males were then removed; females, identifiable by means of notched wings, were dissected to determine the proportion that had mated successfully. The results are given in Table 7-3. Only at 25° is

TABLE 7-3

TEST FOR MATING PREFERENCES BETWEEN FLIES CARRYING *ST* AND *SR* (SEX-RATIO) X CHROMOSOMES IN D. PSEUDOOBSCURA.

"Homozygous (like)" females are those homozygous for the same gene arrangement as the males; "homozygous (unlike)" are homozygous for the other gene arrangement.

		FEMALES					
		HOMOZYGOUS (LIKE)		HETEROZYGOUS		HOMOZYGOUS (UNLIKE)	
TEMPERATURE, °C	MALES	+	TOTAL	+	TOTAL	+	TOTAL
25	*ST*	129	182	73	167	81	159
	SR	70	152	85	174	77	172
		199	334	158	341	158	331
		Chi-square (2 d.f.) = 14.2; $p < 0.001$					
16.5	*ST*	129	201	151	193	127	167
	SR	99	167	114	189	105	185
		228	368	265	382	232	352
		Chi-square (2 d.f.) = 4.6; $p > 0.05$					

there a marked departure from randomness in mating pattern. The chief contribution to this departure is made by the high proportion of *ST/ST* females inseminated by *ST* males. As was the case in the preceding example, there is no obvious reason why *ST* males should or should not prefer *ST/ST* females. These gene arrangements do effect such general features of flies as their average longevity and competitive ability as larvae but otherwise are wildtype in the sense that they do not alter the external morphology of their carriers. Our understanding of the behavioral aspects of mating is so unsatisfactory, though, that we are unable to advance explanations for most observations concerning the distribution of mating frequencies.

The two cases of positive assortative mating described above have been dealt with immediately because, for reasons that will become clear in dealing later with polymorphisms (page 294), negative assortative mating is more interesting—interesting enough, in fact, that examples of negative assortative mating tend to be singled out for emphasis (see Ford, 1964). Consequently, the data described immediately above can serve to maintain a balanced perspective of mating behavior in general; preferences are not always for the unlike.

Two examples of nonrandomness in mating behavior have been chosen to illustrate negative assortative mating. Rendel (1951) in studying the matings of *ebony* (body color) and *vestigial* (small winged) males and females of *D. melanogaster* obtained the data shown in Table 7-4. In

TABLE 7-4

TEST FOR RANDOMNESS OF MATINGS BETWEEN *ebony* AND *vestigial* MUTANTS OF D. MELANOGASTER.

(Rendel, 1951.)

	FEMALES		
MALES	*ebony*	*vestigial*	TOTAL
ebony	75 (14.8%)	119.5 (23.6%)	194.5
vestigial	174 (34.3%)	138.5 (27.3%)	312.5
Total	249	258	507

contrast to the choice of two types of females offered to a single type male in the study on sex-ratio, Rendel placed two types of males with one type of female; matings were classified by examining the progeny of each inseminated female. From the table it is apparent that the distribution of successful matings is not random. *Vestigial* males were much more successful than *ebony* males in mating with *ebony* females; the proportions of matings with *vestigial* females are about even for the two types of males. These data are highly significant (chi-square with one degree of freedom greater than 13; probability near zero); deviations from expectations are positive, of course, in the diagonal cells (*vestigial* ♀ ♀ × *ebony* ♂ ♂) and (*ebony* ♀ ♀ × *vestigial* ♂ ♂). Deviations of this nature are those typical of negative assortative mating.

The second example illustrating the tendency for "unlikes" to mate concerns the moth *Panaxia dominula* of which there are three forms: *dominula, medionigra*, and *bimacula. Dominula* and *bimacula* are homozygous for alternative alleles whereas *medionigra* is the heterozygous form (see page 71). Sheppard (1952) studied 150 cases in which two females were offered a single male, or two different males were placed with one female. In each case, the three individuals included one male and one female of the same genotype. The genotypes of the individuals performing the first copulation were scored. The data (Table 7-5) show

TABLE 7-5

TEST FOR RANDOMNESS OF MATINGS BETWEEN THE *dominula* (*d*), *medionigra* (*m*), AND *bimaculata* (*b*) FORMS OF PANAXIA DOMINULA.

Each trial involved three individuals, two of one sex and one of the other. (Sheppard, 1952.)

MALES		FEMALES		LIKE	UNLIKE	TOTAL
d	—	*d*	*m*	8	20	28
m	—	*d*	*m*	12	14	26
d	*m*	*m*	—	13	14	27
d	*m*	*d*	—	11	22	33
m	—	*m*	*b*	2	0	2
b	—	*m*	*b*	0	1	1
m	*b*	*b*	—	3	15	18
m	*b*	*m*	—	2	10	12
d	—	*d*	*b*	2	1	3
Total				53	97	150

that 97 of the 150 copulations involved unlike pairs; this deviation from 75 : 75 is much greater than one would expect on the basis of chance. An analysis of the six combinations in which the total number of tests exceeded ten reveals that the results are homogeneous. Individuals of unlike genotypes seem to prefer each other in this series of tests.

The final example of nonrandom mating involves the apparently positive assortative mating of nonmimetic males and females of the butterfly species *Papilio glaucus.* Burns (1966) has studied the number of spermatophores carried by the two types of females of this species: the dark form which mimics another butterfly species, *Battus philenor,* and the nonmimetic yellow form. Each spermatophore represents, presumably, a single mating; the males of *P. glaucus* are all yellow (light), nonmimetic. The results of these observations are given in Table 7-6. The

TABLE 7-6

NUMBER OF SPERMATOPHORES FOUND IN LIGHT AND DARK FORMS OF PAPILIO GLAUCUS AND IN BATTUS PHILENOR, AN UNPALATABLE MODEL OF THE DARK PAPILIO.

(Burns, 1966.) The observed distribution of spermatophores has been compared in each instance with the Poisson distribution.

	NO. SPERMATOPHORES PER FEMALE						TOTAL	
	0	1	2	3	4	5	FEMALES	
Mt. Lake, Virginia								
B. philenor	0	17	11	3	1	1	33	
$\bar{x} = 1.73$	5.9	10.3	8.9	5.1	2.2	0.8		Poisson
P. glaucus (dark)	0	33	30	8	0	1	72	
$\bar{x} = 1.69$	13.0	21.9	18.5	10.4	4.4	1.5		Poisson
P. glaucus (light)	0	6	2	2	1	1	12	
$\bar{x} = 2.08$	1.5	3.1	3.2	2.3	1.2	0.5		Poisson
Baltimore Co., Maryland								
P. glaucus (dark)	0	8	3	2	0	0	13	
$\bar{x} = 1.54$	2.7	4.2	3.2	1.7	0.6	0.2		Poisson
P. glaucus (light)	0	4	10	2	0	0	16	
$\bar{x} = 1.88$	2.4	4.5	4.2	2.7	1.3	0.5		Poisson

average number of spermatophores carried by females of the two species are nearly identical: 1.73 (57/33) for *B. philenor* and 1.74 (197/113) for *P. glaucus.* However, it appears that the number of spermatophores carried by the light form of *P. glaucus* is higher than that carried by the dark form: 1.96 (55/28) for the light form and 1.67 (142/85) for the dark. These data suggest nonrandomness in mating preference, but the number of observations is quite small.

The data in Table 7-6 are interesting in connection not only with mating preferences of these butterflies but with their mating behavior generally. One can imagine, for example, that there are a number of encounters between male and female butterflies, a short courtship of some sort, and finally either a successful mating or the termination in failure of the courtship. Do the data support this view? Not quite. If encounters take place by chance and if each encounter had a *fixed* probability of leading to copulation, then the spermatophores should be distributed among females according to the Poisson distribution. This is the distribution that described, for instance, the number of visible mutations recovered in individual cultures of *Drosophila* (Table 3-1).

For each of the five groups of females examined by Burns, an expected distribution of spermatophores among females based on the Poisson distribution has been calculated. These expected figures are shown in Table 7-6. Combining all observations, one finds that approximately 25 virgin (no spermatophores) females were expected among the 146 females examined; actually, there were no females without spermatophores. On the other hand, the Poisson distribution predicts that some 35 females should have carried three or more spermatophores; only 22 females had that many. The range in spermatophore number among females is less than that expected. Indeed, this can be shown by computing variances for these observations. In the case of the Poisson distribution, the variance equals the mean; the observed means and variances are as follows:

BUTTERFLY	AVERAGE	VARIANCE
B. philenor	1.73	0.955
P. glaucus (dark)	1.69	0.601
P. glaucus (light)	2.08	1.902
P. glaucus (dark)	1.54	0.603
P. glaucus (light)	1.88	0.383

In contrast to the expectation based on the Poisson distribution, each variance is smaller—sometimes considerably smaller—than its corresponding mean. Thus, it appears that virgin females are more receptive or, at least, more likely to be fertilized if encountered (and perhaps more likely to be encountered as well) than are nonvirgins. Presumably, too, females that have mated once (or twice) become increasingly nonreceptive. The receptiveness of the nonvirgin light females is somewhat greater, apparently, than that of the dark females.

Mating Behavior and Zygotic Frequencies

When choosing marriage partners, people tend on the average to choose mates with similar physical characteristics. This is certainly true in respect to physical characteristics which are regarded (erroneously, for the most part) as diagnostic in racial classification. It is also true, as we saw earlier, for other physical characteristics such as height. However, the observed correlations were far from absolute. From the folklore of romance we hear that "opposites attract." Now, occasional attractions of unlikes are not ruled out by a correlation coefficient of 0.30 or 0.20; consequently, the blonde Italian girl and the rare brunette in the mass of blonde Russians should have excellent matrimonial opportunities.

Experimental data that reveal an influence of rarity on mating success in *Drosophila* have been published by Ehrman et al. (1965). In these tests, a number of pairs of flies (*D. pseudoobscura*) of two different genotypes were placed together in a small chamber; the data obtained consist of the distribution of all matings between the two types of males and the two types of females. The actual distribution can be compared with a theoretical one based on the relative proportions of the two types of flies.

The results obtained by Ehrman and her colleagues have been summarized in Table 7-7. In the different experiments, three proportions of

TABLE 7-7

EFFECT OF RARENESS ON THE MATING SUCCESS OF *AR/AR* AND *CH/CH* INDIVIDUALS OF D. PSEUDOOBSCURA.

The (*) calls attention to the high ratio of observed to expected numbers of matings for both males and females of rare genotypes. (Ehrman et al., 1965.)

NO. OF PAIRS			FEMALES		MALES	
AR/AR	*CH/CH*	NO. OF MATINGS	*AR/AR*	*CH/CH*	*AR/AR*	*CH/CH*
12	12	265	137	128	131	134
		Expected	132.5	132.5	132.5	132.5
		Obs./Exp.	1.03	0.97	0.99	1.01
20	5	207	138	69	136	71
		Expected	165.6	41.4	165.6	41.4
		Obs./Exp.	0.83	1.67*	0.82	1.72*
5	20	209	69	140	105	104
		Expected	41.8	167.2	41.8	167.2
		Obs./Exp.	1.65*	0.84	2.52*	0.62

flies homozygous for the Arrowhead (*AR*) and Chiricahua (*CH*) gene arrangements of chromosome three were studied: 1 : 1, 4 : 1, and 1 : 4. The ratio of the observed number of matings in which the *AR* and *CH* flies were involved to the number expected is very nearly 1.00 when the flies are in equal proportions. This ratio tells us that *CH* and *AR* females and *CH* and *AR* males are about equally efficient at obtaining mates. This similarity of efficiences is destroyed, however, when flies of one type are much more common than the other; under these conditions, the flies of the rare type—males or females—have a pronounced advantage. The implications of these observations for both population genetics and for insect behavior are tremendous. For example, if carriers of rare genotypes tend to be more successful at finding mates than are other individuals of a population, then rare genes would tend to become stabilized at low frequencies rather than lost. A stabilization of this sort would at least slow down the rate at which genes are lost from populations through genetic drift. Much more needs to be done before the full implications of these findings will become clear. In the meantime, it should be noted (1) that the work of Buri (1956) as well as that of Kerr and Wright (1954) has shown that the frequencies of certain mutant genes do reach 0 and 100 percent much as theory predicts, and (2) that not all rare genotypes have a "mating advantage" in mixed cultures (Ehrman et al., 1965). The rare genotype effect was not observed, for example, in studies on *D. paulistorum.*

Summary

Mating preferences can scarcely be discussed without reference to natural selection. Nonrandom mating either exposes otherwise hidden genes to selective processes (positive assortative mating), decreases the frequencies of homozygotes and augments that of heterozygotes (negative assortative mating), or leaves some individuals without mates and, hence, without offspring (selection by virtue of differential reproduction). The discussion is difficult, too, because there seems to be no systematic bias toward positive or negative assortative mating; matings that do not occur at random may be biased in either direction. One can guess, however, that in the case of processes as complex as courtship and mating, randomness represents in reality a canceling out of many opposing biases. Thus, as all possibilities of mating preferences do seem to exist, populations appear ready and poised to respond to nearly any demand involving mating behavior that environmental conditions may make.

8

MUTATION

General Remarks

The biochemistry of gene mutation is a substantial area within the field of molecular genetics. It owes its existence to (1) the Watson-Crick model of DNA structure and (2) Sanger's demonstration that a given protein consists of a specific sequence of amino acids. From these two concepts, either alone or together, emerge the notion of a point by point correspondence of amino acids in polypeptide chains and base pairs in the DNA molecule, the notion of a genetic code, a basis for predicting the consequences in the protein molecule of given errors in the replication of DNA, and an understanding of the role of these errors in altering enzymatic activity or other function of the protein.

An understanding of gene-protein relationships together with the apparent universality of the genetic code have given rise to studies of molecular evolution (see, for example, Bryson and Vogel, 1965, and Jukes, 1966). These studies relate changes in the amino acid composition of related proteins of different species to the phylogenetic relationships of these species. They include, too, analyses of the duplication of genes and the divergence of gene products that have occurred during the evolution of life. Studies of this sort will eventually form the molecular foundation of evolutionary biology.

Molecular evolution at the moment does not greatly concern the population geneticist. Genetic changes in populations can be traced primarily to mutations that alter the relative fitnesses of individuals within the population. These mutations change genes defined by their physiological action; for the most part these would be changes in the cistrons and operons of molecular genetics.

The mutations to which we shall refer repeatedly are those that affect the fitnesses of their individual carriers. This is so because our basic concern is the evolutionary dynamics of populations in which fitness plays one of the basic roles. Mutations with striking morphological effects are interesting whenever they help in our understanding of the means by which natural selection operates; their fate in populations is determined by natural selection rather than by their phenotypic abnormalities. When the question arises, Are two mutations the same?, we limit our selves to the alternative question, Do they behave as if they were the same? The test for allelism of lethals is based upon the survival of $lethal_1$/$lethal_2$ heterozygotes; if the heterozygotes die, the two lethals are said to be the same. In a population, you see, such lethals would *behave* as if they were the same.

The mutation rate used in population genetics, especially in the population genetics of higher organisms, is a proportion rather than a rate in the true sense. It is the proportion of gametes that should be of one sort but that are, in fact, of a different sort. Thus we cite as a mutation rate the proportion of *a*-bearing gametes produced by *AA* homozygotes or, conversely, of *A* gametes produced by *aa* homozygotes. (This quantity differs from that measured, for example, by microbial geneticists and which is the rate of change per cell division.)

The Roles of Mutation

Mutation is the *only* source of new variation. Whenever we observe two alleles, *A* and *a*, for example, we must assume that one of these alleles has arisen by mutation from the other or that both have arisen from an earlier common form; different allelic forms of a given gene arise only by mutation. Within a local population of limited size, migration acts as an alternative source of new variation. Without mutation somewhere in the past, however, natives and immigrants would be identical. Similarly, recombination can give rise to an enormous array of recombinant types, as students who have studied three- or four-point genetic tests can vouch. Nevertheless, without the initial divergence of allelic types at various loci, there would be nothing to recombine. Migration and recombination are effective only as secondary sources of genetic variation; mutation remains the sole ultimate source.

Mutation plays a special role in the genetics of populations as the spoiler of genetic fixation, as the means by which alleles once lost can be regained. In Chapter 6 we saw that no matter how large the population, there is always a finite probability that one allele will be lost while its alternative form will be fixed at 100 percent. These frequencies —100 and 0 percent—were treated as terminal frequencies in the analysis of the results obtained by sampling red and white beans. Games that reached 0 percent red or 100 percent red were terminated. In natural populations mutation changes the nature of these terminal frequencies. They are no longer traps. A population that carries only allele *A* can accumulate the alternative form, *a*, by mutation; similarly, *A* can accumulate in a population consisting entirely of allele *a*. The rate of fixation and loss of genes in a relatively large, randomly mating population is very small as we have seen; only a small rate of mutation is required to counterbalance this genetic loss.

Mutations: Descriptive

Types of Mutations. The types of mutation that can be detected and subjected to study depend upon the experimental organism and the interests of the experimenter. Microbial geneticists have been interested in biochemical pathways and their genetic control; by far the most common mutation picked up for analysis, then, is the biochemical mutant, the mutant that needs a specific supplement added to the minimal medium in order to survive and grow. There are morphological mutants in lower organisms but, until very recently, relatively few persons have worked with them. On the other hand, higher organisms do not live on chemically defined media and so—with certain notable exceptions that include several metabolic disorders in man—workers have dealt primarily with morphological mutations. Finally, population geneticists are concerned with the flow of mutant genes in and out of populations. This continual flux of mutant genes involves natural selection, so these geneticists have dealt largely with mutations whose effects are measured in terms related to the measurement of selection itself. These are the "viability" mutations which extend from those that kill their carriers (lethals) to those with extremely small deleterious effects (subvitals) and (at least in theory) to those that improve the viability of their carriers (see Figure 3-10 and Table 3-6). We shall see later in this chapter that under certain experimental conditions the latter can be demonstrated quite readily.

Morphological Mutations. A number of techniques exist for studying mutation rates for mutations with gross morphological effects on their carriers. Of these, two will be mentioned here: the attached-X technique in *D. melanogaster*, and the multiple-mutant technique that has been used extensively in *Mus musculus*, the house mouse.

An early study by Timofeeff-Ressovsky (1932) can illustrate the type of information obtained by using the attached-X technique. The details of the inheritance of attached-X's can be obtained from any general genetics text; here we shall merely point out that sons of $\overline{XX}Y$ females obtain their X chromosomes from their fathers and their Y chromosomes from their mothers. Consequently, each son of a male *melanogaster* that has been mated to an attached-X female serves as a test for the occurrence of a mutation in his father's X chromosome. One can look, for example, for *white* sons of red-eyed fathers; red-eyed sons of *white* fathers; or, for that matter, any one of a large number of mutant sons of nonmutant (wildtype) fathers.

Timofeeff-Ressovsky studied mutations at the *white* locus. His studies included analyses of two strains of flies, one obtained in America and the other in Russia. Further, by the use of genetic markers other than *white,* he was able to transfer the *white* locus of the American strain into the Russian strain and, conversely, the *white* locus of the Russian strain into the American strain. Then, with the four stocks (American in American, American in Russian, Russian in American, and Russian in Russian), he determined (1) the rate at which the wildtype allele mutated and (2) the frequency with which it mutated to *white* rather than to one of the many known intermediate alleles of the *white* locus.

The results of Timofeeff-Ressovsky's studies are listed in Table 8-1. Altogether, some 135,000 X chromosomes were tested (one son tests

TABLE 8-1

MUTATIONS FROM WILDTYPE TO MUTANT ALLELES AT THE *white* LOCUS IN TWO GEOGRAPHIC STRAINS, AMERICAN (A) AND RUSSIAN (R), OF D. MELANOGASTER.

(Timofeeff-Ressovsky, 1932.)

ALLELE	BACKGROUND	NO. OF TESTS	NO. OF MUTATIONS	FREQUENCY
W^A	American	31,000	27	0.00087
W^A	Russian	28,200	28	0.00100
Total		59,200	55	av. 0.00093
		41 of 55 mutations were *white*		
W^R	Russian	49,200	26	0.00053
W^R	American	26,100	14	0.00054
Total		75,300	40	av. 0.00053
		19 of 40 mutations were *white*		
		difference: 0.00040 ± 0.00013		

one chromosome) for the presence of new mutations. From the data listed in the table, it is quite clear that the American wildtype allele mutated more often than did its Russian counterpart. It is also clear that the mutation rate of either allele was a property of the allele, not of the background in which it was placed. It is also clear that the higher mutation rate of the American allele resulted from a higher rate of mutation to *white,* itself; the frequency with which the two alleles mutated to non-*white* alleles were nearly identical—0.00024 (14/59,200) for the American allele and 0.00028 (21/75,300) for the Russian. These data tell us, then, that the wildtype alleles at a locus need not be identical. Beneath the phenotypic uniformity of seemingly identical alleles may reside considerable genetic diversity. This fact is an extension of our earlier observation that beneath the uniformity of the wildtype individuals of a species there may exist a wealth of concealed mutant genes.

The study of Timofeeff-Ressovsky was a pioneer effort; our understanding of the *white* "locus" has since improved. Green (1959, 1963) has shown that the *white* locus of *D. melanogaster* can be resolved by refined genetic procedures into four subloci. Furthermore, he has shown that two often-used strains of flies (both American) carry two different wildtype "genes" at the *white* "locus." The findings of Timofeeff-Ressovsky, although improved upon, still form a classic study of evolutionary genetics.

The multiple-locus technique of the mouse geneticists has proved to be an excellent way to study mutation rates in mammals. Large numbers of females homozygous for seven recessive mutations each of which causes a distinct morphological change in the mouse (agouti versus nonagouti, black versus nonblack, color versus noncolor, dense color versus dilute color, dark eye versus pink eye, solid color versus spotting, and short ear versus normal ear) are crossed with males homozygous for the wildtype allele at each of the seven loci. The wildtype alleles are dominant to the recessive mutations, so the F_1 offspring are normally wildtype in appearance. A small number of offspring showing one or the other of the seven mutant traits are observed, however. These presumably represent mutations at the corresponding loci in the wildtype males. Each F_1 individual examined is a test of the seven loci carried by a single sperm.

Results obtained by William Russell in studies carried out at the Oak Ridge National Laboratory are listed in Table 8-2. Over ½ million offspring of wildtype fathers were examined; these represent tests of some 3½ million gene loci. Of these, 32 were mutant; the average mutation rate, consequently, is 0.0000084 (0.84×10^{-5}). A corresponding test in which the seven-mutant animals were fathers while the mothers were wildtype gave a single mutation among nearly 100,000 individuals (700,-000 loci) examined; this corresponds to a mutation rate of 0.0000014 (0.14

TABLE 8-2

SPONTANEOUS MUTATIONS AT SEVEN LOCI IN THE MOUSE, MUS MUSCULUS.

(Russell, after U.N. Report, 1962, p. 106.)

Spermatogonia	
Number of offspring	544,897
Number of mutations	32
Mutation/locus/gamete	0.84×10^{-5}
Oocytes	
Number of offspring	98,828
Number of mutations	1
Mutation/locus/gamete	0.14×10^{-5}

$\times\ 10^{-5}$). These, then, are estimates based on seven loci, of the rates at which mutant genes enter mouse populations; they represent, that is, the proportion of gametes that carry a mutant allele at a given locus.

Viability Mutations: Sex-Linked Lethals. Sex-linked lethals are the favorite material of Drosophila geneticists. Does a given treatment influence mutation rate? Test it by studying sex-linked mutations. To what extent are "recessive" lethals dominant? Find out by studying sex-linked lethals. There are several reasons for the popularity of these lethals: Changes at many loci are lethal to their carriers; hence cumulative lethal mutation rates are relatively high. Techniques, such as the original *ClB* technique, that are based on the analysis of a large block of genes as a single unit are available in many *Drosophila* species. Lethality can be described quantitatively; lethal-nonlethal decisions are frequently based on the presence or absence of easily recognized classes of flies. Decisions regarding other levels of viability can be based on the relative frequencies of these same classes.

The *ClB* technique has been largely superseded by other, more efficient ones such as "*Basc,*" a technique utilizing a doubly inverted X chromosome carrying the mutant genes *apricot* (one of the *white* alleles, a recessive) and *Bar* (a semidominant mutation affecting eye shape). In a series of tests made at Cold Spring Harbor under the direction of Dr. M. Demerec, some 130,000 X chromosomes were tested by means of the *Basc* technique (known, too, as the Muller-5 technique); the rate of mutation to sex-linked lethals observed in these tests was 0.0030. Since the X chromosome of *D. melanogaster* is about the same size as the third chromosome of *D. pseudoobscura,* it would seem that some 300 loci are involved in determining the observed mutation rate; in this case, an average mutation rate per locus would be 0.00001, or 10^{-5}.

Viability Mutations: Autosomal Lethals. A large part of the popularity of sex-linked lethals among Drosophila geneticists lies in the hemizygous nature of male fruit flies; a live male must carry a nonlethal X chromosome. Lethals recovered among sperm must arise, then, during the development of the tested male. The same cannot be said of autosomal lethals; sufficient sperm must be tested per male to decide whether he carries two, one, or no lethal autosomes. Furthermore, isolating individual autosomes from a diploid individual and then bringing replicates of these together once more in homozygous individuals requires one generation more than the equivalent test of the X chromosome.

A summary of a variety of tests of lethal (plus semilethal) mutation rates for the second chromosome in *D. melanogaster* is presented in Table 8-3. The first entry is based on a test of nearly 4000 chromosomes.

TABLE 8-3

SEVERAL ESTIMATES, BASED ON DIFFERENT TECHNIQUES, OF THE MUTATION RATE FOR LETHALS IN THE SECOND CHROMOSOME OF D. MELANOGASTER.

	NO. OF TESTS	NO. OF LETHALS	MUTATION RATE
Direct test	3,778	23	0.0061
Accumulation	6,764	33	0.0049
Accumulation within a once-lethal-free population	41,103	181	0.0044

These were obtained from males, rendered homozygous, and classified as lethal or nonlethal. Males yielding sufficient numbers of lethals among tested sperm were recognized as lethal heterozygotes and were eliminated from the data. Among the remaining tests, twenty-three lethals were recovered. The total mutation rate of second chromosome lethals estimated from these data is 0.0061.

The second entry is based on two studies of lethals which arose in chromosomes that were maintained in heterozygous individuals balanced over genetically marked chromosomes. The number of tests in this case is expressed as chromosome-generations—the product of the number of chromosomes tested in homozygous condition and the number of generations these were allowed to accumulate lethal mutations. Thirty-three lethals were recovered in tests that represent 6764 chromosome generations; the estimated mutation rate in this case is 0.0049.

Finally, an estimate of the mutation rate for second chromosome lethals can be obtained from a series of tests (Table 8-4) involving chromosomes recovered from a large laboratory population. This population was started

TABLE 8-4

ACCUMULATION OF SPONTANEOUS LETHALS AND SEMILETHALS IN SUCCESSIVE "GENERATIONS" OF AN INITIALLY LETHAL-FREE POPULATION OF D. MELANOGASTER.

A "generation" in this and other laboratory populations maintained at Cold Spring Harbor is really a 2-week interval between samples; the actual generation time would be somewhat longer.

GENERATION	TESTS	LETHAL + SEMILETHAL	FREQUENCY, %
1	133	1	0.8
2	52	0	0
3	183	4	2.2
5	212	3	1.4
7	263	15	5.7
9	285	10	3.5
11	283	16	5.7
13	386	24	6.2
15	377	23	6.1
17	408	33	8.1
19	409	26	6.4
22	289	26	9.0
	3280	181	

Total chromosome-generations tested: 41,103

with flies carrying second chromosomes known to be free of lethals. Tests made at various times after the start of the population can be looked upon as a variant of the "accumulation" technique described immediately above. In this case, the estimate of mutation rate based upon more than 40,000 chromosome-generations is 0.0044. In reality, this is the rate at which lethals accumulated in the population every two weeks—the interval between samples of tested chromosomes. Since the generation time in a population cage is somewhat longer than 2 weeks, the true mutation rate should be somewhat higher than that estimated. An examination of the three values listed in Table 8-3 suggests that 0.005 is a reasonably accurate estimate for the cumulative mutation rate of lethals on the second chromosome of *D. melanogaster*; this conclusion agrees with that reached by Crow and Temin (1964) on the basis of additional experimental data. Again, on the basis of allelism tests such as those described on page 45, it would appear that this chromosome has some 400 to 500 loci at which lethals can occur; thus, the mutation rate per locus for this autosome (as for the X chromosome) appears to be about 0.00001, or 10^{-5}.

Viability Mutations: Subvital Mutations. The effort of most geneticists interested in mutations has been spent on the study of lethal mutations. These can be defined in such a way that most cultures containing a nonlethal chromosome can be identified by inspection alone. Viability mutations with less drastic effects on their carriers can be studied only by counting the flies of the different genotypes in each of hundreds of cultures. Counts of individual cultures alone are not sufficient, because the frequency of wildtype flies in any culture is subject to sampling error. Distributions of frequencies based on large numbers of control and experimental cultures are needed to have a basis for *estimating* (in a rather crude way, at that) the frequency of subvital mutations.

Several attempts to estimate the rates of mutation to viability mutants other than lethals have been made. The results of a study by Timofeeff-Ressovsky (1935b) based on irradiated (6000 r of X rays) X chromosomes of *D. melanogaster* are listed in Table 8-5 and are shown diagrammatically in Figure 8-1. The males in these experimental cultures were carrying irradiated chromosomes; the effect of these on viability

TABLE 8-5

DETECTION OF VIABILITY MUTATIONS, INCLUDING LETHALS, INDUCED ON THE X CHROMOSOME OF D. MELANOGASTER BY AN EXPOSURE TO X RADIATION.

The classification is based on the ratio of males hemizygous for an irradiated X chromosome to females (sisters) carrying the same chromosome in heterozygous condition. (Timofeeff-Ressovsky, 1935b.)

	CONTROL		IRRADIATED	
MALES/FEMALES	NO. OF CULTURES	FREQUENCY	NO. OF CULTURES	FREQUENCY
0	1	0.001	107	0.123
0.05	—	—	7	0.008
0.15	—	—	10	0.012
0.25	—	—	7	0.008
0.35	—	—	15	0.017
0.45	2	0.002	31	0.036
0.55	—	—	53	0.061
0.65	—	—	49	0.056
0.75	9	0.011	53	0.061
0.85	46	0.055	58	0.067
0.95	655	0.783	383	0.441
1.05	110	0.131	84	0.097
1.15	14	0.017	11	0.013
Total	837	1.000	868	1.000

is expressed as the ratio of wildtype males to wildtype females in these cultures. Because of the radiation, the frequencies of mutations are high. The frequency of lethals and semilethals (male: female ratio smaller than 0.45) among the X-rayed chromosomes is 16.8 percent (146 among 868 tests). In the next four classes (0.45 to 0.75, inclusive) are 186 cultures in the case of the X-rayed chromosomes; this represents over 25 percent of nonlethal chromosomes. In the case of the control tests, only 11 cultures (1.3 percent) are found in these classes. Clearly, then, radiation is inducing viability changes other than lethals and semilethals. Clear evidence that these effects are associated with the X chromosome has also been provided by a further test of four "normal" and eight "subvital" cultures; the results of these additional tests are given in

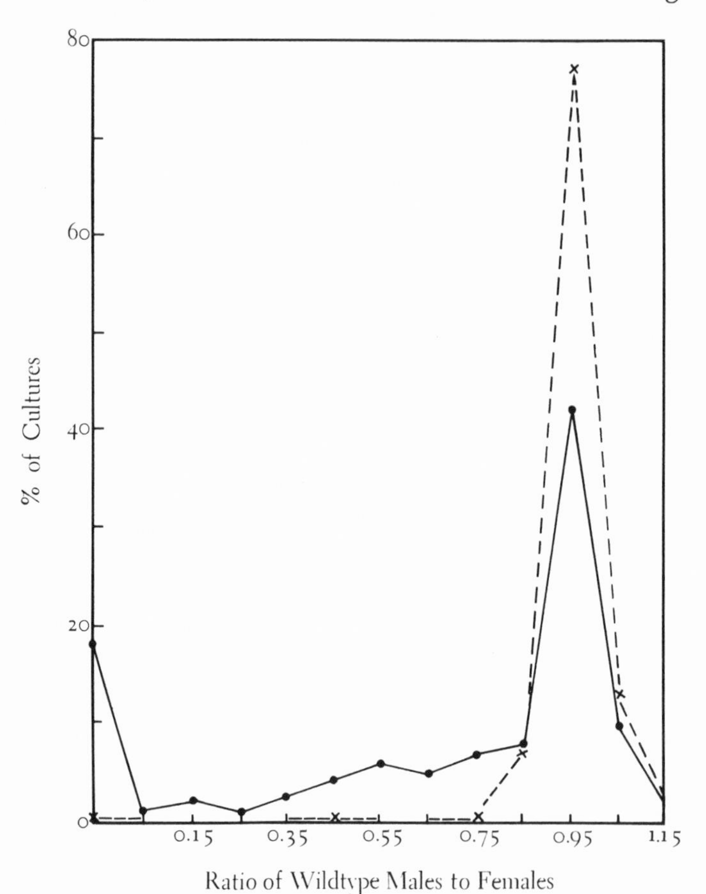

FIGURE 8-1. Viability mutations induced on the X chromosomes of *D. melanogaster* by a large (approximately 6000 r) dose of X radiation. Dashed line: control; solid line: irradiated. (After Timofeeff-Ressovsky, 1935b.)

Table 8-6. We see that the males classified as subvital are consistently subvital in the later, replicated tests; normals, too, are consistently normal. It should be mentioned in connection with this work, however, that following the radiation dose used by Timofeeff-Ressovsky (about 6000 r)

TABLE 8-6

TESTS SHOWING THAT NON-LETHAL VIABILITY MUTATIONS ARE HERITABLE.

(Timofeeff-Ressovsky, 1935b.)

CULTURE	ORIGINAL	1	2	3	4	5	6	7	8	9	10	AV.	NO. OF FLIES
R12	96.8	89	96	99	92	85	96	101	94	88	97	93.7	2247
R73	94.3	96	97	91	102	93	104	90	92	95	94	95.4	2374
R161	98.2	90	98	96	93	98	99	97	100	95	95	96.1	2014
R253	97.4	106	97	90	96	95	91	93	92	98	90	94.8	2116
R41	65.1	61	65	69	60	57	60	56	63	54	62	60.7	1818
R69	51.8	49	55	60	54	62	52	51	59	56	50	54.8	1619
R132	66.9	71	64	78	73	68	72	69	77	70	75	71.7	2189
R214	77.2	72	80	89	78	83	87	85	76	82	88	82.0	2260
R307	60.3	68	57	63	70	65	59	66	64	60	67	63.9	1754
R341	63.5	72	64	73	59	66	75	78	63	70	72	68.2	2410
R379	75.1	75	83	91	67	86	76	79	81	77	—	79.4	1733
R422	58.2	58	69	57	60	67	62	65	54	63	61	61.6	1895

extensive structural damage to the chromosomes would be expected. The ratio of lethals to subvitals observed in these radiation experiments (2 : 3) should not be taken as a ratio that applies as well to spontaneous mutations.

A study similar to that described above has been made on the growth rate of yeast cells by James (1959); again, a high level of radiation (30,000 r of X-rays) was used to induce the mutations. The distribution of growth rates shown in Figure 8-2 is remarkably similar to that of viabilities of male flies carrying irradiated X chromosomes (Figure 8-1).

An alternative procedure for estimating the extent of nonlethal genetic damage has been utilized by Temin (1966); this procedure is based on the conversion of relative viabilities into lethal equivalents. Using mating techniques similar to those shown in Figure 3-6, Temin obtained cultures containing the following classes of flies:

TYPE OF CULTURE	GENOTYPES
Homozygous	$Cy/bw^D : Cy/+_i : bw^D/+_i : +_i/+_i$
Heterozygous	$Cy/bw^D : Cy/+_i : bw^D/+_j : +_i/+_j$

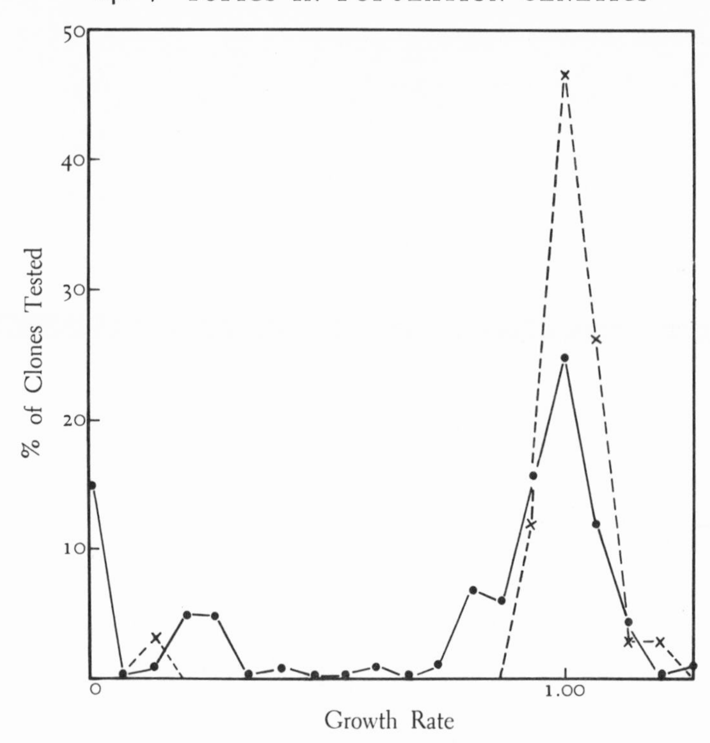

FIGURE 8-2. "Viability" mutations induced in yeast by exposure to 30,000 r X radiation. Viability in this case was measured as the growth rate of colonies of haploid individuals that carried irradiated genomes. (After James, 1959.)

Cultures in which the wildtype flies were homozygous were classified as lethal or nonlethal; the former had fewer than 10 percent of the mean survival of heterozygotes. Viability ratios were computed by dividing the number of wildtype flies of each culture by the average number of $Cy/+$ and $bw^D/+$ flies in the same culture. These ratios for heterozygotes, *all* homozygotes, and *nonlethal* homozygotes were 1.022, 0.679, and 0.882. If the viability ratio of heterozygotes is taken as 1.000 (standard), those of the other two types are 0.664 and 0.863. Now, these values can be thought of as the proportion of chromosomes that would be absolutely free of deleterious mutations and, consequently, equal in viability to the heterozygotes themselves if all deleterious mutations could be concentrated into lethal genes. And, if all deleterious mutations could be concentrated into lethals, the remaining, nonlethal chromosomes would equal the "zero" term of the Poisson distribution. That is, e^{-m} equals 0.664, where m is the proportion of lethal equivalents represented by *all* deleterious genes carried by the tested chromosomes. Similarly, e^{-m}

equals 0.863, where *m* in this case is the proportion of lethal equivalents carried by nonlethal chromosomes. Upon solving for *m* in these two equations, we find the following:

Total lethal equivalents (total load)	0.409
Lethal equivalents for subvitals (detrimental load)	0.147
Difference (lethal load)	0.262

The ratio of the detrimental load to the lethal load (0.147/0.262) equals 0.563. Among the chromosomes tested in this analysis, detrimental mutations *as a group* have an effect on viability 50 to 60 percent as great as lethal mutations themselves.

The procedure given above has been illustrated by using data on chromosomes sampled from large populations. For the present discussion, however, we are really interested in mutation rates rather than in frequencies of genes accumulated over many generations. In her paper, Temin discusses a number of published reports (including some of the data given in Table 8-5 and Figure 8-1) and has found detrimental load : lethal load ratios (D : L ratios) which vary from 1.000 to 0.125.

Variation in the D : L ratio is not unexpected. Lethal mutations, to a large extent, are mutations that alter enzymatic processes or critical stages in the development of the individual; these mutations have their effects largely (but not absolutely, as we shall see shortly) independently of the external environment. The same is not true of subvital mutations. The proportion of homozygous subvital individuals that survive to adulthood is very sensitive to culture conditions. Many of the standard visible mutations used for classroom exercises are decidedly deleterious in overcrowded cultures but only mildly so (or not so at all) in less crowded ones. Temperature and other environmental conditions can alter the viability effects of these genes. Consequently, a ratio of detrimental mutations to lethals that is based on proportions of survivors will vary from study to study unless experimental conditions are rigorously standardized. If the conditions *are* standardized, then the calculated ratio will apply only to that particular set of conditions.

Viability Mutations: Summary. A summary of the array of viability and other effects of newly arising mutations is presented in Figure 8-3; the frequencies shown in this figure are those estimated by Muller (1950b). The figures shown apply to mutations that are expected as the result of an exposure of *Drosophila* sperm to 150 r of X radiation. In the case of recessive lethal mutations, an exposure to 100 r induces as many lethals as occur "spontaneously"; therefore, a rough estimate of spontaneous changes can be obtained by dividing each entry by 2.5. The ratio of four detrimental mutations to one recessive lethal could be a gross under-

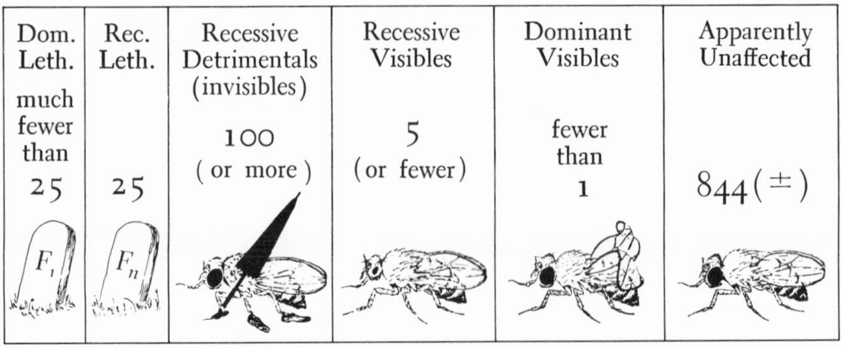

Dom. Leth.	Rec. Leth.	Recessive Detrimentals (invisibles)	Recessive Visibles	Dominant Visibles	Apparently Unaffected
much fewer than 25	25	100 (or more)	5 (or fewer)	fewer than 1	844(±)
F_1	F_n				

FIGURE 8-3. Frequency with which gene mutations of various types may be expected among spermatozoa of *D. melanogaster* following an exposure to 150 r of X radiation. Numbers refer to mutations per thousand spermatozoa. (After Muller, 1950b.)

estimation. Subvitals may conceivably have very slight effects on survival; it is obviously impossible to estimate the frequencies of mutational changes that are imperceptible.

MUTATIONS: INTERACTIONS

Except for the very brief comment that the subvitality of subvital mutations depends to a large extent on environmental conditions, we have treated mutations as if they were rather constant in their physiological effects. This could be seriously misleading. Mutant phenotypes are subject to modification, as are all aspects of life. In some instances, the effects of a given mutation are undone only by the most specific type of suppression; mutant alleles at the same locus which appear to have identical phenotypic effects are often suppressible only by entirely different suppressor mutations. In other instances, it appears that any gross alteration of the organism's environment has its effect on the viability characteristics of the mutant phenotype.

Some notion concerning the interaction of mutant genes can be obtained from a study made by Timofeeff-Ressovsky (1934) in which the relative viabilities of carriers (*D. funebris*) of a number of mutant genes were obtained first with single mutant characters (Table 8-7), and then with the same mutants in paired combinations (Table 8-8 and Figure 8-4). Each mutant gene has a general (and rather pronounced) depressing effect on the viability of its carriers; *miniature* (*m*), for example, lowers the viability of its carriers by 30 percent or more. An exception to this general rule is found in *eversae* (*ev*); under the culture condi-

TABLE 8-7

VIABILITY EFFECTS OF VISIBLE MUTATIONS IN D. FUNEBRIS.

(Timofeeff-Ressovsky, 1934.)

Eversae	*ev*	104
Singed	*sn*	79
Abnormes	*ab*	89
Miniature	*m*	69
Bobbed	*bb*	85
Lozenge	*lz*	74
Wildtype	+	100

tions used, carriers of this mutant gene survived in larger numbers than did the wildtype flies.

When the mutant genes are combined in paired combinations, an expected viability can be calculated as the product of the viabilities of flies carrying the mutations individually. As a general rule, again, the viabilities of the combinations are related to the expected ones. An examination of Figure 8-4 reveals, however, that there is a systematic bias in the experimental results; the viabilities of the combinations tend to be greater than those expected on the basis of the individual tests.

Experimental results similar to those of Timofeeff-Ressovsky have been reported by Pearl et al. (1923) on the duration of life in *D. melanogaster*. These data have not been summarized because they contain an obvious

TABLE 8-8

INTERACTIONS OF MUTANT ALLELES AFFECTING VIABILITY IN D. FUNEBRIS.

(Timofeeff-Ressovsky, 1934.)

		OBSERVED	EXPECTED
er	*sn*	103	82
er	*ab*	84	93
er	*bb*	86	88
sn	*ab*	77	70
sn	*m*	67	55
ab	*m*	83	61
ab	*lz*	59	66
ab	*bb*	79	76
m	*bb*	97	59
lz	*bb*	69	63

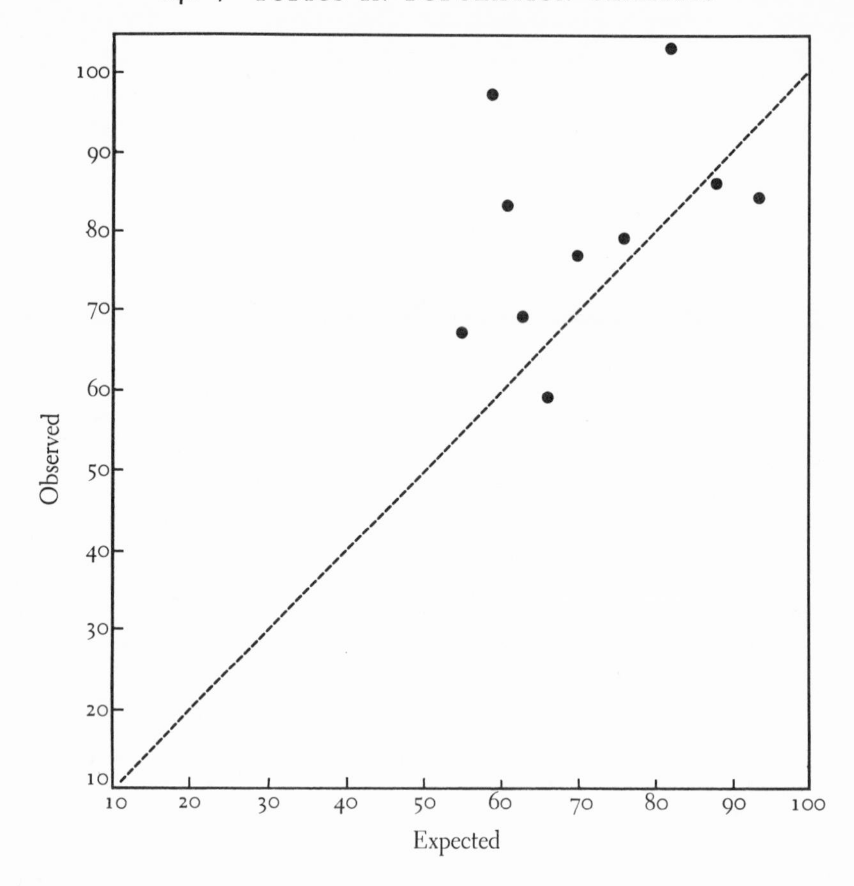

FIGURE 8-4. Comparison of observed and expected interactions of viability mutations in *D. funebris*. (After Timofeeff-Ressovsky, 1934.)

flaw: Pearl and his colleagues assumed that aside from the mutations with obvious phenotypic effects, the wildtype genes of different stocks of flies are identical. Consequently, the various gene combinations they synthesized in the laboratory are accompanied to an unknown extent by hybridity for wildtype genes. The viability effect of the hybridity and that of the visible mutations cannot be disentangled. The interesting point about Timofeeff-Ressovsky's work is that he was well aware of the role of the "genetic milieu" and went to great lengths in his experiment to eliminate differences in background genotypes. It appears that his efforts failed since it is much easier to explain the upward bias of the viability of mutant combinations on hybrid vigor than to claim that mutant genes in combination with one another tend to have lessened effects on viability. Indeed, such a claim would be quickly demolished as a general rule because Dobzhansky and his colleagues (Dobzhansky et al., 1965;

Spassky et al., 1965) have shown for *D. pseudoobscura* that deleterious genes tend to exhibit epistatic interactions such that combinations are less viable than one would predict from individual effects alone.

An extremely laborious study of the interactions of viability mutations and their genetic background was made by Dobzhansky and Spassky (1944) using *D. pseudoobscura.* In this case, chromosomes rather than individual gene loci were studied. Furthermore, aside from the effects of the chromosomes on viability—effects that are determined by counting flies of different genotypes in test cultures—no visible mutations were involved. The tested chromosomes as well as the background chromosomes were wildtype.

A number of second and fourth chromosomes obtained from collections of flies at different localities on Mt. San Jacinto (Figure 4-6) were subjected to viability tests. The viabilities of flies homozygous for some of these chromosomes are listed in the second column ("original") of Table 8-9. These same chromosomes (California origin) were carefully

TABLE 8-9

EFFECT OF BACKGROUND GENOTYPE ON THE VIABILITY OF FLIES (D. PSEUDOOBSCURA) HOMOZYGOUS FOR VARIOUS WILDTYPE SECOND AND FOURTH CHROMOSOMES.

The background genotypes consist of wildtype chromosomes of various geographic localities. (Dobzhansky and Spassky, 1944.)

CHROMOSOME	ORIGINAL	WASHINGTON	COLORADO	CALIFORNIA	MEXICO	GUATEMALA	AV.
AA1003	34.6	32.1	33.9	33.2	28.7	29.9	31.5
AA1015	14.2	12.8	31.2	30.8	24.7	24.3	24.8
AA1178	34.8	35.0	31.0	33.6	31.9	32.1	32.7
KA667	33.6	34.0	29.9	32.9	29.8	33.4	32.0
KD745	24.4	33.6	31.6	32.7	29.9	32.3	32.0
AA955	32.8	28.4	31.8	31.1	30.5	28.5	30.1
AA1035	23.2	28.0	32.5	23.4	28.4	29.4	28.7
PA851	19.0	25.9	25.7	33.0	28.2	24.5	27.5
PA998	19.4	29.3	27.2	28.3	29.8	25.3	28.0
Average	26.2	28.8	30.5	31.0	29.1	28.9	29.7

transferred into stocks in which the other chromosomes were of various origins: Washington (state), Colorado, California (northern), Mexico, and Guatamala. Once more their effects on the viability of homozygous individuals were determined. Each of the original tests involved counts of at least several hundred but more often of a thousand or more flies; the other entries in Table 8-9 represent counts of from 1000 to 5000 flies each. The data of Table 8-9 are shown graphically in Figure 8-5. There is a marked tendency for the five chromosomes that had the lowest viabilities in the original tests to undergo a systematic improvement in the other genetic backgrounds. Conversely, the four chromosomes that originally had the highest viabilities show a systematic tendency to have lower viabilities when the background genotype is changed. In a sense, the data illustrate a well-known statistical phenomenon: regression toward the mean. Children of tall parents tend to be tall, on the average, but not as tall as their parents. Similarly, children of short parents tend to be short but, on the average, not as short as their parents. In the case of the viability effects of the nine chromosomes studied by Dobzhansky and Spassky, however, the basis for the "regression toward the mean" lies in gene interactions and not in chance variations of observed frequencies. The statistical errors in the determination of each point shown in Figure 8-5 are extremely small because of the enormous numbers of flies counted in determining each viability. It appears, then, that tests which reveal that a chromosome tends to lower the viability of its carriers (or tends to raise it) are very often revealing genetic interactions peculiar to a given culture or to the particular experimental material tested.

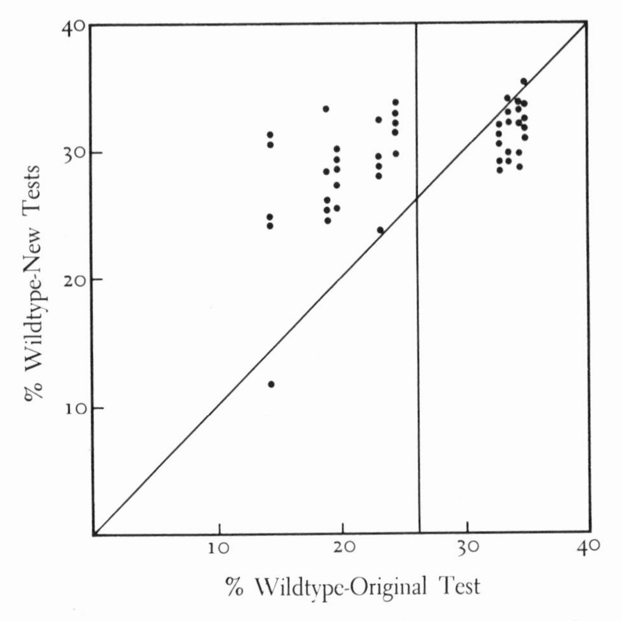

FIGURE 8-5. Comparison of the viability of nine different chromosomes of *D. pseudoobscura* in the original tests (horizontal axis) and in novel genetic backgrounds (vertical axis). (After Dobzhansky and Spassky, 1944.)

Earlier in this chapter the possibility was mentioned that mutations which are beneficial rather than detrimental to their carriers might occur. Ordinarily such mutations would hardly be expected, because in the course of any species' existence they should already have been incorporated into the genetic fabric of the species. To show, then, that such mutations occur and, indeed, are incorporated into the gene pool of a population, one must place a sample of individuals in an aberrant, suboptimal circumstance. We do this to natural populations, of course, when we employ insecticides, herbicides, and antibiotics on a massive scale. We see the results, too, as flies, mosquitos, and lice develop resistance to insecticides and as hospital after hospital becomes infected by dangerously resistant strains of bacteria.

The results of an experiment on the changes in the viability effects of subvital chromosomes when these are kept in homozygous cultures are given in Table 8-10 (Dobzhansky and Spassky, 1947). Four chromosomes,

TABLE 8-10

PROGRESSIVE INCREASE WITH TIME IN THE FREQUENCY OF WILDTYPE FLIES HOMOZYGOUS FOR WHAT WERE INITIALLY SUBVITAL OR SEMILETHAL CHROMOSOMES.

Note that much of the increased viability of these flies is lost following the transfer of the tests from 21 to 25°C. (Dobzhansky and Spassky, 1947.)

	PA748		PA851		PA784		AA1035	
	% +/+	*n*	% +/+	*n*	% +/+	*n*	% +/+	*n*
Original—21°	13.6	750	19.0	1529	21.4	891	23.2	2514
2	23.3	2973	26.2	6028	36.6	4085	24.1	7393
6	24.7	4284	29.2	5373	36.2	6461	22.9	3908
10	25.1	3483	30.2	4284	42.4	3846	25.9	6121
14	29.6	4110	30.9	3866	31.2	4560	26.1	4932
24	28.2	2975	30.0	4110	35.0	1511	29.6	3632
30—25°	19.2	1803	24.9	2396	31.7	710	7.7	1660
Original—25°	9.9	766	26.4	1594	20.5	796	10.2	1808

each with a decidedly deleterious effect on viability, were chosen for study. Some of these chromosomes had such marked effects on their carriers at 25° that the experiment had to be carried out for the most part at 21°, a more favorable temperature for *D. pseudoobscura*. Periodically, samples of chromosomes were withdrawn from the homozygous cultures and were tested in homozygous condition once again. From the data listed in Table 8-10 it is clear that the chromosome in each culture underwent a systematic improvement in its effect on viability. The precise nature of these changes is unknown. It seems unlikely that they represent reverse mutations of the deleterious alleles carried by these chromosomes; on the contrary, it is much more likely that the gradual improvement represents a series of changes scattered throughout the chromosome.

The two lower lines of Table 8-10 reveal that whatever changes were occurring (and the consolidation of beneficial genetic interactions seems the most likely candidate) these were relatively specific for the temperature (21°) at which the experiment was conducted for the first twenty-five generations. After the twenty-fifth generation, the viability effects of the four chromosomes were judged sufficiently "normal" to allow a transfer of the experiment to 25°, a less favorable temperature for this *Drosophila* species. Five generations after the transfer (a period of time sufficient to allow marked viability changes in the early part of the experiment), the viability effects of the chromosomes were much more detrimental than they had been at generation 24 at 21°; they were nearly as bad, in fact, as the original tests at 25° had been. The genetic modification of viability was specific for the temperature at which the modification occurred.

The preceding study by Dobzhansky and Spassky illustrates not only interactions between genes (the improvement of viability through genetic interactions) but gene-environment interactions as well. Results of an earlier study (Dobzhansky and Spassky, 1944) designed specifically to determine the effect of temperature on the viability of flies (*D. pseudoobscura*) homozygous for various autosomes are shown in Table 8-11.

TABLE 8-11

GENE-ENVIRONMENT INTERACTION: THE VIABILITY OF WILDTYPE *D. PSEUDOOBSCURA* HOMOZYGOUS FOR VARIOUS SECOND CHROMOSOMES AND TESTED AT THREE DIFFERENT TEMPERATURES.

(Dobzhansky and Spassky, 1944.)

	TEMPERATURE, °C		
CHROMOSOME	16.5	21	25.5
AA958	0	13.1	1.0
AA966	26.0	12.6	2.4
AA995	31.7	27.2	22.7
AA1003	31.6	34.6	26.9
AA1015	29.7	14.2	0
PA736	27.8	18.4	9.0
PA748	23.0	13.6	9.9
PA784	26.9	21.4	20.5
PA841	32.2	26.7	11.5
PA858	31.5	37.0	30.0

The data of this table (selected from a much larger series of published tests, incidentally) show how temperature can affect viability of homozygotes in a most pronounced manner: Chromosomes that are lethal at one temperature can be nearly normal in their effect on viability at another. Although there are exceptions, as a rule 25° exaggerates the inviability of *pseudoobscura* homozygotes; this temperature is near the upper limit tolerated by the species.

The point is frequently made in conjunction with genetic or gene-environment interactions that we cannot speak of the nature of a gene's effect on viability unless we speak simultaneously of the environment. The lethality of nonresistant alleles carried by bacteria living in an environment containing antibiotics serves as an excellent illustration. Nevertheless, despite the nature and the magnitude of known interactions, the following generalization must be recognized: The effect that a given mutant appears to have in one environment is our best guide concerning the effect it will have in a second, untested one. We must adopt this rule because it is impossible to test more than a few environments in studying the action of a mutant gene. If the studies that have been made do not lead us closer to the truth, on the average, concerning an untested situation than would a random guess, then there exists no reason for carrying out experiments. This is not to say that experiments yield data that will always lead to extremely accurate predictions concerning novel situations (see, for example, Table 8-9 and Figure 8-5). The more complex the problem under study, the less accurate the predictive ability gained by limited experimentation. Even in Figure 8-5, however, one can detect a tendency for chromosomes with high original viabilities to have higher than average viabilities in the new genetic backgrounds.

Gene-environment interactions are well known to plant and animal breeders. Many varieties of agriculturally important plant species are valuable in certain geographic regions while all but worthless in others. Flowering, seed set, germination and other aspects of a plant's existence are regulated to a large extent by light intensity, length of day, and temperature; plants adapted to the long growing seasons of southern latitudes may fail to develop in the short growing reasons of northern areas. Humidity, type of soil, and local pests also determine the success or failure of a given variety in a given locality. Table 8-12 gives the results of a classical experiment by Harlan and Martini (1938) as reported by Stebbins (1950). A mixture of seeds from six varieties of barley were sown annually at a number of localities. Each year's harvest consisted of a mixture of these different seeds. The proportions of seeds of the different varieties were determined by the amount of seed produced by each variety in each locality. After ten years or more, a single variety contributed the bulk of all seeds in each of the different localities. Fur-

TABLE 8-12

SURVIVAL (PERCENT OF TOTAL HARVEST) OF BARLEY VARIETIES FROM THE SAME INITIAL MIXTURE AFTER REPEATED (10 TO 12) ANNUAL SOWINGS AT DIFFERENT LOCALITIES.

(Stebbins, 1950.)

VARIETY	NEW YORK	MINNESOTA	MONTANA	OREGON
Coast	13	20	20	73
Hannchen	7	75	4	7
White Smyrna	0	1	56	14
Manchuria	78	0	5	0
Getami	2	4	14	0
Meloy	0	0	1	6

thermore, a different variety was the leading producer at each of the different localities.

INTERACTION: FACILITATION

Genes do not act in a vacuum in bringing about their effects on their carriers. The biochemical steps controlled by different genes are parts of complex, interconnected pathways. It is impossible, in theory, to conceive of a gene whose action would be entirely independent of those of other genes. And as a matter of fact as well, genetic interactions are commonly observed. Nor do genes have to be carried by the same individual to interact; individuals of different genotypes can influence each other's survival if they develop in mixed cultures.

Lewontin (1955) studied the survival of nineteen homozygous wildtype strains of *D. melanogaster* as well as the homogeneous inbred *Swedish-b* and *white* strains. The culture conditions for these studies were highly standardized. In one series of tests Lewontin determined the survival to adulthood of forty larvae when these were placed on a small amount of culture medium. In another series he studied the survival of forty larvae, twenty wildtype and twenty *white*, under similar conditions. A summary of these two tests is presented in Table 8-13 and Figure 8-6. In fourteen of the twenty tests, the proportion of survivors was higher in the mixed cultures than in the average of the "pure" wildtype and *white* cultures. The increased survival was sufficiently large that the average difference, 0.058, is significant at the 2 percent level (the standard error of the difference equals 0.022; t equals 2.64 with 19 degrees of freedom). Since the average survival in pure cultures equaled 0.550, the increase in mixed cultures is 10 percent or more.

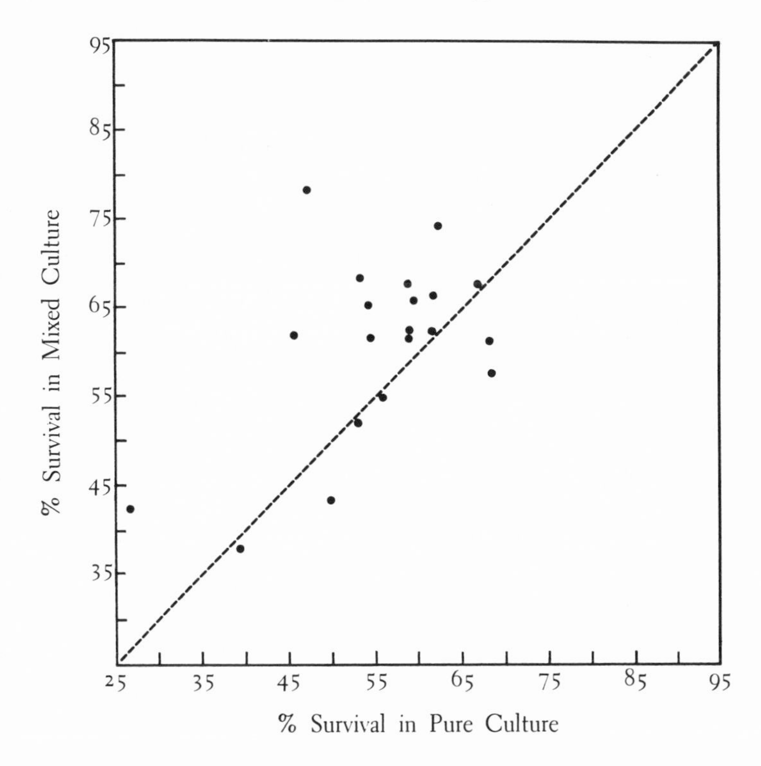

FIGURE 8-6. Proportion of survivors of *white* and homozygous wildtype flies in pure culture and in cultures containing one half *white* and one half wildtype larvae. The initial density was 40 larvae per vial. (After Lewontin, 1955.)

Facilitation, the interaction between individuals of different genotypes such that the survival of each is increased, has been used by Jensen (1965) in increasing the yield of oats. Five excellent varieties of oats were tested in pure stand at each of ten localities. Their yields were compared to that of a synthetic "variety" obtained by simply mixing the seeds of the five varieties. The results of this test are shown in Table 8-14; the mixture gave a higher yield than the average of the pure varieties in eight of the ten localities. The average increase in yield in this case, as in the preceding example, was approximately 10 percent.

Not all intergenotypic interactions serve to increase the survival of the individuals involved. Weisbrot (1966) studied the interaction of wildtype and mutant individuals of *D. melanogaster* and of wildtype *D. pseudoobscura.* The survival of larvae of different sorts was determined when these developed on a certain culture medium and on the same medium "conditioned" by the presence of killed larvae of the same or of a different sort.

TABLE 8-13

COMPARISON OF THE SURVIVAL OF LARVAE OF D. MELANOGASTER IN OVERCROWDED CULTURES.

Each of the various numbered strains was homozygous for a particular wildtype second chromosome; Swedish-B is a laboratory stock. Under "pure" are listed the average percentage survivals of the wildtype and *white* strains tested separately; under "mixed" are those of wildtype and *white* growing together in the same culture. Under "difference" are listed the individual differences together with the mean difference ($\bar{d}$), error of the mean ($s_{\bar{d}}$), t, and probability (p).

STRAIN	PURE	MIXED	DIFFERENCE
1	0.392	0.380	−0.012
11	0.530	0.685	0.155
18	0.265	0.425	0.160
22	0.543	0.620	0.077
24	0.619	0.745	0.126
25	0.554	0.552	−0.002
28	0.616	0.623	0.007
33	0.497	0.433	−0.064
34	0.679	0.615	−0.064
41	0.541	0.653	0.112
45	0.681	0.580	−0.101
46	0.589	0.628	0.039
47	0.469	0.783	0.314
48	0.614	0.668	0.054
50	0.588	0.618	0.030
54	0.529	0.523	−0.006
62	0.669	0.678	0.009
67	0.453	0.622	0.169
74	0.592	0.660	0.068
Swedish-B	0.585	0.680	0.095

$\bar{d} = 0.0583$; $s_{\bar{d}} = 0.0218$; $t = 2.67$ (19 d.f.); $p = 0.01–0.02$

A number of interactions were detected in this study (see Table 8-15). Killed larvae of wildtype strain A2 increased the proportion of survivors among additional A2 larvae but lowered the survival of another wildtype strain (A6) and of the *ebony* strain. Killed larvae of wildtype strain, A6, did not influence the survival of A2 or A6 larvae but lowered that of the *ebony* strain. Killed *ebony* larvae decreased the survival of *brown-ebony* larvae while having relatively little effect on the other strains. *Brown-ebony* larvae when killed in the medium tended to reduce the survival of both *ebony* and *brown-ebony* larvae. Finally, killed *pseudo-*

TABLE 8-14

YIELD IN BUSHELS PER ACRE OF FIVE VARIETIES OF OATS AND THEIR COMPOSITE MIXTURE AT 10 LOCALITIES.

(See Jensen, 1965.)

VARIETY	1	2	3	4	5	6	7	8	9	10	AV.
Garry	135	90	111	128	125	95	111	94	78	72	104
Russell	123	95	120	116	130	111	105	80	83	72	103
Tioga	125	79	104	109	136	109	110	86	75	75	101
Select-51	120	82	117	111	112	106	104	102	70	67	99
Select-70	101	77	113	116	134	110	113	75	72	71	98
Average	121	85	113	116	128	106	109	87	75	71	101
Mixture	128	100	129	114	124	109	116	95	87	84	109

TABLE 8-15

EFFECT OF DEAD LARVAE OF ONE STRAIN OF DROSOPHILA (MELANOGASTER OR PSEUDOOBSCURA) ON THE SURVIVAL OF OTHER LARVAE OF THE SAME OR OTHER STRAINS.

The average numbers of surviving larvae of 40 tested per vial are listed; the total numbers of larvae tested are given in parentheses. (Weisbrot, 1966.)

	A2	A6	*e*	*bw, e*	*pseudoobscura*
Control	13.8 (400)	27.0 (400)	28.2 (360)	23.4 (360)	2.3 (360)
Containing dead larvae					
A2	17.5 (720)	17.6 (1040)	12.0 (440)	—	0.6 (800)
A6	15.3 (880)	25.4 (720)	11.0 (1600)	6.0 (800)	2.6 (640)
e	16.1 (600)	20.3 (720)	24.0 (440)	11.0 (400)	—
bw, e	—	25.7 (560)	19.4 (280)	16.2 (160)	—
pseudoobscura	22.3 (1600)	28.2 (1600)	—	—	3.7 (680)

obscura larvae increased the survival of the A2 wildtype strain of ***melanogaster*** but had no effect on that of the A6 strain. It appears as if inhibitory interactions (presumably through the action of metabolic waste products) are about as frequent as are those recognized as facilitation. At least, they are as common when diverse strains are picked arbitrarily for testing; it is possible that facilitation among naturally co-existing genotypes would, as the result of natural selection, be more common.

MUTATIONS: QUANTITATIVE

The bulk of this chapter has been devoted to a description of mutations and of the interactions of mutations with one another and with the environment. The description of interactions was not limited to obvious mutations but was extended to cover genetic and gene-environment interactions generally. The few remaining pages will be spent on the purely formal aspects of mutation as it affects the genetics of populations.

Forward Mutation. Assume that an allele A mutates to an allele a at a rate u per generation. Assume, too, that at a given moment the frequency of A in the population is p while that of a is q. Following mutation, the frequency of A is reduced to $p(1 - u)$. If there is an equilibrium frequency, it must be such that the frequency of A before and after mutation is the same:

$$p = p(1 - u)$$

Upon solving, it appears that an equilibrium exists if

$$u = 0 \qquad \text{or} \qquad p = 0$$

That is, an equilibrium exists if there is no mutation or if the allele A is completely removed from the population.

In the absence of reverse mutation from a to A, the eventual loss of A is assured. However, the loss is not necessarily a rapid one. The frequency of A in any generation t is given by the equation

$$p_t = p_0(1 - u)^t$$

Now, $(1 - u)^t$ can be written as $[(1 - u)^{1/u}]^{ut}$ or as e^{-ut}. Consequently, $p_t = p_0e^{-ut}$. Thus to calculate the number of generations required to reduce p_0 by one half, for example, one proceeds as follows:

$$\tfrac{1}{2}p_0 = p_0e^{-ut}$$

$$e^{-ut} = 0.5$$

$$ut = -\log 0.5 = 0.7$$

$$t = \frac{0.7}{u}$$

If u equals 10^{-5} as did many mutation rates we encountered earlier, approximately 70,000 generations would be required to reduce the frequency of an allele by one half its initial frequency.

Forward and Reverse Mutation. Suppose that the frequencies of A and a are p and q. Suppose, too, that A mutates to a at a rate u per generation while a mutates to A at a rate v. The frequency of A after mutation has occurred equals

$$p - up + vq$$

or, at equilibrium,

$$up - vq = 0$$

At equilibrium, $\hat{p}$ equals $v/(u + v)$ and $\hat{q}$ $(= 1 - \hat{p})$ equals $u/(u + v)$. ($\hat{p}$ and $\hat{q}$ represent equilibrium values of p and q and are read p-hat and q-hat.)

This is the first *stable* equilibrium we have encountered. Unlike the Hardy-Weinberg "equilibrium," the equilibrium established under the influence of forward and reverse mutation is stable. If it is displaced, it tends to return to a given value, $\hat{p}$. In fact, if a population contained only allele A at the outset, a would accumulate as a result of repeated mutation. Conversely, if the population consisted entirely of a, the allele A would accumulate. The tendency to return to $\hat{p}$ exists even when the frequency of either allele is raised to 100 percent while the other is "lost" (see Figure 8-7).

The accumulation of lethal chromosomes in a population consisting initially only of lethal-free chromosomes is illustrated by the data in Table 8-4. These are the data upon which the mutation rate of 0.0044 given in Table 8-3 was based. Table 8-4 shows the steady increase in the frequency of lethal chromosomes in succeeding generations. How can mutations occurring at an estimated rate of 1×10^{-5} accumulate so rapidly? Because the data pertain to lethal chromosomes which have been estimated to contain some 400 to 500 loci at which lethal mutations can occur. Consequently, the rate of accumulation should be about 400×10^{-5} or 500×10^{-5}; that is, about 0.004 to 0.005. More precisely, the loss of nonlethals should be such that the frequency of nonlethals, P, at generation t should be about $(0.996)^t$ or $(0.995)^t$. As long as lethal chromosomes are relatively rare, these two ways of calculating their theoretical accumulation are nearly equivalent.

Although forward and back mutation give rise to a stable equilibrium, the forces that tend to restore this equilibrium are very weak indeed. The rate at which the frequency of A returns to $\hat{p}$ once it has been displaced is exceedingly low. Thus, although stable equilibria are vital in maintaining genetic variability in populations, it is unlikely that persistent equilibria can be ascribed to opposing mutation pressures. Other causes must be sought. An exception to the preceding statement must be made for mutations to alleles such as lethals where the same generalized phenotype (death) results from mutations at a large number of loci. In this case, the cumulative mutation pressure, as we have seen, is not negligible and mutation alone can cause marked changes in chromosomal frequencies.

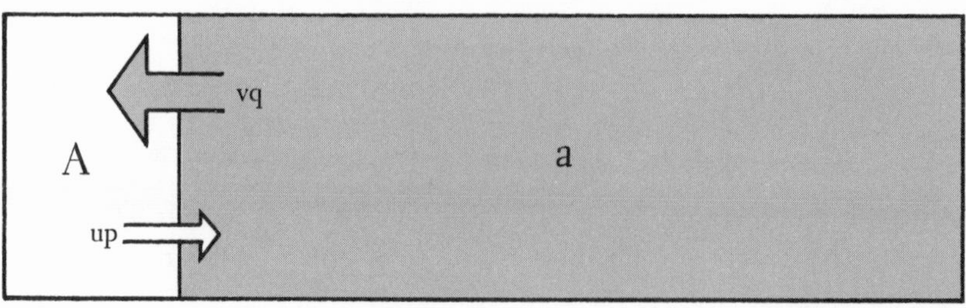

A. Frequency of "A" is increasing.

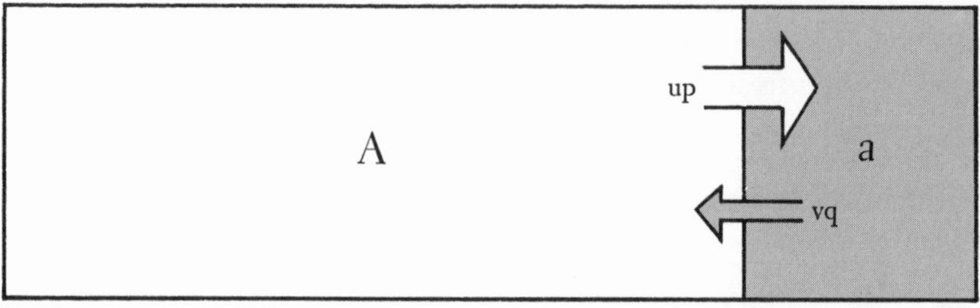

B. Frequency of "a" is increasing.

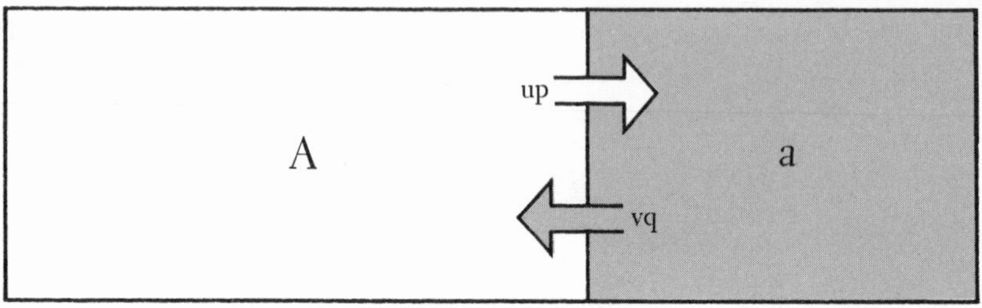

C. Frequencies of "A" and "a" are constant.

FIGURE 8-7. Diagrams showing the establishment of equilibrium gene frequencies under the opposing influences of forward and back mutation. (After Wallace, 1964.)

9

SELECTION

According to the Hardy-Weinberg equation, gene frequencies remain constant from generation to generation within populations of cross-fertilizing diploid individuals, and, furthermore, gene and zygotic frequencies are related through the binomial rule. These statements are based, however, on certain assumptions. In the foregoing chapters we have considered these assumptions one by one, together with the consequences of their rejections. We have discussed the consequences of a finite rather than an infinite population size. We have discussed non-random mating patterns. We have seen some of the consequences of mutation pressure. Now we take up what is probably the most important phenomenon of all: selection. In one way or another, the remaining chapters will deal with selection. Indeed, it has been impossible to completely defer a discussion of selection until now; mention of selection crept into the discussions of inbreeding, random mating, and mutation. This was because natural selection in the form of differential survival or reproduction is an unavoidable consequence of life itself.

What are the consequences of selection? Look about you! Visit a zoo or botanical garden. Thumb through a number of zoological and botanical atlases. Scan some taxonomic reviews. Look into some accounts of the paleontological record. Consider the tremendous variety of life. The conquences of selection are so enormous that we can not hope to do more than discuss them at the very simplest levels: (1) changes in gene frequency and (2) interactions associated with changing gene frequencies.

Despite the apparent modesty of our treatment of selection, the treatment rests nevertheless upon a wholly immodest assumption. Suppose that a group of shadowy, immaterial beings of limited (that is, human-like) intelligence lived in an ethereal world before life as we know it arose. Suppose, too, that these beings were asked whether the chemicals of the real universe could be combined in such a manner that life—the self-reproduction of necessarily rather complex chemical systems—would arise. I am not sure that these shades would have agreed that life would be possible. I am not sure either that modern physical chemists and biochemists can draw up a convincing blueprint for a self-sustaining, living system. Knowing that such a system can exist, it is a relatively simple matter to outline the general characteristics to which the system must conform. But, to predict in advance that known elements will combine so as to set up the necessary stable equilibria, will combine so that some at least can make accurate copies of themselves, and will persist so that system complexes reproduce faster than they are destroyed by accident—this is an entirely different matter.

Granted that life does exist, however, and granted that sequential environmental changes are correlated so that abrupt changes are the exception rather than the rule, I think the human mind can grasp the nature of evolutionary changes, the origins of adaptations of organisms to their environment, the role of reproductive isolation between groups of organisms, and the diversity of living forms. Furthermore, to understand these things it is sufficient to know (1) the types of chemical processes utilized by living organisms, (2) the basic nature of genetic systems, (3) the extent to which variation in the reproductive success of genetically different individuals exists, and (4) the conditions for setting up stable rather than unstable or wildly oscillating communities of organisms. An understanding of the elementary action of natural selection on the genetic endowments of populations goes a long way toward an understanding of evolutionary changes in general. This, then, is the immodest assumption upon which the following discussions rest.

The Nature of Selection

Selection is the result of differential reproductive success of genetically different individuals. As a result of selection, the relative proportions of genes change from generation to generation.

Differential reproductive success can result from differences in the probability of survival of different types of individuals or from differences in the success these individuals have in producing surviving offspring.

The selective differentials are rarely all-or-none; slight differences are sufficient to bring about substantial changes. Certain molds produce antibiotics which destroy bacteria growing nearby. On bacterial plates contaminated by these molds, the molds have full possession of certain restricted areas; the bacteria are unable to grow closer to a "colony" of mold than a centimeter or more. Two strains of bacteria, in contrast, may grow together in a culture tube without either inhibiting the other. Nevertheless, if the generation time of one is shorter than that of the other by even a small fraction, the culture will eventually contain only the faster growing strain. The replacement of one bacterial strain by the other because of a small difference in division rates may be slower than outright destruction, but it is scarcely less certain.

A number of studies have been made in an effort to demonstrate the action of natural selection on alternative forms of various species. One of the early studies of this sort was made with the praying mantis. Cesnola (1904) tethered green and brown individual praying mantises (*Mantis religiosa*) on patches of green and brown grass. Within nineteen days on the green grass, thirty-five of forty-five brown insects had been eaten by birds; none of the twenty green ones had been killed. On brown grass the situation was reversed; all twenty-five greens had been killed within eleven days while none of twenty browns had been even by the nineteenth day. Under the proper circumstance, the green or brown color of mantises is clearly protective.

A study similar to that of Cesnola's was carried out with gray and black forms of the mosquito fish, *Gambusia patruelis,* which were placed in a pale or black tank and exposed to predation by penguins. (Mosquito fish adjust their color to match the background but can do so only rather slowly. The gray and black fish used in this experiment were forms that had been kept in white and black tanks for one or two months just prior to the predation experiment.) The results of one of these experiments are given in Table 9-1. Examination of the figures in this table shows that, in the case of the pale tank, low numbers correspond to gray fish eaten and black fish surviving. The reverse is true in the black tank; low numbers correspond to grays surviving and to blacks eaten. A color appropriate for its background clearly helps these fish to survive predation even though the protection is not absolute by any means.

A number of observations similar to those given above have been made for various species of animals. Ford (1964, pp. 155ff.) describes the predation of the snail (*Cepaea nemoralis*) by the song thrush (*Turdus philomelos*). These birds carry the larger snails to convenient rocks ("thrush anvils") to crack open their shells. Examination of the shell fragments at these anvils reveals that snails of different banding patterns are caught differentially during different seasons of the year as the background foliage of the woodlands changes. Finally, the most famous case of all is that of industrial melanism, which has affected at least eighty species

TABLE 9-1

EVIDENCE FOR THE PROTECTIVE VALUE OF CHANGEABLE COLORATION IN THE FISH, GAMBUSIA PATRUELIS, WHEN EXPOSED TO PREDATION BY PENGUINS.

(Sumner, 1935.)

TANK	FISH	EATEN	NOT EATEN	TOTAL
Gray	Gray	176	352	528
	Black	278	250	528
		454	602	1056
Black	Gray	217	118	335
	Black	78	257	335
		295	375	670

of cryptically colored moths in Britain during the past one hundred years. Ford (1964) has given an excellent review of this phenomenon. It is sufficient for the moment to say that the chief selective force seems to have been differential predation by birds; dark forms of these moths are relatively inconspicuous on soot-covered tree trunks and, consequently, are overlooked by birds. Light-colored individuals, on the contrary, are highly conspicuous on darkened tree trunks (trunks that are denuded of rough-textured lichens, as well) and, as a result, suffer a high mortality.

SELECTION AGAINST RECESSIVE GENES

The consequences of differential predation or other sorts of differential mortality as these affect individuals of different genotypes can be calculated rather easily. Suppose that alleles A and a have frequencies of p and q, and that at the onset of a given generation the three genotypes have frequencies p^2, $2pq$, and q^2. Suppose too, that between the formation of zygotes (fertilization) and the attainment of sexual maturity, only $(1 - s)$ aa individuals survive for every AA or Aa individual. Note that we have not said that all or even most AA and Aa individuals survive; mortality may be extremely heavy as it must be, for example, in the case of the oyster where each female produces millions of eggs. But, for every AA individual that survives the trials of growing up, one Aa individual survives, but only $(1 - s)$ aa individual. The quantity s is known as the selection coefficient, while $1 - s$ is known as the adaptive value or the relative Darwinian fitness of aa individuals. What are the effects of our suppositions on the genetics of this hypothetical population?

The description given in the preceding paragraph can be summarized in algebraic terms as follows:

GENOTYPES	AA	Aa	aa
Initial frequencies	p^2	$2pq$	q^2
Adaptive value	1	1	$1-s$

Average adaptive value or
adjusted total population: $p^2 + 2pq + q^2 - sq^2$ or $1 - sq^2$

Frequency of a after selection: $\dfrac{q - sq^2}{1 - sq^2}$ or $\dfrac{q(1 - sq)}{1 - sq^2}$

The frequency of a has changed from q to $q(1 - sq)/(1 - sq^2)$ as the result of selection.

The steps involved in these calculations may appear less difficult if the problem were restated somewhat differently. Suppose a lecture room were filled with students, one half of them men and one half women. Suppose, too, that a certain number of students leave the room but for every male student that remains an average of one half a female student remains ($1 - s$ equals 0.5 in this case). What is the frequency of female students among the remaining students? Clearly it is no longer one half. The adjusted total of the student population in the lecture room is no longer 1.00; instead it is $(1 \times ½) + (½ \times ½)$, or ¾. Hence, the frequency of female students is ¼ ÷ ¾, or ⅓.

Two variations of the algebraic expression given above are important for us. The first is the special case in which s equals 1.00, the case of recessive lethals. The new frequency of a in this case equals $q(1 - q)/(1 - q^2)$. Since $1 - q^2$ equals $(1 - q)(1 + q)$, the new frequency can be written $q/(1 + q)$.

Those who prefer fractions to decimals can substitute $1/n$ for q; in this case, the new frequency is $1/(n + 1)$. In fact, in a series of successive generations the frequencies of a would be

$$\frac{1}{n} \qquad \frac{1}{n+1} \qquad \frac{1}{n+2} \qquad \frac{1}{n+3} \qquad \frac{1}{n+4}$$

or, reverting to decimals,

$$q \qquad \frac{q}{1+q} \qquad \frac{q}{1+2q} \qquad \frac{q}{1+3q} \qquad \frac{q}{1+4q}$$

These terms tell us that the elimination of a recessive lethal proceeds at a rapid rate only if its frequency is high; otherwise, elimination is slow. For example, n generations are required to reduce the frequency of a lethal by one half, from $1/n$ to $1/2n$. Two generations suffice to reduce a recessive lethal from a frequency of ½ to ¼; 1000 generations are required to reduce it from 1/1000 to 1/2000. In the case of nonlethal recessive genes—genes for which s is less than 1—the same general statement is true except that the rate of elimination is even slower because not all *aa* homozygotes are eliminated from the population each generation.

The calculations made above reveal the inefficiency of negative eugenic measures proposed so vigorously three or four decades ago. The cure-all proposed for genetic ills was sterilization. Unfortunately, the ills against which the measures were proposed and which were really genetic in nature (many of the syndromes under attack were really the result of environment, not heredity) were, separately considered, rare abnormalities. Furthermore, the afflicted persons were generally severely handicapped by the disease itself. Consequently, sterilization merely reduced further an already impaired reproductive ability. The gene frequencies involved were low to start with; consequently, the increased rate of elimination of the gene through the sterilization of homozygous carriers was negligible. Only the most fanatic or the least informed of these early eugenists could detect any value in the proposed programs.

The second useful alteration in the foregoing calculations is a conversion of the original cumbersome expression into a statement concerning the change in the frequency of $a(\Delta q)$ during a single generation. We saw that the frequency of a changed from q to $q(1 - sq)/(1 - sq^2)$. The change in frequency, Δq, equals $q(1 - sq)/(1 - sq^2) - q$ or $-sq^2(1 - q)/(1 - sq^2)$. If q is small (and generally alleles that are opposed by selection are rare), this expression can be written (approximately)

$$\Delta q \approx -sq^2$$

The above expression seems reasonable because it claims that the change in the frequency of a rare, deleterious allele equals the product of the degree to which it is deleterious (selection coefficient, s) and the frequency of homozygous carriers (q^2).

Having learned that Δq is approximately $-sq^2$, one can test for possible equilibrium conditions for a gene opposed by selection. At equilibrium, gene frequencies remain constant from generation to generation and, consequently, Δq equals zero. In that case, sq^2 equals zero as well. And so, if selection against homozygous recessives does occurs, the ultimate result (slow as it may be) is the complete elimination of the recessive allele from the population.

Some Selection Experiments

One of the simplest possible selection experiments consists of observing the elimination of a lethal gene from a population. Because the gene is lethal, s is known to be 1.00. Therefore, the expected frequencies of the gene in successive generations are terms in the sequence $1/n$, $1/(n+1)$, . . . Ordinarily, a considerable amount of work would be needed to determine the frequency of a lethal in each generation. Fortunately, however, some lethals in *Drosophila* arise at loci for which recessive visible mutations are known; individuals heterozygous for the recessive allele and a particular lethal may survive and show the recessive phenotype. One such lethal was found in *D. melanogaster*; it was allelic to *light* (*lt*), a recessive eye-color mutation carried by the *CyL* balancer chromosome.

The fate of the *lt*-lethal in a freely-breeding laboratory population was followed over a period of ten generations (Wallace, 1963a). The parental flies were all heterozygous for the *lt*-lethal and so the starting frequency was 0.50 or ½. Each generation many adult males were mated individually with *CyL/Pm* virgin females. Some of these males produced *CurlyLobe* offspring, all of which had the brick red eye of *CyL/+* individuals. Other males produced *CurlyLobe* offspring one half of which had red eyes while the other half had the bright orange eye typical of homozygous *light* individuals. The males that produced two types of *CyL* offspring were lethal heterozygotes. The frequency of the *lt*-lethal in the population equaled the number of heterozygotes divided by twice the total number of males tested. The results of this experiment are shown in Table 9-2 and Figure 9-1. The data, especially as they appear in the figure, show that the lethal decreased in frequency even more rapidly than our calculations led us to expect. Obviously, s cannot be

TABLE 9-2

ELIMINATION OF AN AUTOSOMAL LETHAL FROM AN EXPERIMENTAL POPULATION OF D. MELANOGASTER. (Wallace, 1963a.)

GENERATION	n	OBSERVED	EXPECTED
0	—	0.500	—
1	454	0.284	0.333
2	194	0.232	0.250
3	212	0.189	0.200
4	260	0.188	0.167
5	290	0.090	0.143
6	398	0.085	0.125
7	366	0.082	0.111
8	382	0.065	0.100
9	388	0.054	0.091
10	394	0.041	0.083

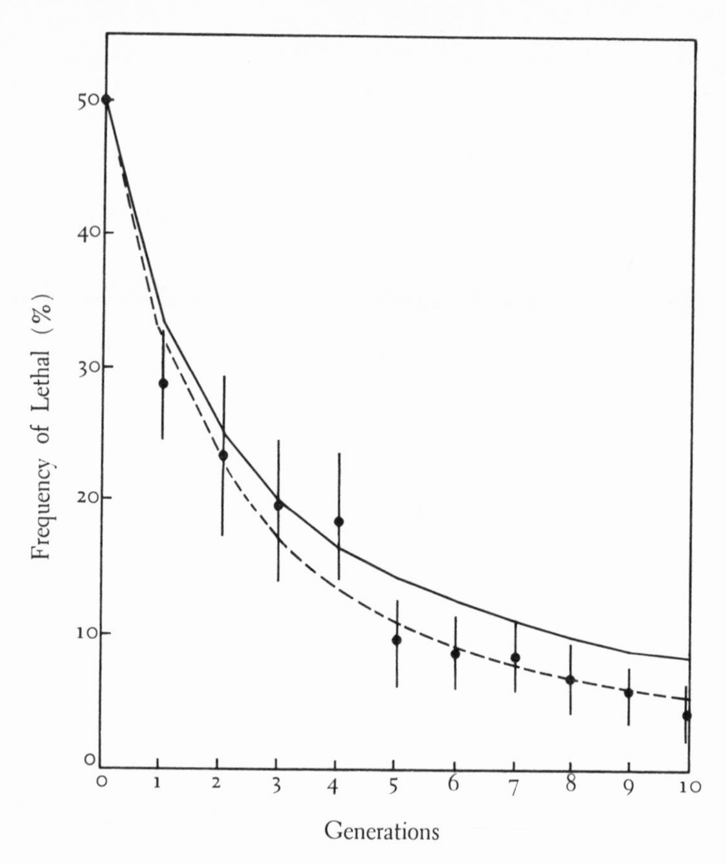

FIGURE 9-1. Elimination of an autosomal lethal from an experimental population of *D. melanogaster.* Closed circles and vertical lines represent observed lethal frequencies together with their 95 percent confidence intervals. Solid line: theoretical expectation for a completely recessive lethal; dashed line: theoretical expectation for a lethal that reduces the fitness of heterozygotes by 10 percent. (After Wallace, 1963a.)

greater than 1.00; the greater-than-expected rate of elimination indicates, then, that a loss of the lethal occurs also in heterozygous carriers.

The gene, *lt*-lethal, was chosen for study in the above example because the selection coefficient of a lethal is known, 1.00. As a result, any discrepancy between observed and expected rates of elimination can be assigned to heterozygous individuals. Clear-cut decisions of a comparable nature cannot be made in following the fate of a recessive visible mutation. Gordon (1935), for example, released a large number of *ebony* heterozygotes (*D. melanogaster*) in an area of England normally devoid of this species. After a period estimated to be six generations at most, the frequency of the *ebony* gene was only 11 percent. Hence, *ebony* homozygotes were either effectively lethal in this introduced population or *ebony* was eliminated through the action of selection on both homozygotes and heterozygotes.

Although it has been stated above that selection can result from either differential mortality or differential fertility (or both), we have restricted our discussion and our examples to the former—differential mortality. A study by Reed and Reed (1950) on the elimination of *white*, a recessive sex-linked eye-color mutant, from populations of *D. melanogaster* illustrates the selective effect of differential reproductive ability. The observed elimination of the *white* allele from the experimental population is shown in Figure 9-2; the gradually decreasing frequencies resemble very much those of Figure 9-1. [The data on *white* are complicated somewhat because of the inheritance pattern of sex-linked genes; this point need not concern us here, however. Interested persons can consult either Li (1955) or Falconer (1960) for a discussion of the population genetics of sex-linked genes.]

It is impossible to assign a selective coefficient to *white* on purely theoretical grounds; if selection operates against white-eyed males or females, this fact and its magnitude must be determined empirically.

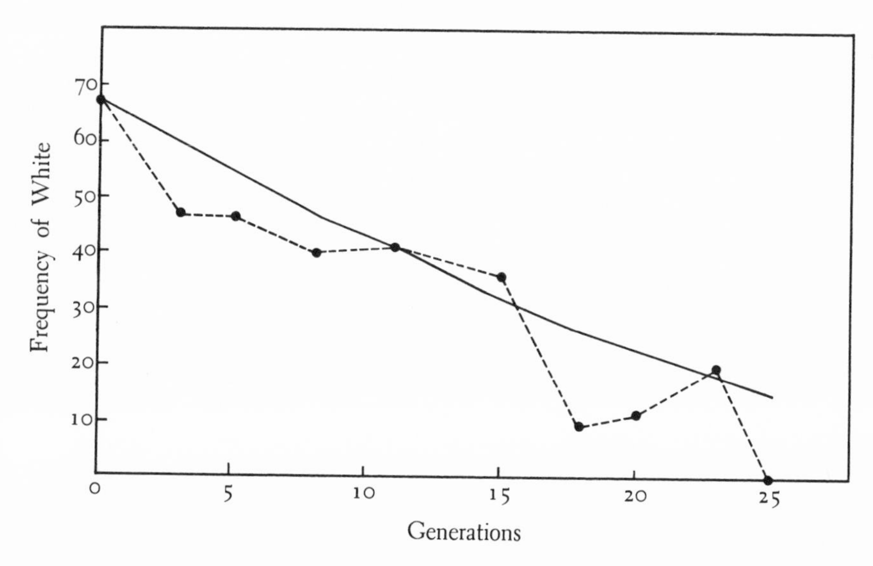

FIGURE 9-2. Elimination of the sex-linked mutant allele *white* from a laboratory population of *D. melanogaster*. Frequency of *white* is calculated as two thirds of its frequency in females plus one third of its frequency in males. The solid line shows the elimination expected as a result of the decreased mating success of white-eyed males. (After Reed and Reed, 1950.)

Tests made to determine the relative viability (as an estimate of adaptive value, $1 - s$) of white-eyed flies are summarized in Table 9-3. These data reveal, in fact, that the viabilities of white- and red-eyed flies in this particular experimental material are equal. The viabilities of the two sexes differ, but the sex ratio in *Drosophila* cultures commonly deviates considerably from the expected 1 : 1 ratio. Within each sex, however, the proportions of the two phenotypic classes are equal just as one expects from the nature of the test cross used (*w*Y♂♂ × *Ww*♀♀). No evidence exists, then, to suggest that the elimination of *white* results from differential preadult mortality.

In addition to their viability, the mating success of white- and red-eyed flies was also tested. The results of these tests, too, are given in Table 9-3. From these data we see that *white* males were only about 75 percent as fortunate at obtaining mates as their red-eyed sibs. Although *white* males appear to survive larval competition in crowded cultures as well as the wildtype males, their mating behavior is such that about one quarter of them are not really members of the breeding population.

TABLE 9-3

RELATIVE VIABILITIES AND MATING ABILITIES OF *white* AND WILDTYPE D. MELANOGASTER.

(Reed and Reed, 1950.)

Viability

	FEMALES		MALES	
	Ww	*ww*	*W*Y	*w*Y
Observed	6947	6844	6032	6083
Expected*	6896	6896	6058	6058

* "Expected" assumes a 1 : 1 ratio within sexes but not a 1 : 1 ratio of sexes.

Mating ability

	FEMALES		
MALES	*ww*	*Ww*	TOTAL
*w*Y	90	64	154
*W*Y	100	105	205
Double	6	4	10

Successful *white* males: 154 + 10 = 164 (0.76)
Successful wildtype males: 205 + 10 = 215 (1.00)

As a result, the *white* allele disappears from the population. The solid line shown in Figure 9-2 has been calculated on the basis of the lessened mating success of *white* males; the observed elimination of *white* from the laboratory population can be accounted for rather well by this one item alone.

Evidence for Selection in Human Populations

Human beings are afflicted with a variety of genetic ailments—sex-linked and autosomal, recessive and dominant. Some of these hereditary diseases involve the most crippling developmental disturbances; some cause embryonic death, others prove to be lethal after birth, still others cripple their carriers—physically, mentally, or both—for life.

In addition to genetic disorders, man exhibits a series of polymorphisms which we take for granted without thinking in terms of physical damage: blue and brown eyes; black, brown, and red hair; tasters and nontasters; A, B, AB, and O blood groups; and many others. Genes whose alternative alleles appear harmless are the source of much of the variation exhibited by mankind.

A number of studies have been made in an effort to determine whether selection in the form of differential mortality or fertility operates on the carriers of these seemingly harmless widespread mutant genes in human populations. These studies have met with some success. Ford (1965) has summarized many of the more important studies.

The results of two studies concerning the possible interactions of anti-A antibodies of type O mothers with unborn type A children are reported in Table 9-4. The search for this type of selection was initiated because it seems plausible that blood of type O mothers, containing anti-A and anti-B antibodies as it does, might destroy tissues and blood cells of embryos of types A and B if the mother's blood were to cross the placenta into the embryo's circulatory system. The type of data collected and its analysis are clever indeed. From long experience with the ABO blood group system, it is known that the proportions of O, A, B, and AB individuals of the two sexes are the same. The proportions of type O and type A children born to type O mothers with type A husbands should equal those born to type A mothers with type O husbands. There is no theoretical basis for expecting destructive interactions between antibodies of type A mothers (anti-B) and the tissues of either type O or type A embryos.

The relative proportions of type O and type A children born of O $\times$ A matings depend upon the frequencies of I^O, I^A, and I^B genes in the population. Knowing these frequencies, however, it is possible to calculate the frequencies I^O and I^A alleles among type A individuals. Using the

TABLE 9-4

EVIDENCE FOR (HOKKAIDO) AND AGAINST (OHDATE) AN INTERACTION BETWEEN TYPE O MOTHERS AND TYPE A EMBRYOS.

(Matsunaga, 1955; Matsunaga et al., 1962.)

			CHILDREN		
	MOTHER	FATHER	O	A	TOTAL
Hokkaido					
	A	O	282 (278)	430 (434)	712
	O	A	320 (269)	369 (420)	689
Ohdate					
	A	O	289 (290)	401 (400)	690
	O	A	258 (291)	435 (402)	693

Gene frequencies used in calculating expected numbers

	I^A	I^B	I^O
Hokkaido	0.30	0.15	0.55
Ohdate	0.23	0.18	0.59

frequencies listed in the first example of Table 9-4 we can calculate the following expected frequencies:

I^OI^O 0.3025
I^OI^A 0.3300
I^OI^B 0.1650
I^AI^A 0.0900
I^AI^B 0.0900
I^BI^B 0.0225

Furthermore, we can calculate the frequencies of I^O and I^A alleles among type A persons:

GENOTYPE	FREQUENCY	I^O	I^A
I^OI^A	0.33	0.165	0.165
I^AI^A	0.09	—	0.090
Total	0.42	0.165	0.255
Relative frequencies (÷ 0.42)		0.39	0.61

Since matings of type A and type O persons are, in a sense, test crosses, the proportions of A and O children will equal the frequencies of I^A

and I^O alleles carried by type A parents. Therefore, among the 712 children born to A mothers and O fathers in the first example, one expects to find 278 O children and 434 A children; these are almost exactly the numbers observed. In the case of O mothers and A fathers, one expects to find 269 type O children among the total of 689 while 420 are expected to be type A. These expectations differ markedly from the actual observations.

There appears to be a shortage of A children born to type O mothers in the first example given in Table 9-4. Just how large is the deficit? We can estimate it by recalling that the 369 A children, according to the assumptions regarding incompatibility of A children and their type O mothers, are a residue of a larger unknown number, N. We can let 369 equal kN, where k is a fraction whose value we want to calculate. We also know that $282/430 = 320/N$ or, substituting and multiplying,

$$\frac{282 \times 369}{k} = 430 \times 320$$

The value of k yielded by this calculation is 0.76; apparently, 24 percent of type A zygotes initially carried by type O mothers are lost before birth. These embryos are no better off than the gray fish in the black tank of Table 9-1.

The loss of nearly one quarter of type A embryos carried by type O mothers raises a number of interesting points. First, it shows that selection can operate at relatively high intensities in human populations. Second, it shows how intense selection can be, even in human populations, without leaving obvious signs. Finally, this loss is interesting in respect to the lack of evidence for selection in the second example listed in Table 9-4. There is no need to assume because of the discrepancy that one of the two sets of data is in error. It may be that the difference represents different patterns of selection in different populations. It must be admitted, however, that the basis of these different patterns is not at all clear at the moment.

Selection against Heterozygotes

In discussing both the elimination of the *lt*-lethal from a laboratory population and of *ebony* from a "wild" population, we mentioned that the frequency of the mutant gene dropped somewhat faster than was expected. This was true even if homozygous *ebony* flies were assumed to be lethal or sterile. To "explain" this overrapid elimination, we mentioned that additional mutant alleles might be lost from the population because of selection against their heterozygous carriers. In short, the effect on fitness of these genes may not be completely recessive.

The algebraic aspects of selection against deleterious genes not only in homozygous but also in heterozygous condition are relatively simple. They can be made even simpler if we assume that such genes will generally be *very* rare in populations. In this manner we shall be able to ignore the even rarer homozygotes (whose frequency will be q^2, the square of a small amount) and concentrate entirely on the elimination of mutant genes by selection against heterozygotes. As in the earlier calculations (page 162) we shall assume that homozygous *aa* individuals have fitness $1 - s$. Now, however, we shall assign a fitness $1 - hs$ rather than 1.00 to *Aa* heterozygotes. A fraction h of the deleterious effect of the gene a is expressed in heterozygous condition; h, then, is a measure of the dominance of a in respect to fitness.

The conditions we have described can be summarized as follows:

GENOTYPES	*AA*	*Aa*	*aa*
Frequencies	p^2	$2pq$	q^2
Fitnesses	1	$1 - hs$	$1 - s$
Average fitness:	$1 - 2hspq - sq^2$, or approximately $1 - 2hspq$ when q is very nearly 0		
Frequency of a after selection:	$(q - hspq)/(1 - 2hspq)$, or, since q is very nearly 0 while p is very nearly 1.00, $q(1 - hs)$		

The new frequency of a is given by $q(1 - hs)$. In this case, the change in gene frequency, Δq, equals $q(1 - hs) - q$, or $-hsq$.

The point that emerges from these calculations is that the loss of a *rare* mutant gene through the action of selection on its heterozygous carriers is directly proportional to its frequency, not to the square of its frequency. This is true only for mutant genes with very low frequencies, to be sure. Rare deleterious genes are lost much more rapidly from populations if they affect heterozygotes than if they affect homozygotes only. Needless to say, such mutations, unless they are retained for other reasons, are eliminated completely from populations.

Attempts to Estimate h Experimentally

By assigning adaptive values to the three genotypes *AA*, *Aa*, and *aa* of 1, $1 - hs$, and $1 - s$, we have removed an earlier restriction that we had placed on the fitness of heterozygous individuals. In discussing recessive mutations (page 163), the heterozygotes were assigned an adaptive value of 1.00 just as were the homozygous *AA* individuals. In effect, the value of h was said to be zero in this earlier calculation, because $1 - hs$ equals 1.00 if h equals zero. Were h to have a value of 1.00, *Aa* individuals would have a fitness of $1 - s$ just as *aa* individuals; in this case, *a* would be completely dominant to *A* in respect to fitness. Although one can imagine values of *h* greater than 1.00 or less than zero (negative values), for the moment we shall allow *h* to vary only from zero to 1.00.

Many mutations are known which affect their heterozygous carriers. Even if we neglect mutations with visible morphological effects and limit ourselves to mutations which affect viability (an important component of fitness), we find many examples of dominance and partial dominance. If one irradiates *D. melanogaster* males with some 7000 r of X radiation and mates these males with untreated females, about 90 percent of all eggs laid by these females will fail to hatch. Ordinarily, the mating of untreated males and females results in eggs of which about 5 percent or fewer do not hatch. By varying the radiation dose, one can show that the proportion of unhatched eggs is related to exposure. Since the radiation exposure described has been limited to one parent, the dying zygotes are merely heterozygous for induced changes; consequently, these changes are known collectively as "dominant lethals." According to the algebraic terminology used above, both *h* and *s* equal 1.00 in the case of dominant lethals. (A cytological examination of the early nuclear divisions of affected eggs reveals that gross chromosomal abnormalities and aberrant division patterns are responsible for much of the observed dominant lethality; such observations have been made by Sonnenblick, 1940, and Sonnenblick and Henshaw, 1941.)

In both *Drosophila* and man many genes are best known by their marked effects on heterozygous carriers; many of these genes are lethal in homozygous condition. In man, the genes causing chondrodystrophic dwarfism and brachydactyly are supposed to be lethal when homozygous. Other recessive lethals in man have rather harmless effects on heterozygotes. In *Drosophila*, many of the useful dominant laboratory mutations—*Curly, Plum, Stubble, Hairless,* and *Ultrabithorax,* to mention a few mutants of *D. melanogaster*—are alleles that are lethal when homozygous. One class of mutations in particular, the *Minutes,* are outstanding in the effect they have on their heterozyogus carriers; each *Minute* is lethal when homozygous, each prolongs the development time of heterozygotes and generally reduces its size considerably.

In addition to mutations which have rather pronounced effects on heterozygotes, there are numerous others which at least superficially appear to be completely recessive. A small degree of dominance is, of course, easily overlooked. The *lt*-lethal whose elimination from a population was described in Table 9-2 and Figure 9-1 appeared to be a "recessive" lethal. Still, it disappeared from the population at a rate faster than that expected for a recessive lethal. The dashed line shown in Figure 9-1 represents the rate of elimination of a lethal that lowers the fitness of heterozygous carriers (in addition to killing homozygotes) by 10 percent; this line fits the observed points considerably better than the one representing the elimination of a completely recessive lethal.

A large number of tests have been made in an attempt to measure the extent to which lethal mutations in *Drosophila* affect their heterozygous carriers. To a large extent these tests involve comparisons of lethal and nonlethal chromosomes manipulated by means of crosses with genetically marked stocks; the tests are tests of chromosomes and not of genes at individual loci.

A study of the partial dominance of seventy-five six-linked lethals recovered during the course of low-level radiation experiments of *D. melanogaster* were made by Stern et al. (1952). Thirty-six of these lethals had arisen in control males, the remaining thirty-nine in males exposed to 50 r of gamma (radium) radiation. Since 100 r induces lethals with about the same frequency as that with which they arise spontaneously, only thirteen of the thirty-nine lethals arising in the exposed males should be radiation-induced; the remainder should be spontaneous lethals no different from the control lethals themselves. Therefore, the two series of lethals have been combined for the following discussion.

To determine whether or not lethal chromosomes lowered the survival of their heterozygous carriers, Stern and his colleagues crossed wildtype females known to be heterozygous for a given lethal to males carrying the genetically marked (w^aB) *Basc* chromosome. These crosses were arranged deliberately so that the resultant progeny would develop under severely crowded conditions. Daughters, heterozygous for the *Basc* chromosome, were collected in large numbers from these cultures and were tested individually to determine whether they were $w^aB/+$ or w^aB/le. If the lethals were completely recessive, the two types of daughters would be expected in equal numbers; if lethals are partially dominant, however, one would expect fewer w^aB/le daughters to survive under the severe conditions imposed by overcrowded cultures.

The results suggest that the recessive lethals of this series of tests were indeed partially dominant *on the average*. The mean frequency of w^aB/le females observed in the first test of each lethal was 0.487; the average dominance of these "recessive" lethals was, consequently, about 2.5 percent (this value is lower than that calculated on all tests which include repeat tests on many lethals, 3.9 percent). Among the tested lethals, some were found that gave more than the expected 50 percent w^aB/le females, while others gave significantly fewer than the expected 50 percent. Between these extremes were the bulk of all lethals tested; for these, taken lethal by lethal, the results were neither significantly below nor above expectation. One can say that different lethals behave differently; most of them tend to reduce the viability of their heterozygous carriers (hence the lowering of the *average* proportion of w^aB/le females) but a few tend to increase the survival of their carriers.

A summary of the results obtained by Stern and his colleagues is presented in Table 9-5 and Figure 9-3. A procedure analogous to that used by Wallace and Madden (see page 49) was used in analyzing the original data. First, the variance in the proportions of w^aB/le daughters among the seventy-five different lethals was calculated. Second, an expected binomial sampling variance was computed; this is the statistical variance that is related to the number of females tested per lethal. The variance actually observed among the seventy-five lethals was considerably larger than that expected on the basis of numbers of flies counted

TABLE 9-5

ESTIMATING THE BETWEEN-LETHAL VARIATION IN THE VIABILITY EFFECTS OF SEX-LINKED "RECESSIVE" LETHALS ON THEIR HETEROZYGOUS CARRIERS IN D. MELANOGASTER.

(After Stern et al., 1952.)

Total variance	0.00134263
Sampling variance	0.00053813
Experimental variance	0.00022250
Between-lethal variance (residue)	0.00058200
Between-lethal standard deviation	0.0241

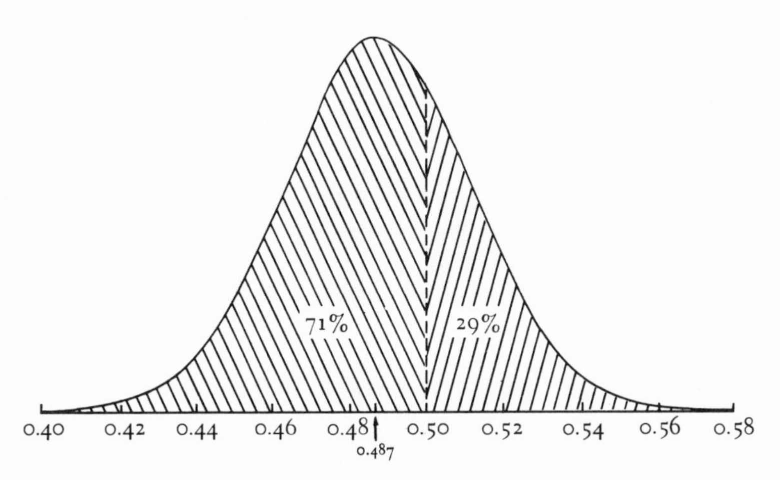

FIGURE 9-3. Frequency of w^aB/le females among the daughters of $+/le$ females and w^aB males (le = lethal). The curve represents variation remaining after binomial and experimental variances are removed; consequently, it represents variation between the frequencies of w^aB/le heterozygotes carrying different lethals. (After Stern et al., 1952.)

per lethal alone; we have already suggested this by saying that some lethals gave significantly more w^aB/le females than expected while others gave significantly fewer. About one third of the lethals were tested more than once for one reason or another; in some cases the replicate tests were carried out by different persons working in different laboratories. The variation from test to test of the same lethal can be compared with that expected by sampling (binomial) error alone. It seems as if some lethals upon repeated testing did give results more variable than those which can be explained by numbers of daughters tested. Consequently, the excess variance ("experimental" variance) can be calculated.

Now, neither the sampling (binomial) nor the experimental variance should bias the mean number of w^aB/le females per culture; therefore, these variances can be removed from the total variance observed. The variance that remains is an estimate of the variation between mean proportions of w^aB/le females from lethal to lethal (genetic variance, if you wish) unencumbered by contributions made by extraneous sources of variation. The genetic variance can be converted into a genetic standard deviation and now the distribution of proportions of w^aB/le females can be calculated. The results appear in Figure 9-3. From the analysis given here, one can see that nearly 30 percent of all lethals give proportions of w^aB/le daughters greater than the expected fifty percent while the remaining 70 percent or more give proportions below this expected frequency. Presumably, many of those which give too many or too few w^aB/le daughters give so nearly the correct number that tests of enormous size would be needed to demonstrate the slight deviations involved.

Another series of tests of newly arisen lethals has been made (Wallace, 1965, and unpublished) as an offshoot of an experiment that was carried out for a slightly different purpose. The question originally posed concerned the viability of individuals carrying homologous second chromosomes that were identical except for mutations that had arisen during five or ten generations in which the two homologues had been perpetuated in separate lines of descent. In a large series of such tests, it was inevitable that lethal mutations would arise in the homologue of one line of descent but not in that of the other. As a result, the data include estimates of the viabilities of flies that carry (1) a lethal chromosome or, alternatively, (2) a nonlethal chromosome otherwise identical to the lethal. A comparison of these two groups reveals, first, that the lethals obtained for testing were not uniform in their effects on the viabilities of heterozygous carriers; two lethals of thirty-two tested were markedly deleterious in heterozygous condition, one of these lowered the viability of its heterozygous carriers over 30 percent. The elimination of the two obviously semidominant lethals from those studied left thirty others; on the average, these appeared to lower viability by about 1 percent.

The lethals tested in the preceding two sets of experiments constitute a reasonably random sample of newly arisen lethals. Stern and his colleagues accumulated lethal stocks as they arose during the course of a study on the mutagenic effects of low-level radiation. The lethals in my own experiments arose while chromosomes were perpetuated in balanced heterozygotes for five or ten generations; their presence was not evident until counts of test cultures were well underway (which cultures contained lethals and which contained nonlethal heterozygotes was not known until after the full counts of test cultures had been completed). The tests to be described in the following paragraphs will serve as a transition between tests of newly arisen lethals and lethals sampled from long-established populations.

The transition is needed to illustrate that lethals obtained from existing populations need not be identical as a group with those newly obtained by mutation. This point can be illustrated, too, by a hypothetical example. Suppose newly arisen lethals consist of two types in equal frequencies—those that are completely recessive and those that lower the fitness of their heterozygous carriers by 10 percent. The average fitness of individuals heterozygous for a large sample of newly mutated lethals would be (0.50)(1.00) plus (0.50)(0.90), or 0.95; the average dominance ($\bar{h}$) of these lethals would be 5 percent. Suppose now that these same lethals are maintained for ten generations in a large, freely breeding population but maintained at such low frequencies that their elimination by homozygosis is negligible. Those which lower the fitness of their heterozygous carriers will have their frequencies reduced. While the complete recessives maintain their original frequencies unchanged, the partially dominant lethals will be reduced to $(0.90)^{10}$, or to some 35 percent of their original frequencies. Of all lethals remaining in the population 26 percent $[(0.35 \times 0.50) \div 0.675]$ will be partially dominant while 74 percent will be completely recessive. The average dominance ($\bar{h}$) of lethals in the population at the end of ten generations will be (0.26)(0.10) plus (0.74)(0.00), or 2.6 percent. The value of $\bar{h}$ will have been reduced by one half because of ten generations of selective elimination of incompletely recessive lethals.

An actual example illustrating what has been described above in hypothetical terms can be found in the behavior of lethals in certain artificial populations of *D. melanogaster*. These are populations that will be described in greater detail in a subsequent chapter. For the moment, it will be sufficient to say that each population was started initially with lethal-free second chromosomes. Lethal chromosomes accumulated in all populations—slowly in the control population (see Table 8-4) but rapidly in others that were exposed to chronic gamma radiation.

At regular intervals throughout the existence of the artificial populations, samples of eggs were taken to obtain (eventually) chromosomes which were then tested in homozygous and heterozygous condition by means of the techniques outlined in Figures 3-3 and 3-6. The early tests were made by means of the *CyL*-test of Figure 3-3; the proportion of wildtype flies in heterozygous cultures of these tests approached 33⅓ percent. Later, the *CyL-Pm*-test of Figure 3-6 was used exclusively so that the relative viabilities are expressed as ratios (with the *CyL/Pm* class serving as the standard with viability of 1.00).

The results obtained by the *CyL* tests are summarized in Table 9-6.

TABLE 9-6

FREQUENCIES OF WILDTYPE FLIES HETEROZYGOUS FOR RANDOM COMBINATIONS OF DIFFERENT CHROMOSOMES FROM EACH OF FIVE EXPERIMENTAL POPULATIONS.

Three types of combinations have been identified: those in which both wildtype chromosomes are lethal or semilethal (2-drastic), those in which one chromosome is lethal or semilethal (1-drastic), and those in which neither chromosome is lethal or semilethal (0-drastic). The data listed in the top part of the table were collected over a 2-year period and are included in those of the bottom portion, which represents a study extending over a 5-year period. (Wallace and King, 1952; Wallace, 1962.)

	2-DRASTIC		1-DRASTIC		0-DRASTIC	
POPULATION	$\bar{x}$	n	$\bar{x}$	n	$\bar{x}$	n
1	35.3	38	34.7	223	34.8	352
3	33.7	28	34.4	216	34.5	292
5	31.6	239	32.4	235	32.0	54
6	32.5	231	33.2	243	34.5	54
7	32.7	29	33.0	163	33.8	339
1	35.2	121	34.8	878	34.7	1575
3	33.1	242	33.5	1137	33.9	1377
5	32.9	1329	33.4	817	33.1	147
6	33.5	1406	33.8	691	34.7	90
7	32.6	235	33.2	934	33.7	1080

The tests include those made on five experimental populations. The top part of the table gives the very early results reported by Wallace and King (1952). The figures are the average frequencies of wildtype heterozygotes that carry no lethal second chromosome, one, or two. It is quite clear from these data that lethals had an adverse effect on their carriers. If population 1 is excluded from the calculations, it appears that the frequency of heterozygotes carrying one lethal is decreased about 1.4 percent, while that of heterozygotes carrying two (nonallelic) lethals is decreased by some 3.2 percent. Now, it can be shown that a change in the frequency of wildtype flies in this type of culture can be expressed as a change in viability as follows:

$$\frac{2(f_w + \Delta f_w)}{f_c - \Delta f_w} - \frac{2f_w}{f_c}$$

or, approximately, $(2\Delta f_w)/(f_c)^2$, where f_w is the frequency of wildtype flies, f_c is the frequency of *CurlyLobe* flies, and Δf_w is the change in frequency of wildtype. For example, a lowering of the frequency of lethal-bearing wildtype heterozygotes by 1.5 percent corresponds to a reduction in the relative viability of these flies of about 7 percent, an estimate of h for these lethals.

Following the analysis of Wallace and King given above, Wallace (1962) summarized all available data on these populations obtained by the *CyL* technique. These data are given in the second section of Table 9-6. Once more it appears that in general lethal chromosomes lower the viability of their heterozygous carriers. However, there are differences between the early and total (including the early) data. Tests of chromosomes from the control population (#3) are consistent in showing a regular increasing sequence of wildtype frequencies in the three categories: 2-lethal, 1-lethal, and 0-lethal. For this population, a single lethal decreases the frequency of wildtype heterozygotes by 0.4 percent; this corresponds to a change in viability of approximately 1 percent. It is now quite obvious that population 1 adheres to its contradictory pattern; there is a corresponding difference in the history of this population as compared with the others. The chromosomes of population 1, as were those of all other populations, were initially lethal-free; however, because of its radiation history this population had, at the moment of its inception, a high frequency of radiation-induced lethals which were then gradually (and presumably selectively) eliminated by natural selection. All other populations accumulated their lethals more or less gradually and continuously. Excluding population 1 again, we can calculate that in the total data a lethal chromosome lowered the frequency of its wildtype carriers by an average of approximately 1.2 percent; this corresponds to a reduction in viability of about 5 percent. This is considerably less than the earlier estimate, especially when one recalls that the earlier data are part of the total. Removing the earlier data from the total leaves a residue within which the average dominance of lethal chromosomes is about 2 percent; this is indeed a considerable reduction from the initial effect for which h was estimated to be 7 percent.

The data listed in Tables 9-7 and 9-8, data that are based on the *CyL-Pm* technique, were collected for the most part later than those listed in Table 9-6. During a transitional period, while the technically more difficult *CyL-Pm* technique was being introduced for the first time,

TABLE 9-7

COMPARISON OF THE RELATIVE VIABILITIES OF *CurlyLobe* AND *Plum* FLIES THAT CARRY (1) QUASI-NORMAL OR (2) LETHAL OR SEMILETHAL WILDTYPE CHROMOSOMES.

(Wallace, 1962.)

POPULATION	*CyL/N*	*n*	*CyL/D*	*n*
5	1.062	122	1.063	483
6	1.091	70	1.091	523
7	1.065	395	1.064	317
17	1.068	361	1.068	367
18	1.060	497	1.057	315
19	1.067	204	1.081	521
Average	1.065	1649	1.073	2526
	Pm/N	*n*	*Pm/D*	*n*
5	1.157	124	1.114	481
6	1.180	71	1.165	522
7	1.126	397	1.141	315
17	1.134	360	1.140	368
18	1.134	497	1.126	315
19	1.123	201	1.161	524
Average	1.134	1650	1.143	2525

TABLE 9-8

COMPARISON OF THE RELATIVE VIABILITIES OF WILDTYPE FLIES (D. MELANOGASTER) HETEROZYGOUS FOR TWO DIFFERENT QUASI-NORMAL CHROMOSOMES (*N/N*), ONE QUASI-NORMAL AND ONE LETHAL OR SEMILETHAL (*N/D*), OR TWO DIFFERENT LETHAL OR SEMILETHALS (*D/D*).

(Wallace, 1962.)

POPULATION	*N/N*	*n*	*N/D*	*n*	*D/D*	*n*
5	1.028	28	1.109	190	1.050	363
6	1.101	8	1.125	125	1.135	444
7	1.047	219	1.048	354	1.058	132
17	1.075	162	1.060	397	1.076	163
18	1.078	301	1.086	392	1.052	116
19	1.078	66	1.097	273	1.133	377
Average	1.067	784	1.079	1731	1.097	1595

samples of each population were tested alternately—one sample by the *CyL* technique, the next by the *CyL-Pm* technique. After a short period of alternation of this sort, all tests were made by means of the *CyL-Pm* technique. The data in the two tables confirm the suggestion made in the previous paragraph; after a long period of time, 120 generations or more, lethals remaining in laboratory populations are no longer partially dominant on the average. If the average effects of lethal chromosomes on their carriers deviate from that of nonlethals of the same populations, they deviate toward an enhancement rather than a lowering of viability.

It has been shown that the selective elimination, generation by generation, of lethal chromosomes with the most pronounced deleterious effects on their heterozygous carriers eventually leaves a residue in a population which, on the average, are completely recessive or, perhaps, have a slight enhancing effect on viability in heterozygous condition. This is, however, not accepted by everybody. Hiraizumi and Crow (1960), Crow and Temin (1964), and Temin (1966) have published evidence that appears contradictory to conclusions reached above. In part, the contradictions are spurious. Not all populations are alike. We showed above that for the combined data on the control population (#3), lethal chromosomes appeared to lower the viability (not the frequency of wildtype flies) of their heterozygous carriers by approximately 1 percent. No *CyL-Pm* tests are available for the final generations of this population. However, regressions of the frequencies of wildtype flies heterozygous for 0-, 1-, and 2-lethals give no evidence for any systematic change in dominance relationships within this population. We can see from the analyses of this population that lethals can consistently exert a deleterious effect on the viability of heterozygotes.

Hiraizumi and Crow (1960) made a systematic study of preadult viability, rate of development, fertility, and longevity of flies heterozygous for the genetically marked *cn bw* (both recessive eye-color mutations) and wildtype chromosomes that were (1) lethal, (2) semilethal, or (3) normal. The classification of wildtype chromosomes had been made during tests involving 735 chromosomes obtained from wild flies trapped near Madison, Wisconsin.

The test for preadult viability consisted of mating two *cn*/+ males with four *cn bw* females in each of twenty vials for each tested wildtype chromosome. After 5 to 6 days, the parental flies of each vial were transferred to a new one. On the average, the two vials of each replication yielded some 115 flies so that about 2200 to 2300 flies were counted for each chromosome. These tests covered 53 lethals (no wildtype flies in the original test cultures), 64 semilethals (0.0 to 0.167 wildtype flies), and 50 controls (0.250 to 0.416 wildtype), or 177 wildtype chromosomes in all; the total number of flies counted in these tests was nearly 400,000.

The results of the tests on preadult viability have been summarized in

Figure 9-4. The mean frequencies of wildtype flies observed for each type of chromosome tested were:

Control	0.5087 ± 0.0021
Semilethal	0.4998 ± 0.0023
Lethal	0.5026 ± 0.0021

Both the lethal and semilethal wildtype heterozygotes have mean frequencies that differ significantly from that of the control flies. On the other hand, lethals and semilethals do not differ significantly from one

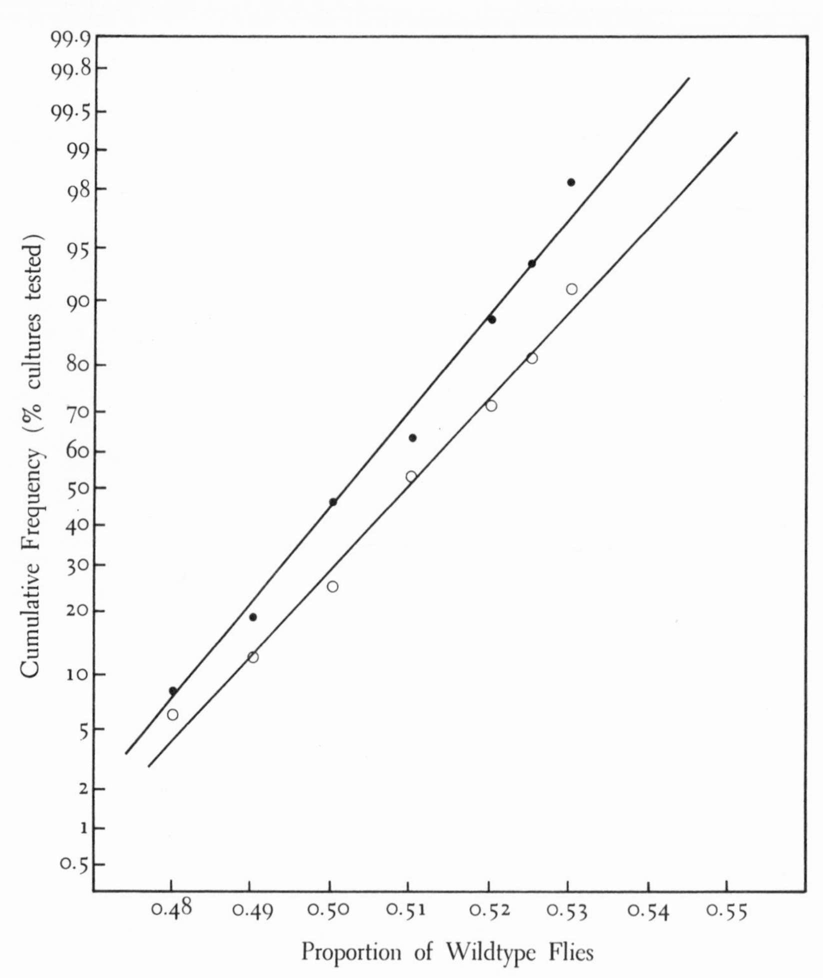

FIGURE 9-4. Cumulative frequency distributions of cultures containing wildtype flies in different proportions where the expected (Mendelian) proportion is 50 percent. Solid circles: wildtype flies were *cn bw/le;* open circles: wildtype flies were *cn bw/+*. (After Hiraizumi and Crow, 1960.)

another. As a result, the frequencies of lethal and semilethal heterozygotes can be combined (0.5010 ± 0.0016); this frequency differs significantly from that of the control (difference $= 0.0026$; $t = 2.96$; $P = 0.003$).

The difference in the frequency of wildtype flies in these test cultures is not a direct measure of the difference in viability of these flies. If x and y represent the frequencies of *cinnabar* and wildtype flies, the viabilities of wildtype flies in the control cultures can be represented as y_c/x_c and that of the lethal and semilethal heterozygotes as y_l/x_l. From these terms it can be seen that the viability of lethal-bearing heterozygotes relative to that of the control flies equals $(y_l/x_l) \div (y_c/x_c)$ or $(y_l x_c)/(y_c x_l)$. For the material tested by Hiraizumi and Crow, this ratio equals 0.9697. The average effect of lethal and semilethal chromosomes on the preadult viability of their carriers is, according to these tests, a reduction of approximately 3 percent.

Tests of the relative performances of control and lethal-bearing wildtype flies in respect to other components of fitness such as speed of development and longevity revealed that the lethal heterozygotes were either equal to the controls or inferior to them. There was no evidence that lethal heterozygotes exceeded the controls in any of the fitness components studied.

Hiraizumi and Crow estimate from their data that the relative difference (selection coefficient) for total fitness between lethal heterozygotes and controls is about 10 percent, although they admit that this estimate has a large error. For a number of theoretical reasons they conclude that the estimate of 10 percent is an overestimate of the true selective difference. In Chapter 10 we shall encounter a compelling reason for believing that this estimate of selective difference is indeed too large; if the true selective difference were as large as 10 percent, the frequency of lethals (together with semilethals) among the sampled chromosomes would have been about ten times larger than mutation rate itself. Since the mutation rate of lethals and semilethals is only 0.005, the expected frequency of lethals and semilethals combined would be 5 percent. But the frequency observed in the first two hundred or more sampled chromosomes was about 25 percent. One month later the frequency was more than 35 percent. An *increase* in the frequency of lethal chromosomes of 10 percent in some two generations, together with an overall frequency considerably in excess of the expected 5 percent, cannot be easily reconciled with the supposed 10 percent lowering of the fitness of lethal heterozygotes.

These difficulties were appreciated by Hiraizumi and Crow. To account for the observed increase in lethal frequency, they suggested that migrants had entered the sampled population from another one in which lethal frequencies were still higher. The likelihood that this suggestion is correct is remote indeed. As we have seen, migrants constitute a minute proportion of any one population. But for immigrants to raise the frequency of lethal chromosomes by 10 percent, either they must be exceedingly numerous or the population from which they come must have an unprecedentedly high frequency of lethals. Were the frequency of lethal chromosomes among migrants as high as 75 percent (a frequency higher than any observed for the second chromosome of *D. melanogaster* in nature), one fly in five of the sampled population would have been an immigrant. From what we learned of migration in Chapter 2, this does not seem to be a reasonable proportion of new immigrants. Even if it were, however, it would not resolve the problem confronting us; it merely postpones it. If the 10 percent disadvantage of lethal heterozygotes is to be regarded as a *general* phenomenon, how can the high frequency of lethals among migrants be explained? Specifically, how can one justify postulating a frequency of lethals approaching 75 percent among migrant individuals when in the population from which they came the expected frequency of lethals is also only 5 percent?

There is a built-in bias associated with the type of analysis made by Drosophila geneticists that may serve, at least in small part, as an alternative explanation for the partial dominance of "recessive" lethals. First, it is an empirical fact that a combination of two different chromosomes which yields fewer than one half the expected number of wildtype flies consists, in the overwhelmingly large proportion of cases, of two lethals, one lethal and one semilethal, or two semilethal chromosomes. Among the exceptional cases that I can recall, in which one of the two chromosomes of a semilethal combination was classified as quasi-normal (that is, neither lethal nor semilethal), the quasi-normal chromosome was very nearly semilethal in its effects on homozygous individuals.

Now, in comparing the effects of lethal and control chromosomes on viability, these chromosomes are tested in combination with a standard, genetically marked chromosome. In Table 9-7 the standard chromosomes were *CyL* and *Pm*. Both of these chromosomes are lethal when homozygous. During the many years that these chromosomes have been maintained in balanced lethal stocks, both have had an opportunity to accumulate lethals at several loci other than *Cy* or *Pm*, the known lethals. If the lethals on the genetically marked chromosomes lower viability by an occasional interaction with mutations carried by the wildtype chromosomes, they are much more likely to do so in combination with a lethal or semilethal than with a normal chromosome. In the case of the tests based on the *CyL-Pm* technique, lethal wildtype chromosomes that are allelic to either *CyL* or *Pm* are eliminated from the array of tested chromosomes. In the case of tests utilizing *cn bw*, eight *cn bw* chromosomes (four females) were involved in each test culture. Deleterious interactions of lethals with *cn*, *bw*, or any other deleterious mutation existing in this strain of flies are handicaps automatically assigned to lethals and semilethals as "partial dominance." It must be admitted, however, that the observations on preadult viability made by Hiraizumi and Crow are not easily explained by this type of interaction. Such interactions should affect only a fraction of all lethals tested; the remainder (the majority of lethals tested) should resemble the controls. Their data suggest, on the contrary, that lethals and semilethals appear to differ systematically from the control chromosomes.

More recently, Temin (1966) has reexamined the question of the partial dominance of seemingly "recessive" lethals. The test cultures Temin employed were four-class cultures analogous to those of the *CyL-Pm* technique (Figure 3-6). The four classes obtained in her experiments were

$$Cy/bw^D : Cy/+_i : bw^D/+_i : +_i/+_i$$

and

$$Cy/bw^D : Cy/+_j : bw^D/+_i : +_i/+_j$$

Because the Cy/bw^D class varied in number from experiment to experiment, $\frac{1}{2}(Cy/+ + bw^D/+)$ was used as the standard of comparison, thus converting the experimental procedure into one analogous to the *CyL* technique of Figure 3-3. Altogether, four experiments were performed. In two of these the wildtype chromosomes were taken from a wild population near Madison, Wisconsin; in two others, the chromosomes were from an experimental population whose ancestry included my own population 3.

The results of Temin's experiments are presented in Table 9-9. The data have been subdivided so that the results based on chromosomes from the Wisconsin population are separate from those of the laboratory population. There is no obvious difference between these two sources of wildtype chromosomes. The mean selective disadvantage of lethal heterozygotes according to these data is 2.9 percent with an error of 1.4 percent; consequently, it appears reasonably certain that the tests reflect a partial dominance of lethals whose true value lies somewhere between 0.1 and 5.7 percent. The average frequency of lethals in these populations—about 25 percent for the Wisconsin population and slightly less for the laboratory population—suggests that selection against average lethal heterozygotes is not as high as 5 to 6 percent, because (as we shall see in Chapter 10) the equilibrium frequency should then be 16 to 20 times greater than the mutation rate or, using 0.005 as the rate of mutation for second chromosome lethals in *D. melanogaster*, approximately 8 to 10 percent. The frequency actually observed in the populations is considerably greater.

In her experiments, Temin also obtained an estimate of the partial dominance of lethal chromosomes from cultures in which the wildtype

TABLE 9-9

COMPARISON OF THE VIABILITIES OF FLIES OF VARIOUS GENOTYPES HETEROZYGOUS FOR EITHER A QUASI-NORMAL OR LETHAL SECOND WILDTYPE CHROMOSOME.

H is the estimation of the dominance of the lethal chromosome; a positive value means the lethal lowers the viability of its heterozygous carriers while a negative value means it enhances the viability of these individuals. (Temin, 1966.)

GENOTYPE	LETHAL		NONLETHAL		H
	n	$\bar{x}$	n	$\bar{x}$	
+/+ I	93	1.213	116	1.291	0.060
IV	136	1.911	206	2.071	0.077
$Cy/+$ I	54	1.187	155	1.178	−0.008
IV	79	1.997	263	1.976	−0.011
$bw^D/+$ I	51	1.284	158	1.228	−0.046
IV	77	1.759	265	1.941	0.094
					av. 0.028
+/+ II	98	1.143	133	1.156	0.011
III	75	2.599	184	2.792	0.069
$Cy/+$ II	55	1.190	176	1.201	0.009
III	38	2.856	221	2.907	0.018
$bw^D/+$ II	54	1.139	177	1.109	−0.027
III	43	2.217	216	2.457	0.098
					av. 0.030
					Grand average 0.029 ± 0.014

flies were homozygous. These are the cultures that permit the initial classification of chromosomes into "lethals" and "nonlethals." The estimation of dominance is based in this case upon the ratio of singly marked flies ($Cy/+$ or $bw^D/+$) to the doubly marked flies (Cy/bw^D) in the lethal as opposed to the nonlethal cultures.

Calculations of dominance based upon the genetically marked classes of two sets of cultures which are classified according to the proportions of wildtype homozygotes can be biased because of unsuspected correlations. To illustrate this effect, one hundred "cultures" of flies yielding four "genotypes" each were simulated by the use of a table of random digits (Dixon and Massey, 1951, Table 1). The digits in the table were taken in pairs to form two-digit numbers. In the cases of genotypes *A*, *B*, and *C* (equivalent to CyL/Pm, $CyL/+$, and $Pm/+$, or, in the case of Temin's experiment, Cy/bw^D, $Cy/+$, and $bw^D/+$) if the first digit was odd, it was scored as three, if even it was scored as four. Consequently, the numbers of flies in these classes were restricted to the 30s and 40s. Class *D*, which was to represent the homozygous wildtype flies, was given greater latitude: If the first digit of the pair was 0, 5, or 6 it was scored as 0; if it was 1, it was scored as 1; if it was 2 or 7 it was scored as 2; if it was 3 or 8 it was scored as 3; and if it was 4 or 9 it was scored as 4.

When the numbers of flies in all 100 cultures are summed, the results are:

A	*B*	*C*	*D*	TOTAL
3978	3975	3898	2527	14,378

As expected, the total number of *A*, *B*, and *C* are very similar, because they represent samples drawn from the same table of random digits. The total for class *D* is substantially smaller than the other three because the individual values of this class included numbers smaller than 30.

The 100 "cultures" are now subdivided into two groups: "lethals" in which the number of wildtype "flies" is less than 10 percent of the total, and "nonlethals," the remainder. The results of this subdivision can be tabulated as follows:

	NO.	*A*	*B*	*C*	*D*	TOTAL
Lethals	31	1240	1221	1199	167	3,827
Nonlethals	69	2738	2754	2699	2360	10,551

We can now use *A* as a standard for comparison and compute the ratios *B*/*A* and *C*/*A* for lethals and nonlethals. Since *A* is independent of *B*, *C*, and *D*, it can safely serve as a standard of viability—or so it would seem. The results of the calculations are as follows:

	LETHALS	NONLETHALS	*H*
B/*A*	0.985	1.006	0.021
C/*A*	0.967	0.986	0.019

The calculations reveal that the "lethal" chromosomes depress the viability of their heterozygous carriers (*B* and *C*) by approximately 2 percent. This, of course, is nonsense in the case of these "cultures," since there were no chromosomes involved. The data for each class—*A*, *B*, *C*, and *D*—were drawn from the table of random digits independently of all other classes; nothing analogous to dominance is present in the sampling procedure. The apparent "dominance" comes from the systematic bias introduced by the definition of lethality. Because of this statistical bias, Wallace and Dobzhansky (1962, p. 1029) omitted such cultures from calculations of dominance. Nor were cultures yielding homozygous wildtype flies included in the material presented in Table 9-2.

This section dealing with the detection and measurement of the partial dominance of lethal chromosomes must be terminated on an unsatisfactory note. Some tests involving large numbers of flies and cultures have failed to reveal significant differences between viabilities of (otherwise comparable) flies with and without lethal chromosomes; other tests, also involving large numbers of flies and cultures, suggest that lethal chromosomes have a slight depressing effect on the preadult viability of their carriers.

The General Effect of Selection on a Population: Fisher's Fundamental Theorem

The usual effect of natural selection is the elimination from populations of alleles that lower the fitness of their carriers, homozygotes or heterozygotes. In the absence of new mutation, each generation the average fitness of the population would be higher than it was in the previous one because fewer of the deleterious mutant genes would remain. Ultimately, as we have seen, the deleterious genes would be completely removed from the population.

Fisher (1930) described a very important relationship that exists between change in fitness and the genetic variance in fitness that exists within a population. He stated this relationship in the form of a theorem called the fundamental theorem of natural selection: *The rate of increase in fitness of any organism at any time is equal to its genetic variance in fitness at that time* (Fisher, 1958).

In the following paragraphs we shall give a simplified account of this theorem. We shall consider only those alleles whose action on fitness is such that heterozygotes (*Aa*, for example) are exactly intermediate between the two homozygotes (*AA* and *aa*); in this way we avoid the complications of dominance. We shall assume, too, that no interactions occur between alleles at different loci (epistatic interactions). In short, the alleles can be shuffled and reshuffled into any combination without distortions in fitness changes arising from the combinations themselves.

Under these simplifying restrictions, sexual reproduction does not affect the calculations we are about to make. Consequently, we will eliminate sex and treat the population as if it were a collection of asexual clones; for convenience, imagine that our population is a bacterial culture.

The arithmetic procedures for relating increase in fitness to genetic variance for fitness are outlined in Table 9-10. First, we say that the population consists of individuals that can be classified into genotypes ($g_1, g_2, g_3, \ldots, g_n$) each of which has a certain frequency ($f_1, f_2, f_3, \ldots, f_n$) and a certain fitness ($w_1, w_2, w_3, \ldots, w_n$). The average fitness of this population equals $f_1w_1 + f_2w_2 + f_3w_3 + \cdots + f_nw_n$, or $\overline{w}$.

The variance in fitness equals (by the definition of variance)

$$f_1(w_1 - \overline{w})^2 + f_2(w_2 - \overline{w})^2 + f_3(w_3 - \overline{w})^2 + \cdots + f_n(w_n - \overline{w})^2$$

TABLE 9-10

SUMMARY OF FREQUENCIES (f) AND RELATIVE FITNESSES (w) OF VARIOUS GENOTYPES (g) IN TWO SUCCESSIVE GENERATIONS OF A POPULATION FROM WHICH FISHER'S FUNDAMENTAL THEOREM CAN BE DEMONSTRATED.

First generation			
Genotypes	g_1	g_2	$g_3 \ldots g_n$
Frequencies	f_1	f_2	$f_3 \ldots f_n$
Fitnesses	w_1	w_2	$w_3 \ldots w_n$
Average fitness $= f_1w_1 + f_2w_2 + f_3w_3 + \ldots + f_nw_n = \bar{w}$			
Second generation			
Genotypes	g_1	g_2	$g_3 \ldots g_n$
Frequencies	$\frac{f_1w_1}{\bar{w}}$	$\frac{f_2w_2}{\bar{w}}$	$\frac{f_3w_3}{\bar{w}} \ldots f_n\frac{w_n}{\bar{w}}$
Fitnesses	w_1	w_2	$w_3 \ldots w_n$
Average fitness $= \frac{f_1w_1^2 + f_2w_2^2 + f_3w_3^2 + \ldots + f_nw_n^2}{\bar{w}}$			

By squaring each parenthetic expression, multiplying, and regrouping the terms, the following three series of terms emerge:

$$f_1w_1^2 + f_2w_2^2 + f_3w_3^2 + \cdots$$
$$- 2\bar{w}(f_1w_1 + f_2w_2 + f_3w_3 + \cdots)$$
$$+ \bar{w}^2(f_1 + f_2 + f_3 + \cdots)$$

Now, $f_1w_1 + f_2w_2 + f_3w_3 + \cdots$ equals $\bar{w}$ and $f_1 + f_2 + f_3 + \cdots$ equals 1.00; thus the three series reduce to

$$f_1w_1^2 + f_2w_2^2 + f_3w_3^2 + \cdots - \bar{w}^2$$

This is the variance among genotypes of the population in respect to fitness.

To calculate the increase in fitness brought about in a single generation by natural selection, it is necessary to remember that the fitness of a given genotype determines the frequency of that genotype in the following generation; that, after all, is the definition of fitness. It follows, then, that after one generation of selection, the new frequencies of genotypes $g_1, g_2, g_3, \cdots$ will be proportional to $f_1w_1, f_2w_2, f_3w_3, \cdots$ The new average fitness of the population equals, as before, the sum of the frequencies of fitnesses times the fitnesses themselves. With the correction needed to keep the sum of frequencies equal to 1.00, the average fitness equals

$$\frac{f_1w_1^2 + f_2w_2^2 + f_3w_3^2 + \cdots}{f_1w_1 + f_2w_2 + f_3w_3 + \cdots}$$

or

$$\frac{f_1w_1^2 + f_2w_2^2 + f_3w_3^2 + \cdots}{\overline{w}}$$

The increase in fitness equals the new average minus the old average, or

$$\frac{f_1w_1^2 + f_2w_2^2 + f_3w_3^2 + \cdots}{\overline{w}} - \overline{w}$$

or

$$\frac{f_1w_1^2 + f_2w_2^2 + f_3w_3^2 + \cdots - \overline{w}^2}{\overline{w}}$$

The w's of these calculations are completely arbitrary; they represent the relative fitnesses of the different genotypes. We shall now say that the original values were chosen in such a way that $\overline{w}$ equals 1.00. And with that stipulation we see that the increase in fitness is in fact identical to the genetic variance in fitness of the population.

10

SELECTION *VERSUS* MUTATION

The last two chapters have dealt with two opposing processes. The discussion of mutation dealt largely with mutation from normal alleles to mutant alleles with deleterious effects on their carriers; one of the examples cited was the mutation from red to *white* (eye color) in different geographic strains of *D. melanogaster*. The discussion of selection dealt almost exclusively with the elimination of deleterious mutations from populations; again, one of the examples was the elimination of the *white* mutant genes from some experimental populations of *D. melanogaster*. The present chapter will deal simultaneously with both processes, the accumulation of mutant alleles as the result of mutation and their elimination from the population by selection. The result will be the second *stable* equilibrium we have encountered for gene frequencies; the first, we may recall, arose as the result of opposing forward and reverse mutations.

Algebraic Calculations

In contrast to some earlier procedures, we shall proceed immediately to examine the algebraic aspect of mutation versus selection. We do this because we have already dealt with the components that we need for the combined treatment.

In the case of mutation from A to a, the change in the frequency of A equals $-up$, the rate at which A mutates to a each generation, times p, the frequency of A. If a is a rare allele, p is very nearly equal to 1.00 and so Δq equals u ($\Delta p = -u$).

In the case of selection, Δq can be given either of two values depending upon whether or not the allele a exhibits an appreciable degree of dominance. If a is completely recessive to A, Δq equals $-sq^2$ when q is small. If, on the other hand, a is dominant even to a relatively slight extent, the change in its frequency, Δq, equals $-hsq$—again, when q is small.

Under the joint influence of mutation and selection, an equilibrium will be established when the gain of an allele by mutation equals the loss by selection. Consequently, for a recessive allele

$$u = sq^2$$

or, at equilibrium,

$$q = \sqrt{u/s}$$

If s equals 1.00 (a is a lethal mutation), $q = \sqrt{u}$. For a partially dominant allele, however, an equilibrium arises when

$$u = hsq \qquad \text{or} \qquad q = \frac{u}{hs}$$

The values for $\hat{q}$ (q-hat), the equilibrium frequency of the mutant allele a, under various assumptions regarding mutation rate (u), selective disadvantage (s), and dominance (h) have been listed in Table 10-1. In the case of a recessive allele, u and s are the only variables concerned. The striking feature brought out in the table is the surprisingly high frequency obtained by mutant genes under low rates of recurrent mutation. Thus, a lethal that arises once in a million gametes is at equilibrium when one gamete in a thousand carries it; the one sperm in a thousand meets the one egg in a thousand with a frequency of one per million—precisely the frequency needed to counterbalance mutation.

The examples in Table 10-1 also show the extent to which the frequency of partially dominant alleles fall short of that of comparable, completely recessive mutations. A lethal gene that affects its heterozygous carriers so that their fitness is only 99 percent that of homozygous normals reaches an equilibrium frequency of only 1 in 10,000 rather than 1 in 1,000 reached by a complete recessive ($u = 10^{-6}$).

Calculations such as these are the ones which earlier led us to question the suggestion that lethals in *Drosophila* lower the fitness of their heterozygotes by 10 percent. The equilibrium frequency of such a lethal at a single locus would be 10^{-5} (mutation rate per locus) divided by 1.00 (s) times 10^{-1} (h), or 10^{-4}. For a collection of some 500 loci such as that

TABLE 10-1

EQUILIBRIUM FREQUENCIES ATTAINED BY RECESSIVE AND PARTIALLY DOMINANT MUTANT ALLELES IN LARGE POPULATIONS GIVEN VARIOUS VALUES FOR MUTATION RATE (u), SELECTION COEFFICIENT (s) OF HOMOZYGOTES, AND DEGREE OF DOMINANCE (h).

COMPLETE RECESSIVES

	u		
s	10^{-4}	10^{-6}	10^{-8}
1	10^{-2}	10^{-3}	10^{-4}
0.1	3.2×10^{-2}	3.2×10^{-3}	3.2×10^{-4}
0.01	10^{-1}	10^{-2}	10^{-3}

PARTIAL DOMINANTS

		u		
s	h	10^{-4}	10^{-6}	10^{-8}
1	0.1	10^{-3}	10^{-5}	10^{-7}
1	0.01	10^{-2}	10^{-4}	10^{-6}
0.1	0.1	10^{-2}	10^{-4}	10^{-6}
0.1	0.01	10^{-1}	10^{-3}	10^{-5}

carried by the second chromosome of *D. melanogaster,* the frequency of lethals per chromosome would be 0.0001×500, or 5 percent. We arrived at this same value earlier by saying that the equilibrium frequency would equal mutation rate to lethal chromosomes (0.005) divided by 0.10, or 5 percent. This same line of reasoning rules out a more extreme suggestion of Goldschmidt and Falk (1959) that the fitness of lethal heterozygotes are reduced by nearly 20 percent; in this case, the equilibrium frequency of second chromosome lethals should be less than 3 percent.

THE USE OF GENE FREQUENCIES AND FITNESS VALUES TO ESTIMATE MUTATION RATES

The obvious use of the calculations we have just presented is for the estimation of mutation rates. This is the obvious use, that is, when one studies material for which mutation is difficult to estimate but for which estimates of fitness are relatively easily arrived at. Human beings are such material. And it is no surprise that mutation rates in man have been estimated to a large extent by the use of these (or more sophisticated) calculations.

Mutation to dominant mutant alleles in man can be measured by direct count. One can obtain from the proper authorities (from hospital or civil records, for example) data on all children born to parents both of whom are normal in respect to the trait in question. Information on each child is obtained so that the number afflicted with the trait can be ascertained. Thus, of 94,075 children born in one hospital in Copenhagen, 8 were chondrodystrophic dwarfs born to normal parents. The mutation rate equals 8 (the number of dominant genes for dwarfism) divided by 188,150 (the number of alleles at the given locus carried by 94,075 children); this is very nearly equal to 4×10^{-5}. A number of assumptions are involved in this and similar calculations. Some of these are (1) that the mutant gene is fully penetrant, (2) that only one locus is involved, (3) that no other source of dwarfism—genetic or environmental—exists, and (4) that correct parentage is assured.

An indirect method for estimating mutation rates, a method which is based upon known gene frequencies and calculated fitnesses, also exists. The following example, like the one in the preceding paragraph, has been adapted from Stern (1960). Records of 108 chondrodystrophic dwarfs revealed that they had produced 27 children. These dwarfs had had 457 normal sibs who in turn produced 582 children. The average fitness of these dwarfs can be calculated as $(27/108) \div (582/457)$ or $(27 \times 457)/(108 \times 582)$. The fitness of the dwarfs, then, equals 0.196; since the dwarfs are all heterozygotes, hs equals 0.804. The frequency of dwarfs can be estimated from 94,075 births, 10 of which were dwarfs; thus $2pq$ equals 0.000106. Since p is very nearly 1.00, q equals 0.000053. The mutation rate, u, is known to equal hsq, or 0.804×0.000053. The mutation rate equals 0.000043 (4.3×10^{-5}). This value is nearly identical to that calculated above by the direct method.

(There is still another indication in this material that the fitness of chondrodystrophic dwarfs is about 20 percent. Ten dwarfs were born among 94,075 recorded births; 8 of these were born to normal parents, the other 2 to a dwarf parent. It appears, then, that only 20 percent of the dwarf genes in a population are there because they were passed on from parent to offspring; the remainder are lost and are replaced by new mutations.)

Procedures similar to the indirect method illustrated above can be used for calculating the mutation rate of recessive genes. Such procedures are sensitive, however, to the fitness of seemingly normal heterozygotes and, consequently, are subject to considerable error.

The Use of Mutation Rates and Gene Frequencies to Estimate the Dominance of "Recessive" Genes

Unlike the situation faced by human geneticists, Drosophilists find it rather easy (although tedious) to measure mutation rates. Unlike human geneticists, too, they find it difficult if not impossible to follow all the individual members of a population; mankind is unique in the sense that each of us has a name and nearly everything of importance that happens to us is known and is recorded by others. In studying flies, then, we use known mutation rates and known gene frequencies to estimate the degree of dominance of deleterious mutations. In doing this we assume, of course, that the gene frequencies observed at the moment are equilibrium frequencies.

The Comparison of the Input and Outflow of Lethal Genes. If the frequency of lethal chromosomes in a population is constant from generation to generation, then the lethals entering the population by mutation must exactly counterbalance those which are eliminated by selection. Furthermore, if lethals are completely recessive, the rate of elimination is given by IQ^2, where I is the probability that two lethal chromosomes are allelic (that they carry lethals at the same locus) and Q is the frequency of lethal chromosomes in the population. It seems (Dobzhansky, 1939, p. 351) that A. H. Sturtevant first made this type of calculation.

Data suitable for comparing the gain and loss of lethals have been published by Dobzhansky and Wright (1941). Within populations of *D. pseudoobscura* inhabiting the Death Valley region of California, the frequency of lethal third chromosomes was 0.153 (131 lethals among 857 chromosomes tested). The probability that two lethals taken in the same locality were allelic was 0.0311 (24 cases of allelism observed among 772 intercrosses of different lethals). Lethals are eliminated by allelism, then, at a rate equal to $(0.153)^2 \times 0.0311$, or 0.0007. The mutation rate to lethals equals 0.0030 (40 lethals among 13,472 chromosome-generations tested). Hence, it appears that lethals arise by mutation at a rate that exceeds their elimination by homozygosis. It also appears (Figure 10-1) that an elimination of lethals from the population at an additional rate of 0.0023 lethal per gamete is needed in order that the 0.153 really be an equilibrium frequency.

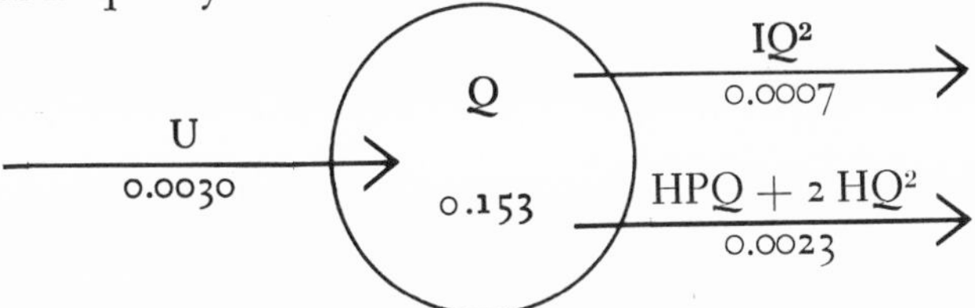

FIGURE 10-1. Balance attained by the input and outflow of lethals in a population. The experimentally determined value of I in this case was 0.0311; the value of H required to establish the balance is 0.013.

If the excess elimination is effected by means of the partial dominance of lethals and the decreased fitness of their heterozygous carriers, the decrease H can be estimated as follows:

$$\tfrac{1}{2}\,(2PQ)\,H + 2Q^2H = 0.0023$$

P equals $(1 - Q)$ or 0.847, so H equals 0.0023/0.1764, or 0.013. The calculation suggests that an average lethal chromosome lowers the fitness of its heterozygous carriers by roughly 1 to 2 percent.

A General Statement Concerning Input and Outflow of Lethal Genes. In the preceding paragraphs we dealt with the elimination of lethals from a population in two, somewhat artificial, steps. First, we asked about the elimination of lethals by homozygosis. Having seen that this elimination appeared inadequate to counterbalance mutation, we proceeded to calculate the partial dominance required to eliminate the remaining lethals through the reduced fitness of their heterozygous carriers.

A more general approach to the problem of estimating H, the degree of dominance of lethal chromosomes, has been given by Crow and Temin (1964). They start with the following items: U, the lethal mutation rate per chromosome; Q, the proportion of chromosomes carrying one or more lethal genes; P, $1 - Q$; and A, the probability that any two chromosomes (lethal or not) are lethal in heterozygous combination. The term A can be expressed, too, as IQ^2, where I is the probability that two *lethal* chromosomes are allelic and Q^2 is the probability that both chromosomes of a pair are in fact lethal.

At equilibrium, the frequency of lethals does not change from generation to generation. This leads to the expression

$$Q = \frac{Q - HPQ - 2HQ^2 - A + UP}{1 - 2HPQ - 2HQ^2 - A}$$

or, upon solving, to

$$H = \frac{U - A}{Q}$$

If inbreeding is taken into account, this equation is replaced by

$$H + F = \frac{U - A}{Q}$$

where F is the coefficient of inbreeding resulting, in this case, from the mating of close relatives.

Using this equation, Crow and Temin analyzed data from a large number of sources and found that the average value for $H + F$ is 0.0174 in the case of second chromosome lethals in *D. melanogaster*.

Although repeated tests made by various investigators appear to support the notion that lethals in natural populations are, as a rule, appreciably deleterious to their heterozygous carriers, there are good reasons for accepting this conclusion with caution (Wallace, 1966a, 1966b).

The observational data that make caution necessary have already been presented in Chapters 2 and 5. In Chapter 5 we saw that the frequency of allelism of lethals is related to geographic distance in such a way that lethals taken at distances of one hundred meters from one another have a detectably lower frequency of allelism than do those taken from a single collection site. The allelism of lethals taken simultaneously at one site in a Colombian population had a probability of allelism of nearly 5 percent; those taken at distances separated by 90 meters had an allelism frequency of only about 2.5 percent. The average frequency of allelism observed in intercrosses of all the Colombian lethals was very nearly 3 percent; these lethals, although they had been collected in an effort to study the relation of distance and allelism, were all taken within smaller distances of one another than have been most lethals of previous studies.

A further point regarding past tests of allelism is the insistence by many workers that, in order to be considered alleles, two lethals must permit *no* heterozygous individuals to survive. A summary illustrating the oversight by this insistence of what may be considered "effective" alleles is presented in Table 10-2. In these tests, lethals were detected by the *CyL* technique (Figure 3-3) and were defined as chromosomes which permitted fewer than 10 percent of the expected number of wildtype flies to survive (about 3.2 percent wildtype flies among the total counted

TABLE 10-2

Analysis of 97 Lethal/Lethal Heterozygous Combinations Where the Combinations Themselves Were Lethal or Semilethal.

The lethal chromosomes tested have been classified "absolute" (no wildtype flies in test culture) and "near" (some wildtype homozygotes but fewer than 10 percent of the expected number). The combinations, too, have been classified; *A*, three or fewer wildtype heterozygotes in test culture; *B*, more than three wildtype flies but fewer than one half the expected number.

	A	*B*	TOTAL
Absolute × absolute	58	20	78
Absolute × near-lethal	2	6	8
Near-lethal × near-lethal	3	8	11
	—	—	—
Total	63	34	97

in half-pint culture bottles). For our purposes, these have been divided into "absolute" (no wildtype flies) and "near" lethals (at least one wildtype fly but fewer than 3.2 percent). A total of ninety-seven cases in which the lethal heterozygotes themselves were lethal or semilethal has been subdivided into (*A*) cultures in which there were 3 or fewer wildtype flies, and (*B*) cultures in which there were more than three but fewer than one half the expected number of wildtype flies. The important feature for our point is that nearly one quarter (20 of 78 cultures) of the absolute $\times$ absolute combinations, which were allelic in the sense that the heterozygotes were semilethal or worse, yielded more than three wildtype flies and hence would not have been counted as allelic in many earlier studies. In a locality where the frequency of allelism is really 5 percent, a poorly sampled collection of lethals may yield an estimate of only 2.5 percent, while a rigid adherence to absolute allelism may lower the final estimate of allelism to something less than 2 percent.

The other criticism of the conclusion that lethals found in natural populations exhibit a systematic partial dominance is based on the neglect of *F*, the inbreeding coefficient. This quantity is hard to measure directly in free-living populations. The analysis of dispersion of *Drosophila* presented in Chapter 2, however, emphasized very much the local nature of breeding populations. A large proportion of any *Drosophila* population seems to have its origins within meters of the place where the flies are actually found. The proportions are so large and the distances involved are so small that it seems unwise to arbitrarily decree that inbreeding is negligible and, hence, that the excess of mutation over elimination is a measure of dominance alone.

Observing the Accumulation of Lethals in Artificial Populations

This chapter can be concluded by taking up again the main point of the discussion—that mutation and selection jointly establish stable gene frequencies under which the opposing tendencies are in a stable equilibrium.

Natural populations are not suitable for the study of an *approach* to equilibrium because, presumably, they are at or near equilibrium conditions at all times. Much better are artificial populations from which lethals have been eliminated at the start and which then proceed to accumulate new ones. The populations discussed in the following paragraphs were exposed continuously to rather high levels of gamma radiation in order that mutations within them would occur at higher than normal frequencies.

The accumulation of sex-linked lethals in an irradiated population has been illustrated in Figure 10-2. Sex-linked lethals do not tend to accumulate extensively in *Drosophila* populations because males are hemizygous

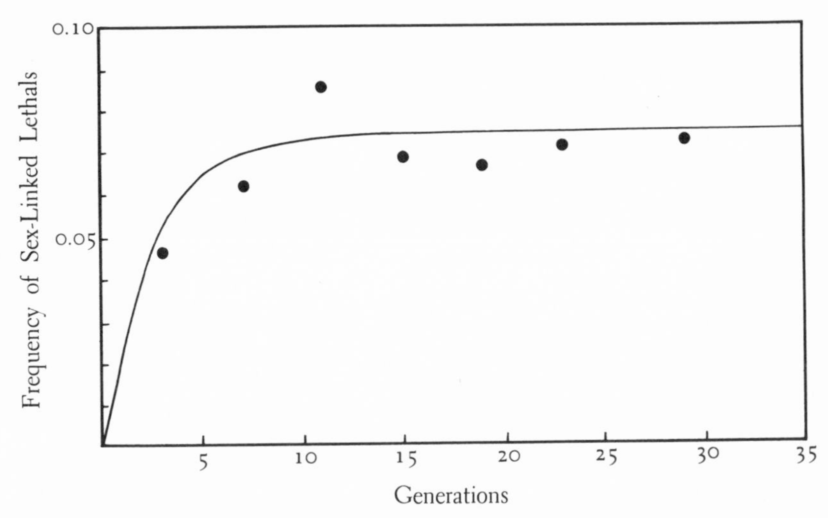

FIGURE 10-2. Accumulation of sex-linked lethals in an artificially irradiated population (#6) of *D. melanogaster*. (Wallace, unpublished.)

for the X chromosome. Of all X chromosomes in a *Drosophila* population, one third are found in males (XY); two thirds are found in females (XX). Sex-linked lethals kill their male carriers, so one third of all sex-linked lethals are "washed out" of the population each generation. If the mutation rate to such lethals is U, then when Q, the frequency of lethals, reach $3U$ the rate of elimination ($\frac{1}{3} \times 3U$) equals mutation rate (U). This frequency, Q equals $3U$, is an equilibrium frequency. In Figure 10-2 the frequencies of sex-linked lethals in an irradiated population over a period of nearly thirty generations are shown, together with the accumulation curve expected if U were 2.5 percent. The points and the curve agree reasonably well.

(The radiation dose delivered to this population was approximately 2000 r per generation. Had this dose been applied to mature spermatozoa, one would expect to induce about 5 or 6 percent sex-linked lethals; 2.5 percent is much less than this higher frequency. Much of the continuous radiation was delivered, however, to immature germ cells of young larvae. Many sex-linked lethals, especially in males, suffer from germinal selection. Finally, lethals are induced at lower frequencies in females than in males. For the point we are making about the equilibrium between mutation and selection, we can regard 2.5 percent as a reasonable mutation rate for this population.)

The accumulation of second chromosome lethals under the impact of continuous irradiation has been illustrated in Figure 10-3; the data on which the figure is based are listed in Table 10-3. These data are com-

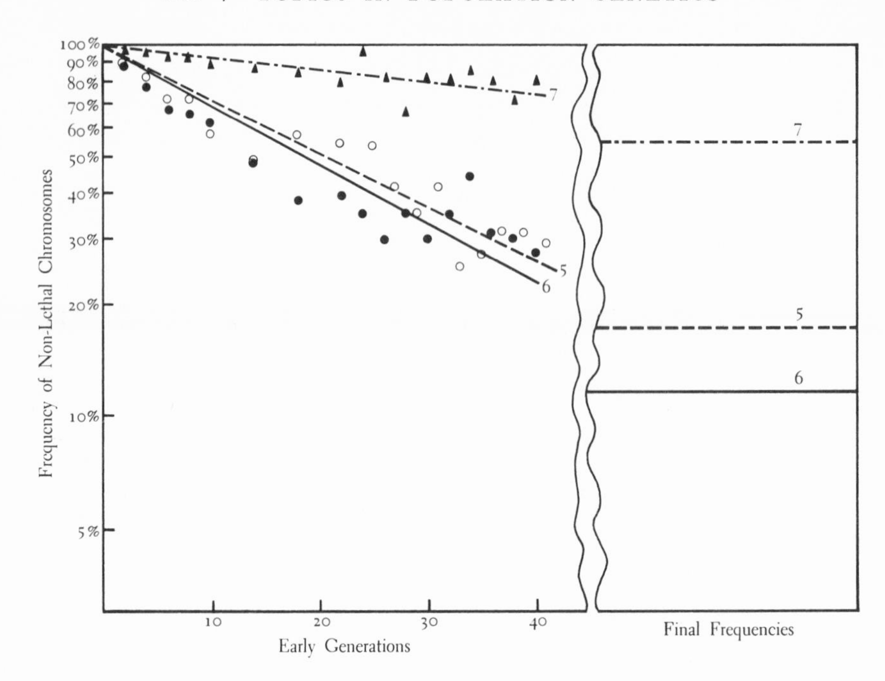

FIGURE 10-3. Loss of non-lethal (or, conversely, the accumulation of lethal) autosomes in three irradiated populations of *D. melanogaster*. On the left are curves illustrating the early generations; on the right are the equilibrium frequencies that were eventually established. Population 5: open circles and dashed line; population 6: closed circle and solid line; population 7: triangles and dot-dash line. (After Wallace, 1956, and unpublished.)

parable to those listed in Table 8-4 except that those were spontaneous lethals while the ones in the irradiated populations are very largely radiation-induced.

The curves in Figure 10-3 show, not the frequencies of lethals, but of nonlethals. It will be recalled that under continuous mutation, the frequency of the nonmutated allele (chromosome in the present case) decreases exponentially:

$$p_n = p_0\,(1 - u)^n = p_0 e^{-un}$$

where p_0 is the initial frequency of the nonmutated allele, p_n is its frequency after n generations, and u is the mutation rate of this allele. Within the experimental populations, the lethals were eliminated by homozygosis (and by partial dominance) so that they did not accumulate at a constant rate. That this elimination was already appreciable by the fortieth generation can be detected by comparing the straight lines representing the average loss of nonlethals from populations 5 and 6 with the early points themselves. In the early generations the rates of decrease

TABLE 10-3

FREQUENCIES OF LETHAL AND SEMILETHAL CHROMOSOMES IN THE EARLY GENERATIONS OF CHRONICALLY IRRADIATED POPULATIONS OF D. MELANOGASTER.

	POPULATION 5		POPULATION 6		POPULATION 7	
GENERATION	*n*	FREQUENCY	*n*	FREQUENCY	*n*	FREQUENCY
0	—	0	—	0	—	0
2	275	0.113	270	0.119	260	0.042
4	261	0.180	263	0.217	265	0.053
6	250	0.292	258	0.333	242	0.083
8	89	0.292	272	0.353	278	0.079
10	269	0.428	255	0.380	291	0.107
14	45	0.533	246	0.524	262	0.141
18	196	0.434	272	0.621	264	0.174
22	203	0.458	179	0.609	170	0.206
24	—	—	20	0.650	20	0.050
25	19	0.474	—	—	—	—
26	—	—	182	0.698	182	0.187
27	180	0.594	—	—	—	—
28	—	—	20	0.650	20	0.350
29	20	0.650	—	—	—	—
30	—	—	79	0.696	78	0.192
31	80	0.588	—	—	—	—
32	—	—	79	0.646	78	0.205
33	80	0.750	—	—	—	—
34	—	—	80	0.563	80	0.163
35	80	0.725	—	—	—	—
36	—	—	80	0.688	80	0.213
37	80	0.688	—	—	—	—
38	—	—	80	0.700	80	0.300
39	80	0.688	—	—	—	—
40	—	—	80	0.725	80	0.213
41	80	0.713	—	—	—	—

in these two populations were systematically greater than the rates suggested by the average taken over forty generations.

These populations were maintained and sampled periodically for about 160 generations. By the end of the experiment, lethal frequencies in the three populations had become stable. The average frequencies of nonlethal chromosomes in them at this time have been indicated by the horizontal lines on the right in Figure 10-3. As in the case of the sex-linked lethals, autosomal lethals can be seen under proper circumstances to accumulate in populations and then to stabilize at equilibrium frequencies. The equilibrium frequencies, in turn, are dependent upon mutation rates. Finally, it should be noted that the equilibrium frequency for second chromosome lethals in population 6 is much greater than that of sex-linked lethals of the same population (Figure 10-2). In the one case the elimination involves the square root of mutation rate while in the other it involves a simple (and low) multiple of this rate.

11

SELECTION IN FAVOR OF HETEROZYGOTES

This is the third of four chapters dealing with the role of selection in determining gene frequencies in populations. First, we saw that selection tends to eliminate from populations those alleles which lower the fitness of their carriers. As the allele becomes rare, its elimination continues unabated if it lowers the fitness of heterozygotes; otherwise, selection is remarkably slow in eliminating recessive mutations—even recessive lethals. Next we considered the interplay between selection against deleterious mutations and their origin by mutation. We saw that a certain frequency of a mutant allele is eventually reached where selection is so inefficient that mutation (small as mutation rates generally are) counterbalances the loss of the mutant allele. As a result, stable equilibria arise whose values depend upon the effect of the gene on fitness, the extent to which it exhibits its effect in heterozygotes, and its rate of origin through mutation.

In the present chapter, instances in which maximum fitness is the property of the heterozygotes rather than of one of the homozygotes will be examined. The notion that such situations might exist is so unexpected to beginning genetics students that two examples will be cited as illustrations.

The *CyL/Pm* stock of *D. melanogaster*, which is so useful and which has entered into so many of our earlier discussions, can serve as the first example. This stock of flies is typical of dozens of balanced lethal stocks which are known for various *Drosophila* species. Each such stock is true-breeding in the sense that all surviving adults of each generation are phenotypically uniform and similar to their parents. They are uniform, however, only because the homozygotes die. One half of the sperm produced by *CyL/Pm* males carry the *CyL* chromosome, the other half carry *Pm*. Eggs, too, are one half *CyL*-bearing and one half *Pm*-bearing. One half of the progeny each generation consist of *CyL/CyL* and *Pm/Pm* homozygotes. These individuals die because of the lethal action of *Cy* and *Pm*. Only the heterozygotes survive. If their fitness is said to be 1.00, that of the two homozygotes must be set at zero, each. The true-breeding stock is an example of selection in favor of heterozygous individuals.

Another example is not as artificial as the one just cited but concerns what approaches a balanced lethal condition within a natural population. The case in point is a population of *D. tropicalis* found near Lancetilla, Honduras. Inseminated adult females from this population were permitted to produce offspring in the laboratory; the latter were tested as larvae (one larva per female) to determine what gene arrangements were present in the original population. Seventy-four larvae were examined (a test of chromosomes from 74 females and 74 unknown male parents). Fifty-two of the seventy-four (70.3 percent) were heterozygous for a particular inversion. Since 50 percent is the maximum frequency of heterozygotes expected under the Hardy-Weinberg equilibrium, and since it is improbable that preferential mating of unlike homozygotes is involved, it appears that heterozygotes survive in larger proportions than do homozygotes in this population. A laboratory population of this same material yielded the distribution 3 $A/A : 134\ A/D : 3\ D/D$ which was cited in Chapter 4 (page 67) as a distribution that obviously does not fit the Hardy-Weinberg distribution. The heterozygous individuals in this example, as in the case of the *CyL/Pm* stock of *D. melanogaster* cited above, appear to survive in greater proportions than do homozygotes.

ALGEBRAIC CALCULATIONS

If a superior fitness resides in heterozygous individuals, two possibilities for calculating gene frequencies may be considered. We could, possibly, adhere to our earlier symbols but allow h to become negative; in this case $1 - hs$ would become $1-(-h)s$ or $1 + hs$, a fitness for heterozygotes greater than that of homozygous "normals." Under this scheme, however, a balanced lethal system can be represented only by an infinitely large value for $-hs$. It is more convenient to represent the heterozygote itself as having fitness 1.00 and let the fitnesses of the two homozygotes approach zero.

Following the second scheme outlined above, we can set up the following outline:

GENOTYPE	AA	Aa	aa
Frequency	p^2	$2pq$	q^2
Fitness	$1-s$	1	$1-t$
Average fitness:		$1 - sp^2 - tq^2$ or $\overline{W}$	
Frequency of A after selection:		$(p - sp^2)/\overline{W}$	
Frequency of a after selection:		$(q - tq^2)/\overline{W}$	

If two alleles have frequencies p_0 and q_0 in one generation and frequencies p_1 and q_1 in the next, then an equilibrium has been reached if $p_1/p_0 = q_1/q_0$. Upon solving this equation, it can be seen that the equality holds only if $p_1 = p_0$.

The values of p_1 and q_1 in terms of the original p and q have been set down in the outline given above. Consequently, we can solve for equilibrium frequencies by setting:

$$\frac{p - sp^2}{p\overline{W}} = \frac{q - tq^2}{q\overline{W}}$$

or

$$1 - sp = 1 - tq$$

or

$$sp = tq$$

From the last equation, we can find that at equilibrium

$$p = \frac{t}{s + t}$$

$$q = \frac{s}{s + t}$$

The change in gene frequency each generation can be calculated as the difference between the new frequency and the old:

$$\Delta q = \frac{q - tq^2}{1 - sp^2 - tq^2} - q$$

Solving the above equation gives

$$\Delta q = \frac{pq(sp - tq)}{1 - sp^2 - tq^2}$$

(Note that Δq equals zero at equilibrium—that is, when $sp = tq$.)

The important point in these calculations is that, for the third time, we have encountered conditions that lead to stable gene frequencies, to a stable genetic equilibrium. Furthermore, since s and t can have any value up to and including 1.00, the equilibria can be very stable indeed.

The equilibrium established as the result of forward and back mutation was an extremely feeble one; changes in gene frequency tending to restore equilibrium conditions following a temporary displacement could not exceed mutation rates themselves. In the extreme case of the selective superiority of heterozygotes, a balanced lethal system, the population cannot be dislodged from equilibrium frequencies.

Experimental Data

Aside from balanced lethal stocks, a number of examples can be cited to illustrate the superiority of certain heterozygotes in respect to fitness—a superiority relative even to what would superficially appear to be the homozygous "normal" genotype. One of these examples was described by Nabours and Kingsley (1934) for the grouse locust, *Apotettix eurycephalus*. The example concerns a recessive lethal gene found in this species. In crosses both of heterozygotes with heterozygotes (F_2) and of heterozygotes with homozygous normals (backcross), the proportion of lethal heterozygotes among the progeny was greater than that expected (Table 11-1). If the average viability of lethal heterozygotes is adjusted to 1.00, that of the homozygous "normals" is about 0.93.

The relative viabilities of the three genotypes in the above example (+/+, 0.93; +/*le*, 1.00; *le*/*le*, 0) can be used to demonstrate that lethal genes, as well as genes with lesser effects on fitness, can be maintained in populations by natural selection. The fitness of +/+ individuals equals

TABLE 11-1

Deviations from Expected Ratios in Crosses Involving a Lethal Allele in the Grouse Locust, *Apotettix eurycephalus*.

(Nabours and Kingsley, 1934.)

TYPE OF MATING	NO. OF HOMOZYGOUS NORMALS	TOTAL OFFSPRING	HOMOZYGOTES EXPECTED	ESTIMATED VIABILITY OF HOMOZYGOUS NORMALS
Heterozygote × Heterozygote	3097	9657	3219	0.944
Heterozygote × Homozygote	3638	7607	3803.5	0.917
				av. 0.931

$1 - s$, so that s equals 0.07. Similarly, t in the case of lethal homozygotes equals 1.00. The frequency of the lethal gene in populations of the grouse locust could be expected to be $0.07/(0.07 + 1.00)$ or very nearly 7 percent; the frequency of the wildtype allele would then be 93 percent. A lethal gene that is maintained at low frequencies in a population by means of selection in favor of its heterozygous carriers has an advantage in these heterozygotes that is approximately the same as the frequency of the lethal allele itself. A similar example in *D. melanogaster* concerns the balancer chromosome marked with the dominant gene *Pm*; after entering an experimental population consisting otherwise of wildtype chromosomes, *Pm* was maintained for nearly a year at a frequency of 10 percent. *Pm* homozygotes are lethal, so *Pm*/+ flies must have had an advantage of 10 percent in this population.

Another example of heterozygote superiority can be obtained from a report of Battaglia (1958) on a marine copepod, *Tisbe reticulata,* that inhabits the lagoon of Venice. Three distinct phenotypes found in these organisms were shown to result from three alleles:

$V^V V^M$	combined *violacea* and *maculata* phenotypes
$V^V V^V$, $V^V v$	*violacea*
$V^M V^M$, $V^M v$	*maculata*
vv	*trifasciata*

Crosses between individuals heterozygous for both "dominant" alleles ($V^V V^M$) consistently yielded progeny in which the heterozygotes outnumbered the two types of homozygotes combined (Table 11-2). The

TABLE 11-2

DISTORTED F_2 RATIOS REFLECTING DIFFERENT VIABILITIES OF THREE GENOTYPES OF THE MARINE COPEPOD TISBE RETICULATA; THE DISTORTIONS ARE GREATER AT HIGH THAN AT LOW LEVELS OF CROWDING.

(Battaglia, 1958.)

CROWDING	n	$V^V V^V$	$V^V V^M$	$V^M V^M$
High	1751	353	1069	329
		0.660	1.000	0.615
Medium	1743	343	1015	385
		0.675	1.000	0.758
Low	3839	904	2023	912
		0.893	1.000	0.901

degree to which the heterozygotes exceeded expectation (or homozygotes fell short of it) varied with the degree to which experimental cultures were crowded. This dependence of fitness on crowding (and hence on mortality) is shown clearly in the table, where fitness of homozygotes varies from about 60 to 90 percent that of the heterozygotes. In the absence of differential mortality or of differential reproductive ability, there can be no selection.

Probably the most extensive study of the existence of and reasons for the superiority in fitness of certain heterozygotes has been carried on by Dobzhansky and his colleagues using the various gene arrangements of the third chromosome of *D. pseudoobscura.*

Simply to illustrate the equilibrium that is established when heterozygotes exceed homozygotes in fitness, we have chosen four experimental populations studied by Dobzhansky and Pavlovsky (1953). The gene arrangements involved in these experiments were Chiricahua (*CH*) and Standard (*ST*). These had been obtained from flies trapped at Piñon Flats, California (Figure 4-6). The four populations were started with flies collected from the same culture bottles; these flies were chosen so that the starting frequency of *ST* was 20 percent in all four laboratory cages. These were subsequently sampled periodically (and simultaneously) for nearly one year; the frequency of *ST* was determined for each sample for each population through the examination of larval salivary smears.

The results of this study have been listed in Table 11-3 and shown graphically in Figure 11-1. The frequency of *ST* increased in all populations. It did so in an orderly way so that samples taken simultaneously from the four different populations had nearly identical frequencies. Although the frequency of *ST* increased steadily, it did not do so at a constant rate; instead, the increase became less with time until finally there was scarcely any change in gene frequency at all.

Within these laboratory populations, twenty five days is equivalent to one generation. By using changes in gene frequency and numbers of elapsed generations, one can estimate the adaptive values of the three genotypes. These calculations are rather laborious (see Wright and Dobzhansky, 1946, as well as Dobzhansky and Pavlovsky, 1953); in essence, they take advantage of both rate of change of gene frequency and of the seeming approach to an equilibrium. Calculations of fitness based on the data in Table 11-3 yield the following:

	ST/ST	*ST/CH*	*CH/CH*
Fitness	0.90	1.00	0.41

Selection coefficient $s = 0.10$; $t = 0.59$

From these fitnesses, the expected equilibrium frequency of *ST* can be calculated as $0.59/(0.59 + 0.10)$, or 86 percent. Furthermore, using the

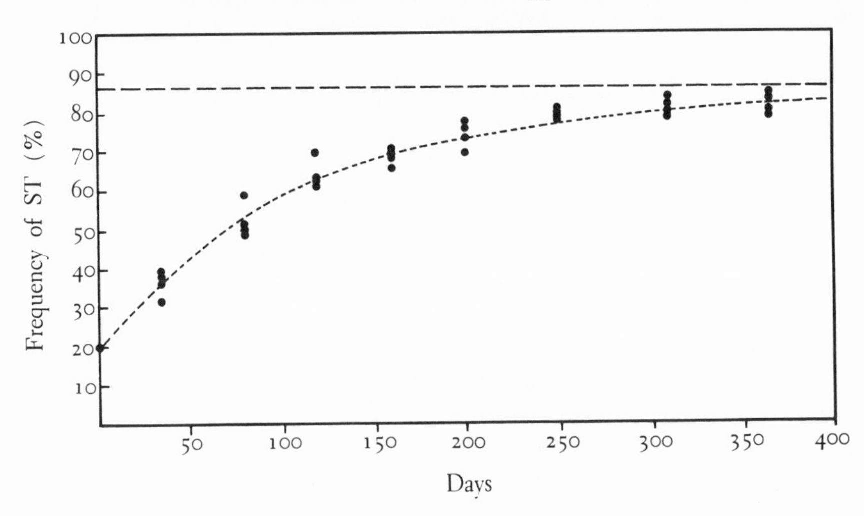

FIGURE 11-1. Frequency of *ST* chromosomes in four replicate populations of *D. pseudoobscura* that were started initially with 20 percent *ST* and 80 percent *CH* obtained originally from Piñon Flats, California. (After Dobzhansky and Pavlovsky, 1953.)

TABLE 11-3

INCREASING FREQUENCIES OF *ST* CHROMOSOMES IN FOUR EXPERIMENTAL POPULATIONS OF D. PSEUDOOBSCURA THAT WERE POLYMORPHIC FOR *ST* AND *CH* GENE ARRANGEMENTS FROM PIÑON FLATS, CALIFORNIA.

(Dobzhansky and Pavlovsky, 1953.)

	% *ST*				
DAYS	1	2	3	4	AV.
0	20	20	20	20	20
35	38	37	38	32	36
80	50	59	53	52	53
120	63	70	61	62	64
160	69	71	69	66	69
200	75	76	69	76	74
250	78	80	81	79	79
310	79	84	80	83	82
365	83	84	81	80	82
Equilibrium (calc.): 86%					

equation for change in gene frequency, the entire sequence of changing frequencies for the four populations can be reconstructed. This reconstructed curve, approaching 86 percent as a limit, has been drawn in

Figure 11-1; the observed changes in gene frequency agree with the calculated ones remarkably well.

Before leaving the studies on *D. pseudoobscura,* it is worth noting that in one of the first reports on the superiority of inversion heterozygotes in this species, Dobzhansky (1948) showed that an equilibrium frequency can be approached from either side as it should be. If a population is set up so that the frequency of a given arrangement is lower than the final equilibrium while another is arranged so that it is higher, the frequencies in the two populations converge toward the same equilibrium frequency (Figure 11-2).

Still another set of data concerning the higher fitness of heterozygous indivdiuals is represented graphically in Figure 11-3; these are results obtained by Oshima (1961) in a study on the persistence of some natural lethal chromosomes in experimental populations of *D. melanogaster.* The lethals (second chromosome) were in this case obtained from populations of flies in Japan. Six populations were set up in each of which all flies were initially heterozygous for one of six lethal chromosomes and for wildtype nonlethals also obtained from natural populations.

In contrast to the results reported earlier (Table 9-2 and Figure 9-1) which concerned the faster than expected elimination of a "recessive" lethal from a population, the six lethals studied by Oshima maintained themselves at higher than expected frequencies. In comparing Figures

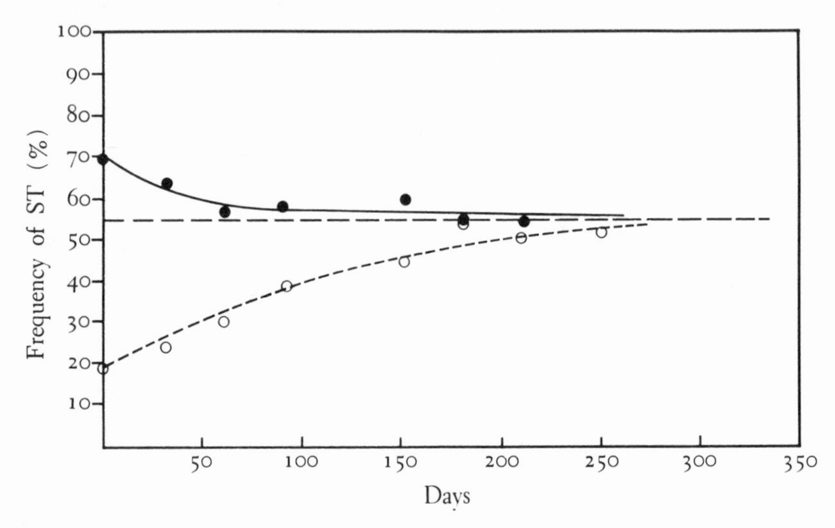

FIGURE 11-2. Demonstration that an equilibrium frequency of *ST* chromosomes can be approached from either side. The populations are of *D. pseudoobscura* containing *ST* and *AR* chromosomes from Mather (Yosemite), California. (After Dobzhansky, 1948.)

9-1 and 11-3, one should note (1) that the duration of Oshima's study is 40 generations rather than 10 as was the earlier one, and (2) the frequencies plotted in Figure 11-3 are frequencies of heterozygotes $[2pq/(1 - q^2)]$ rather than gene frequencies. Because these known lethals ($t = 1.00$) were not eliminated from the populations as fast as recessive lethals are expected to be, the explanation must lie in the higher fitness of lethal heterozygotes. In effect, the lethals studied by Oshima resemble the one found by Nabours and Kingsley in the grouse locust (Table 11-1). Lethals whose heterozygotes have equilibrium frequencies of 20 percent or so must confer an advantage of about 10 percent on those heterozygotes. The advantage of the heterozygous carriers of some of the lethals in Oshima's study were much larger than 10 percent.

The Superiority of Heterozygotes: One Locus of Many?

The knowledge that hybrid organisms often possess qualities useful to man is much older than the science of genetics. Mules have been known since biblical times; various features of these animals make them preferable to both horses and asses for some types of work. Hybrid corn—that is, the double-cross hybrids which account for nearly all corn production in the United States at the present time—was adopted for general use about 1930. Hybrid poultry is of much more recent origin. Much research in plant breeding consists of devising efficient means for making useful hybrids of many different plant species.

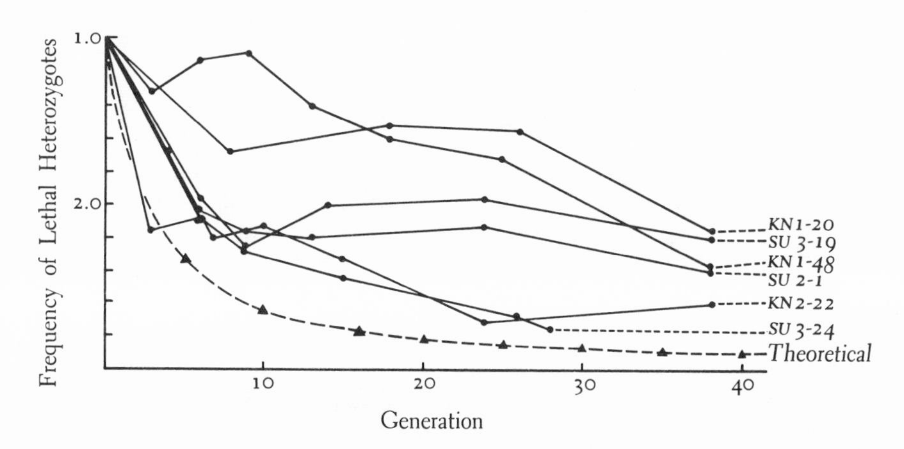

FIGURE 11-3. Decreasing frequencies of individuals heterozygous for autosomal lethals in laboratory populations of *D. melanogaster*. One lethal was tested per population (an identifying number for each has been listed on the right). The dashed line shows the decline expected in the case of completely recessive lethals. (After Oshima, 1961.)

Despite the enormous amount of labor that has been spent in developing and utilizing hybrids, the genetic basis of hybrid vigor is inadequately known. There are several possible genetic causes of hybrid vigor; the evaluation of their respective roles in the production of superior hybrids has not been completed.

A number of difficulties, some arising from confusing terminology, beset a discussion of and an understanding of the genetic bases of hybrid vigor. Hybrid vigor, a term we have used without definition, means vigor associated with the hybrid product of interline, intervariety, interstrain, or interpopulation crosses. Vigor, in this sense, means vigor in respect to a given trait such as yield, stamina, weight, or other measurable quantity. For students of natural populations hybrid vigor in respect to fitness is especially important. To some people the hybrid must exceed both parents to exhibit hybrid vigor; to others, it is sufficient that the hybrid exceed the average of the two parents. Heterosis is synonymous with hybrid vigor.

There are two commonly discussed explanations for heterosis: dominance and overdominance. Different strains of an organism (or individuals of different natural populations) differ in genes at many loci. Strains that are at all inbred have deleterious alleles fixed at a number of loci as the outcome of chance events alone. Two different strains, however, are unlikely to have deleterious alleles fixed at the same loci; consequently, hybrids between them will have most of the deleterious alleles that have been contributed by one parent hidden by the dominant normal alleles of the other. Even if the parents of the hybrids were drawn from rather large, genetically heterogeneous populations, the hybrid progeny they produce are heterozygous at more loci on the average than either of the parents.

Overdominance is an alternative explanation of hybrid vigor. The term overdominance describes the relationship between the phenotypes of *aa*, *AA*, and *Aa* individuals in which *Aa* individuals lie outside the range extending from *aa* to *AA*. Although the notion of overdominance is relatively simple to describe, its experimental study is surprisingly difficult. In the first place, experimental manipulations of genes are not so refined that single genes can be handled; blocks of genes are manipulated in crosses and such blocks, if they exhibit hybrid vigor, may do so because of dominance rather than overdominance. In the second place, through their efforts to understand all sources of phenotypic variation, quantitative geneticists have virtually stripped the term "overdominance" of any biological meaning. Overdominance in the rigorous sense of these workers requires that the heterozygotes fall outside the range of the two homozygotes despite the genetic background in which the alleles are found or the environment in which the individuals develop. This requires, of course, that alleles at one locus be completely independent of those at all other loci and of external influences as well. Since genes do not function alone, overdominance has for our purposes been defined out of existence. For this reason Dobzhansky and other evolutionary geneticists have used the terms "heterotic mutants" and "heterotic effects" in referring to the superiority of heterozygotes. In the hands of the quantitative geneticist, "heterotic effects" vanish into such categories as "epistatic interactions" and "gene-environment interactions."

A term—marginal overdominance—was once used during an informal discussion of geneticists at North Carolina State College; I like this term. Its use is illustrated in Table 11-4. In this table we have listed the fitnesses

TABLE 11-4

GENERAL SCHEME ILLUSTRATING HOW HETEROZYGOUS INDIVIDUALS MAY COME TO EXHIBIT "MARGINAL OVERDOMINANCE" EVEN THOUGH OVERDOMINANCE, STRICTLY DEFINED, IS LACKING UNDER EACH INDIVIDUAL "CONDITION."

(After Wallace, 1959a.)

	GENOTYPE		
CONDITION	*AA*	*Aa*	*aa*
1	1	1	$1 - s_1$
2	$1 - s_2$	1	1
3	1	1	$1 - s_3$
4	$1 - s_4$	1	1
.	.	.	.
.	.	.	.
.	.	.	.
n	$1 - s_n$	1	1
Margin	$1 - S_{AA}$	1	$1 - S_{aa}$

of individuals of the three genotypes *AA*, *Aa*, and *aa* under a variety of "conditions." Examination of the table reveals that under no single condition does the heterozygote exceed both homozygotes; thus, strictly speaking, there is no overdominance for fitness ("euheterosis" of Dobzhansky, 1952b) shown in this table. At the lower margin of the table, however, we have listed the overall fitness of the three genotypes calculated over the entire gamut of conditions; now we see that the heterozygotes are clearly superior to both homozygotes. This is "marginal overdominance."

The conditions listed in Table 11-4 can be assigned various names. They may represent different time periods. If these periods are long relative to the generation time, gene frequencies in the population will undergo cyclic changes as the result of changing selective forces. If, on the other hand, these periods are short in comparison to the generation time, the effect will be to confer a selective superiority on the heterozygotes.

Instead of time, we can substitute space for the listed conditions. Again, if the spaces numbered 1, 2, 3, . . ., and n are large compared to the mobility of the organism, selective changes will occur within each area much as we learned in Chapters 9 and 10. On the other hand, if these spaces are small so that individuals are constantly encountering different ones in quick succession, the effect once more will be to confer an advantage on the heterozygotes.

A third view of these conditions is that they apply to various aspects—morphological, behavioral, physiological, and others—of individuals, aspects that are related to survival and reproduction. In this case we can see that the effect of this mosaic of pleiotropisms is an overall superiority of heterozygotes.

Still another view of these conditions is that they represent a series of background genotypes present in the population which modify the selective properties of the three genotypes in question. Averaged through all such genotypes in the population, the heterozygotes are found to possess the highest fitness.

Finally, these conditions may apply to a special temporal sequence, the sequence of developmental stages leading to the mature individual. In this case it would be fruitless to talk about condition 4 as if it were isolated from preceding and subsequent embryonic or juvenile stages. The net effect is overdominance even in the most rigorous sense. This example shows, I believe, the futility of an overzealous insistence on a too rigorous definition of overdominance. Time—whether it marks the developmental stages during the growth of an individual, the hour by hour change in temperature and humidity, or cyclic climatic changes of larger duration—does not come in logically distinct types. The effort to separate the fifth example from the rest as the only example of overdominance in the "true sense" is largely wasted.

The term "marginal overdominance" appears to include most if not all situations covered by the (undefined) term "heterotic effects." Furthermore, it has a sufficiently rigorous definition to be useful in quantitative treatments of gene action. Perhaps it will manage to find a place in population genetics.

This chapter can be concluded with a short calculation which, in the special case of lethal mutations, bears on the question of one locus or many in conferring heterosis. We shall assume that lethal chromosomes in heterozygous combination with nonlethals appear to be completely recessive; that is, experimental tests suggest that the viability of $+/+$ and $+/le$ individuals are identical. Nevertheless, we suspect that the lethals are really partially dominant to the extent h but that this fact is hidden by the greater heterozygosity of genes in $+/le$ heterozygotes than in the $+/+$ homozygotes.

Among the loci closely linked to those of the lethals are deleterious mutations that in homozygous condition depress fitness by an amount s each. These, like the lethals, are partially dominant; each depresses the viability of its heterozygous carrier by an amount hs. Altogether there are n loci that may be considered closely linked to a given lethal.

As we have seen in Chapter 10, the frequency of a partially dominant deleterious allele in a population equals u/hs. Within a chromosomal segment containing n loci, there are nu/hs such genes. In flies homozygous for nonlethals $(+/+)$, these genes are also homozygous and so depress the viability of these flies by an amount equal to $(nu/hs) \times s$ or nu/h. In the lethal heterozygotes, on the contrary, there are no loci homozygous for these "minor" mutations but counting those on both homologous chromosomes there are $2nu/hs$ genes in this segment of n loci each of which lowers the viability of these flies by an amount hs. Therefore, these genes lower the viability of $+/le$ heterozygotes by an amount equal to $2nu$. This is a smaller effect than that calculated for the $+/+$ homozygotes.

To account for the apparent recessiveness of the "recessive" lethals, the following relationship must be true:

$$\frac{nu}{h} - 2nu = h$$

From this we calculate

$$n = \frac{h^2}{u(1 - 2h)} \quad \text{or} \quad \text{approximately, } \frac{h^2}{u}$$

Consequently, for genes that mutate at a frequency of 10^{-6} and have a partial dominance of 2 percent (2×10^{-2}), the chromosomal segment needed to conceal the partial dominance of the lethal mutations would contain $4 \times 10^{-4}/10^{-6}$, or 400 loci. This number is very nearly as great as the estimated number of lethal bearing loci on the entire V-shaped second chromosome in *D. melanogaster* (page 138). Admittedly, we know nothing about the physical nature of the loci whose number we have just calculated; nevertheless, the large number suggests that the association of a sizable chromosomal segment with each lethal-bearing locus is required in order that the covering up of deleterious mutations at other loci account for the *apparent* recessiveness of partially dominant lethals.

In reviewing the examples of heterosis described in this chapter—color patterns in copepods, the lethal in the grouse locust, and the various lethals in *D. melanogaster*—it seems that overdominance, especially in the marginal sense outlined in Table 11-4, serves as a reasonable explanation for the facts observed. At least, the alternative explanation, which is based on the dominance of normal alleles, leads to equally implausible conclusions. We shall return to this problem again in later chapters.

General Summary

This seems to be the appropriate place to summarize the effects of chance events, mutation, and selection on gene frequencies in populations. These effects have been shown schematically in Figure 11-4. Chance events are dispersive; under their impact gene frequencies move slowly but inexorably toward limits of 0 and 100 percent. Mutation, infrequent as it may be, is sufficient to keep the limiting frequencies, 0 and 100 percent, from being absorbing barriers. If an allele is lost from a population by chance, it can be replaced by mutation. Most mutations are deleterious in respect to fitness. Selection tends to eliminate such alleles from populations. Eventually, equilibria are established between mutation and selection. These equilibria are always near the margin of the chart (0 percent generally), because only at low frequencies of the mutant allele can mutation counterbalance selection. Finally, selection in favor of heterozygotes establishes equilibria at intermediate frequencies; the pressures maintaining these equilibria can be powerful indeed. For practical purposes, such equilibria are independent of chance events and of mutation; in theory, however, even alleles that take part in such balanced systems can be lost or fixed by accident.

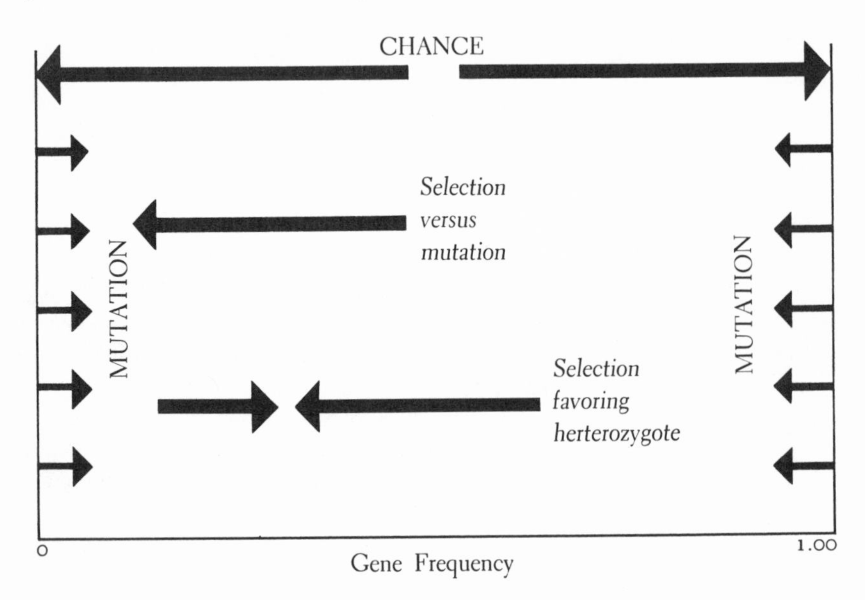

Figure 11-4. Diagram summarizing the roles of mutation, chance, and selection on gene frequencies in a population. (After Wallace, 1964.)

12

NONADAPTIVE SELECTION

Selection of the sort with which we have become acquainted in the last three chapters can be looked upon as "beneficial" for populations. It serves as a means for eliminating deleterious mutations from the gene pools of populations. If it were not for recurrent mutation, all mutations that lower the fitness of their carriers would eventually be purged. Until recently, natural selection was looked upon as the only means by which the consequences of gene mutations could be repaired: the "repair" consisted of their elimination from the breeding population. It is now known that certain mistakes in one strand of double-stranded DNA can be excised by enzymes and a strand with the correct sequence of base pairs can be synthesized under the direction of the remaining strand.

We shall now discuss two systems that illustrate another facet of natural selection: the alteration of gene frequencies by natural selection with no accompanying "improvement" in the content of the gene pool. The two cases that are to be considered in detail are both examples of gametic selection in animals. Under "gametic selection" we shall include factors that disturb gene frequencies by their action at the gametic rather than the zygotic level. Hence, disturbances during meiosis that lead to gametic frequencies other than 50 : 50 in the germ cells of heterozygous individuals as well as differential ability of gametes bearing different alleles to function (true gametic selection) are both included in this chapter.

Sex Ratio

The *sex-ratio* agent that serves for the first illustration is that which has a genetic basis and is associated with the X chromosome of numerous *Drosophila* species of the *obscura* group. Gershenson (1928) described males of *D. obscura* which had only female progeny but with no apparent reduction in total progeny size. He found that the responsible gene was sex-linked, and predicted that such a gene should, in theory, lead to the extermination of the species that possessed it. The same type of *sex-ratio* gene was found in *D. pseudoobscura*; Sturtevant and Dobzhansky (1936b) studied its distribution in natural populations as well as its genetic and cytological nature. Their conclusion was that the *sex-ratio*-bearing X chromosome (*SR*) underwent two divisions during meiosis in males rather than one and, simultaneously, that the Y chromosome was eliminated. Consequently, all four sperm derived from each primary spermatocyte were functional and X-bearing. Novitski et al. (1965) have questioned whether two divisions of the X chromosomes occur in *SR* males; rather, they report that the Y-bearing sperm of these males are nonfunctional. This point would modify the following discussion only if females inseminated by *SR* males produce but one half the number of offspring that those inseminated by normal males produce. There is no evidence that this is the case although the point should be reinvestigated. Certainly, there is no evidence that one half of the eggs laid by females inseminated by *SR* males are sterile.

At the moment we are interested in calculations of the sort which led Gershenson to conclude that *SR* should cause the extinction of a species. The scheme illustrating these calculations is shown in Figure 12-1. The frequencies of *ST* (the Standard X chromosome) and *SR* equal p and q. These two types of chromosomes have no effect on the meiosis of females and so the proportions of eggs carrying *ST* and *SR* are also p and q. In the case of sex-linked genes, the frequencies of the two types of males equal the corresponding gene frequencies because each male has but one X chromosome. Normal (*ST*/Y) males have progenies consisting of sons and daughters in approximately equal numbers. The *SR*/Y males, however, produce only daughters and, presumably, in numbers equal to the sum of sons and daughters that they would have produced if they did not contain the *sex-ratio* condition.

From Figure 12-1 it can be seen that the frequency of *SR* among sons equals q, the frequency of this chromosome among their mothers. The frequency of *SR* among daughters is greater than q; the new frequency is

$$\frac{q(\frac{3}{4}\ p + q)}{\frac{1}{2}p + q}$$

If q were very small and p were nearly 1.00, the frequency of q in daughters would be nearly 1½ times that in their mothers. This is an appreciable increase. Furthermore, the new and higher value of q would be passed on to sons in the following generation while daughters would once again come to possess new and even higher frequencies of *SR*. In theory, this process would continue until the new and old frequencies of q were equal (the equilibrium value of q); this happens only when q equals 1.00. At this frequency, there would be no sons produced (because all males would be *SR*/Y males) and so the population of flies could no longer reproduce. Unless there were simultaneous selection for parthenogenesis, the species would become extinct.

In reality, however, many species of the *obscura* group of *Drosophila* contain *SR* chromosomes. It seems (Wallace, 1948) that *SR*/*SR* females of *D. pseudoobscura* suffer a series of defects such as reduced larval viability and adult longevity, as do *SR*/Y males. *SR*/*ST* females, on the other hand, surpass *ST*/*ST* in many respects. Thus there are selection pressures other than the gametic selection of *sex-ratio* itself that influence the relative frequencies of *ST* and *SR* chromosomes in natural populations.

The existence of *SR* chromosomes in any *Drosophila* species poses a

♂♂ \ ♀♀	X (p)	X^{SR} (q)
X Y (p)	X X daughters - - - - - X Y sons	X X^{SR} daughters - - - - - X^{SR} Y sons
X^{SR} Y (q)	X X^{SR} daughters	X^{SR} X^{SR} daughters

FIGURE 12-1. Inheritance of *SR* in *D. pseudoobscura*.

problem that has not yet been satisfactorily answered: Why is the species (*D. pseudoobscura*, for example) not extinct? Four possibilities exist:

1. The species will become extinct eventually; *sex-ratio* has arisen only recently and has not yet had time to spread. This can be ruled out because, first, the enormous geographic range over which *SR* is found speaks against a recent origin, and, second, observations on the frequencies of *SR* in natural populations of *D. pseudoobscura* and *D. persimilis* over several decades have failed to reveal any perceptible increases.

2. *Sex-ratio* is intrinsically harmful physiologically and hence tends to be limited in its ability to spread through a population. This explanation seems unlikely because a reasonably intricate set of selection coefficients is required in the two sexes to ensure a stable equilibrium.

3. *Sex-ratio* frequently spreads through local populations and causes their extinction. These, in turn, are replaced by flies from elsewhere. In the end, only those populations in which *sex-ratio* happened, in association with genes at other loci, to conform to conditions required for stable equilibria are those in which *sex-ratio* is found. This is a plausible explanation. It involves interpopulation ("group") selection. This selection, in turn, is that which accomplishes feats for a species that when described in unguarded terms appear teleological.

4. Any chromosome gives rise to homozygous individuals which are inferior in fitness on the average to those that are heterozygous for two different chromosomes. This certainly is true for the large autosomes in various *Drosophila* species. Thus *sex-ratio,* having arisen within a species on a single chromosome, is limited in its spread because of associated deleterious genes that become homozygous as the *sex-ratio*-bearing chromosome increases in frequency. Although this explanation sounds plausible in the case of females (it would certainly be true for an autosome), it fails to explain the low fitness of *sex-ratio* males. Nor does it account for the balance of selective forces that is needed to establish a stable equilibrium.

The explanation which I prefer, then, is that which involves interpopulation or group selection; the failure of some local populations to find acceptable solutions to the presence of *SR*, their elimination, and their replacement by migrants from other populations that were more fortunate in the chance association of *SR* with deleterious mutations at linked loci.

The *t* Locus in the Mouse

Our second case of gametic selection involves mutations at a particular locus, the *t* locus, in the mouse, *Mus musculus.* "Mutations" at this locus arise with relatively high frequency (one gamete in 500) in laboratory strains that already contain a *t* allele; in other non-*t*-bearing strains they are as infrequent as mutations generally are. Virtually every sample of mice captured in the "wild" (houses, barns, and farm buildings) has yielded *t* alleles.

The *t* alleles have two properties that make them interesting material in population studies: Homozygous *t/t* mice are either dead or sterile. The proportions of *t* and + sperm produced by +/*t* males is grossly distorted; as many as 95 to 98 percent of all functional sperm produced by these heterozygotes can be *t*-bearing.

The fate of *t* alleles in wild (and computer-simulated) populations has been studied by Lewontin and Dunn (1960) and Lewontin (1962). The alleles normally found in populations have an average segregation ratio in heterozygotes of 95 percent *t* : 5 percent +. Populations generally contain from 35 to 50 percent heterozygotes. In large populations where chance events would be unimportant, segregation ratios of 95 : 5 would counterbalance the lethality of *t/t* homozygotes and establish much higher equilibrium frequencies of heterozygotes.

By using a computer to carry out the routine calculations, Lewontin found that small populations (two males and six females each) duplicate field observations. Within any one population, a *t* allele can be lost despite gametic selection in its favor. A population devoid of *t* can have it reintroduced, however, by a migrant male from another population. In the case of lethal *t* alleles, a high segregation ratio for the heterozygotes decreases the probability that the wildtype allele will become fixed by accident. Furthermore, the lethality of *t/t* individuals prevents the fixation of that allele; its frequency within a population cannot exceed 50 percent. Consequently, those lethal *t* alleles which are retained in wild populations should be those with very high segregation ratios; the loss of such alleles by chance from one small population is counterbalanced by their reintroduction by migrants from nearby populations. Evidence that this is an accurate description of the pattern of selection can be obtained by the comparison of segregation ratios of newly arisen alleles and of those alleles recovered from wild populations (Table 12-1).

Several *t* alleles are known that are not lethal when homozygous; these

TABLE 12-1

Distribution of Segregation Ratios (Percent *t* Sperm Produced by +/*t* Males) for Various *t* Alleles of the Mouse, *Mus musculus.*

The alleles have been grouped as newly arisen (new), lethals obtained from natural populations (wild-lethal), and male sterility alleles (wild-sterile).

	99–90	89–80	79–70	69–60	59–55	(50)	44–40	39–30	29–20	TOTAL
New	3	1	0	0	2	8	3	2	0	19
Wild-lethal	15	1	0	0	0	0	0	0	0	16
Wild-sterile	1	2	0	0	0	0	0	0	0	3

cause sterility of t/t males rather than the death of t/t homozygotes. The fate of such sterility alleles in small wild populations should differ from that of the lethal t alleles. In the case of the sterility alleles (in marked contrast to the lethal ones), populations can lose the wildtype allele by chance; when both males of the small simulated populations are t/t, the population becomes extinct. Thus alleles that possess high segregation ratios should possess as well high probabilities for bringing about their own extinction. As a result, one would expect sterility t alleles recovered from natural populations to have lower segregation ratios than their lethal counterparts. This difference is borne out to some extent by the limited data in Table 12-1.

Summary

Mutant alleles found in Mendelian populations possess characteristics which, among those exhibited by an array of newly mutated alleles, enhance their persistence within populations. This was true, at least, in the case of the lethal genes and other deleterious mutations studied earlier. Alleles whose frequencies are regulated by gametic selection possess characteristics that, in contrast, enhance the continued *survival of the population* that possesses the allele.

The lessened dominance of deleterious mutations found in populations as compared with newly arisen mutations of the same sort can be accounted for by the different persistences of these mutations in populations. Recessive mutations, once they become rare, persist in populations for enormous periods; equally rare, partially dominant mutations are eliminated at a newly constant rate.

Mutant alleles that are potentially dangerous to the population and are subject to gametic selection tend to be associated with properties (also genetically determined) that permit the continued existence of populations containing these alleles. *Sex-ratio* chromosomes, for example, interfere seriously with the fitnesses of their hemizygous and homozygous carriers. Sterility-causing t alleles have lower segregation ratios than those t alleles which are lethal when homozygous. The differential persistence that explains these associations is the differential persistence of populations. If we refer to the bed-of-nails picture of local populations within a restricted area (Figure 2-7), we can see that interpopulation selection consists first of removing certain nails from the board. The empty places are then filled once more by the progeny of migrants from persistent populations. Thus the extermination of a local population means, too, the extermination of an "uncontrolled" allele. Eventually, if the allele is found at all, it is found to be under the control of associated properties that make its existence compatible with the continued existence of the populations in which it is found.

PART II

We now begin the second half of *Topics in Population Genetics*. The emphasis in the treatment of problems now changes considerably. Mutation, selection, migration, hybrid vigor, and, of course, the Hardy-Weinberg equilibrium remain with us. In contrast to the earlier chapters in which we did our best to separate the various agencies of change free of one another and to examine them as independent phenomena, we shall now examine populations to see what sorts of evolutionary changes occur under the joint action of these agencies. We shall ask not whether heterozygotes have superior fitness but how often they have it. We shall examine the notion that although populations can, in theory, approach perfection, in reality they do not; and we shall try to relate existing imperfections ("genetic loads") to the continued survival and reproduction of these populations. In short, we shall be discussing aspects of populations related to those introduced in the first dozen chapters but related in a somewhat complex and obscure way. And so our task is clear: Simplify that which seems complex, clarify that which is obscure.

13

IRRADIATED *DROSOPHILA* POPULATIONS

Cold Spring Harbor Populations

The relation between mutation and the average fitness of a population was first indicated by Haldane (1937). This relationship is simple indeed.

The equilibrium frequency for a recessive allele that arises repeatedly by mutation equals $\sqrt{u/s}$ (see page 192); consequently, the frequency of homozygous mutant individuals, q^2, equals u/s. The fitness of each of these homozygotes is impaired by an amount s. Therefore, the impairment suffered by the population as a whole equals sq^2 or $s(u/s)$, or u. The amount by which the mutant gene impairs the fitness of each of its homozygous carriers does not enter into the measure of the total impairment of the population; only mutation rate remains. The equilibrium frequencies of mutant genes possessing different values for s are such that s drops out of the product of s times the frequency of homozygous mutants. The average fitness of the population is lowered by an amount u; high mutation rates lower fitness more than low ones.

Confirmation of the above relation between fitness and mutation rate was the purpose for establishing several irradiated populations of *D. melanogaster* at Cold Spring Harbor during 1949 and 1950. These populations have been mentioned previously; data obtained through their study have been cited in Tables 8-3, 9-6, 9-7, 9-8, 10-1, and 10-2 as well as Figures 10-2 and 10-3. A detailed discussion of these populations is now in order. First, we shall give some information concerning their origins. Next, we shall discuss some of the information their study has yielded. Of this information, we shall restrict the discussion to the evidence (1) that the fitness of a population is not always correlated with the frequency of "deleterious" mutations it contains, (2) that drastic alterations in the composition of gene pools of populations may occur with scarcely any outward sign, and (3) that the adaptation of populations to the crowded conditions of laboratory population cages can be demonstrated experimentally.

The populations with which we shall be concerned are five in number; they are identified as populations 1, 3, 5, 6, and 7. Each was started with a mixture of flies carrying sixteen different lethal-free second chromosomes isolated initially from a culture of Oregon-R flies that had been maintained for a decade or more by the mass transfer of parents every two or three weeks. The times at which the populations were started, together with information concerning their irradiation histories, are listed in Table 13-1. During the interval between the start of populations 1 and 3 and that of 5, 6, 7, the sixteen chromosomal strains were maintained separately; lethal-free chromosomes were reextracted from each strain to set up the last three populations.

The Effects of Chromosomes on the Fitness of Homozygous and Heterozygous Carriers. The calculations of Haldane (1937) apply in a strict sense only to equilibrium populations. The experimental populations

TABLE 13-1

SHORT HISTORIES OF FIVE EXPERIMENTAL POPULATIONS STUDIED AT COLD SPRING HARBOR.

(Wallace, 1952.)

POPULATION	APPROX. SIZE (NO. ADULTS)	STARTING DATE	FIRST SAMPLE NO.	IRRADIATION
1	10,000	7/25/1949	1	Original ♂♂ 7000-r X ray Original ♀♀ 1000-r X ray
3	10,000	7/25/1949	1	None
5	1,000	4/ 1/1950	20	2,000 r/gen., chronic, gamma
6	10,000	4/15/1950	21	2,000 r/gen., chronic, gamma
7	10,000	4/15/1950	21	300 r/gen., chronic, gamma

studied at Cold Spring Harbor were not necessarily at equilibrium. Nevertheless, the following reasoning applies even to nonequilibrium populations: The greater the frequency of deleterious mutations (whether they are equilibrium frequencies or not), the greater the frequencies of their homozygous carriers, and, consequently, the greater the anticipated reduction in average fitness of the population. The argument applies to incompletely recessive mutations as well as to those that are completely recessive; the depressing effect of a partially dominant (that is, incompletely recessive) deleterious allele on the average fitness of a population should be very nearly proportional to its frequency. At equilibrium, the effect of a partially dominant mutant allele at any one locus equals $2(1 - u/hs)(u/hs)(hs)$, or roughly $2u$.

How shall the fitness of a population be measured? It is impossible (or impossibly laborious) to measure fitness directly, so it is necessary to measure instead an important component of fitness and to assume that, as a rule, other components of fitness are correlated with the one actually measured. With *Drosophila* it is possible (Figures 3-3 and 3-6) to manipulate chromosomes so that one can estimate the preadult viabilities of flies carrying random combinations of different chromosomes from a given population. The estimates are based on the relative frequencies of genetically marked and wildtype flies obtained from crosses such as $CyL/+_1 \times CyL/+_2$. The survival of the $+_1/+_2$ flies through egg, larval, and pupal stages relative to that of the $CyL/+$ flies can be calculated from the deviation of genotypic ratios from the $2CyL/+ : 1+/+$ expected on the basis of Mendelian segregation. A large number of cultures of this sort yields information on a large number of chromosomal combinations within which different types are formed with frequencies proportional to those with which they arise in the populations themselves. Thus, the average survival obtained from a study of a large number of combinations can be taken as a measure (in a relative, not an absolute, sense) of the average fitness of flies in the population itself.

The first indication that the frequency of deleterious genes in a population does not bear a direct relationship to the relative fitness of that population came from a comparison of populations 1 and 3. Population 1 had been subjected at the outset to so heavy an exposure of X radiation that about 95 percent of first generation zygotes were killed by dominant lethals. This population simulated the consequences of an atomic disaster. What happened after this exposure can be seen in Tables 13-2 and 13-3. Table 13-2 lists the frequencies of lethal and semilethal chromosomes observed in the irradiated (#1) and control (#3) populations over the first two years of their existence (a sample interval in these studies was two weeks; time throughout the course of the experiments was measured in these 2-week intervals). Lethals gradually accumulated in population 3 since there were none present initially (some of these same data are listed

TABLE 13-2

OBSERVED FREQUENCIES OF LETHALS AND SEMILETHALS IN POPULATIONS 1 AND 3.

SAMPLE	POPULATION 1 n	POPULATION 1 % $L + SL$	POPULATION 3 n	POPULATION 3 % $L + SL$
1	131	0.214	133	0.008
2	156	0.173	52	0
3	173	0.156	183	0.022
4–5	178	0.118	212	0.014
6–7	202	0.158	263	0.057
8–9	248	0.177	285	0.035
10–11	226	0.199	283	0.057
12–13	289	0.218	386	0.062
14–15	367	0.207	377	0.061
16–17	389	0.272	408	0.081
18–19	402	0.284	409	0.064
20–22	246	0.207	289	0.090
24–26	275	0.273	434	0.120
28	63	0.254	—	—
30	284	0.257	278	0.144
34	252	0.302	254	0.197
38	278	0.284	253	0.206
42	180	0.244	179	0.162
44	20	0.300	20	0.300
46	180	0.244	177	0.186
48	80	0.275	80	0.275
50	79	0.266	78	0.141
52	79	0.215	80	0.263

in Table 8-4); lethals in relatively high frequencies were present in population 1 from the beginning because of the X radiation.

Not all deleterious mutations are lethal or semilethal in their effects on homozygous individuals; many lower the viability of their carriers by smaller amounts. Amounts, in fact, small enough that it would take a tremendous amount of work to identify every deleterious mutation of this sort. Although they cannot be identified individually in the study of populations 1 and 3, the presence of these deleterious mutations as a group can be readily demonstrated. In Table 13-3 are listed the average frequencies of wildtype flies in cultures containing quasi-normal homozygous wildtype flies, wildtype homozygotes whose frequencies exceed those of lethals or semilethals. The lower the frequency of such wildtype flies, the higher the frequency of deleterious mutations among the tested chromosomes. Of the twenty-two comparisons listed in Table 13-3, twenty-

TABLE 13-3

AVERAGE FREQUENCIES IN F_3 TEST CULTURES OF QUASI-NORMAL WILDTYPE HOMOZYGOTES IN POPULATIONS 1 AND 3.

SAMPLE	POPULATION 1	POPULATION 3	#3 − #1 (SIGN ONLY)
1	0.3066	0.3208	+
2	0.3194	0.3165	−
3	0.3111	0.3146	+
4–5	0.3039	0.3149	+
6–7	0.3079	0.3214	+
8–9	0.2963	0.3164	+
10–11	0.2985	0.3140	+
12–13	0.3081	0.3179	+
14–15	0.2999	0.3084	+
16–17	0.2990	0.3087	+
18–19	0.2926	0.3225	+
20–22	0.3033	0.3277	+
24–26	0.3063	0.3204	+
28	0.3104	—	No test
30	0.3083	0.3277	+
34	0.2972	0.3019	+
38	0.2960	0.3100	+
42	0.3107	0.3159	+
44	0.2906	0.3158	+
46	0.3135	0.3317	+
48	0.2987	0.3137	+
50	0.3078	0.3127	+
52	0.3138	0.3139	+

one show that the average frequency of quasinormal homozygotes of population 3 (the control) is higher than that of similar flies of population 1. There are, consequently, higher proportions of subvital mutations, as well as lethals and semilethals, in the irradiated population than in its control. These findings agree with expectations; radiation does indeed induce mutations and these, in general, are deleterious.

The effect of deleterious mutations on Mendelian populations is manifested in individuals that come from random mating, not in highly inbred individuals or individuals made homozygous by special laboratory techniques. In Table 13-4 are listed the average frequencies of wildtype flies in "heterozygous" cultures in which these flies are heterozygous for two different wildtype chromosomes picked at random from a population. Knowing that the chromosomes of population 1 carried the higher frequencies of lethals, semilethals, and other deleterious chromosomes, it

TABLE 13-4

AVERAGE FREQUENCIES IN F_3 TEST CULTURES OF WILDTYPE FLIES HETEROZYGOUS FOR TWO DIFFERENT WILDTYPE CHROMOSOMES OBTAINED EITHER FROM POPULATION 1 OR 3.

	ALL		NORMAL		ALL	NORMAL
SAMPLE	#1	#3	#1	#3	3–1	3–1
28	0.3533	—	0.3533	—		
32	0.3419	0.3312	0.3438	0.3324	–	–
36	0.3431	0.3356	0.3439	0.3356	–	–
40	0.3690	0.3482	0.3690	0.3482	–	–
42	0.3514	0.3484	0.3514	0.3484	–	–
44	0.3547	0.3407	0.3547	0.3426	–	–
46	0.3529	0.3652	0.3529	0.3652	+	+
48	0.3505	0.3487	0.3505	0.3487	–	–
50	0.3471	0.3546	0.3515	0.3546	+	+
52	0.3489	0.3492	0.3489	0.3492	+	+

was expected that random combinations of these chromosomes would exhibit the greater effects of these deleterious mutations. This is not the case, however. As the data listed in Table 13-4 show, the frequency of wildtype flies in population 1 was frequently higher than in the control, population 3. Table 13-5 shows that the average frequencies of these flies ($\overline{y}$) in a much larger series of samples were 0.34799 (population 1) and 0.33748 (population 3); the difference between the two means (0.01051) is highly significant ($p < 0.0002$).

Despite the greater numbers of mutant alleles in population 1, alleles that were undoubtedly in the population as a consequence of the initial radiation exposure, the chromosomes found within the population after the lapse of nearly thirty generations are those which in random combination with one another confer high fitness on their carriers. The chromosomes remaining are not a random sample of those originally present; this can be seen, for example, in the rapid early drop in the frequency of lethals in this X-rayed population. Rather than a random sample of all chromosomes originally present, population 1 either retained or developed a collection that conferred high fitness in random combination. The fitness of homozygotes, such as those made by means of the *CyL* technique, is of little importance in a large population of randomly mating individuals because such homozygotes arise rarely if ever.

There remains one point that should be cleared up before the conclusion outlined in the preceding paragraph is accepted. The test cultures cited contained only two classes of flies—*CurlyLobe* and wildtype. It is conceivable that the wildtype chromosomes of population 1 as a group

TABLE 13-5

DETAILED ANALYSIS OF THE FREQUENCIES IN F_3 TEST CULTURES OF WILD-TYPE FLIES HETEROZYGOUS FOR TWO DIFFERENT CHROMOSOMES FROM EACH OF FIVE EXPERIMENTAL POPULATIONS.

n, number of samples studied; $\bar{y}$, average frequency of wildtype heterozygotes in all samples; *b*, the slope of the regression of *y* on generation; s_b, error of the slope; *p*, level of significance ($B = 0$); *d*, difference between control population (#3), and other populations; and s_d, error of the difference.

	POP.	*n*	NO. OF COMB.	$\bar{y}$	*b*	s_b	*p*
All	1	37	3832	0.34799	−0.000154	0.000060	<0.02
	3	37	3762	0.33748	−0.000298	0.000073	<0.001
	5	39	3176	0.31972	0.000001	0.000078	~ 1
	6	37	3318	0.33171	−0.000018	0.000058	>0.5
	7	37	3391	0.33261	−0.000116	0.000047	<0.03
Normal	1	37	3825	0.34842	−0.000159	0.000060	<0.02
	3	37	3755	0.33798	−0.000283	0.000070	<0.001
	5	39	3090	0.32900	0.000144	0.000060	<0.02
	6	37	3287	0.33483	0.000012	0.000019	>0.50
	7	37	3378	0.33392	−0.000077	0.000046	0.09

		d	s_d	*p*
All	1–3	0.000144	0.000095	0.10–0.15
	3–5	0.000299	0.000107	0.005
	3–6	0.000280	0.000094	0.003
	3–7	0.000182	0.000087	0.04
Normal	1–3	0.000124	0.000092	0.15–0.20
	3–5	0.000427	0.000092	0.0001
	3–6	0.000295	0.000092	0.001
	3–7	0.000206	0.000084	<0.02

yielded *CurlyLobe* flies with low viabilities. Hence, the wildtype flies of that population may only *appear* to have high fitness. This possibility was ruled out by the more sophisticated *CyL-Pm* technique (Figure 3-6). In this test, the *CyL/Pm* flies are the standard of comparison for both populations 1 and 3; the *CyL/+* flies of the two series did not differ, while the proportion of wildtype heterozygotes of population 1 exceeded that of population 3. This test confirms the conclusion reached on the basis of data presented in Tables 13-2, 13-3, and 13-4.

Populations exposed continuously to gamma radiation underwent changes in average fitness which resembled that observed in the case of population 1. The accumulation of lethals and semilethals in these populations has already been illustrated in Figure 10-3; this figure shows not only the initial rate of increase of lethals and semilethals (more accurately, the initial rate of decrease of nonlethals) but also the final equilibrium frequencies. Lethal chromosomes in these populations had considerably higher frequencies than those in populations 1 and 3. The average frequency of quasinormal wildtype homozygotes offers a basis for estimating relative frequencies of chromosomes with only slight detrimental effects on the viability of their carriers. The average frequency of quasinormal wildtype flies in the homozygous test cultures of populations 5 and 6 combined was 0.2986 (samples 20 through 71), while that of population 3 during the same time interval was 0.3161. Presumably, then, populations 5 and 6 had higher frequencies of subvital mutations. Furthermore, the frequency of these flies decreased with the passage of time (−0.00030 per sample interval) in populations 5 and 6. The mean frequency of quasinormal homozygotes in population 7 during the interval from samples 21 through 71 was 0.3154; this is not significantly lower than that of the control population.

Despite the obvious increase in the frequencies of lethals and detrimental genes in the continuously irradiated populations, the average viability of individuals carrying random combinations of these same chromosomes increased, rather than decreased, with time. This change was noted in Table 9-6 where comparable data from early and late in the history of these populations were compared. The same change is revealed even more clearly in Table 13-5 where regressions have been computed for the viabilities on sample interval over some 200 weeks, or nearly four years. In studies of this sort, only relative changes can be discussed meaningfully, since there are no absolute values to serve as reference points. Consequently, every regression slope has been compared to that observed in the control population (#3). First, we can see that population 1 had a higher grand mean frequency of flies heterozygous for random combinations of wildtype chromosomes than did the control. This is the point we made immediately above (page 230). Furthermore, the regression of the mean frequency in population 1 seems to be the same as that in the control; the slopes of the regressions in the two populations do not differ significantly from one another. As far as we can tell, then, heterozygotes in population 1 always survived slightly better in test cultures than did those of population 3.

In the case of the continuously irradiated populations—5, 6, and 7—the mean frequencies of flies heterozygous for random combinations of wildtype chromosomes were lower than those of the control population. Rather startling, though, are the slopes of the regressions; relative to the control, each is positive. The lowest estimates of viability for individuals carrying chromosomal combinations representative of these populations occurred in the early samples when the cumulative effects of radiation-induced mutation were small. As the cumulative effects of continuous radiation grew, the estimates of fitness also grew. This is contrary to what seemed to be the most likely expectation based on the theoretically expected behavior of deleterious mutant genes as described in algebraic terms in Chapter 10.

Faced with the conflict between expectations and observations, an alternative scheme upon which to base expectations was erected. According to this scheme (Wallace and King, 1952; Wallace and Madden, 1953; Wallace, 1956) selection for new mutations in populations proceeds on the basis of the fitness of their heterozygous carriers. The decision concerning the retention of a mutation in or its rejection from a population is made when the mutation is rare and is found almost entirely in heterozygous condition. Only later, when the retained mutation becomes more frequent, does the fitness of homozygous individuals become important. Even then, the viabiltiy of the new homozygotes affects only the equilibrium frequency for the new mutation, not the question of its being in the population. As a result of shifting the initial responsibility for retention of mutant alleles onto heterozygous carriers, the lack of correlation between the fitness of homozygotes and of heterozygotes becomes reasonable. A more quantitative treatment of this type of selection has been developed by Bodmer and Parsons (1960) and Parsons and Bodmer (1961). Underlying the argument is the implied assumption that the superiority of heterozygotes over the two corresponding homozygotes is not a rare phenomenon.

Evidence for a Turnover in the Genetic Composition of a Population. Except to an expert, the flies from one population of *D. melanogaster* seem to be very much like those from another. Careful observations will reveal phenotypic differences between individuals within populations; more extensive ones will reveal average differences between individuals of different populations (Wallace, 1954). An interpretation of observed morphological differences, beyond that they have arisen, is difficult to make. The following analysis of lethals within population 5 shows the extent to which the genetic composition of a population can change with scarcely a hint to the casual observer.

Two types of tests were made routinely with the experimental populations. In one series of test cultures, individuals homozygous for the various chromosomes were obtained so that the effects of these chromosomes in homozygous condition could be measured; these tests lead to the estimates of frequencies of lethals, for example. In the second series, individuals for various (random) combinations of these same chromosomes were obtained; these tests revealed the effects of known chromosomes on their heterozygous carriers. Occasionally, individuals carrying two different chromosomes failed to survive or survived in small numbers only. Almost always these combinations were combinations of lethals or semilethals with one another. These are the cases of allelism; as a group they represent the differences between the numbers of cultures tested per population in "all" and "normal" tests of Table 13-5.

Population 5 was a small population (the number of adults was kept at approximately 1000) that received a high chronic radiation exposure. The frequency of allelism among random chromosomal combinations in this population was higher than that for the other, larger populations. At various times, this frequency changed; these changes are recorded in Figure 13-1. We are concerned here with the extended period during which exceptionally high frequencies of allelism occurred.

During a period of eight generations (interval from sample 85 to 93), four samples were taken from population 5. Among 294 random combinations of all chromosomes tested, 27 were lethal. This is a frequency of allelism of nearly 10 percent. In earlier samples, the average frequency of allelism was 2 percent or less. Obviously, one particular lethal had suddenly increased in frequency within this population. Just how common this lethal had become can be seen in Figure 13-2. The frequency of lethal chromosomes in population 5 at this time was very nearly 80 percent; consequently, among the random combinations of all chromosomes,

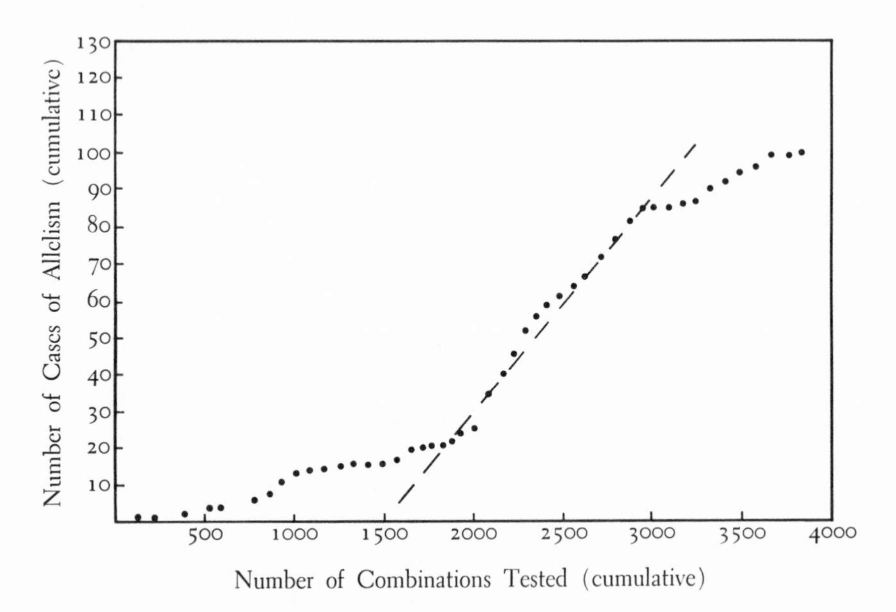

FIGURE 13-1. Frequency of allelism of lethals in laboratory population 5 expressed as the cumulative cases of allelism (vertical axis) per cumulative number of tests (horizontal axis). Clearly, the frequency of allelism is not constant through time. (Wallace, unpublished.)

4 percent were *N/N* (*N* stands for "quasinormal"; *L* for lethal), 32 percent were *N/L*, and 64 percent were *L/L*. If 10 percent of all chromosomal combinations consisted of allelic lethal combinations, then some 16 percent of all *L/L* combinations were allelic. Thus, 40 percent of all lethal chromosomes were in reality one particular lethal or, alternatively, about one third of all chromosomes in the population carried this lethal.

Is it possible that the turnover in the chromosomal constitution would have been detected by routine tests of lethal frequencies alone? Scarcely, at least with samples of the size we were testing at that time. In Table 13-6 are listed the frequencies of lethals and semilethals in the period just preceding and just following the sharp increase in allelism frequency. By consolidating the data for the preceding seven and the following six samples, one can show that the average frequency of lethals was about 9 percent higher in the later ones. These data are shown diagrammatically in Figure 13-3. The change in the overall frequency of lethals is not nearly as striking as the change observed in allelism frequency.

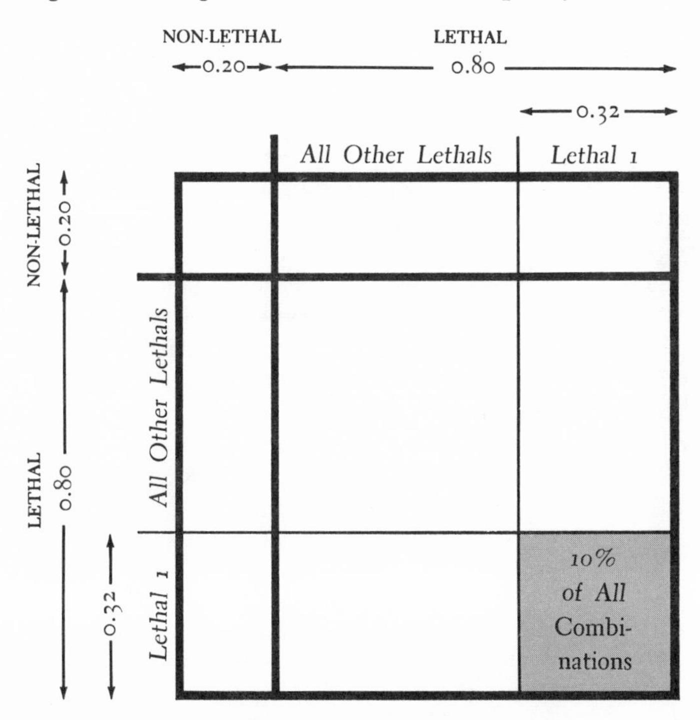

FIGURE 13-2. Elaborate "Hardy-Weinberg" diagram which shows that in a population where lethals have a total frequency of 80 percent and where 10 percent of all chromosomal combinations are lethal (allelic), then some 40 percent of the lethal chromosomes are actually one particular lethal. (Wallace, unpublished.)

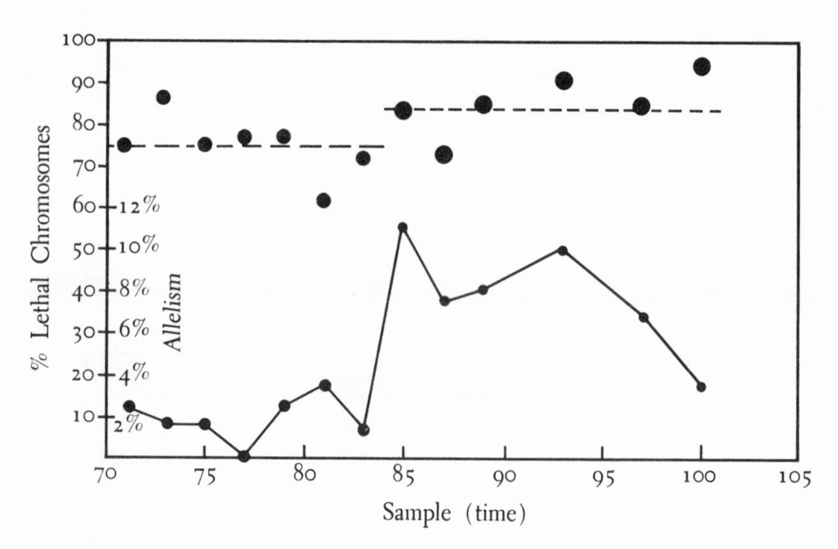

FIGURE 13-3. Changes in the frequency of lethals (dashed line) and in the frequency of allelism (solid line) in population 5 at the 85th sample ("generation"). (Wallace, unpublished.)

A little—precious little, though—is known about why this one lethal spread within population 5. Approximately forty generations after it first appeared, and then persisted, in population 5, two subpopulations (#'s 17 and 18) were set up with flies from #5. One of these (#17) was maintained under the same stringent conditions of scant food and no yeast supplement that was required to maintain a small population; the other (#18) was given a standard amount of enriched food, so that it quickly expanded to a population of 10,000 or more adults. The high frequency of allelism tended to be maintained in the smaller population. Five cases of allelism were observed in the first 268 combinations tested in population 17 as opposed to the need to test over 1200 combinations in population 18 before five cases of allelism were observed. The lethal was seemingly favored either by the stringent diet or small population size.

Evidence for the Adaptation of Flies to Life in Population Cages. In a sense, the selection for a particular lethal, apparently in response to a meager diet, represents the adaptation of a fly population to the conditions surrounding it in a laboratory cage. To the best of our knowledge, however, it represents a special case. If possible, additional evidence should be presented to show that adaptive changes are the rule, not the exception.

Problems arising from efforts to maintain population 5 as a small popu-

TABLE 13-6

ANALYSIS OF THE NUMBERS OF QUASI-NORMAL, LETHAL (L), AND SEMI-LETHAL (SL) CHROMOSOMES IN POPULATION 5 DURING THE PERIOD FROM SAMPLE 71 TO 100, ROUGHLY 1 YEAR.

SAMPLE	QUASI-NORMAL	$L + SL$	TOTAL	% $L + SL$
71	21	59	80	73.8
73	8	51	59	86.4
75	15	44	59	74.6
77	15	51	66	77.3
79	9	30	39	76.9
81	22	35	57	61.4
83	21	54	75	72.0
85	14	66	80	82.5
87	21	59	80	73.8
89	11	63	74	85.1
93	5	53	58	91.4
97	9	51	60	85.0
100	6	51	57	89.5
Consolidated: Before and after sample 84				
71–83	111	324	435	74.5
85–100	66	343	409	83.9
Total	177	667	844	av. 79.0

lation of approximately 1000 adults illustrate how flies can become adapted to their immediate environment. The standard regime for feeding the experimental populations called for the replacement of an exhausted food cup by a fresh one containing about 27 cubic centimeters of enriched fly food on each of five days every week. Each cage had spaces for fifteen food cups so that each cup remained in the cage for three weeks before it was removed.

To maintain the smaller number of adults in population 5, the brewer's yeast supplement was omitted from the food for that one cage. This was excellent at the beginning of the experiment but eventually it appeared that the number of flies in cage 5 had become nearly the same as that in the other populations. After that, one half of the food was removed from each food cup; rather than a cup of food one inch in diameter and two inches deep, population 5 received (nonenriched) food only one inch deep. Later it appeared again that the number of flies in population 5 was increasing and so the schedule for feeding was reduced to four days each week. Subsequently, it was reduced to three days. Finally, it seemed that feeding the population only twice a week would be sufficient; this schedule proved to be too severe, however. And so, step by step, the continual adaptation of the flies to ever more stringent conditions had forced us to reduce the quantity of medium from five full cups of unenriched medium per week (sufficient to reduce the population size to the desired level when first used) to three half-cups of the same, unenriched medium per week—one third the original amount.

The increased ability to survive in the face of adversity that is illustrated so well by the flies in experimental population 5 can be detected in all the experimental populations by noting the numbers of flies hatching in the F_3 cultures of the routine *CyL* tests. The numbers of flies hatching in the heterozygous test cultures have been listed in Table 13-7. These numbers have been computed for intervals of approximately ten generations. The data show that the average number of flies counted per culture increased approximately 2½ times from samples numbered in the 30s to those taken nearly four years later (numbered in the 120s). This increase is shown graphically in Figure 13-4. It is highly unlikely that these observations represent a gradual and systematic improvement in food making or "fly handling." It is remarkable that the relative frequencies of wildtype and *CyL* flies in these cultures remained virtually constant through all these years; not only must a good deal of the increased number of progeny flies per culture stem from genes lying on the third chromosome, but

TABLE 13-7

AVERAGE TOTAL NUMBER OF FLIES PER F_3 TEST CULTURE CALCULATED FOR INTERVALS OF 10 SAMPLES (20 WEEKS) FOR FIVE EXPERIMENTAL POPULATIONS.

Test cultures included in these data are those yielding wildtype flies heterozygous for two different second chromosomes.

SAMPLE	POPULATION				
PERIOD	1	3	5	6	7
30s	164.8	147.1	165.4	154.2	157.2
40s	237.1	221.1	276.3	273.5	287.2
50s	203.4	215.1	205.9	204.6	204.6
60s	323.7	314.7	301.6	306.9	295.4
70s	397.8	374.5	374.4	379.8	347.5
80s	340.8	397.9	296.6	362.8	336.8
90s	373.2	408.3	388.5	430.1	392.0
100s	420.6	399.5	353.7	376.4	357.9
110s	463.1	477.4	404.4	418.4	394.8
120s	404.6	455.2	374.7	383.2	364.8

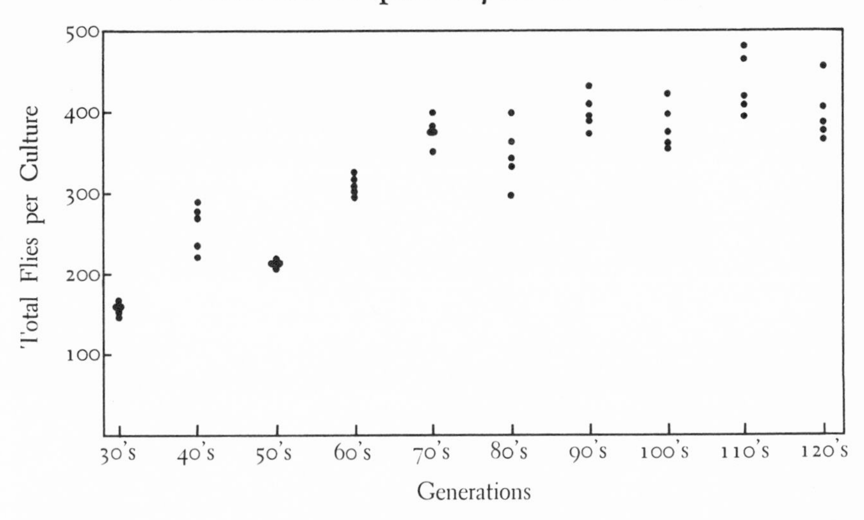

FIGURE 13-4. Average number of flies per F_3 test culture at various times during the analysis of five experimental populations of *D. melanogaster* at Cold Spring Harbor. (Wallace, unpublished.)

those which are on the second must be very nearly dominant in their effects so that both *CyL*/+ and +/+ flies were affected.

Tabulations similar to that given in Table 13-7 for heterozygous combinations of wildtype chromosomes have been listed in Tables 13-8 and 13-9 for cultures yielding lethal ("good" lethals, not semilethals) and quasi-normal homozygotes. These, too, show the same trend toward increased numbers of flies per test culture as time passed. Perhaps the most interesting aspect of these data is emphasized in Figure 13-5, where the ratio of total flies in homozygous cultures to that in heterozygous ones is plotted. In the case of lethals, the flies in the homozygous test cultures are nearly all *CyL*/+ and in theory should be only two thirds as numerous as the total of all flies in heterozygous cultures. Instead, we see that during the early samples, the number of *CyL*/+ flies in homozygous lethal tests were as numerous as the combined number of *CyL*/+ and +/+ flies in the heterozygous tests. The ratio drops but seemingly levels off at nearly 75 percent instead of the expected 67 percent. These ratios suggest that the increased yield of test cultures was caused by an increased ability of larvae to survive in these culture bottles, and that from sample 60 onward, practically all eggs that were laid and capable of hatching gave rise to adults. Thus there is no method by which lethal homozygous tests can yield as many flies as the heterozygous ones. In the early tests, on the contrary, only a fraction of all larvae were able to survive; in lethal cultures, there were sufficient numbers of *CyL*/+ larvae to produce virtually a full number of flies despite the lethality of all wildtype flies. For each

TABLE 13-8

AVERAGE TOTAL NUMBER OF FLIES PER F_3 TEST CULTURE CALCULATED FOR INTERVALS OF 10 SAMPLES (20 WEEKS) FOR FIVE EXPERIMENTAL POPULATIONS.

Test cultures included in these data are those in which the homozygous wildtype flies were lethal.

SAMPLE	POPULATION				
PERIOD	1	3	5	6	7
30s	157.3	166.5	161.5	160.1	151.5
40s	228.8	223.0	195.3	206.4	225.6
50s	172.7	166.2	181.2	178.9	164.9
60s	219.4	252.1	241.6	253.6	220.1
70s	300.2	288.1	292.5	283.5	257.6
80s	264.6	330.4	244.3	283.4	265.0
90s	299.2	316.4	318.0	332.3	285.1
100s	308.1	326.7	301.3	291.5	256.9
110s	322.5	359.5	298.0	337.4	291.4
120s	317.5	333.1	286.2	300.8	270.2

TABLE 13-9

AVERAGE TOTAL NUMBER OF FLIES PER F_3 TEST CULTURE CALCULATED FOR INTERVALS OF 10 SAMPLES (20 WEEKS) FOR FIVE EXPERIMENTAL POPULATIONS.

Test cultures included in these data are those in which the homozygous wildtype flies were quasi-normal.

SAMPLE	POPULATION				
PERIOD	1	3	5	6	7
30s	174.2	178.0	167.4	171.8	166.0
40s	261.1	264.5	215.7	229.0	259.3
50s	197.7	207.0	200.1	197.0	198.0
60s	311.1	304.7	299.8	289.7	287.6
70s	375.5	359.6	354.6	336.5	312.8
80s	343.6	377.0	283.9	330.0	325.7
90s	370.3	370.7	318.5	406.6	375.4
100s	395.7	381.9	353.3	362.6	338.7
110s	443.2	450.2	379.1	420.9	362.1
120s	375.8	351.7	344.3	369.8	367.6

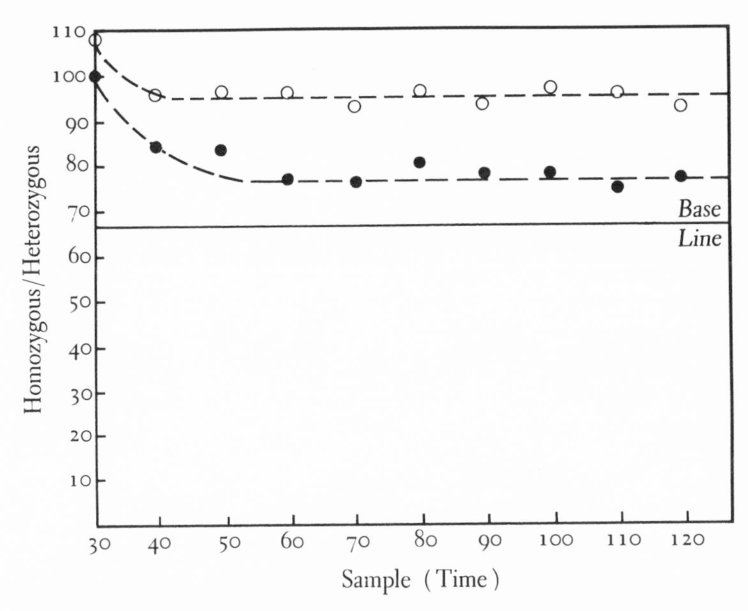

FIGURE 13-5. Ratio of culture size of homozygous test cultures to that of heterozygous test cultures during various samples of the Cold Spring Harbor experiments. Solid circles: lethal homozygotes; open circles: nonlethal (quasinormal) homozygotes. The base line, 67 percent, represents the proportion of *CyL*/+ heterozygotes per culture; if all wildtype flies were to die, homozygous cultures on the average should not be smaller than 67 percent of the average total flies found in heterozygous cultures. (Wallace, unpublished.)

dead wildtype larva, there was an otherwise doomed *CyL*/+ larva ready to take its place. Recalling the results obtained by Weisbrot (see Table 8-15) on the interactions between developing larvae, it appears that immunity to inhibitory waste products may be an adaptive change favored within a severely overcrowded population cage. A second possibility—adaptation to a near-starvation diet—is a real but less likely one because tests involving population 5, the population which was supplied with the least food of any and which adapted to this meager diet, did not out-produce those of the other four populations in the half-pint culture bottles.

BONNIER'S POPULATIONS

Irradiated populations of *D. melanogaster* were studied also by Professor Bonnier and his colleagues at Stockholm (Bonnier et al., 1958). These populations were not maintained in precisely the same manner as those studied at Cold Spring Harbor. Furthermore, the information upon which a judgment as to the "well-being" of the population was based differed from that used to measure fitness in the Cold Spring Harbor populations. Nevertheless, the results of the Swedish study are reminiscent of those observed at Cold Spring Harbor.

The populations studied by Bonnier were maintained as shown in Figure 13-6. In one series (P), adult flies were irradiated (1500 to 2000 r) and were then shaken into a plastic box. Eggs were collected from these flies on petri dishes of culture medium. Following the egg collection, 5000 to 6000 young, freshly hatched larvae were transferred to small culture tubes—25 per tube in the P25 series, 200 per tube in P200.

As the adults emerged in the culture tubes, they were shaken into a

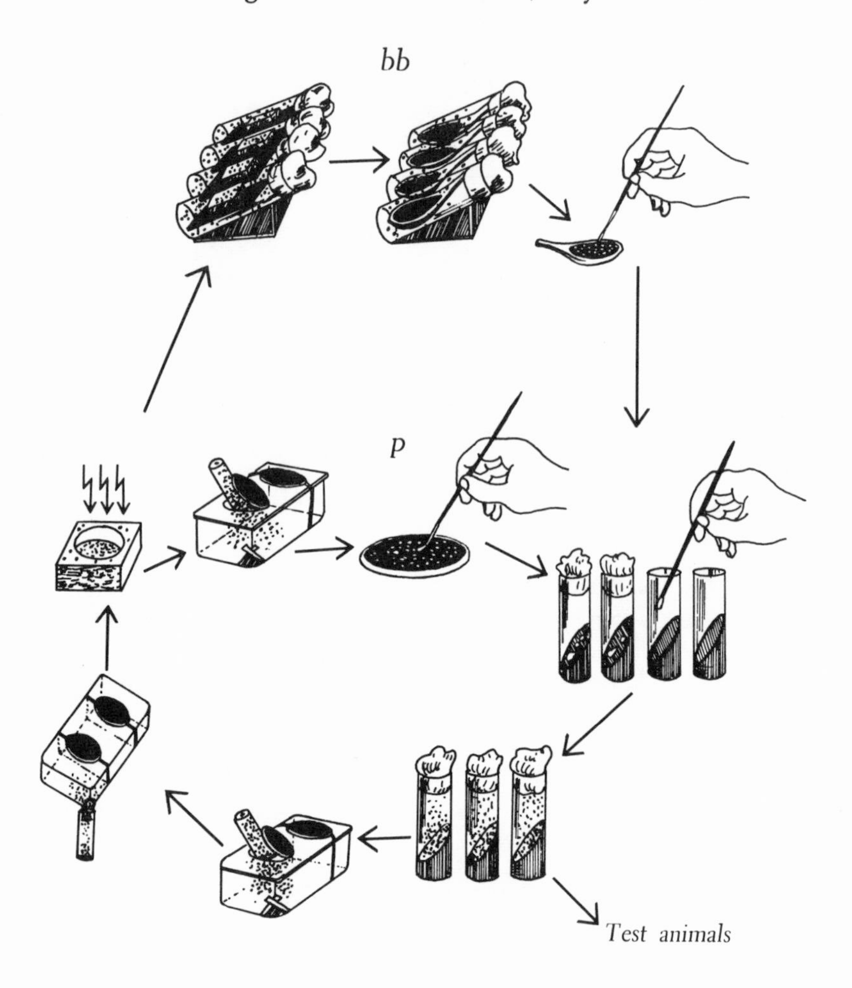

FIGURE 13-6. Procedures by which Bonnier kept his experimental populations of *D. melanogaster* (see the text). (After Bonnier et al., 1958.)

plastic cage for storage (and to allow more or less random mating between flies of different tubes, although no effort was made to shake the adults into the cage as virgins). Three weeks after the irradiation of the original adult flies, the new generation was shaken from the storage cage, irradiated, and the entire process was repeated.

A variation on the above scheme was introduced in an effort to mimic more closely human reproductive patterns. Under the system described immediately above, all 5000 to 6000 transferred larvae might have been the progeny of very few females since no control was exercised over the total number of progeny contributed by any one female. The number of children per couple in man is rather small even under the "best" of circumstances. To imitate man's lower fecundity, Bonnier subdivided the irradiated flies into lots of five pairs each; only twenty-five larvae were collected from any one lot of parents. Consequently, one female could contribute at the very most only twenty-five larvae to the 5000 to 6000 picked for the following generation. The populations maintained in this special manner were designated (bb); as in the earlier series, the larvae were placed 25 (bb25) or 200 (bb200) to a culture tube to develop.

One of the measures of the well-being or fitness of the irradiated populations was the actual number of larvae collected following the irradiation of the parental adults. Quite often fewer than the desired 6000 were actually obtained. The charts presented in Figure 13-7 show that in some populations—especially in bb25—the number of larvae collected was much below the goal. Interestingly, however, the populations improved with time. The effect of radiation was greatest in the early generations; in subsequent generations, despite the cumulative radiation dose and the accumulation of deleterious mutations, the populations tended to recover. This pattern resembles that revealed by the data in Table 13-5. Bonnier and his colleagues suggest, as we did above, that mutations beneficial to their heterozygous carriers arise and spread rapidly in these experimental populations.

The results obtained by the Swedish workers resemble those obtained at Cold Spring Harbor in still another respect: When chromosomes were tested to determine the frequencies of lethals, the average number of flies hatching per culture was related to the larval density of the tested population. This has been shown in Table 13-10, where the average numbers of *CyL*/+ flies per test culture are listed. Both P200 and bb200 yielded considerably more *CyL*/+ flies per culture than did P25 or bb25. Furthermore, populations in which no limit was placed on the contribution of individual females to the following generation (series P) had higher numbers of *CyL*/+ flies in subsequent test cultures than did those of the "bb" series; the difference between P25 and bb25 is highly significant but that between P200 and bb200 is not. These observations parallel those described in discussing Figure 13-4 and its accompanying tables.

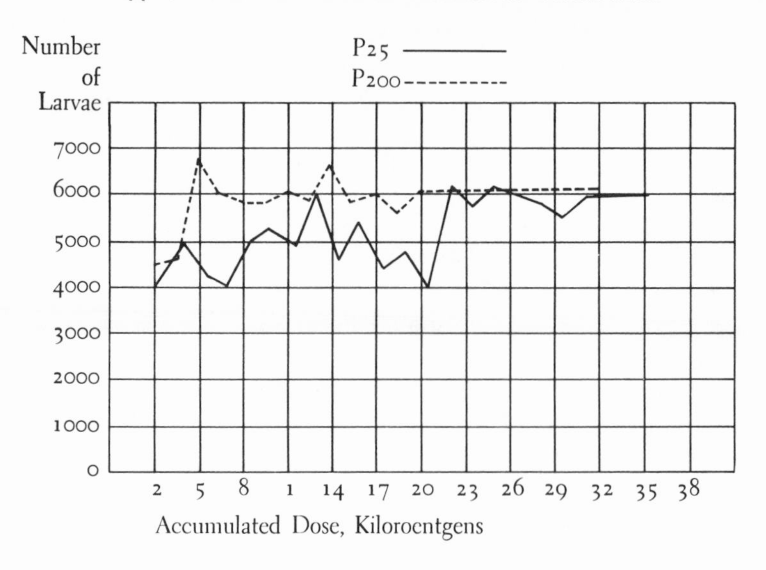

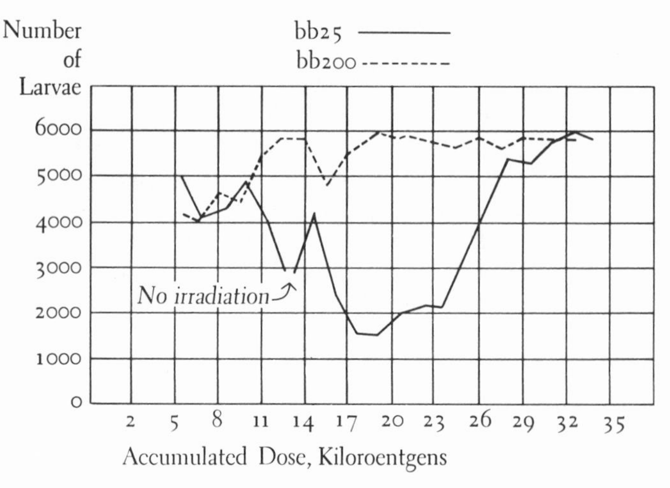

FIGURE 13-7. Total number of larvae collected per generation (an estimate of population fitness) in Bonnier's experimental populations. (After Bonnier et al., 1958.)

THE COLUMBIA-ROCKEFELLER POPULATIONS

The study by Dobzhansky and Spassky (1947) in which the improvement of semilethal chromosomes under natural selection, with or without mutation induced by irradiation, was one of the early studies of irradiated populations (see Table 8-10). More recently, some of Dobzhansky's colleagues have studied irradiated populations for one purpose or another.

TABLE 13-10

AVERAGE NUMBER OF *CyL/+* FLIES IN F_3 TEST CULTURES CONTAINING CHROMOSOMES FROM EXPERIMENTAL POPULATIONS OF D. MELANOGASTER MAINTAINED AT DIFFERENT POPULATION DENSITIES.

(Bonnier et al., 1958.)

NO. OF LARVAE IN EACH MAINTENANCE VIAL	OFFSPRING PER PARENTAL FEMALE	AV. NO. OF *CyL/+* IN TEST CULTURES
25	(P) Unlimited	85.75
25	(bb) Limited	69.88
200	(P) Unlimited	110.35
200	(bb) Limited	103.93

Sankaranarayanan (1964, 1965, 1966) studied the egg-to-adult survival of eggs produced by females from populations receiving 0 r (control), 2000 r, 4000 r, and 6000 r every generation. Under these radiation exposures, egg-to-adult survival quickly drops to a low, constant level. Upon cessation of radiation, much but not all of the control survival is quickly restored. It appears that lethal frequencies within the irradiated populations do not drop immediately to their control values; certain lethals seem to be retained as the result of selection in favor of heterozygous carriers (not necessarily on account of the lethal allele itself).

Ayala (1966) studied some of the more important ecological aspects of irradiated and control populations of *D. birchii* and *D. serrata*. Among measurements made by Ayala were (1) the number of individuals produced per week, (2) the total biomass produced per week, (3) total population size (number of individuals), and (4) total biomass. Table 13-11 records Ayala's data on the number of individuals produced per week in twelve populations. Four different sets of experiments are included among the twelve populations: *D. birchii* at 25° and 19°, and *D. serrata* at 25° and 19°. In each set there was a control and two irradiated populations. (An irradiated population was one in which males were exposed to 2000 r in each of three generations immediately before setting up the experimental populations.)

After an initial lag period of some ten or more weeks, the irradiated populations in three of the four sets took the lead in the number of individuals produced per week (as well as in other measurements). In the fourth set, the irradiated and control populations did not differ significantly; the reason for the similarity in this instance is that the control population itself underwent a steady increase in weekly production. The data obtained by Ayala correspond more closely to the descriptive ma-

TABLE 13-11

AVERAGE NUMBER OF INDIVIDUALS PRODUCED PER WEEK IN LABORATORY POPULATIONS, IRRADIATED AND CONTROL, OF D. SERRATA AND D. BIRCHII.

(Ayala, 1966.)

	INDIVIDUALS PRODUCED PER WEEK
D. serrata (25°C)	
Control	595 ± 21
Experimental 1	941 ± 30
Experimental 2	1133 ± 27
D. serrata (19°C)	
Control	236 ± 28
Experimental 1	499 ± 19
Experimental 2	519 ± 18
D. birchii (25°C)	
Control	533 ± 22
Experimental 1	933 ± 32
Experimental 2	833 ± 31
D. birchii (19°C)	
Control	473 ± 28
Experimental 1	495 ± 16
Experimental 2	557 ± 29

terial presented above (page 237) concerning the tendency for population 5 to expand in size despite efforts to keep it small. Data on the increase in the yield of test cultures (Figure 13-4) suggest that the control as well as the irradiated Cold Spring Harbor populations improved in this respect.

SUMMARY

The studies described in this chapter have dealt with the interplay of mutation and selection in determining some characteristics of experimental populations. The supply of mutations in some populations was augmented by exposure to mutagenic radiation. The result that seems most startling is the *improvement* of estimates of fitness which accompanied *increases* in the frequency of "deleterious" mutations. Our earlier treatment of the role of mutation in respect to fitness had led us to expect the reverse. The observed improvement should not be regarded, however, as evidence that the calculations given earlier are wrong. I recall Dr.

Sewall Wright making this point very clearly when he first learned of the experimental results obtained from the Cold Spring Harbor populations. For genes that behave as those which were described in the earlier algebraic calculations on mutation and selection, the expectations are correct. If the experimental results fail to support the expectations, then there must be mutant alleles that behave otherwise. One of our remaining tasks is to estimate, if we can, the proportion of such mutant alleles.

In discussing the studies of Dobzhansky and Spassky (1947; see page 149), we stressed the improvement that occurred in what were initially semilethal chromosomes when homozygous cultures of these semilethals were maintained in grossly overcrowded cultures. We also stressed the close relationship between the improvement observed in these experiments and the temperature under which the cultures were maintained. Several of the chromosomes that had improved within mass cultures maintained at 19° lost a good deal of their improvement when transferred to 25°. In these studies, the semilethality of the starting chromosomes served as an obvious challenge to natural selection; natural selection "responded" by retaining preferentially those mutations that suppressed or otherwise modified this semilethaltiy.

Changes that occur within experimental populations may, in a more subtle manner, mimic the obvious changes observed by Dobzhansky and Spassky. Populations respond (if they respond at all) to challenges with which they are actually confronted. Flies kept in stock cultures under standard conditions of crowding are not necessarily adapted to the conditions that exist in experimental populations. Consequently, there may be a multitude of "beneficial" genetic changes that can take place in these flies when they are transferred to cages; an example of such a change would be immunity toward waste products as we mentioned earlier. If this is the case, however, we must then admit the futility in trying to discover a universally standard fly—an ideal fly—against which all other flies of that species must be measured. The "standard" fly for any species is the standard that has evolved in its own little locale, in its own little population, and out of the limited genetic materials available to it. Moreover, the "standard" may itself be composed of a collection of genotypes rather than a single one alone. To what extent seemingly deleterious mutations enter into such collections of genotypes remains to be seen.

14

HYBRID VIGOR

The descriptions given in Chapter 13 of events taking place in irradiated populations of flies led to the notion that some mutations are retained in populations and increase in frequency because their heterozygous carriers enjoy a selective advantage over that of other individuals. The result was an eventual improvement of the fitness of some irradiated populations. The improvement does not necessarily lead to a population of superflies; rather, it represents an improvement over the fitness exhibited during the first few generations of radiation exposure.

Heterotic mutations—mutations that are favored in heterozygous individuals, at least within the populations in which they arise—are apparently frequent enough to confound the obvious predictions about population fitness. In predicting the fitness of population 1, for example, it was of no use to know that it contained more deleterious mutations than did population 3; the fitness of random heterozygotes of population 1 was higher than that of population 3. Similarly, it was not enough to know that the irradiation of the Swedish populations had decreased their "fitness" (measured in this case as the ability to produce larvae for the next generation) in early generations; upon continued exposure to radiation the population improved in this respect.

The studies of irradiated populations were started with the notion that high frequencies of deleterious mutations must mean low average fitness. As the frequencies of lethals and other deleterious mutations went up, fitness was expected to go down. This belief was not borne out. Now we have a new problem. We have had to postulate that a certain proportion of mutations are heterotic in the sense that they improve the fitness of their heterozygous carriers but not that of individuals homozygous for them. Having made such a postulate, it is necessary to determine the frequency with which such mutations occur. If it is possible, we would like to estimate "frequency" in terms of both loci and alleles per locus.

The material presented in the sections immediately below constitute a "historical" approach to the question we have just posed. It speaks well for population genetics that this is one of the shortest "histories" in biology; the entire series of arguments has been rendered obsolete by new electrophoretic techniques that allow us, in theory at least, to detect genic variation locus by locus. Nevertheless, the older arguments are still valuable as a background against which to view the newer observations.

The Viability Effects of Random Mutations

Many experiments have been carried out in which radiation-induced mutations have been assessed for their effects on viability. The original experiments of Muller (1927, 1930) can serve as examples. In this case, special techniques were developed for determining the frequency of mutations; moreover, the mutations studied were, for the most part, those that killed their homozygous carriers. Experiments of this sort (see Figures 8-1, 8-2, and 8-3) revealed that radiation-induced mutations are deleterious—especially in homozygous condition.

Still other experiments have been made to determine the effect of known deleterious mutations on their heterozygous carriers. Experiments of this sort have the general form of that by Hiraizumi and Crow (Figure 9-4) or Stern (Table 9-5 and Figure 9-3) in which the effect of the lethal gene is determined in an otherwise uncontrolled background. We are not quite sure in these experiments, what sorts of alleles—identical or different—are present in the $+/+$ homozygous controls; the $+$ symbol means "nonlethal" but it does not necessarily follow that the two $+$'s are identical.

To determine the effects of new mutations in heterozygous condition—and to use the experimental data in an effort to understand the genetics of populations—we shall develop our models around otherwise homozygous individuals; new mutations will be studied in heterozygous condition within the homozygous background.

Model 1: Homozygote Superiority. The first model is one that is consistent with the material presented in Chapters 9 and 10. According to this model, there exists a normal allele at each locus together with a number of relatively rare mutant alleles. The frequencies of these rare mutants can be given in terms of mutation rates and selection coefficients: complete recessives have frequencies (at equilibrium) equal to $\sqrt{u/s}$, while those whose deleterious effects are partially dominant have (lower) frequencies equal to u/hs.

Under this model, the ideal genotype—one achieved by few if any individuals of a population—would be that shown in Figure 14-1A; every locus in this ideal would be occupied by a normal allele.

Within a population of randomly mating individuals, mutations accumulate; consequently, "ideal" individuals of the sort described above most likely do not exist. Instead, individuals carrying mutant genes at various loci make up the bulk of such a population. These mutations are generally carried in heterozygous condition so the phenotype of the "ordinary" individual (Figure 14-1B) is very much like that expected for the "ideal"—somewhat worse, perhaps, because of the partial dominance of some mutant alleles and because of occasional homozygosis for others.

By means of techniques such as those shown in Figures 3-3 and 3-6, we can choose a chromosome at random from a population and by means of special crosses synthesize individuals homozygous for this chromosome (Figure 14-1C). We have seen ample evidence (Table 3-6) that these homozygotes have poor average viabilities relative to control combinations which are equivalent to Figure 14-1B. The poor viability of these homozygotes does not mean that all loci are occupied by deleterious mutations. On the contrary, the few mutant alleles in homozygous condition are the limiting factors in determining the viability of these individuals; a lethal allele at a single locus is sufficient, when homozygous, to kill an otherwise genetically perfect individual.

Population Model Based on Homozygote Superiority

A	Ideal genotype	A B C D E F G H . . . / A B C D E F G H . . .
B	"Ordinary" genotype	A B C D e F G H . . . / a B C D E F g H . . .
C	Artificial homozygote	A B C D e F G H . . . / A B C D e F G H . . .
D	Artificial homozygote heterozygrous for newly induced mutation	A B C D e F G H . . . / A B C D e F G h* . . .

FIGURE 14-1. Population model based on homozygote superiority. (Wallace, 1959a.)

At this point we should consider the frequency of mutant loci among all loci in the artificially synthesized homozygote. Lethals do not concern us here because we shall deal experimentally only with viable homozygotes. If we consider mutations for which s equals 0.1 and u equals 10^{-5}, then $\sqrt{u/s}$ equals 1 percent; 1 percent of all loci are expected to be occupied by alleles of this sort while the remaining 99 percent would be more normal. If these genes were partially dominant (say, $h = 0.05$), then their expected frequency, u/hs, equals 20×10^{-4}; in this case, only 2 loci of 1000 would be occupied by such mutations. These examples could be extended but the point is already clear: Homozygotes under this model carry very few mutant alleles; the bulk of all loci are occupied by "normal" alleles.

The rarity of mutant alleles is important in the discussion that now deals with the individuals represented in Figure 14-1D. The crosses by which this type of individual can be arrived at have been described (Wallace, 1958, 1959a). For our present discussion it is enough to say that by a judicious use of standard mating procedures such as those given in Figures 3-3 and 3-6, one can irradiate the wildtype chromosome in one line and then proceed to synthesize individuals identical to those of Figure 14-1C except that one (not both) wildtype chromosome has been irradiated. Presumably the radiation induces mutations at one locus or another in a number of exposed chromosomes. Presumably, too, these mutations involve the normal alleles since these occupy 99 percent or more of all loci. And, because the model is specific on this point, each mutant allele must be a deleterious form of the unmutated normal allele.

Can we predict the effect of the newly induced mutations on these otherwise homozygous individuals? Recall that the control (Figure 14-1C) and experimental (Figure 14-1D) individuals are homozygous for precisely the same mutant alleles but, in addition, the experimental ones carry additional, newly induced mutations. There is no alternative under this model; the newly induced mutations should have a depressing effect on viability.

Model 2: Heterozygote Superiority. As an alternative to model 1, we can devise a scheme such as that shown in Figure 14-2. Remember that our task (page 249) is to estimate the frequency with which alleles show an advantage in heterozygous condition. For purposes of this model we shall assume that large numbers of alleles exist at every locus such that the viability of homozygotes for the various alleles is somewhat lower than that of heterozygotes carrying two different alleles. We are now in a position to set up four types of individuals in Figure 14-2 which are analogous to the same four discussed for the preceding model.

According to model 2, the "ideal" genotype (Figure 14-2A) would be any one of a great number of genotypes in which two different alleles are found at every gene locus. This would be the "ideal" since the model

Population Model Based on Homozygote Superiority

A	Ideal genotype	$\frac{A_1\ B_9\ C_2\ D_7\ E_4\ F_5\ \ldots}{A_7\ B_2\ C_8\ D_1\ E_9\ F_3\ \ldots}$
B	"Ordinary" genotype	$\frac{A_1\ B_6\ C_5\ D_2\ E_5\ F_8\ \ldots}{A_1\ B_7\ C_8\ D_4\ E_5\ F_3\ \ldots}$
C	Artificial homozygote	$\frac{A_1\ B_6\ C_5\ D_2\ E_5\ F_8\ \ldots}{A_1\ B_6\ C_5\ D_2\ E_5\ F_8\ \ldots}$
D	Artificial homozygote heterozygrous for newly induced mutation	$\frac{A_1\ B_6\ C_5\ D_2\ E_5\ F_8\ \ldots}{A_1\ B_6\ C_5\ D_*\ E_5\ F_8\ \ldots}$

FIGURE 14-2. Population model based on heterozygote superiority. (Wallace, 1959a.)

states that individuals homozygous for any one allele are inferior in fitness to heterozygotes.

Within a freely breeding population, the most frequently encountered individual (Figure 14-2B) would of necessity be homozygous for a number of alleles. This would be so because (1) the number of alleles per locus is presumably limited and (2) Mendelian inheritance provides no mechanism for retaining large numbers of alleles in populations.

Through the use of special mating procedures, homozygotes can be synthesized (Figure 14-2C) which would be expected to have viabilities inferior to those individuals that carry two different chromosomes (such as Figure 14-2B).

The consequences of irradiating a single chromosome of an otherwise homozygous combination (Figure 14-2D) could be quite different from that postulated for model 1. Under the present model, all loci are homozygous in the absence of an induced change; consequently, *every* induced mutation leads to heterozygosity at the locus involved.

What are the expected consequences of the radiation-induced changes? Unfortunately, we cannot be sure because the model does not offer a basis for making predictions. The overall effects of the new mutations on viability depend upon (1) the proportion of all possible alleles that enter into advantageous heterozygous combinations with the one carried by the chosen chromosome, (2) the proportion of all loci at which alleles behave in the manner described by the model, and (3) the average increase in viability expected from the substitution of heterozygosity for homozygosity at the mutated locus.

To discriminate between the two models experimentally, we proceed just as we do when we perform a chi-square or any other statistical test. We temporarily accept a model that leads to a definite prediction, we compare the data with the results expected, and we reject the model if the data do not fit. In the present case, model 1 is the only one that makes a prediction; it says that the induced mutations will reduce the viability of their carriers. We accept this model and then proceed with the experiment.

The Experimental Data. The viabilities of flies are determined experimentally by noting deviations from Mendelian proportions that arise under somewhat overcrowded culture conditions. One genotypic class is taken as a standard against which to compare the others (Figure 3-6; see also page 35).

Unfortunately, we are interested at the moment in comparing two types of wildtype flies: those homozygous for a given chromosome and those identical in all respects but heterozygous for some newly induced mutations as well. These two types of flies cannot be allowed to develop in the same culture because they could not be separated for counting; the flies that are to be compared must be grown in separate cultures and compared through a common standard.

The experimental procedure used to test the effect in heterozygous condition of radiation-induced mutations on their otherwise homozygous carriers culminates in two series of cultures that yield flies as shown in Figure 14-3. In both the X-ray and control series, *CyL/Pm*, *CyL/+*, *Pm/+*, and +/+ flies develop. The *CyL/Pm* flies serve as the control (viability is said to equal 1.000) so that the viabilities of the other classes can be measured in relative terms. The only differences between the two series are (1) the wildtype flies of the X-ray series carry one irradiated chromosome and (2) one other genotypic class (*CurlyLobe* or *Plum*, depending upon the nature of the cross) also carries an irradiated chromosome. The relative viabilities of the corresponding genotypic classes can be determined by comparing their viabilities which have both been determined relative to that of *CyL/Pm* (1.000).

The results of the first large series of tests of the sort described above are given in Tables 14-1, 14-2, and 14-3. All in all, seven tests were made, each of which involved from 600 to 800 control and a similar number of experimental cultures. More than three and a quarter million flies were counted in these tests.

Table 14-1 lists the relative viabilities of those genotypes (*Plum* in most tests, *CurlyLobe* in others) that carried nonirradiated wildtype

	1		1-r		1-s		1-t
Control:	$\frac{CyL}{Pm}$	:	$\frac{CyL}{+_1}$	:	$\frac{Pm}{+_1}$	:	$\frac{+_1}{+_1}$
Xray:	$\frac{CyL}{Pm}$	:	$\frac{CyL}{\textcircled{+_1}}$	:	$\frac{Pm}{+_1}$	:	$\frac{+_1}{\textcircled{+_1}}$
	1		1-r′		1-s′		1-t′

FIGURE 14-3. Four genotypes produced by the final crosses made to test the effect of newly induced mutations on their (otherwise homozygous) heterozygous carriers. Circles indicate irradiated wildtype chromosomes.

TABLE 14-1

VIABILITY EFFECTS OF IDENTICAL, NONIRRADIATED WILDTYPE CHROMOSOMES CARRIED BY GENETICALLY MARKED HETEROZYGOTES (D. MELANOGASTER) IN CONTROL AND EXPERIMENTAL TEST CULTURES.

(Wallace, 1959a.)

	Pm/+	*Pm/+*
A	1.146 (766)	1.137 (764)
B	1.139 (676)	1.140 (672)
C	1.137 (636)	1.145 (637)
D	1.143 (596)	1.136 (598)
G	1.215 (839)	1.222 (837)
	CyL/+	*CyL/+*
E	1.127 (639)	1.125 (637)
F	1.189 (499)	1.201 (496)

chromosomes even in the experimental series of cultures; a glance at Figure 14-3 shows that there are such genotypes. The data in Table 14-1 reveal that the viabilities of these flies in the two series do not differ; the viability shown in the "Control" column is higher than that in the "X ray" column in three of seven experiments. The probability of seeing the overall difference in viability as the result of chance alone is greater than 0.50.

The comparisons of the viabilities of wildtype flies in the control and experimental series of cultures are listed in Table 14-2. These results are quite different from those of the preceding table. In all six of the large experiments involving homozygous control flies and their counterparts in the X-ray series, the viabilities of the wildtype flies of the X-ray series are

TABLE 14-2

VIABILITY EFFECTS OF NEW MUTATIONS IN HETEROZYGOUS CONDITION IN FLIES (D. MELANOGASTER) HOMOZYGOUS FOR WILDTYPE SECOND CHROMOSOMES.

The irradiated wildtype chromosome is marked (′). (Wallace, 1959a.)

	$+_1/+_1$		$+_1/+_1'$	
A	1.008	(766)	1.033	(764)
B	1.000	(676)	1.007	(672)
C	0.989	(636)	1.015	(637)
D	0.979	(596)	0.989	(598)
E	0.983	(639)	0.990	(637)
F	0.992	(499)	1.002	(496)
		3812		3804
	Av. diff.:	1.5%		
	Error:	0.5%		
	Probability:	0.002		

TABLE 14-3

VIABILITY EFFECTS OF NEW MUTATIONS ON THEIR GENETICALLY MARKED HETEROZYGOUS CARRIERS (D. MELANOGASTER).

The irradiated wildtype chromosome is marked (′). (Wallace, 1959a.)

	CyL/+		*CyL*/+′	
A	1.094	(766)	1.115	(764)
B	1.093	(676)	1.108	(672)
C	1.105	(636)	1.110	(637)
D	1.100	(596)	1.108	(598)
G	1.185	(839)	1.182	(837)
	Pm/+		*Pm*/+′	
E	1.127	(639)	1.125	(637)
F	1.189	(499)	1.201	(496)

higher than those of the corresponding control. In all, the probability that the observed difference (0.015) is the result of chance alone is no greater than 0.002. These results cast serious doubt upon model 1 for the model is incapable of accounting for an increased viability of flies carrying an irradiated chromosome. Model 2 could not predict that this increase would be seen but an increase in viability is compatible with this model.

Not all the data obtained in these tests agree with expectations based on model 2; the discrepant data are given in Table 14-3. Why are these data unexpected? We said that the results (Table 14-2) of the viability analyses in the case of wildtype flies did not agree with model 1 but were compatible with model 2. Fine; so we can assume for the moment that model 2 as described in Figure 14-2 is correct. We must admit, then, that individuals carrying two different chromosomes (Figure 14-2B) are heterozygous at many loci. In that case, however, the *CyL*/+ and *Pm*/+ flies of our test cultures must also be heterozygous for genes at many loci. And, if this is so, not every radiation-induced mutation leads to increased heterozygosity in these flies since many of these mutations must fall at loci already occupied by dissimilar alleles. However, the results shown in Table 14-3 reveal that the *CurlyLobe* or *Plum* flies bearing the irradiated wildtype chromosome have higher viabilities than their controls in five of the seven series. The overall probability for the observed difference is about 0.05; this level of significance would not be sufficient to claim that the two series differ, but it is too low to ignore in the present case.

Experiments essentially similar to those described above have been made by other investigators. Mukai and Yoshikawa (1964) carried out tests in which the effects of newly induced mutations were studied in a homozygous background (as in the experiments described above) and in a heterozygous (wildtype chromosomes of two different origins) background. In one test, both the second and third chromosomes were manipulated simultaneously; in the other, the study was limited to the second chromosome alone. The results (Table 14-4) agree very well with those expected on the basis of the results described immediately above. Newly induced mutations in heterozygous condition but carried by otherwise homozygous individuals tend to increase viability; those carried by individuals already heterozygous at many loci tend to lower viability. A more extensive test made by Wallace (1963b) also tended to confirm the original results: In this second test, *CurlyLobe* flies carrying irradiated wildtype chromosomes had slightly but significantly lower viabilities than those with a control chromosome; this observation helps dispense with the problem we encountered above.

In the second test by Wallace, irradiated chromosomes tended to increase the viabilities of otherwise homozygous wildtype individuals; they had no pronounced effect on wildtype individuals carrying two different chromosomes obtained originally from the same population; and they tended to lower the viability of wildtype individuals carrying chromosomes from widely separated localities. This sequence of differing results agrees well with the decreasing probability that a newly induced mutation

TABLE 14-4

EFFECT ON VIABILITY OF RADIATION-INDUCED MUTATIONS IN HETEROZYGOUS CONDITION IN FLIES (D. MELANOGASTER) HOMOZYGOUS (*AA* VERSUS *AA′*) OR HETEROZYGOUS (*AB* VERSUS *AB′*) FOR THEIR BACKGROUND GENOTYPE.

The irradiated chromosomes are indicated by (′). (Mukai and Yoshikawa, 1964.)

Experiment 1: 2nd and 3rd chromosomes, 150 r

GENOTYPE	NO. OF CHROMOSOMES	NO. OF FLIES	VIABILITY
AA	330	51,500	0.984
AA′	330	51,700	1.002
AB	430	71,300	1.283
AB′	430	80,100	1.252
Experiment 2: 2nd chromosome, 500 r			
AA	292	224,000	1.015
AA′	291	230,000	1.041
AB	324	394,000	1.147
AB′	312	371,000	1.142

will fall at an otherwise homozygous locus. A mutation at a locus for which the control can be represented as a_1/a_2 does not increase the total amount of heterozygosity; it merely substitutes a new allele (a_*) for a preexisting allele (a_1/a_*).

Finally, an experiment with a highly inbred strain of Tribolium has yielded results similar to those obtained by Wallace and by Mukai and Yoshikawa; females heterozygous for an irradiated genome appear to have a higher egg-laying capacity than their "homozygous" controls (Crenshaw, 1965).

Attempts to repeat the experiments described above have not always succeeded. Muller and Falk (1961) and Falk (1961) have discussed at length theoretical reasons why the first model is the more realistic one and have gathered experimental data that appear to bear out their arguments. These arguments and data were discussed by Wallace (1963b). In the meantime, Falk and Ben-Zeer (1966) have described additional experiments, the results of which seem to resemble the earlier ones by Wallace.

The experimental data gathered by various workers to test the alternative models have left us in a somewhat confused state. On top of the confusion has emerged still a third model, one which states that there is an optimal proportion of heterozygous loci. This possibility was discussed by Wallace (1959a) and discarded as unlikely; Mukai et al. (1966), on the basis of exceptionally fine experimental data, have recently argued in its favor once more.

Using Inversions to Study Heterosis

Inversions give us a convenient means for analyzing the frequency and location of genes whose alleles are involved in the development of hybrid vigor. Dobzhansky and Rhoades (1938) suggested that inversions might be used for this purpose in maize; very little has been done with their suggestion, however (see Sprague, 1941; Chao, 1959). Vann (1966) has made a systematic study of the hybrid vigor that accompanies inversion heterozygosity in *D. melanogaster*. Although his techniques do not lead directly to an estimation of the proportion of loci at which heterotic alleles occur (indeed, they cannot distinguish between the two possible causes of hybrid vigor—dominance and overdominance), the experimental results are certainly pertinent to any serious discussion of the causes and magnitude of heterosis.

Vann's experiment was built around a number of disparate facts: (1) X radiation can be used to induce chromosomal aberrations whose positions in the genome are determined by more or less randomly occurring breakage points. (2) Crossing-over is suppressed in inversion heterozygotes within the limits of the inversion. (3) Inversions with average deleterious effects on fitness tend to be eliminated from freely breeding populations; inversions whose heterozygotes have superior fitness (heterosis in respect to fitness or "euheterosis," Dobzhansky, 1952b) will be retained in these populations. (4) The genetic similarity of a number of strains of flies can be governed somewhat by a careful choice of geographic origins combined with various inbreeding regimes.

The experimental material of Vann's study consisted initially of two stocks of *D. melanogaster* descended from different females captured at Riverside, California, two from Raleigh, North Carolina, and one from Bogotá, Colombia. Highly inbred lines were obtained from each of these stocks. Two inbred lines were obtained from each of the California stocks (RC-3 and RC-9); furthermore, after ten generations of brother-sister matings each of these was subdivided so that eight inbred lines (four pairs of sublines) were available for this material (Figure 14-4).

Inversions were obtained by irradiating males (4000 r to 5000 r of X rays) of the inbred California lines and mating them with females of the *same* inbred line. From the progeny of this cross, many *single-pair* matings were set up; the male and female of each pair carried single irradiated genomes. To find out whether either of these genomes contained a radiation-induced aberration, a dozen larval offspring were sacrificed so that their salivary chromosomes could be examined cytologically.

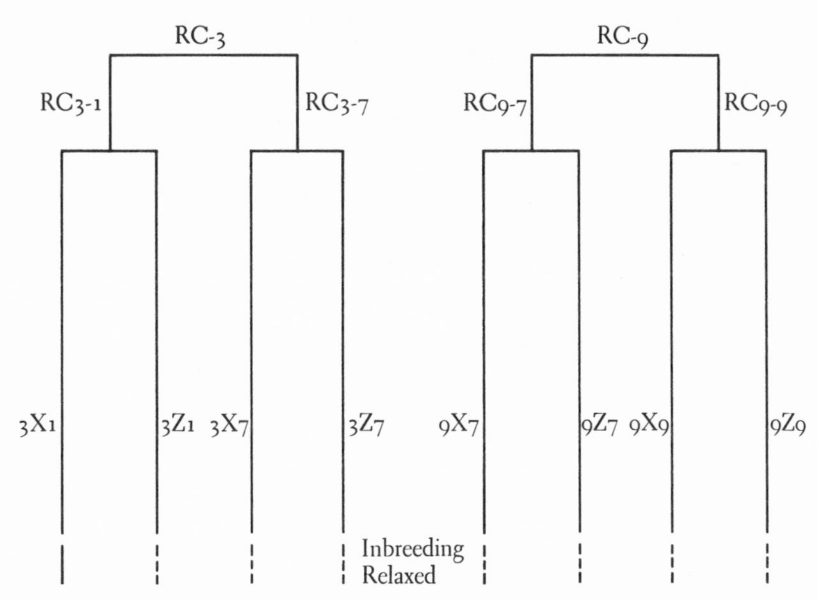

FIGURE 14-4. Derivation of the various inbred lines used to study the relation of hybridity to heterosis. (After Vann, 1966.)

If no aberration was found, the culture was discarded; if one was observed, the culture was saved. The adults that hatched were used in a series of matings that can be described most easily by examples based on the symbols given in Figure 14-4.

1. *Same culture.* This cross was made between males and females taken from the aberration-containing culture itself. These parental flies were very similar genetically for they were brothers and sisters produced by a single-pair mating of highly inbred material.

2. *Same subline.* If the aberration culture occurred in a series obtained by irradiating males of RC3X1, for example, this cross would involve flies (males or females) from the aberration culture and others from the line RC3X1. Following fifty generations of inbreeding, the different strains were kept by subculturing two nonvirgin females and a single additional male; the genetic difference between the parents of this cross was somewhat greater than that of the "same culture" cross.

3. *Other subline.* Using the example given above, this cross would consist of matings between RC3X1 and RC3Z1, the two sublines of RC3-1. The parental flies of this cross differed only in respect to factors that were still segregating after the ten generations of brother-sister matings which preceded the isolation of the two sublines.

4. *Intrastrain.* This cross would consist of matings between RC3X1 and RC3X7 or RC3Z7 (two crosses). These parental flies were related because

they were descended from the same mass culture of flies; they differed in respect to factors that were segregating in that mass culture.

5. *Interstrain.* This type of mating would involve, as an example, RC3X1 with either RC9X7, RC9Z7, RC9X9, or RC9Z9 (four crosses). These flies were related only in the sense that both stocks came originally from flies captured at Riverside, California.

6. *Unrelated.* This mating would involve RC3X1, for example, and flies from the Raleigh, North Carolina, or Bogotá, Colombia, inbred strains. The geographic origins of these flies are extremely remote; if genetic similarity is related to geographic proximity, these flies were the most dissimilar of all that were used in the experiments.

The cultures that were set up according to the scheme outlined above were allowed to produce many offspring. These were transferred in large numbers (as many as several hundreds in some instances) to fresh cultures. This mass transfer of adults each generation was continued for some nine or ten generations. Within the overcrowded cultures, larval mortality was extremely high. To the extent that hybridity for genes lying within the inverted gene segment enhanced survival, the structural heterozygotes had a selective advantage. Recombination outside the limits of the inversion removed any persistent hybridity in noninverted chromosomal segments.

Following nine or ten generations during which the populations were maintained by the mass transfer of adults each generation, thirty or more larvae of each culture were examined to determine the fate of the induced aberrations. The results are summarized in Table 14-5. Translocations and pericentric inversions are listed in the table but they really do not

TABLE 14-5

AVERAGE FREQUENCIES (%) OF INDIVIDUALS HETEROZYGOUS FOR VARIOUS TYPES OF RADIATION-INDUCED CHROMOSOMAL ABERRATIONS RETAINED IN EXPERIMENTAL POPULATIONS OF D. MELANOGASTER.

The degree of genetic dissimilarity between the structurally altered and normal chromosomes depended upon the original cross. (Vann, 1966.)

ORIGINAL CROSS	INVERSIONS: PARACENTRIC	INVERSIONS: PERICENTRIC	TRANSLOCATIONS
Same culture	1.0	0	0
Same subline	5.3	0	0
Other subline	7.4	0	0
Intrastrain	14.1	0.2	0
Interstrain	22.9	3.7	1.7
Unrelated	28.8	5.6	1.6

concern us. Both of these aberrations are known to affect adversely the reproductive ability of heterozygous individuals; both types are extremely rare in natural populations. Only a few remained in the experimental ones. Paracentric inversions, on the contrary, were not necessarily rare after this series of transfers. Furthermore, there is a definite relationship between their final frequencies and the genetic dissimilarity of the original parents; the more dissimilar the parents, the higher the frequency of inversion heterozygotes found in the populations.

The frequencies of structural heterozygotes listed for paracentric inversions in Table 14-5 can be converted into "gene" frequencies. From (1) the known starting frequencies (25 percent for "same culture" crosses; 12.5 percent for the others), (2) the observed final frequencies, and (3) the number of elapsed generations, one can estimate the effect of the inversions on the fitness of their heterozygous carriers. This calculation (and the earlier calculation of gene frequencies as well) is based on the reasonable assumption that induced inversions are lethal when homozygous (as a great many such inversions are even under optimal conditions). The relationship between final gene frequencies and H, the amount by which the fitness of heterozygotes is raised or lowered, is shown in Figure 14-5.

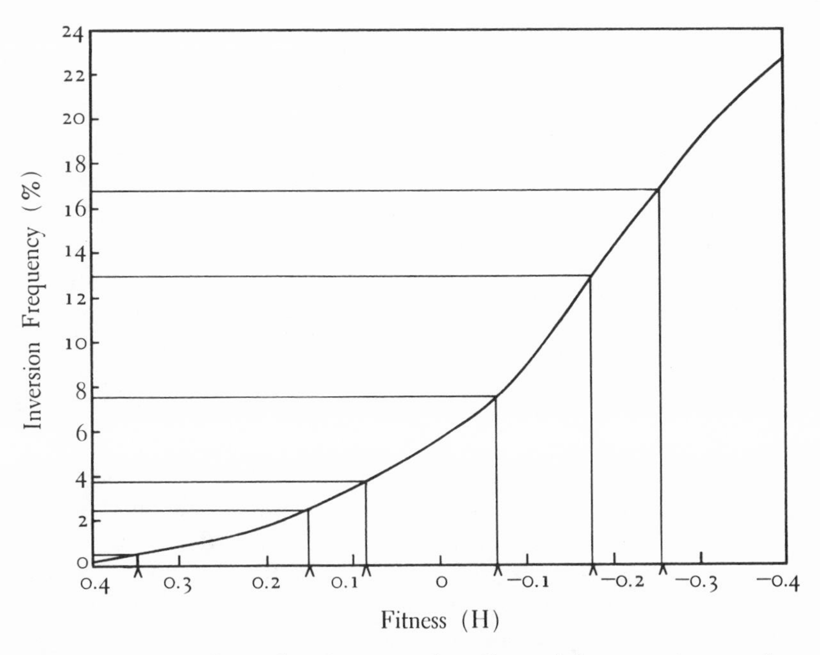

FIGURE 14-5. Relationship between the effect of the inversion on the fitness of its heterozygous carriers and inversion frequency in Vann's study. (After Vann, 1966.)

The estimated average selection coefficients, *H*, of inversion heterozygotes for the various types of crosses are listed in Table 14-6. Following the usual convention, positive values of *H* mean a lowering of fitness while negative values mean an enhancement. Now, there are either no or very few gene differences between inverted and uninverted chromosomes in the "same culture" populations. Hence, the value of *H* (0.35) observed in these crosses represents the average debilitating effect of the two radiation-induced chromosomal breaks on their heterozygous carriers under the conditions of Vann's experiment. This effect should be manifest in all types of populations, therefore the calculated values of *H* can be adjusted to reveal the contributions to heterosis made by the same collection of inversions but within populations where the genetic differences between the inverted blocks increases from no difference ("same culture") to an unknown but presumably considerable difference ("unrelated").

TABLE 14-6

ESTIMATED AVERAGE EFFECT (*H*) OF RADIATION-INDUCED PARACENTRIC INVERSIONS ON THE RELATIVE FITNESSES OF THEIR HETEROZYGOUS CARRIERS IN VARIOUS TYPES OF POPULATIONS OF D. MELANOGASTER.

(Vann, 1966.)

	CALCULATED	ADJUSTED
Same culture	0.35	0
Same subline	0.15	−0.20
Other subline	0.08	−0.27
Intrastrain	−0.07	−0.42
Interstrain	−0.18	−0.53
Unrelated	−0.25	−0.60

Vann also determined the relation between *H* and inversion length within "intrastrain," "interstrain," and "unrelated" populations. The inversions he studied were an unselected sample of newly induced ones so they came in a variety of lengths. Furthermore, they were scattered about the four major autosomal arms. Nevertheless, the negative value of *H* increased regularly with increasing lengths of the individual inversions. More precisely, the negative value of *H* increased linearly with approximately the cube root of inversion length. The contribution of added increments of heterozygosity to heterosis decreases as the total amount increases; this is an important point.

Still another observation by Vann on the relation between *H* and inversion length is an extremely interesting one. The linear regressions of H on inversion length (cube root of length, actually) are not parallel but converge at a single point on the Y axis (zero inversion length). The value of H that represents the conversion point is 0.35, the identical value of H observed in the "same culture" populations. The meaning of this convergence is not hard to see. No matter how genetically different two strains of flies may be, the number of gene differences included in smaller and smaller gene blocks must become less and less. The limit is reached when the block of genes has zero length; inversions of zero length can contain no genes; hence they cannot exhibit hybrid vigor. The value of H (0.35) that represents the point of convergence is the same 0.35 that represents the harm caused by radiation-induced breaks in "same culture" populations. The "same culture" populations, too, had no genetic differences which could contribute to hybrid vigor; for these populations this was true no matter how large the inverted segment may have been.

Of the fifteen paracentric inversions studied by Vann, only one was eliminated from every population; the rest were retained at least in the three types where the parental flies differed at many gene loci. Obviously, then, genes entering into heterotic combinations are located throughout the genome. The role of gene differences in contributing to hybrid vigor seems to be similar from place to place within the genome; this statement is based on (1) the relatively large contribution to heterosis made by small inversions despite the strain in which they were induced or their position in the genome and (2) the decreasing contribution made by additional amounts of heterozygosity.

Do Vann's data help to discriminate between our two models? Keep in mind that inbred lines represent—to the extent permitted by chance events—the best possible collection of alleles, in terms of viability, from the material from which they were originally derived. The contributions made to heterosis by small chromosomal segments appears to be too great, then, to be accounted for by the mere covering up of deleterious recessives. Vann measured the lengths of his inversions in units representing 1/120 of a chromosome arm. For inversions that are ten units long, the rate of increase of H per unit in the case of "unrelated" populations is approximately 2 percent. On page 215 we saw that under the standard mutation versus selection model (the model that was presented in Figure 14-1), a block of genes containing about 400 loci are needed to explain a consistent contribution of 2 percent to hybrid vigor. Forcing Vann's data to fit this model would require that 400 loci be found in less than 1 percent of a single chromosome arm. This would require that there be at least 80,000 gene loci per chromosome. Consequently, although his experiment was not designed to differentiate between the two models described above, an attempt to explain Vann's observations by means of model 1 leads to seemingly absurd conclusions.

Electrophoretic Analyses of Gene-Enzyme Systems

Early in this chapter we said that we would present a historical account of the problem of measuring the amount of hybridity or heterozygosity that occurs within populations. Of necessity, the arguments used during the "historical" era (all eight years of it) were tenuous and the experimental approaches were devious. In many respects these early procedures —both theoretical and experimental—resembled the ones required for solving the child's puzzle concerning the King who had to choose a new Prime Minister. Uncertain arguments are no longer needed. The problem can be and, in time, will be solved by direct observation.

The solution depends upon the unique relation between the amino acid sequence in a polypeptide chain and the base-pair sequence in the DNA segment (the gene) that is responsible for its synthesis. Identical base-pair sequences lead to the synthesis of polypeptide chains with identical amino acid sequences. Alter a base pair in a gene or delete several of its base pairs and you alter the structure of the synthesized protein as well.

Now, proteins can be separated from one another by forcing them to migrate through starch gel (or other supporting matrix such as paper, agar, or acrylimide gel) under the influence of a high-voltage electrical field (electrophoresis). Two proteins that differ in electrical charge, size (number of amino acid residues), configuration (a reflection, supposedly, of the order of amino acids), or amino acid content are unlikely to have identical migration rates.

Enzymes are proteins. The final location of enzymes in the gel following electrophoresis can be revealed by appropriately concocted developing solutions. Basically, the appropriate solution for a given enzyme contains a substrate upon which the enzyme can work, together with the colorless form of a dye that will combine with one of the enzymatic reaction products to form a colored, insoluble precipitate. The precipitate is deposited in the gel at the site of enzymatic activity. The result is one or more colored bands that mark the location of the enzyme within the gel (Figure 14-6).

All that remains, then, is to determine directly the proportion of heterozygous loci within individuals or of the proportion of loci at which two or more common alleles are segregating within populations. This requires the development of techniques for revealing more and more enzyme systems together with electrophoretic studies of the structural variation of these enzyme molecules. Indirect arguments and hypothetical models are no longer required.

Lewontin and Hubby (1966) have produced the first survey of gene-enzyme systems in *D. pseudoobscura.* Of eighteen loci studied, 40 percent were polymorphic in one or more geographic localities from which their

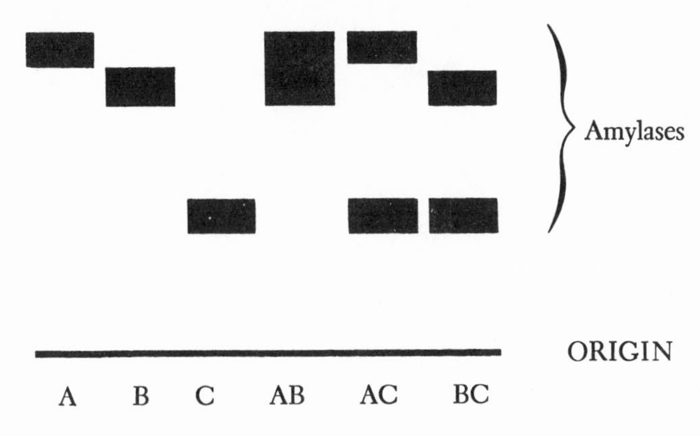

FIGURE 14-6. Six cattle amylase phenotypes to which three alleles—Am^A, Am^B, and Am^C—can give rise. (After Ashton, 1965.)

material was drawn, and 30 percent were polymorphic within any one population. MacIntyre and Wallace (unpublished) observed, too, that of eight gene-enzyme systems of *D. melanogaster* known to possess allelic variation, three or four were polymorphic in individual collections made at a number of places in Europe, South America, and the United States. A similar finding was reported for *D. ananassae* by Johnson et al. (1966).

In brief, these studies, which presumably offer minimal estimates of the frequencies of segregating loci (minimal because some amino acid substitutions cannot be detected by electrophoretic techniques, and because for the most part small numbers of individuals have been tested per population), suggest that genes at 30 percent or more of all loci are represented by two or more common alleles within any one *Drosophila* population. The study of the viability effects on their heterozygous carriers of mutations induced at random by X radiation had led to the conclusion that more than one half of all gene loci on a given chromosome are occupied by alleles that are somewhat deleterious when homozygous. It remains to be seen if the conclusion based on radiation-induced mutations has any bearing on the direct observations of gene-enzyme variation.

WHERE TO GO FROM HERE?

The data reported in this chapter, especially the beautifully clear ones of Lewontin and Hubby (1966) on electrophoretic variation in protein structure, bring us to a fork in our road. At this point it would be reasonable to pass directly to a discussion of naturally occurring polymorphisms together with an account of their evolution. At the same time, however, it is appropriate to discuss the problem raised by the notion that individuals heterozygous for different alleles at a great many loci are favored by natural selection. This problem is part of a larger one: the genetic load. We have not yet considered this important concept.

It will be convenient to devote the next few chapters to a discussion of genetic load and what Haldane (1957) called "the cost of natural selection." Following these we shall return once more to our present position and then proceed anew by discussing striking instances of polymorphism in natural populations.

15

GENETIC LOAD

> and behold I saw a man clothed with rags, standing in a certain place, with his face from his own house, a book in his hand, and a great burden upon his back.
>
> —BUNYAN, *Pilgrim's Progress*

The terms "genetic load" and "load of mutations" were first used by Muller (1950a) in his presidential address to the second annual meeting of the American Society of Human Genetics. The terms were purposely chosen for their emotional impact. They were intended to bring home the dangers of man's hereditary defects. By means of calculations similar to those presented in Chapters 9 and 10, Muller proceeded to show that recurrent mutation leads to an accumulation of deleterious and debilitating alleles in human populations. By utilizing a series of estimates concerning numbers of gene loci and mutation rates per locus, he estimated that between 10 and 50 percent of all germ cells contain at least one new mutation. From data on the partial dominance of mutant alleles he estimated further that 20 percent or more of the human population must undergo elimination each generation for genetic reasons to offset the input of deleterious genes through mutation. Finally, he calculated that each individual is heterozygous for at least eight genes (possibly for scores) which are deleterious in homozygous, if not in heterozygous, condition.

Each of us, according to Muller, tends to carry his burden unconsciously at first, then zealously even as Christian in *Pilgrim's Progress,* but finally we became weighted down by these harmful genes as our powers fail in later years.

The picture painted by Muller is a bleak one indeed. None of the obvious escape routes seems to help us throw off our burden. Relax natural selection? No, because that is the source of our difficulty; we live our "three score years and ten" only because of the intense selection that weeded out the weakest of our primitive ancestors. Improve the environment? Inexorably gene frequencies will approach equilibrium values under which mutation and elimination are balanced. Once more, No! As long as mutations occur, men are doomed to suffer genetic deaths. And since with the advent of atomic and radiological devices mutation rates promise to increase, man's genetic debilitation must increase correspondingly.

Muller's paper "Our Load of Mutations" is without doubt a classic. It was one of the first efforts to understand the impact of genetic phenomena on the well-being (or fitness) of a population. There are few papers dealing with any species that approach this one in the detail with which various aspects of the overall argument are presented and supported.

Our immediate interest in genetic load is based on problems that confront natural populations of lower organisms rather than on a consideration of man's ultimate fate. This restriction is applied in an effort to simplify the problems confronting us. These will prove to be complicated enough! If man were to be included along with other organisms, we would need to consider not only "What is" but also "What might be." This book is not the place to debate what would be an ideal existence for man. Suffice it to say that we can imagine (without claiming that it would in fact be an ideal situation) each human couple having two and only two children, all children living to an advanced age, and elderly citizens passing away peacefully at some proper moment. Presumably, such an orderly course of events could occur either in the actual absence of genetic variation (no mutation) or in the effective equality of all alleles because of advanced techniques in environmental control. In such a population there would be no genetic load, but for organisms other than man such a blissful state of affairs cannot be even a dream.

THE GENETIC LOAD: QUALITATIVE ASPECTS

The qualitative aspects of genetic load have been presented in a number of earlier chapters. The summary given in Table 3-6, for example, lists the frequencies of lethal and semilethal, subvital, supervital, and sterility genes found in three species of *Drosophila.* For all practical purposes, there were no supervitals. The other mutations listed in the table represent the genetic loads of the three species. Throughout Chapter 8, in the treatment of mutation, we discussed mutant alleles that harmed their carriers in one manner or another. The *t* alleles of the mouse (Chapter 12) are alleles that are either lethal in homozygous condition or bring about sterility of male homozygotes. Again, these are part of the genetic load of mice in a qualitative sense; this part of the load, because of the segregational impetus gained by mutant-bearing sperm, affects more individuals than one would expect from mutation rates alone.

Muller's treatment of genetic load, despite the arithmetic calculations he made, was fundamentally a qualitative one. His treatment leaves us with the following picture: Populations contain large stores of mutant genes. Most of these are deleterious in respect to survival or fitness when homozygous; many are partially dominant in respect to their harmful effects on heterozygotes. Consequently, populations do in fact carry—in the sense of John Bunyan—a genetic load, a genetic burden.

The Genetic Load: Quantitative Aspects

"Load" has proved to be an apt and expressive term for describing the array of mutant genes found within the gene pools of populations. Because of its emotional impact, "genetic load" has since 1950 almost completely displaced "concealed variability" in the literature of population genetics.

In addition to its descriptive meaning, however, genetic load has been defined in a formal mathematical sense (Crow, 1958). As defined, the genetic load of a population is the amount by which the average fitness of the population is depressed for genetic reasons below that of the genotype with maximum fitness. Symbolically, the load can be written

$$\frac{W_{max} - \overline{W}}{W_{max}}$$

where W_{max} is the fitness of the best genotype and $\overline{W}$ is the average fitness of the population. Following the convention employed throughout Chapters 9, 10, and 11, W_{max} can be assigned the (arbitrary) value, 1.00; in this case the genetic load can be given as

$$1 - \overline{W}$$

It is important to notice that the genetic load as defined here is quite different from that which we described previously. It is no longer a vast collection of mutant genes; it is no longer a term that can be described by referring to lethals, sterility genes, and morphological mutants. Instead, the genetic load is a quantity that (in theory, at any rate) can be computed accurately according to set mathematical rules. Our attention for the remainder of the chapter will be focused on the following questions:

1. What types of genetic loads are there?
2. How large are they?
3. What are the relative contributions of these different types to the total load?
4. What does the load, as defined, mean to the population itself?

Types of Genetic Loads

The Mutational Load. In Chapter 10 we calculated the equilibrium frequencies for deleterious mutant genes that arise by recurrent mutation. First we discussed completely recessive mutations and then we expanded the discussion to include mutations which are partially (to an extent, h) dominant.

A deleterious mutation that is completely recessive will reach an equilibrium frequency such that q equals $\sqrt{u/s}$. The frequency of homozygous recessives in this case equals q^2 or u/s. Since each homozygote suffers a loss of fitness equal to s, the total loss of fitness to the population—its mutational load—equals $(u/s) \times (s)$, or u.

The mutational load imposed by a recessive mutation equals u, the mutation rate, despite its effect, s, on the fitness of individuals. Lethals, semilethals, or subvitals—it makes no difference; equilibrium frequencies are such that the mutational load equals u itself. For many loci, the collective mutational load is U, the sum of all individual mutation rates.

The equilibrium frequency, q, for a deleterious mutation that is partially dominant equals u/hs. The frequency of individuals heterozygous for this allele equals $2pq$, or, since q is very low and p is very nearly 1.00, $2q$. Thus the frequency of heterozygotes at equilibrium equals $2u/hs$. Each heterozygote suffers a reduction in fitness of hs; consequently, the average fitness of the population is lowered by $(2u/hs) \times (hs)$, or $2u$. The mutational load is $2u$. Mutations that are partially dominant (in the sense that they affect the fitness of their heterozygous carriers) lower the fitness of a population by an amount equal to $2u$; for all such loci, the combined load would be $2U$, where U equals the sum of all individual mutation rates.

The advent of the "atomic age," in which radiological techniques are commonplace in both diagnostic and therapeutic medicine as well as in research and technology, makes the mutational load an important one for man. The consequences of an artificial doubling of mutation rate by radiation exposure have been depicted in Figure 15-1; doubling the mutation rate eventually doubles the mutational load.

The Segregational Load. Mutant alleles enter populations by recurrent mutation; if their heterozygous carriers have a selective advantage, however, "mutant" alleles are retained in populations by natural selection

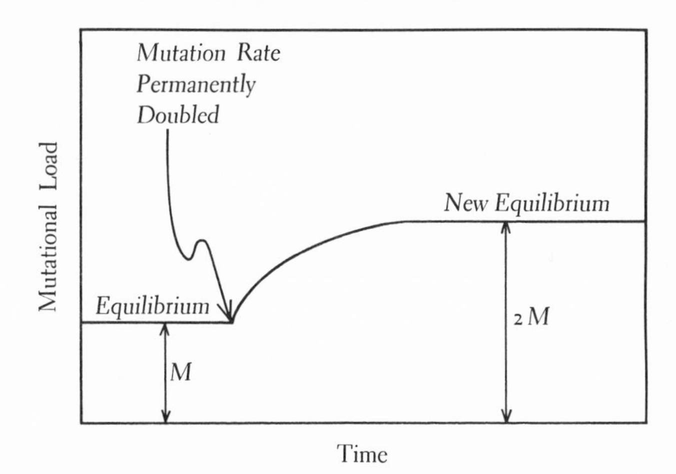

FIGURE 15-1. Effect on mutational load of permanently doubling the mutation rate. (After Crow, *Eugenics Quarterly,* Vol. 4, No. 2, 1957.)

as well. Every individual that is heterozygous for different alleles at a given locus (*Aa*, for example) and that survives as the result of selection, helps retain both alleles in the population.

The algebra of heterozygote superiority was given in Chapter 11. If the adaptive values of *AA* and *aa* individuals are $1 - s$ and $1 - t$ relative to heterozygotes (whose fitness equals 1.00), the equilibrium frequencies of *A* and *a* are $t/(s + t)$ and $s/(s + t)$, respectively. The average fitness of a population in which heterozygotes have an advantage can be calculated as follows:

GENOTYPE	FREQUENCY	FITNESS	PRODUCT
AA	$[t/(s+t)]^2$	$1 - s$	$[t/(s+t)]^2 - st^2/(s+t)^2$
Aa	$2[s/(s + t)]\,[t/(s + t)]$	1	$2[s/(s+t)][t/(s+t)]$
aa	$[s/(s+t)]^2$	$1 - t$	$[s/(s+t)]^2 - ts^2/(s+t)^2$

$$\text{Average: } 1 - st^2/(s + t)^2 - ts^2/(s + t)^2$$

$$1 - st/(s + t)$$

The segregational load (also called "balanced" load) for a given locus equals, then, $st/(s + t)$.

Balanced lethals represent a case in which heterozygotes have superior fitness; *s* and *t* in the case of balanced lethals are both equal to 1.00. The segregational load, then, equals 0.50. One half of the population dies because one half of the population is homozygous for either one lethal or the other. The segregational load in the case of the lethal found in the grouse locust (Table 11-1) can be estimated as 0.07/1.07, or about 6 to 7 percent.

The examples of balanced lethals and of the heterotic lethal of the grouse locust emphasize a very real problem posed by genetic loads. Namely, the genetic load can be too large for the population to bear. Consider man with his twenty three pairs of chromosomes. A balanced lethal system involving one of these would cause the death of one half of all zygotes; one half would survive. A second balanced lethal system on another chromosome would reduce the proportion of survivors to one fourth, the third to one eighth, the fourth to one sixteenth. Very quickly we reach fractions of surviving offspring so small that it becomes impossible for a woman to leave an average of two surviving children, the minimum for population replacement.

The problem outlined above confronts any species. A female *Drosophila* can lay from several hundred to a thousand eggs if pampered in the laboratory; I am not sure that a wild female can produce this many. The data of Lewontin and Hubby (see page 264), however, suggest that *Drosophila* populations (*D. pseudoobscura*) maintain segregating alleles at several thousand loci. If these loci are looked upon as independent (which they are not since this species has only five linkage groups), the sizes of s and t must be very small indeed in order not to overwhelm the population with the segregational load they generate. If, as an example, s and t both equaled 1 percent, the segregational load per locus would be 0.5 percent. The surviving fraction of all zygotes produced each generation would then be $(0.995)^{3000}$ or, approximately, 3 of every 1000. It seems unlikely that *Drosophila* females produce eggs in numbers sufficient to cope with this load. Consequently, we must eventually (by the end of this book, if possible) find a way out of the dilemma posed by Lewontin and Hubby's observations.

The Incompatibility Load. In a population that contains a variety of alleles, it frequently occurs that individuals of one genotype exert a detrimental effect on those of another. Weisbrot (see Table 8-15) showed that culture medium preconditioned with killed larvae of one genotype could depress the survival of larvae of other genotypes. (He showed, too, that preconditioned medium could enhance the survival of still other larvae; a similar effect was noted in oats by Jensen as shown in Table 8-14.) We also saw data (Table 9-4) which suggest that in certain populations at least, babies of blood type B are adversely affected by the anti-B antisera of their type O mothers.

Perhaps the most famous case of incompatibility in man involves alleles at the *Rh* locus. In our account we shall retain the old-fashioned terminology: *Rh* positive and *Rh* negative. *Rh*-positive persons are of genotypes *RR* and *Rr*; they have an "*Rh* substance" in their red blood cells. *Rh*-

negative persons, *rr*, lack this substance. If an *Rh*-negative person is transfused with blood from an *Rh*-positive person, he will develop anti-*Rh* antibodies.

Transfusions are not the only means by which *Rh*-positive blood can be introduced into the blood stream of an *Rh*-negative person. Any woman who is *Rh* negative (*rr*) and who is married to an *Rh*-positive (*RR* or *Rr*) man, will bear children all or one half of whom will be *Rh* positive (*Rr* only in this case). Now, a placenta is not an entirely perfect barrier between the circulatory systems of mother and fetus; blood occasionally does pass from the fetus to mother or from mother to the fetus. Consequently, blood from an *Rh*-positive fetus can induce anti-*Rh* antibodies in its *rr* mother. And the mother's blood, upon leaking into the bloodstream of this or any subsequent *Rh*-positive fetus, is capable of destroying the red blood cells of the fetus.

The incompatibility load for a given locus depends upon the gene frequencies of the alleles involved together with a number which represents the probability that the unwanted interaction will in fact occur when the proper circumstances arise. In the case of the *Rh* locus, for example, interactions based on the ABO system occur so quickly that these interactions serve to prevent the development of anti-*Rh* antibodies by *rr* mothers. If it were not that the ABO interactions are frequently lethal themselves, we could say that they protect us from *Rh* interactions.

The Substitutional Load. At any given instant, members of a species are reasonably adapted to their way of life. But not perfectly so. Neither the way of life nor the environment of a species is constant, and so the average fitness of a population is never as high as one can imagine that it should be. Genes selected at earlier times on the basis of earlier selective forces are continually being replaced by other, newly arisen mutations that more nearly meet the demands of the present—continually, but not in enormous numbers. To the extent that this turnover does occur, however, we have a calculable lowering of an ideal fitness that can be called "the substitutional load."

The Relative Sizes of Mutational and Segregational Loads

A preliminary estimate of the relative sizes of mutational and segregational loads can be made on the basis of known mutation rates on the one hand and polymorphic systems on the other. The known mutation rate for sex-linked lethals (X chromosome) in *D. melanogaster* is between two and three lethals per thousand gametes. The second chromosome, which is roughly twice as large, has a mutation rate that is also about twice as large, five lethals per thousand gametes. If we accept 0.003 as the mutation rate for the X chromosome or any of the five equal-sized chromosome arms, we can estimate that 1.5 percent of all gametes contain a newly arisen lethal. Following Muller's estimates (Figure 8-3), we can guess that detrimental mutations are about four times as frequent as lethals. The total mutation rate for lethals and detrimentals, consequently, is about 1.5 percent $\times$ 5, or 7.5 percent. Finally, if these mutations are dominant to any appreciable extent, the load they impose on a population is $2U$, or 15 percent. The mutational load on a population might be, then, approximately 15 percent, although we must admit that an estimate arrived at in this way contains many possible errors.

The mutational load computed above can be compared with the segregational or balanced load arising from selection in favor of inversion heterozygotes in *D. pseudoobscura* (see Figure 11-1 and Table 11-3). We saw (page 208) that in the four populations of *D. pseudoobscura* studied by Dobzhansky, the adaptive value of *ST/ST* individuals was 0.90, while that of *CH/CH* was 0.41. The values of s and t, then, are 0.10 and 0.59; consequently, the segregational load for this one pair of contrasting gene arrangements is (0.10)(0.59)/0.69, or 8.5 percent. This one system, then, gives rise to a genetic load that is one half the size of that resulting from mutation at all loci.

The B/A Ratio. An alternative method for estimating the relative contributions of mutation and segregation to the total genetic load has been described by Morton et al. (1956). This method depends upon a comparison of the genetic load that is estimated for a population which has been made completely homozygous and that exhibited by the population under a regime of random mating.

If an allele is maintained in a population by recurrent mutation, and if it is somewhat detrimental in heterozygous condition, its equilibrium frequency is u/hs. The genetic load imposed by such an allele on a randomly mating population is $2u$.

We can recall here that the allele described above lowers the fitness of homozygous carriers by s; this fact is unimportant in a randomly mating population because homozygotes are extremely rare. If the population were to be made homozygous by special mating techniques or by inbreeding, then the frequency of homozygous detrimentals would equal their gene frequency, u/hs. The amount by which the fitness of the homozygous population would be lowered would equal $u/hs \times s$, or u/h.

The genetic load of the randomly mating population for use in calculations of this sort has been designated by the letter A; that of the homozygous population has been designated B. The B/A ratio, which is merely a ratio of the two loads, equals in the case of mutation $(u/h) \div 2u$, or $1/2h$. Since h is generally quite small (we saw a number of estimates which were 2 percent or less), $1/2h$ should be quite large—say, 25, 50, or even more.

The B/A ratio in the case of a locus at which two alleles are maintained by the superior fitness of heterozygous individuals is calculated in a similar manner.

The value for A has been calculated above (page 271) as the segregational load—$st/(s+t)$.

The value for B is calculated for a population that has become entirely homozygous. In this case, the equilibrium frequencies of the two alleles, A and a, were $t/(s+t)$ and $s/(s+t)$; therefore, when the population is made completely homozygous, these will also be the relative frequencies of AA and aa homozygotes. The amount by which the fitness of the homozygous population is depressed relative to the maximum (1.00 of the heterozygotes) equals

$$\frac{st}{(s+t)} + \frac{ts}{(s+t)} \quad \text{or} \quad \frac{2st}{s+t}$$

In the case of a locus whose alleles are maintained by the superiority of heterozygotes, the B/A ratio would be

$$\frac{2st}{s+t} \div \frac{st}{s+t} \quad \text{or} \quad 2.00$$

In more general terms, the B/A ratio for loci whose alleles contribute to the segregation load is k, where k equals the number of alleles maintained in this way.

As a rule, one expects a low value for the B/A ratio if the segregational load contributes heavily to the genetic load, while the B/A ratio is expected to be high if the mutational load is the chief contributor. A low value, however, does not mean that the mutational load is negligible; a single locus (or an inversion polymorphism) can bring about a low B/A ratio, thereby masking the high ratio expected for the mutational load at all other loci. (This defect impairs the B/A ratio as an investigative tool because it permits but a single conclusion despite the outcome of the calculations; a discussion of other related problems can be found in Mettler et al., 1966.)

The Genetic Load: a Reexamination

In one of its two senses, as we explained earlier in this chapter, the genetic load is a rigorously defined quantity. Given certain information, a mathematically correct figure can be calculated; this figure *is* the genetic load. In this concluding section we want to look at this quantity. We want to look at it not from the point of view, Can we calculate the genetic load? Having defined it, we can surely calculate it. Rather, we want to examine genetic load as defined with the notions, What does it mean? and Is it a meaningful quantity?

The criticisms we intend to make of "genetic load" or comments concerning its interpretation will be limited in this chapter to several, elementary points. There has been considerable dissatisfaction with the concept of genetic load as it has been defined (Dobzhansky, 1964; Sanghvi, 1963; Li, 1963; but see Crow, 1963). We shall make no attempt to resolve these arguments in this chapter. Here we shall make simple comments because the problem is too large for the intellectual ammunition we have accumulated so far. We shall be satisfied for the moment in showing that the concept of genetic load is not a concept that necessarily serves as a measure of the well-being of a population.

1. *Low genetic load does not mean high fitness.* The genetic load is a measure of the proportionate difference between the average fitness of a population and that of the genotype with maximum fitness. A genetically uniform population might be wholly incapable of maintaining itself in a given environment; it may, in fact, be dwindling with every passing generation. Nevertheless, it would have no genetic load according to the rigorous definition. On the other hand, a population such as that of *D. tropicalis* described on page 204 might thrive despite its balanced lethal system, a system whose load amounts to 0.50. Thus genetic load by itself tells us very little about a population's ability to continue from one generation to the next.

2. *If the genetic load can be relegated to individuals already destined to die without reproducing, it is of no importance to a population.* Through territoriality or other means, populations of many animals are rigorously limited in respect to the total number of breeding pairs in a given locality. The individuals produced in excess of this virtually constant number become "wanderers" and eventually fall victim to predation or otherwise die without leaving offspring. At best, wanderers may set up their territories in submarginal areas where the probability of survival and reproduction is extremely low.

We can imagine two populations of a given animal species of which 10 percent of the offspring survive as parents with territories while 90 percent are excess and die before reproducing. In one of these we shall imagine that the gentic load is 20 percent; in the other, mutation has been miraculously eliminated so a genetic load does not exist. Now, if the genetic load of the one is included within the 90 percent doomed to die, mutation rate is counterbalanced by selection but, nevertheless, no harm befalls the population on account of the genetic load. A geneticist with a talent for bookkeeping would make his entries for these two populations as follows: Population 1; 10 percent survival, 20 percent genetic death, 70 percent environmental death. Population 2; 10 percent survival, 90 percent environmental death.

We might note further that, if either population has any possibility of adapting successfully to presently submarginal territories (say, by developing a more cryptic coat color), the population with the genetic load is the one more likely to succeed.

3. *The largest part of the segregational load falls to the "normal" allele.* If a heterozygote is favored by selection and if one homozygote has a fitness very nearly the same as that of the heterozygote, the bulk of the segregational load can be blamed on this relatively normal homozygote.

Consider a population in which fitnesses are as follows: *AA*, 0.99; *Aa*, 1.00; and *aa*, 0. The frequencies of *A* and *a* will be very nearly 0.99 and 0.01 (see page 205). The portion of the segregational load contributed by the lethal gene equals sq^2, or 0.0001. That contributed by the "normal" allele, *A*, is tp^2, or 0.0098. The normal gene contributes nearly 100 times as much to the total segregational load as does the lethal mutation.

Hiraizumi and Crow (1960) used the above calculation as evidence that the action of natural selection will be aimed primarily at making "normal" alleles even more normal, even more like the heterozygotes. The logic behind this argument contradicts that of Muller (1950a) who predicted that the dominance of slightly detrimental genes should be greater than that of more deleterious ones. Muller's argument states that slight differences in fitness offer no basis for effective selection.

4. *If the value of* h *for a particular allele is a variable, the* B/A *ratio leads to meaningless conclusions.* Ordinarily, when devising indices or when carrying out a calculation involving estimated values, one avoids using as the denominator in a ratio any number that may take on extremely small values; slight variations in estimations of the denominator could lead to enormous variations in the calculated ratio. In the discussion of migration (page 82), for example, the MN blood group system was rejected as a suitable basis for estimating migration of genes from the American white population into that of the American Negro. The usefulness of these genes was compromised by the similarity of gene frequencies in the two races. The denominator used in estimating migration consists of the difference of these frequencies; similar frequencies could easily lead to zero differences by accident and, through these zero differences, to absurd estimations of migration rates.

It seems to me that the use of h in the denominator of the B/A ratio can at times also lead to an absurd or erroneous conclusion. Presumably, h for a given allele within a heterogeneous population is a variable. The degree of dominance exhibited by an allele over its alternative form is certainly influenced by the remainder of the genotype. Presumably, too, the closer the mean value of h approaches zero from the positive side (see Figure 15-2), the greater the proportion of all combinations in which h is actually negative (confers an advantage on its heterozygous carrier). This point is brought out by the family of curves in Figure 15-2. As long as the value $h = 0$ is not a special value representing a discontinuity in the horizontal axis, the fraction of the total area that lies to the

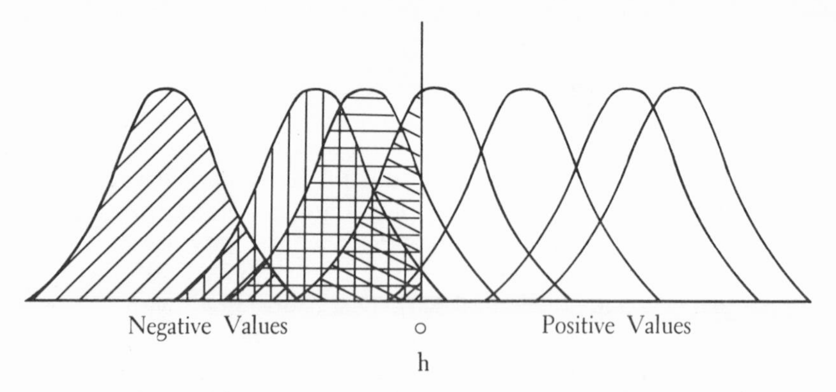

FIGURE 15-2. Relation between the recessivity of an allele in respect to fitness and the proportion of background genotypes in which it enhances fitness.

left of the zero point increases as the mean value of h approaches zero.

The B/A ratio calculated from actual data is interpreted in a manner directly opposed to the argument presented above. Large values, as we saw, are interpreted as evidence that the genetic load is largely mutational; small values, that the segregational load may be important. Thus alleles for which h is nearly zero constitute the best evidence in the use of the B/A ratio that the segregational load is unimportant. We have seen immediately above that, on the contrary, alleles whose mean h is very nearly zero are those which most often have negative values of h, values that reflect the superiority of heterozygotes. The segregational load depends upon the superiority of heterozygotes.

5. *The genetic load concept is a static concept.* The concept of the genetic load is a static one unable to embrace change and, consequently, it leaves the impression that lasting changes do not occur in populations. The lack of escape routes by which man might lighten his genetic load, a lack upon which we commented earlier, is an artifact resulting from an attempt to describe *change* in terms suitable only for *static* (equilibrium) conditions. Figure 15-3 illustrates a specific example. According to the figure, a permanent environmental improvement fails to lead to a lasting reduction in the mutational load of a population. To the extent that a permanent environmental improvement leads to more suitable surroundings for a species and to a more thriving population, the ups and downs of the mutational load are unimportant indeed.

Similarly, the concept of genetic load cannot deal realistically with the origin of favorable mutations. The maximum genetic load suffered by a homogeneous population occurs at the moment a beneficial mutation arises, one that confers high fitness on its carriers. At this moment, almost the entire population is composed of what must be described as old-fashioned and now inferior genotypes (Figure 15-4). When the new mutation has spread through the population, the population has not improved according to the genetic load concept, it has merely recovered.

A Concluding Comment

It is wise, I believe, to remember that the concept of genetic load in its original descriptive sense was devised with man in mind; Muller (1950a) spoke and wrote of *our* load of mutations.

There is no doubt that many of man's ills have a genetic basis. These ills do indeed constitute a burden. They are a tremendous burden to afflicted persons. They are a burden to parents of afflicted children. They are a burden to us all in the sense that they constitute a major public health problem.

In respect to the above matters, "load" is an excellent word. I do not believe, however, that it is a term that deserved the rigorous mathematical treatment it has been given. It is my opinion, for example, that sterility and lethality are scarcely equivalent in human terms. Nor do I think that all forms of lethality are equivalent. I cannot conceive, either, of the equation of the death of one person with the slight physical impairment of many. I do not find the term, as defined for mathematical purposes, to be particularly useful in pondering the status of a given

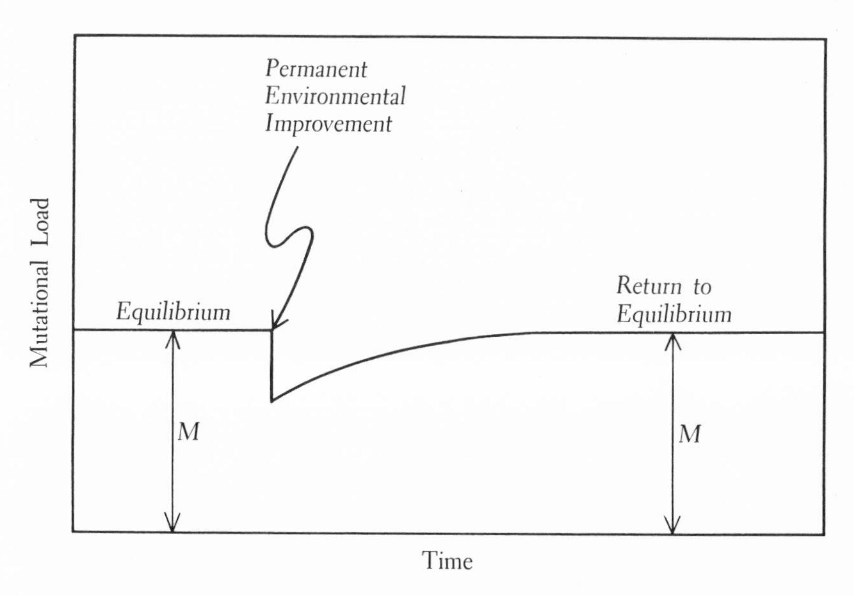

FIGURE 15-3. Effect of a permanent improvement in the environment upon mutational load. (After Crow, *Eugenics Quarterly*, Vol. 4, No. 2, 1957.)

population at a given moment nor in predicting its future. I am inclined to look upon spontaneous mutation as one of the lesser problems faced by populations—human or otherwise; this, of course, is a personal opinion not necessarily shared by others.

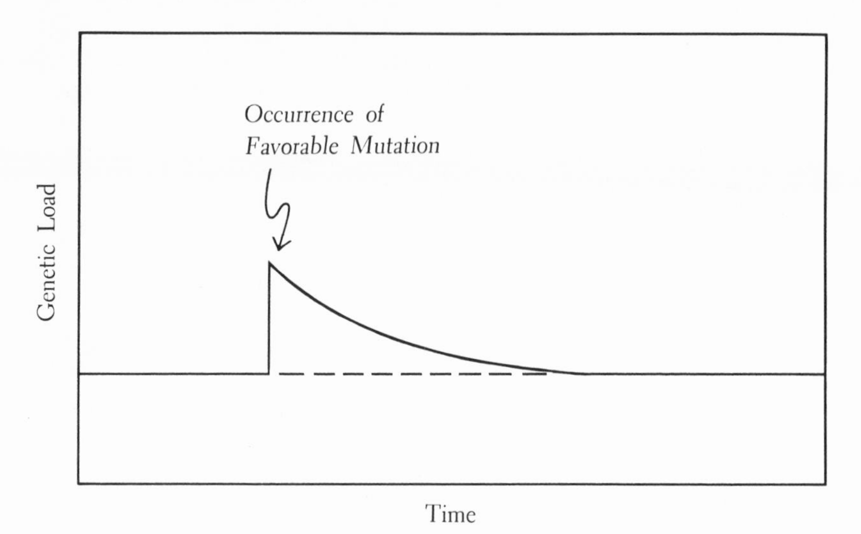

FIGURE 15-4. Effect of a new, favorable mutation upon mutational load.

16

ADAPTIVE "LANDSCAPES"

An adaptive "landscape" is a model in which the average fitness of a population is represented by the altitude of the terrain at a given point. The extent of the landscape is provided by the variety of *all possible* genetic compositions the population might have. The most interesting composition, of course, is the one that the population actually does have. In the following pages we shall attempt to answer two questions: Where on a landscape does a population actually sit? Why does it pick this particular spot rather than some other?

Two-Dimensional Landscapes

The simplest landscapes are those that represent the change in the average fitness of a population brought about by the change in the relative frequencies of two alleles at a single locus. The two examples that are described here involve (1) a recessive lethal allele and (2) a pair of alleles whose heterozygotes have higher fitness than either homozygote. Both examples have been discussed in earlier chapters in respect to equilibrium frequencies and changes in gene frequencies. The emphasis is now placed on the average fitness of the population—on the contours of a very simple landscape.

The Effect of a Recessive Lethal on Average Fitness. Assume that two alleles, A and a, exist at a given locus and that a is lethal when homozygous. If the relative frequencies of these two alleles are p and q, the average fitness of the population equals $1 - q^2$. (In the case of a lethal mutation, s equals 1.00.)

By assigning various frequencies from 0 to 1.00 to the lethal allele, we see (Figure 16-1) that the average fitness of the population takes on values from 1.00 (no lethals present) to 0 (no normal alleles). Frequencies of lethals greater than 50 percent are not realistic because the surviving members of a population can at best be heterozygous for a lethal gene.

We have seen in an earlier chapter (page 164) that lethals are gradually eliminated from randomly mating populations. The change in gene frequency, Δq, equals $-q^2/(1 + q)$; the frequencies of lethals remaining in the successive generations 0, 1, 2, 3, . . . are q, $q/(1 + q)$, $q/(1 + 2q)$, $q/(1 + 3q)$, . . . As the frequency of the lethal gene decreases, the average fitness of the population increases. Looking upon the curve in Figure 16-1 as a "landscape" with no depth, we see that the population climbs higher and higher up the hill on successive generations. Selection acts in this case to maximize the average fitness of the population; with no further mutations, the population would eventually climb to fitness 1.00.

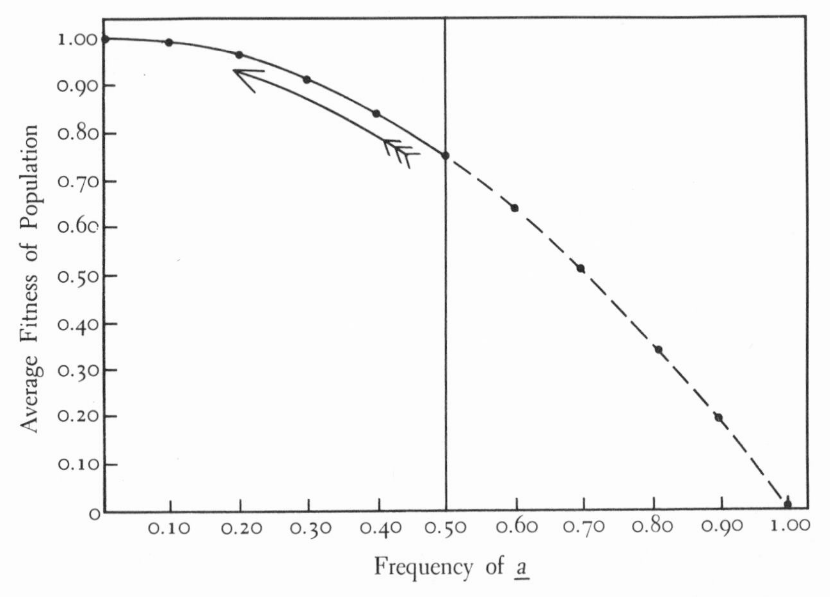

FIGURE 16-1. Relation between the frequency of a recessive lethal (a) and the average fitness of a population. Frequencies beyond 0.50 have no meaning for actual populations.

Heterozygote Superiority and Average Fitness. In an earlier chapter we considered the case in which two alleles, *A* and *a*, produced heterozygotes of high fitness, higher than that of either homozygote. The scheme used to examine the consequences of this type of selection was the following:

GENOTYPE	*AA*	*Aa*	*aa*
Frequency	p^2	$2pq$	q^2
Fitness	$1-s$	1	$1-t$

Average fitness: $1 - sp^2 - tq^2$

We proceeded (page 205) to calculate p and q for each of two successive generations, and then showed that these frequencies remain constant when p equals $t/(s+t)$ and q equals $s/(s+t)$.

The earlier calculation was based solely upon changes in gene frequency. Now we propose to calculate the change in fitness with respect to gene frequency and to determine the value of p, for example, that leads to the highest fitness. We can make this calculation by taking advantage of some elementary differential calculus:

$$\begin{aligned} \overline{W} &= 1 - sp^2 - tq^2 \\ &= 1 - sp^2 - t + 2tp - tp^2 \\ d\overline{W}/dp &= -2sp + 2t - 2tp \end{aligned}$$

When

$$\begin{aligned} d\overline{W}/dp &= 0, \\ p &= \frac{t}{s+t} \\ q &= \frac{s}{s+t} \end{aligned}$$

The second derivative, $d^2\overline{W}/dp^2$, is negative; consequently, the average fitness of the population is at a maximum when the frequencies of *A* and *a* have reached their equilibrium frequencies.

Once more we find that selection maximizes the average fitness of a population (Figure 16-2) as it goes about the business of establishing equilibrium gene frequencies. Once more, as the arrows in Figure 16-2 suggest, the population moves uphill on the adaptive landscape until it sits at the very top.

A Three-Dimensional Landscape

A population that possesses alternative alleles at a single locus only does not exist; literally thousands of loci are occupied by two or more alleles according to the estimate made by Lewontin and Hubby (1966). In making the calculations upon which Figure 16-1 and 16-2 are based, all

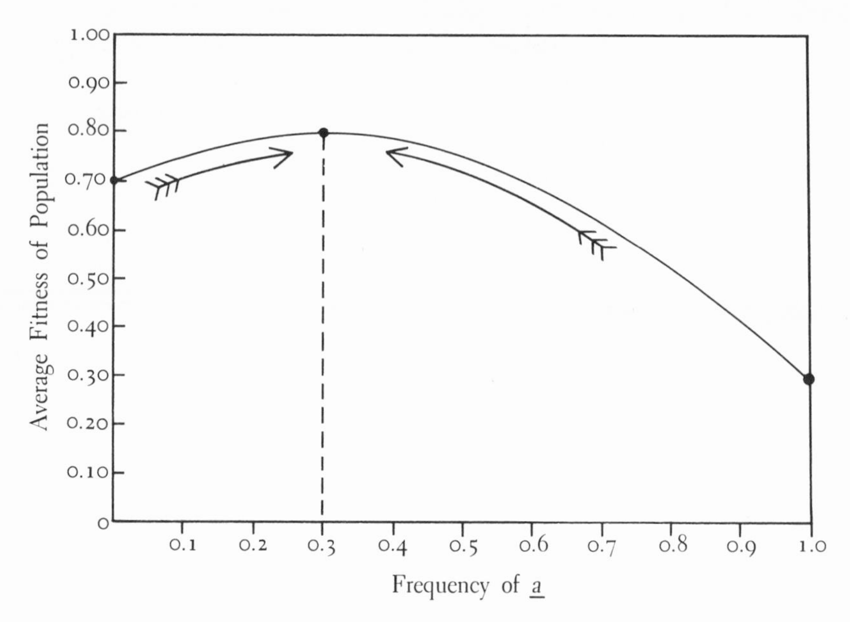

FIGURE 16-2. Relation between the frequency of a heterotic allele (*AA* < *Aa* > *aa*) and the average fitness of a population.

variation except for the locus under discussion was forgotten. The "background" variation can be ignored because one assumes that it affects the three genotypes—*AA*, *Aa*, and *aa*—equally.

We are now about to take a giant stride forward. In this section we shall discuss the adaptive landscape formed by the interaction of two pairs of alleles—*A* and *a*, *B* and *b*. If one considers that there are thousands of segregating loci, it seems ridiculous to refer to the simultaneous treatment of two loci as a "giant stride." But it is, indeed; for without the aid of electronic computers to handle the massive computations involved, we would probably be content with the analysis of the simpler, one-locus cases.

Lewontin and White (1960) completed the first thorough analysis of a two-locus polymorphism. They analyzed the joint effect of two independent polymorphisms on the average fitness of grasshopper populations. The systems involve two chromosomes (the **CD** and **EF** chromosomes) of *Moraba scurra*, an Australian grasshopper. Each of these chromosomes is found in two cytologically distinguishable forms: a V-shaped, metacentric form and a rod-shaped, telocentric form. Presumably, a V is converted into a rod by means of a pericentric inversion. The V-shaped form in each case is regarded as the standard (*ST*) form. The rod-shaped form is designated Blundell (*BL*) in the case of the **CD** chromosome, Tidbinbilla (*TD*) in the case of **EF**.

In Table 4-5 we listed the observed distribution of the zygotic constitutions of 584 individuals in respect to both **CD** and **EF** chromosomes; altogether there are nine possible genotypes. The point made during our earlier discussion of Table 4-5 was that the Hardy-Weinberg equilibrium can be expanded to account for the joint distribution of two or more pairs of alleles. The sample of 584 individuals was adequate to illustrate that elementary point; the observed deviations from the expected distribution based on the expanded Hardy-Weinberg equilibrium (that is, on the product of two separate Hardy-Weinberg distributions) were not significant. This particular sample of 584 individuals was only one of many that Lewontin and White had at their disposal; the distribution of individuals among the nine genotypes in the combined data does differ significantly from expectation. The observed deviations have been used in turn to estimate the relative fitnesses of the different types of grasshoppers; these fitnesses are listed for two different geographic localities in Table 16-1.

TABLE 16-1

ESTIMATED RELATIVE FITNESSES OF NINE GENOTYPES IN THE GRASSHOPPER, MORABA SCURRA.

CD and **EF** represent two nonhomologous chromosome pairs; *ST* represents the standard gene arrangement; *BL* (Blundell) and *TD* (Tidbinbilla) are names given to telocentric forms of these chromosomes (see Table 4-5). (Lewontin and White, 1960.)

	CHROMOSOME **EF**	CHROMOSOME **CD** *ST/ST*	*ST/BL*	*BL/BL*
Wombat	*ST/ST*	1.002	1.000	0.927
Royalla		0.842	1.000	0.808
Wombat	*ST/TD*	0.646	0.849	1.044
Royalla		0.636	0.997	0.974
Wombat	*TD/TD*	0.000	1.054	0.626
Royalla		0.393	0.682	0.916

With the estimates of fitness for each of the nine genotypes provided by the observations themselves, it is now possible to construct the adaptive landscape these fitnesses control. In doing this it is necessary to remember that for a given frequency of *BL* and of *TD*, the relative frequencies of all nine genotypes are fixed. The average fitness of a population with this gene composition is given by the sum of the products of relative frequencies times fitnesses. In Table 16-2 a sample calculation has been made to show the average fitness of a population in which the frequency of *BL* is 20 percent and that of *TD* is 70 percent.

TABLE 16-2

CALCULATION OF A SINGLE POINT ON THE ADAPTIVE LANDSCAPE FORMED BY THE RELATIVE FITNESSES LISTED IN TABLE 16-1 FOR M. SCURRA OF ROYALLA.

The point chosen is that for which the frequency of *BL* is 0.20 and that of *TD* is 0.70.

GENOTYPE		FREQUENCY			
CD	EF	CD	EF	FITNESS	PRODUCT
ST/ST	*ST/ST*	0.64	0.09	0.842	0.0485
ST/ST	*ST/TD*	0.64	0.42	0.636	0.1710
ST/ST	*TD/TD*	0.64	0.49	0.393	0.1232
ST/BL	*ST/ST*	0.32	0.09	1.000	0.0288
ST/BL	*ST/TD*	0.32	0.42	0.997	0.1340
ST/BL	*TD/TD*	0.32	0.49	0.682	0.1069
BL/BL	*ST/ST*	0.04	0.09	0.808	0.0029
BL/BL	*ST/TD*	0.04	0.42	0.974	0.0164
BL/BL	*TD/TD*	0.04	0.49	0.916	0.0180
				Sum (av. fitness) =	0.6497

The adaptive landscape of a two-locus system cannot be represented as a simple curve for it is not a line; it is a surface. It is a surface because the gene frequencies for each of the two systems can vary from 0 to 100 percent; thus two dimensions are required for the representation of all possible combinations of gene frequencies. As a result, the average fitness must be represented as a height of a rod, for example, set upon the proper spot in the gene-frequency field (Figure 16-3). And now one can see why there is so much labor involved in computing this surface. The frequency of *BL* can take any value from 0 to 1.00 and so can that of *TD*. For each pair of values, the frequencies of the nine genotypes must be calculated. And, from the known fitnesses of the different genotypes one must calculate an average fitness for the population. Lewontin and White were satisfied to make the calculation for the 441 combinations obtained by letting each gene frequency increase from 0 to 100 percent by increments of 5 percent; an idealized landscape based on their results, but drawn in contour rather than three-dimensional form, is represented in Figure 16-4.

An examination of the idealized landscape discovered by Lewontin and White reveals that it is saddle-shaped; it is not a simple mound as one might have expected. The saddle-shaped topography covers, remember, all possible frequencies of *BL* and *TD*. The populations of

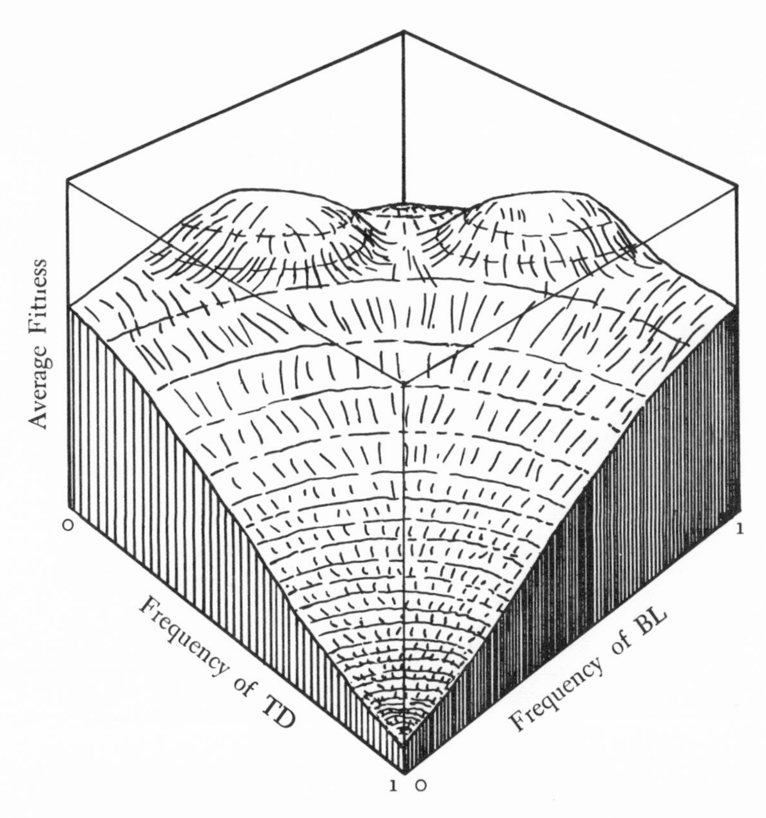

FIGURE 16-3. Cutaway model of the three-dimensional landscape associated with different frequencies of *BL* and *TD* chromosomes in populations of *Moraba scurra.*

grasshoppers sampled in Australia, however, had a particular set of frequencies each; in the 1959 Wombat population, for example, the frequency of *BL* was 87.5 percent and that of *TD* was 14.0 percent. Where on the adaptive landscape does this population fall? Surprisingly enough, each population falls in the seat of the saddle. Although the seat of a riding saddle is the high point of reference to the two sides of a horse (line CD in Figure 16-4), it is the low point relative to horn and back (line AB in the figure). Why has selection not maximized the average fitnesses of these populations? Why have the populations not continued to climb up the slopes of the landscape as we learned earlier that they can?

The answers to the questions raised above have probably been supplied by Allard and Wehrhahn (1964). The grasshoppers studied by Lewontin and White are extremely sluggish and sedentary. They live in small grassy spots (chiefly cemeteries) from which there is little

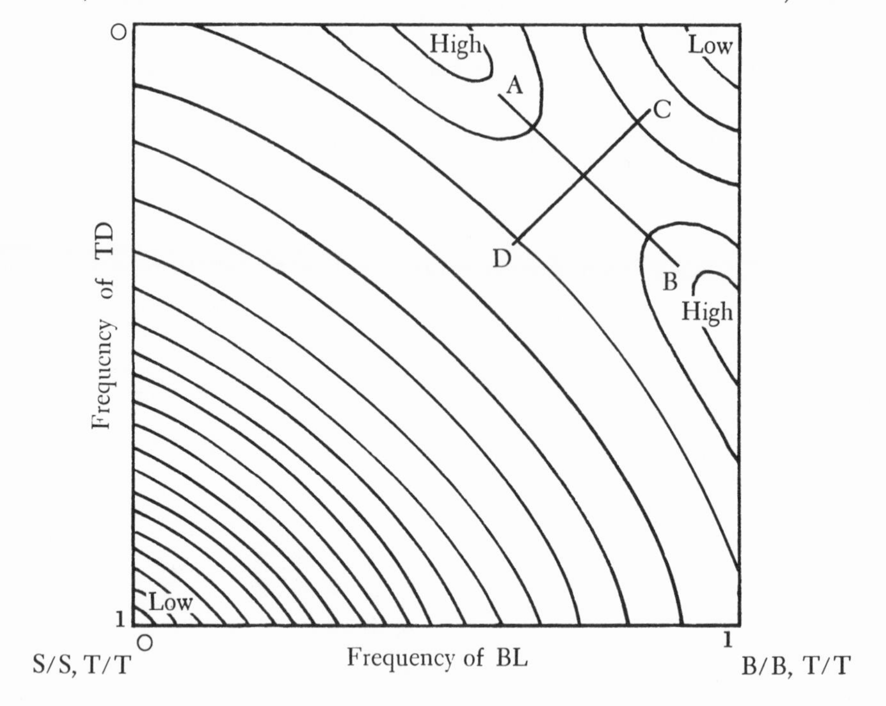

FIGURE 16-4. Diagrammatic representation of a typical contour map representing the adaptive landscape associated with different frequencies of *BL* and *TD* chromosomes in populations of *M. scurra*. (After Lewontin and White, 1960.)

incentive to leave. Allard and Wehrhahn suggest that there may be an appreciable amount of inbreeding in these populations. Inbreeding, it will be recalled, results in homozygosis. If F is a measure of inbreeding, only $1 - F$ of a population behaves as the Hardy-Weinberg equilibrium predicts (see Figure 6-3). We shall not reproduce the new calculations in this book, but we shall outline their nature:

1. By assuming a given level of inbreeding, F, the expected zygotic frequencies for the **CD** and **EF** chromosomes can be represented as follows: For **CD,**

ST/ST	ST/BL	BL/BL
$(1-F)p^2 + Fp$	$(1-F)2pq$	$(1-F)q^2 + Fq$

where the frequencies of ST and BL are p and q; and for **EF,**

ST/ST	ST/TD	TD/TD
$(1-F)r^2 + Fr$	$(1-F)2rs$	$(1-F)s^2 + Fs$

where the frequencies of ST and TD are r and s.

2. The products of the above frequencies (times the number of individuals sampled) lead to the expected numbers for each of the nine genotypes in each of the available samples of grasshoppers.
3. The recalculation of expected numbers leads to revised estimates of fitness because the deviations of observed from expected numbers upon which these estimates are based will have been changed.
4. The individual points of the adaptive landscape, calculated as shown in Table 16-2, will be altered. The alterations will be brought about by new zygotic frequencies and by new estimates of fitness.

The altered landscapes that emerge as the result of including inbreeding in the calculations resemble those shown in Figure 16-5. The figure shows the gradual transformation of the saddle-shaped landscape observed for the collections from Wombat into a simple mound. The transformation is brought about by substituting 0.05, 0.10, and 0.25 as values for F in place of 0, the value originally assumed. When F equals 0.10, the population rests virtually at the peak of a well-defined "hill." Our experiences with the populations illustrated in Figures 16-1 and 16-2 lead us to expect that a population should come to rest at the peak in this manner.

Frequency-Dependent Fitness

In their original analysis, when they saw that the populations lay at saddle points within their calculated adaptive landscapes, Lewontin and White made an excellent point: Populations need not come to rest on the highest point of a theoretical landscape if the fitness of a genotype depends upon its frequency. They illustrated this statement with a well-chosen example.

Let the frequencies of the alleles A and a be p and q, and let the fitnesses of the three types of zygotes be as follows:

AA	Aa	aa
1	1	$1.5 - q$

Mere examination of the fitnesses reveals that the three zygotic types will have equal fitnesses (and, therefore, that the gene frequencies will be at equilibrium) when both p and q equal 0.5. The average fitness of the population is 1.00 at this time. However, when q has values lower than 0.5, the fitness of the population is greater than 1.0 because that of aa individuals is greater than 1.0. Nevertheless, selection within the population would increase the frequency of a from low values to 0.5 even though this change moves the population "downhill" and brings it to rest part way down the slope of the adaptive landscape.

The dependence of fitness on gene frequency is not an abstract theoretical possibility. We saw examples of this dependence in Chapter 7

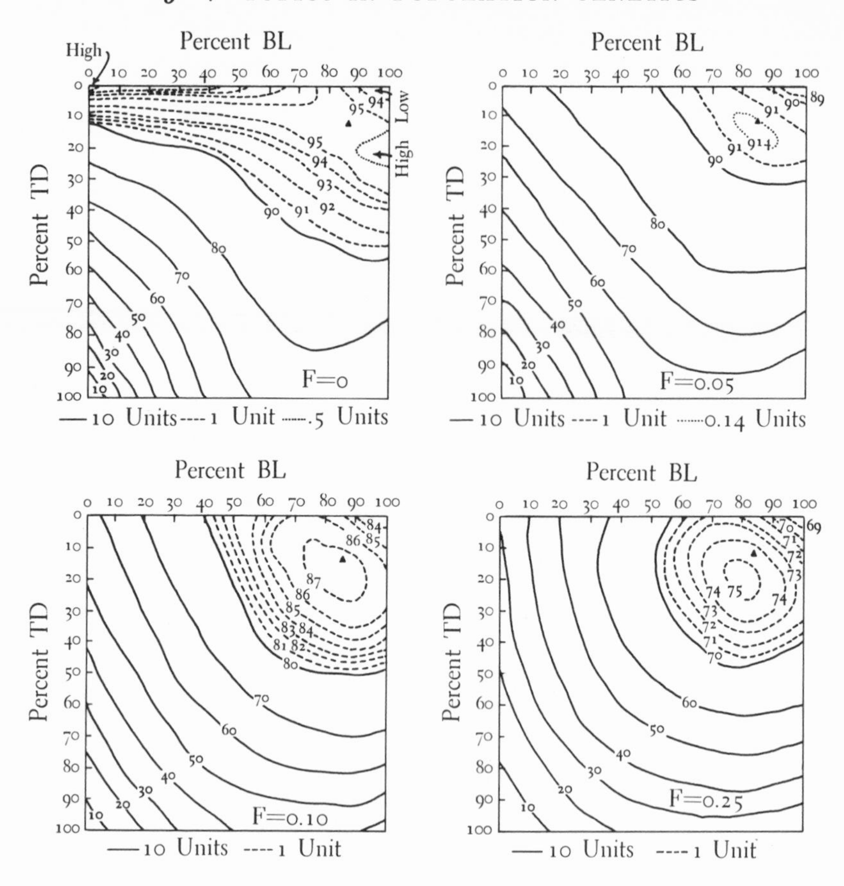

FIGURE 16-5. Contour maps showing the adaptive landscapes for *M. scurra* under different levels of inbreeding. (After Allard and Wehrhahn, 1964.)

where we discussed mating behavior; we saw further examples in Chapter 8 in a discussion of the beneficial interactions that occur at times between individuals of different genotypes. Gene frequencies are also involved in determining the "incompatibility load," one component of genetic load (page 272).

LANDSCAPES INVOLVING MANY LOCI

The discussion of the fitness of populations as determined by two segregating loci has exhausted our ability to use realistic illustrations in representing the relationship between fitness and gene frequencies. We might stretch our artistic and interpretive abilities by showing the frequencies of alleles at each of three segregating loci on the individual axes of a three-dimensional drawing. In this case, however, fitness would have to be represented by the density of stippling within the drawing; populations might move, for example, toward regions of dense stippling.

For complicated landscapes, it seems wise to adopt a technique that has long been used by Sewall Wright (see, for example, Wright, 1932): the contour map containing many peaks and valleys (Figure 16-6). A map of this sort must be used with caution, of course. It is at best an attempt to represent in diagrammatic form a hopelessly complex situation. Lewontin and Hubby (1966) have estimated that 30 percent of all gene loci *in D. pseudoobscura* are occupied by two or more common alleles. If the total number of loci in *Drosophila* equals 10,000 (a number only twice as large as that of cytologically visible bands in salivary chromosomes), this means 3000 segregating loci and the possibility of forming 3^{3000} or more different genotypes. Peaks and valleys representing high and low average fitnesses of populations containing various frequencies of these alleles cannot be represented (or even thought about) in a realistic manner.

Ford (1964, pp. 33-34) finds little merit in the landscape analogy of adaptive values. He points out that peaks and valleys are not permanent; peaks subside while valleys become elevated. Therefore, in

FIGURE 16-6. Adaptive landscape with several hills and valleys.

Ford's view, the analogy, if it is to be useful, must more closely resemble a seascape than a landscape. Ford's objections can be traced to two sources: (1) his view of evolutionary change as *rapid* change and (2) his attempt to take a rigorous rather than an imaginative view of an admitted analogy.

Living organisms evolve but that does not render useless what Haldane (1954b) has called the "statics" of evolution. Indeed, the mountain ranges, hills, and valleys of geology are also evolving continuously, but this does not mean that mapmaking is a useless profession. Some features of a terrain change much more rapidly than others. One can regard the contours in Figure 16-6, for example, as an illustration of the adaptive peaks and clusters of adaptive peaks occupied by species and genera of the order primates or of some other order of mammals; such peaks will remain virtually unchanged for millenia. On the other hand, one may regard these peaks as separate genetic solutions to malaria hit upon by various human populations—Hb^S of sickle cell anemia in one, Hb^C of thalassemia in another, or the gene responsible for favism in still another. Populations hitting upon these solutions climb one or another of the various peaks; by climbing one peak, it may happen that the possibility of climbing a second simultaneously is precluded because of genetic incompatibilities (Allison, 1961). Nevertheless, peaks at this level are relatively fluid. Genetic changes that cause the merging of two peaks into a single one may arise; on the other hand, medical treatment may within a decade cause a valley to rise and peaks to subside. The smaller the genetic basis for the peak under consideration, the more transient is its nature.

One need not regard a contour map such as that shown in Figure 16-6 as if it represents something permanent. It is an analogy whose meaning changes even as one studies it. The events that are being represented are multidimensional and hence incapable of illustration. So it must not be discouraging to find that a population, having climbed to the top of a peak, finds itself exposed to an entirely new dimension. As an example, we can cite the establishment of a stable polymorphism through the selective advantage of heterozygous individuals; once the population avails itself of this peak, vistas involving the evolution of dominance are opened up that were nonexistent previously. We must look at landscapes of this nature with a proper amount of imagination or, as Ford has said, they will indeed mislead us.

Concluding Remarks

The hypothetical peaks and valleys described in this chapter form landscapes that represent the *average* fitnesses of populations under equilibrium conditions. The surfaces that we have represented by contours or otherwise are not fitnesses of individuals or of clusters of individuals artificially arrived at. For example, the heterozygotes in the case illustrated in Figure 16-2 have a fitness of 1.00; the average of an equilibrium population has a maximum fitness of $1 - st/(s + t)$. However, if *AA* males were to be mated to *aa* females, any number of heterozygotes could be obtained all of whom would have fitness equal to 1.00. These hybrids would not represent an equilibrium population; left to themselves, they would form an equilibrium population (F_2) containing the two types of homozygotes as well as the heterozygotes. The average fitness of an F_2 population is not necessarily the highest possible. In a practical sense, we see here why commercial seed companies produce large quantities of hybrid corn for planting, and why the corn growers do not use their own harvest for seed purposes. Similarly, we see why the horse and the ass (each of which would have its peak in a contour map such as that shown in Figure 16-6) can be crossed to obtain a useful hybrid, the mule. The mule is not part of the landscape. Furthermore, we can be assured that an equilibrium population containing equal frequencies of genes drawn from the horse and ass populations would lie in a rather deep valley; of that there is no doubt.

17

BALANCED POLYMORPHISM

In an earlier chapter we described the outcome of selection that favors heterozygotes. We mentioned various studies—but notably that of Lewontin and Hubby (1966)—on the variation found in gene-enzyme systems of *Drosophila*. The techniques available for the study of this variation, techniques that enable us to examine gene products made by individual loci, promise to extend immeasurably our observations on genetic variation. They promise, in fact, to make earlier techniques obsolete. But they have not yet done so. And so there still exist valid reasons for describing cases of polymorphisms known from yesteryears. Some of these are cases that Fisher (1930) had in mind when he showed that the selective advantage of heterozygous individuals led to the retention of both alleles in a randomly mating population.

There is much variation in natural populations. A great deal of it has a genetic basis. Much of it, however, is environmental in origin. For example, isolated strains of *D. melanogaster* differ from one another in mean number of sternopleural hairs. Similarly, individual flies within a single strain also differ. Furthermore, the numbers of hairs on the two sides of the same individual frequently differ as well. If one sums the absolute difference observed between the two sides of a sample of flies (their "asymmetry") and divides this sum by the total number of hairs counted, the result is about 5 percent. Thus the difference between the numbers of bristles on the two sides of a fly is roughly 10 percent that of either side alone. The observed asymmetry presumably reflects an environmental influence on bristle number, because the two sides of a fly do not differ in a systematic way and yet they are genetically identical. The differences between individuals, on the other hand, are not entirely the result of environmental variation because artificial selection is effective in changing the mean number of bristles per fly. Selection progresses in either a plus or minus direction depending upon the choice of parents each generation.

Bristle number, body size, wing length, and many other morphological traits of flies show continuous variation part of which has a genetic basis, part environmental. Physiological traits such as resistance to poisons, high salt concentrations, and other noxious conditions may also exhibit continuous variation. Genetic studies have revealed the presence of genes responsible for such traits on every chromosome or chromosome segment subjected to analysis.

Not all genetic variation is continuous. Classical genetics is founded on the study of discontinuous variation arising from single-gene substitutions. Systems of this sort were used in illustrating the Hardy-Weinberg equilibrium. In human populations such systems are the ABO, the MN, and the Rh blood group systems, the ability or inability to taste PTC, the various hemoglobin disorders such as sickle cell anemia and thalassemia, and various biochemical disorders such as favism. Naturally occurring, discontinuous variations of this sort which have relatively simple genetic bases are known as genetic polymorphisms or, for purposes of our discussion, balanced polymorphisms (see Figure 17-1).

Genetic polymorphism was defined by Ford (1940) as the occurrence together in one habitat of two or more forms of a species in such proportions that the rarest cannot be accounted for by recurrent mutation. How then can one account for the presence of these different alleles? Three alternatives come to mind:

1. *Chance*. An "obvious" explanation for the existence within a single population of several common alleles at a given locus is that these alleles are selectively neutral. This has been the explanation advanced in case after case—including that of the various gene arrangements in *D. pseudoobscura* (see Figure 11-1); and for each such case examined in detail, the obvious "explanation" has proved to be wrong. Shaw (1965) has suggested that some explanation other than selection will best account for the high frequency of gene-enzyme variants; one argument he uses is precisely that used by Dobzhansky and Epling (1944) in referring to the selective neutrality of gene arrangements—namely, we have not yet seen a selective difference. Perhaps we have not yet looked carefully enough! At any rate, the most likely result by far of selective neutrality is the loss from the population of all alleles but one for each locus.

2. *Selection in favor of heterozygotes*. Under this heading, I include "marginal overdominance" (page 213). This covers the shifting selection

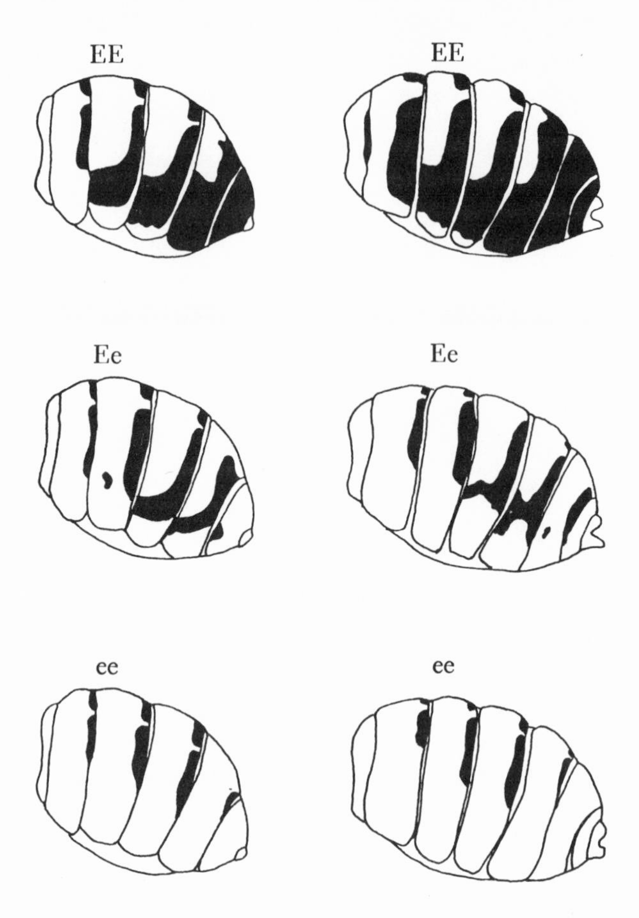

FIGURE 17-1. Phenotypes of dark (*EE*), intermediate (*Ee*), and light (*ee*) color types in *D. polymorpha;* male, left, female, right. (After daCunha, 1949.)

coefficients (even in the absence of overdominance in the strict sense) encountered in a patchy environment or in brief intervals of time. This mechanism by which diverse alleles are retained in a population is, of course, that described in Chapter 11; the selective advantage of heterozygotes results in the retention of two or more alleles in a population.

3. *Selection in favor of rare alleles.* This mechanism for retaining genes in populations was mentioned in the earlier discussion on mating behavior; the data of Ehrman et al. (1965) clearly show that at times this can be a powerful selective influence. Furthermore, we described a mathematical model in Chapter 16 (page 289) that demonstrates the stability of equilibria set up under this type of selection. Teissier (1954a, 1954b) argued that selection which favors rare alleles will perpetuate the

genetic variation within populations; more recently, Kojima and Yarbrough (1967) have obtained evidence that some alleles responsible for electrophoretic variation in *Drosophila* are subject to frequency-dependent selection. It is interesting to consider this type of selection in terms of marginal overdominance as described on page 213; in this case the "conditions" described in that discussion become "gene frequencies."

In the remainder of this chapter we shall examine (1) a polymorphism in which rare alleles are favored, (2) the age of some polymorphisms, (3) genetic changes associated with polymorphic systems, and, briefly, (4) the nature of the selective forces that favor heterozygotes.

Self-Sterility Alleles: Selection in Favor of Rare Alleles

In many plant species, among them various species of the genus *Oenothera*, pollen carrying a given allele at a particular locus (the self-incompatibility locus) is unable to germinate on the stigma, or to grow down the style, of a plant whose genotype includes the same allele. Thus no a_1 pollen can fertilize eggs of a_1a_x plants, where a_x stands for any other allele at the a locus.

A plant species cannot begin to utilize a system such as this unless there are at least three alleles in the population; it is only then that pollen of a given sort (a_2, for example) encounters a plant (a_1a_3) upon which it can function. It seems likely that the advantage of self-incompatibility systems of this sort lies in the protection they afford against breeding.

It is not difficult to see that new self-incompatibility alleles have an advantage over well-established ones in freely breeding populations. Imagine a population in which three alleles—a_1, a_2, and a_3—exist in equal frequencies. Each type of pollen can function in one third of all plants since a_1 can function only on a_2a_3, a_2 on a_1a_3, and a_3 on a_1a_2. Suppose, however, that an allele a_4 arises but is very rare. It can function on all three types of common plants—a_1a_2, a_1a_3, and a_2a_3. Pollen carrying the allele a_4 is three times as likely to encounter a receptive stigma as is pollen carrying any one of the original alleles. Consequently, the new allele has a tremendous advantage over the old ones.

The same argument can be used to show that at equilibrium the various alleles have equal frequencies. Suppose a_3 is rare in the original three-allele system. Pollen of types a_1 or a_2 will rarely encounter a_2a_3 or a_1a_3 plants. Pollen bearing a_3 will frequently encounter the more common a_1a_2 heterozygotes. Thus the number of a_1a_3 and a_2a_3 plants will tend to increase. The advantage of the a_3 allele will vanish when the frequencies of a_1, a_2, and a_3 are equal. Similarly, with the new allele a_4; its advantage will vanish when it is as common as the original alleles (25 percent each).

Each new allele as it arises has an advantage over those which have already been incorporated into the population. Let the new allele be a_{n+1} and let the n previous alleles be at equilibrium frequencies ($1/n$ each). Only heterozygous plants exist and with n alleles there are $n(n-1)/2$ different heterozygous genotypes. Pollen carrying a given allele can fertilize eggs carried by heterozygous genotypes in which it is not involved; there are $n-1$ other alleles so there are $(n-1)(n-2)/2$ genotypes that are receptive to pollen carrying any one allele. Within the original population, then, pollen of each type could function on $[(n-1)(n-2)/2] \div [n(n-1)/2]$ or $(n-2)/n$ of all plants. Pollen carrying the new allele, a_{n+1}, can function on all plants when this allele is rare. Consequently, if the rare allele is assigned fitness 1.00, $(n-2)/n$ represents the fitness of the other alleles. The selection coefficient s of these older alleles equals $2/n$; as the number of existing alleles increases, the disadvantage of the older ones (and therefore the advantage of the new one) becomes less and less.

The mathematical analysis of the number of self-incompatibility alleles that one would expect to find in a population of limited size is not a simple one (see Wright, 1964, and references listed by him). Nevertheless, we can outline the main parts of the problem:

1. The advantage of new alleles become smaller as the number of previously retained ones increases.
2. Although the "old" alleles should occur in equal frequencies, these frequencies will vary as the result of chance events in a finite population; rare alleles will then enjoy an advantage while common ones will be at a disadvantage.
3. Some alleles will be lost by chance each generation; the smaller their advantage, the easier they are lost altogether.
4. If the population consists of N individuals, $2N$ gametes will be "chosen" to form the surviving zygotes of each generation. If the new alleles arise at a rate u, then $2Nu$ new alleles are present in the population each generation.

The number of alleles becomes constant when the number lost each generation equals the number gained through mutation. In his paper, Wright (1964) cites a study by Emerson (1939 and unpublished) in which forty-five self-incompatibility alleles were found in a population of *Oenothera organensis* numbering no more than 1000 and probably fewer than 500 individuals. The system we have described here is capable, then, of maintaining a considerable number of alleles in a population.

The Age of Some Existing Polymorphisms

Interest in the age of polymorphisms can be traced to several questions: How stable are polymorphic systems? How are they eliminated from populations? How do they help in reconstructing evolutionary events?

There is a widespread belief, derived largely from the genetic load concept, that balanced polymorphisms are inefficient, transitory affairs doomed for replacement by equivalent but more efficient monomorphic systems. On a number of occasions, Muller (1950a, 1958) has expressed this view. Hiraizumi and Crow (1960) have also argued that the bulk of the segregational load is borne by the more normal allele and hence that selection will tend to make this homozygote equal in fitness to the superior heterozygote. Although there is some basis for these arguments, they are not as compelling as they may seem. The argument concerning selection for the improvement of the normal allele, as we pointed out previously, actually contradicts that concerning the greater dominance of slightly than of grossly deleterious mutations. The latter argument admits that natural selection is inefficient in making the final small improvements of near-normal phenotypes.

The advantage of a polymorphic system lies, in general terms, in the need for versatile biochemical or developmental systems involving, for example, two gene products, two times of action for a given locus, or two sites of action on membranes within cells. Given this need, it appears that polymorphism is the solution most likely to arise within a reasonable time: the retention of unlike alleles at a given locus because of the selective superiority of heterozygotes. In theory, a polymorphic system involving two different alleles can be made into an equivalent monomorphic one by gene duplication—by the substitution of *Aa*/*Aa* for *A*/*a*. Presumably duplications of this kind are responsible for the many clearly related subunits of hemoglobin that are now found in man and other mammals. However, the duplication of chromosomal segments is not a procedure that is under the control of the organism or one to be called upon if and when the need arises. The "desired" duplication, when it finally occurs, may include many loci other than that involved in the polymorphism. Consequently, gene duplication is such an inefficient solution to problems demanding immediate attention that it permits an accumulation of many of the faster arising polymorphisms.

The gene arrangements of the third chromosome of *D. pseudoobscura* represent one of the best studied cases of balanced polymorphism. They have been subjected to scrutiny since the early work of Sturtevant and Dobzhansky (1936a). As a rule, the patterns formed by the geographic distributions of the various gene arrangements (see Dobzhansky and Epling, 1944) have remained constant during the past three decades. No formerly polymorphic population has become monomorphic; the largely monomorphic (*AR*) populations of Utah and Arizona have remained monomorphic. On the other hand, changes of long-lasting duration have been observed; Pikes Peak (*PP*), an arrangement originally absent from extensive California collections, is now found in these populations with frequencies as high as 15 percent.

Another well-known polymorphic system is that found in many populations of the platyfish, *Platypoecilus maculatus.* Multiple allelic series of dominant genes govern the shape and position of spots formed by the precise groupings of many micromelanophores. Samples of preserved material collected at various localities in Mexico at intervals over a period of seventy years reveal that most populations have remained constant with only slight shifts in gene frequencies (Gordon, 1947).

A third system, the polymorphism of banding patterns in the garden snail, *Cepaea nemoralis,* has also been thoroughly studied (see Ford, 1964). The bands, together with the background color of the shell, afford protection against predation in a number of environments. The use of bands for this purpose is common among snails; it has, for example, also been studied in *Cepaea hortensis.* Still another species of snail that is polymorphic for color patterns is *Limicolaria martensiana,* of Africa. In this case, fossil individuals 8000 to 10,000 years old whose color patterns are still intact have been discovered. The data listed in Table 17-1

TABLE 17-1

FREQUENCY (PERCENT) OF COLOR FORMS IN A FOSSIL AND SEVERAL LIVING POPULATIONS OF THE SNAIL, LIMICOLARIA MARTENSIANA.

(Owen, 1966.)

	FOSSIL,	LIVING		
	KABAZIMU ISLAND	KAYANJA	ISHASHA ROAD	RWENSHAMA
COLOR FORM	($N = 1277$)	($N = 2840$)	($N = 882$)	($N = 841$)
Streaked	61.0	54.6	40.6	33.7
Broken-streaked	5.2	8.9	4.2	9.8
Pallid 1	3.9	24.0	7.0	1.9
Pallid 2	28.3	11.9	43.2	37.3
Pallid 3	1.6	0.6	5.0	17.3

show that the overall proportions of color types in fossil populations thousands of years old are very much like those found in living populations of the same species.

A number of man's polymorphisms exist among the great apes as well. It is a simple matter to show that some chimpanzees taste PTC while others do not (Stern, 1960, p. 708). The great apes, too, have their version of the ABO blood group polymorphism (Stern, 1960, p. 707). Of more interest at the moment, perhaps, is the contrast between the structures encountered in hemoglobin polymorphisms and in hemoglobin subunits. All polymorphic versions of hemoglobin in man such as sickle cell hemoglobin (Hb^S) and Hb^C of thalassemia differ from the normal by a single amino acid substitution in one of the major subunits (most often in the beta chain). On the other hand, even the most similar subunits, the beta and delta chains, differ from each other by some eight amino acid residues of 146. Indeed, the same hemoglobin subunits carried by different species are generally more similar than are the different units found in the same species. A minor exception to the above statement has been reported; C. J. Muller (1961) claims that polymorphic versions of the hemoglobin within sheep and goats differ from one another by several amino acid substitutions. If this is confirmed, it would appear that this is an older polymorphism than those found in human populations. On the other hand, the sharing of various recognizably similar hemoglobin subunits by many species of mammals testifies to the rarity with which gene duplications have occurred during the evolution of mammalian species.

Genetic Changes Associated with Polymorphisms

Polymorphisms involve more than the superior fitness of certain heterozygous individuals and the retention of two or more alternative forms of a given gene in a population. The mere existence of these alleles evokes selective changes at other loci. Interactions of this sort are the basis for the claim that genes do not act in solitary splendor; they act as parts of a network of gene-controlled reactions. The outcome of selective changes taking place within this network will be the topic of Chapters 18 and 19. Since we have already discussed the spotting polymorphisms found in Mexican populations of the platyfish, the outcome of an interpopulation hybridization experiment (Gordon, 1947) can be cited to illustrate the interactions between polymorphic loci and genes at other loci.

In addition to the large spots formed by clusters of micromelanophores, platyfish carry patterns—dorsal-spotted, spot-sided, stripe-sided, and black-sided—caused by the distribution of macromelanophores. A number of examples of each of these patterns are shown in Figure 17-2; it can be seen that there is considerable variation. Furthermore, the frequencies of individuals that show these different patterns vary markedly in different river basins.

Breeding experiments carried out by intercrossing fish from the same locality lead to the conclusion that different patterns are caused by different alleles of genes at a particular locus. That this result depends, however, upon tests of individuals from a single locality is shown by hybridization experiments carried out with individuals from different river basins. The results of hybridization and a subsequent backcross are shown in Figure 17-3. The gene for "dorsal-spotted," when present

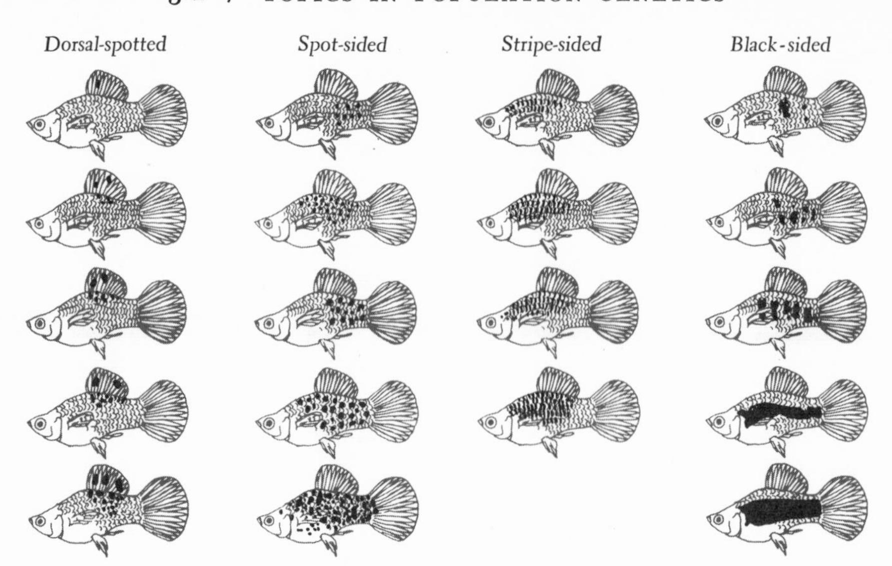

FIGURE 17-2. Variation in the expression of the macromelanophore pattern genes in the Rio Jamapa and Rio Coatzacoalcos populations of the platyfish, *Platypoecilus maculatus.* (After Gordon and Gordon, 1957.)

in the hybrids formed by crosses of individuals from two different river basins, causes heavy melanic spots to appear at various unexpected places on the fish's body. Following a backcross to individuals of the second river basin, the "dorsal-spotted" allele of the first causes enormous numbers of pigmented cells to blacken nearly one half the fish's body. Clearly, within any one population there exists a need (1) (for reasons unknown) for the black patterns formed by macromelanophores and (2) for a precise control over the deposition of these melanic cells. The solution to these two needs lies not entirely with the alleles at the polymorphic locus but with the interplay of these alleles and the background genotype. Remove one of the alleles from its proper background, insert it in another, and potentially dangerous, melanotic growths occur. When high fitness depends upon the proper interactions between genes, we say that the genes in question are "coadapted."

The types of genetic changes that are associated with polymorphic systems—changes that are generally of a coadaptive nature—largely invalidate an earlier argument we presented. We saw (page 277) that if a polymorphism involved two alleles whose homozygotes had dissimilar fitnesses, the seemingly more "normal" allele contributes the greater share to the total genetic load. Consequently, the argument ran, there should be strong selection tending to make this quasi-normal homozygote equal to the heterozygote; in the end this selection should replace the polymorphism by a more efficient monomorphism.

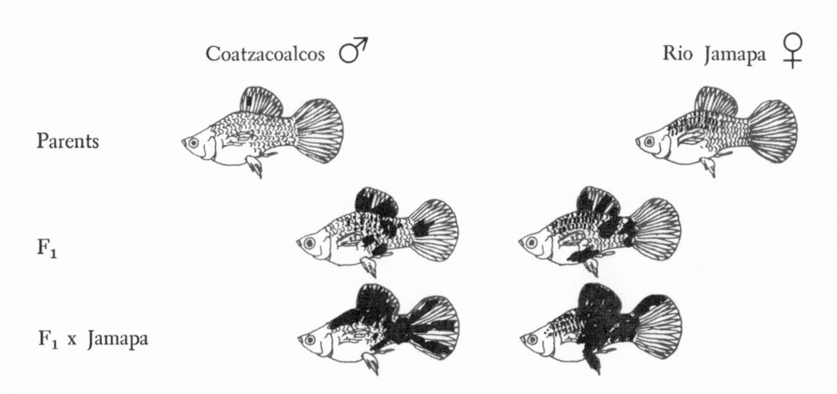

FIGURE 17-3. Parents, F_1 hybrids, and backcross hybrids of platyfish from two isolated river basins, Rio Jamapa and Rio Coatzacoalcos. (After Gordon and Gordon, 1957.)

This argument contains a series of flaws. First, the observed polymorphism must have arisen in response to some preexisting load borne by the population. The balanced load ascribable to polymorphism was not erected *de novo.* Second, response to selection requires (1) that selection pressure be applied and (2) that the genotype under pressure be able to respond. Now, the genetic load as defined can be apportioned between the two types of homozygotes (with the lion's share falling to the "better" of the two!) with none left over; this does not mean, however, that the heterozygote is perfect and free of selection. The total load on a population, genetic or otherwise, consists of those factors which ultimately limit the population's replacement each generation to an average of one adult daughter per mother; this load brings about a natural selection for an improvement in the fitness of heterozygotes just as it does in that of homozygotes. The mere existence of a polymorphic system is evidence that the homozygous quasi-normal individuals of an earlier time either could not respond to selection or were responding too slowly to forestall the origin of a superior heterozygote. And given that the polymorphism does exist, there is no need to believe that the homozygotes will respond more readily to improvement than they did originally or than the heterozygotes. Finally, selection for improvement of the fitness of the heterozygotes results in an even larger genetic load; consequently, the additional improvement is undetectable under the present definition of genetic load. Improvements of this nature are recorded instead as accumulations of deleterious mutations by balanced polymorphisms. It is the use of an internal standard that permits this perverse conclusion; an improvement in the fitness of heterozygotes does not alter their assigned relative fitness (1.00); instead, it is reflected as a deterioration of homozygotes.

The Identification of the Selective Forces That Maintain Polymorphisms

Despite the experimental work done on a great number of polymorphisms in different species, relatively little is known in most instances about the reasons why the polymorphisms exist. There are exceptions to this statement, of course. The banding patterns of snails represent camouflages; frequencies of the responsible genes are capable of responding on demand to changing textures and colors of the ground cover. Polymorphisms are associated with mimicry in butterflies; again, the various alleles through their effects on color patterns offer protection against predation.

Among the really interesting aspects of the superior fitness of heterozygotes are those responsible for the cryptic polymorphisms such as the inversion polymorphisms of Diptera and the gene-enzyme variants of many organisms. Here the problems involved are internal ones concerned with the biochemical reactions that occur within the organism. The problems appear to concern both reactions and their control. The possession of two alleles may aid in the restriction of certain reactions to the right cells and tissues, and to the right times during the development of the individual or during the life of individual cells. The biochemistry of the control of metabolic and developmental systems is still an infant science. The primitive state of this aspect of biochemistry can be illustrated as follows: The active site of some enzymes occupies only about 15 or 20 percent of the entire protein molecule. It has been suggested that the remainder of the molecule represents "fossil" DNA, DNA the organism has not yet managed to remove from its genetic apparatus. This suggestion completely ignores the need for enzymes to be rendered inactive by various inhibitors, as well as the need for certain enzymes to be associated with membranes. The problems associated with biochemical control promise to be infinitely more complex than those associated with performing individual reactions; within this complexity may be found the need for balanced polymorphisms and, in turn, the impetus for gene duplication.

18

COADAPTATION

In Chapter 17 a new term was introduced—coadaptation. Genes are said to be coadapted if high fitness depends upon specific interactions between them. A need for the term arose in the discussion of the genetic control of macromelanophore development in the platyfish. Populations of these small fish contain individuals with a variety of black markings. What function these spots and stripes serve in nature is unknown. They may serve as camouflage, as warning flashes, or as recognition signals during courtship. Whatever the reason, the presence of a variety of gene-controlled patterns in all populations studied, together with the stability of the relative proportions of these patterns over long periods of time, makes it appear that the whole system is one stabilized by selection. Selection in favor of heterozygous individuals appears to be the most likely stabilizing force.

Relatively simple hybridization experiments reveal that the macromelanophore production must be controlled or an excessive growth of melanophores takes place on the surface of the fish. The controlling systems that have been developed are genetic systems. Furthermore, the systems developed in different river basins (in isolated populations of platyfish) differ. The background genotype of Rio Jamapa is quite unable to control the pattern caused by the gene responsible for the dorsal spot found in Rio Coatzacoalcos (Figure 17-3). On the basis of such observations, we claimed that the gene for dorsal spot of one locality was coadapted to the background genotype of the same locality. Interpopulation crosses are able to destroy this coadaptation. Further, we said that coadaptation arises because genes act as parts of a complex network of interrelated gene-controlled processes. Coadaptation arises as the result of selection for a well-functioning network, a network whose genetic basis can be transmitted reliably from one generation to the next. Genes do not act independently of each other and so what is selected is not this or that allele but rather a favorable constellation of alleles.

In the following pages the notion of coadaptation will be examined in greater detail. First, we shall examine Dobzhansky's experiments that prompted him to first use the term. These experiments involve different gene arrangements of *D. pseudoobscura.* Following that, we shall turn to coadaptation that does not involve chromosomal inversions. Finally, we shall attempt to show that coadaptation can at times arise very quickly—during the course of laboratory experiments, for example.

Coadaptation of Gene Arrangements in D. Pseudoobscura

The third chromosome of *D. pseudoobscura* possesses a variety of naturally occurring gene arrangements. We have referred repeatedly to these gene arrangements in this book. They served to illustrate certain features of the Hardy-Weinberg equilibrium; they served, too, to illustrate that a selective superiority can be a property of heterozygous individuals (page 208).

That the gene arrangements found in populations of *D. pseudoobscura* do in fact possess selective significance was shown by Wright and Dobzhansky (1946); proof that these arrangements were subject to selective pressures was given by the systematic changes in their frequencies which occurred in cage populations. In Figure 11-1 we saw the changes in the frequencies of *ST* which occurred in four populations that contained initially 20 percent Standard and 80 percent Chiricahua (*CH*) gene arrangements from Piñon Flats, California. From these data, the relative fitnesses of *ST/ST*, *ST/CH*, and *CH/CH* individuals were estimated as 0.90, 1.00, and 0.41.

A fairly large number of combinations of several gene arrangements have been studied in experimental populations. Among these tests, there have been many that involved two alternative arrangements—*AR* versus *ST*, *AR* versus *CH*, and *ST* versus *CH*, for example. Furthermore, a number of these tests have involved gene arrangements obtained from the same geographic locality. Such tests were made to determine whether the same two gene arrangements would go through identical sequences of frequency changes even though they had come from geographically remote areas.

The results of a number of such "intralocality" populations are listed in Table 18-1. With the single exception of the population containing *ST* and *TL* chromosomes from Mather, the heterozygotes in these tests con-

TABLE 18-1

RELATIVE FITNESSES OF A NUMBER OF HOMOZYGOUS AND HETEROZYGOUS GENOTYPES INVOLVING GENE ARRANGEMENTS OF D. PSEUDOOBSCURA OBTAINED FROM A NUMBER OF LOCALITIES WITHIN CALIFORNIA.

(See the text for references.)

	GENOTYPE				
LOCALITY	1	2	1/1	1/2	2/2
Piñon	*ST*	*CH*	0.85	1.00	0.58
Keen	*ST*	*CH*	0.91	1.00	0.42
Mather	*ST*	*CH*	0.78	1.00	0.28
Piñon	*ST*	*AR*	0.81	1.00	0.50
Keen	*ST*	*AR*	0.79	1.00	0.58
Mather	*ST*	*AR*	0.64	1.00	0.58
Piñon	*AR*	*CH*	0.86	1.00	0.48
Keen	*AR*	*CH*	0.54	1.00	0.60
Mather	*AR*	*CH*	0.81	1.00	0.60
Mather	*AR*	*TL*	0.69	1.00	0.12
Mather	*ST*	*TL*	1.12	1.00	0.33

sistently exhibit a higher fitness than the two corresponding homozygotes. This means that in each of these populations inversion frequencies reached stable equilibria. Furthermore, they approached these equilibria at rates determined by the estimated fitnesses. Not all equilibrium frequencies were identical even for populations involving the same pair of gene arrangements. Thus the equilibrium frequencies of *ST* in different *ST* versus *AR* populations range from 73 to 54 percent, depending upon the geographic origin of the chromosomes involved. The gene content of chromosomes of the same gene arrangement but from different localities is not identical.

As a further test of the geographic differentiation in the gene content of chromosomes with various gene arrangements, a number of "interlocality" laboratory populations were set up. In these, one gene arrangement was taken from one locality while the other was taken from a second, geographically remote locality. The results of such tests differ markedly from those in which the gene arrangements were collected in a single locality.

The contrast between intra- and interlocality populations can be seen by comparing the data listed in Table 18-2 (Figure 18-1) with those given earlier in Table 11-3 (Figure 11-1). The eight populations reported in these two tables were studied simultaneously. The four intralocality populations approached equilibrium frequencies; in three of the four

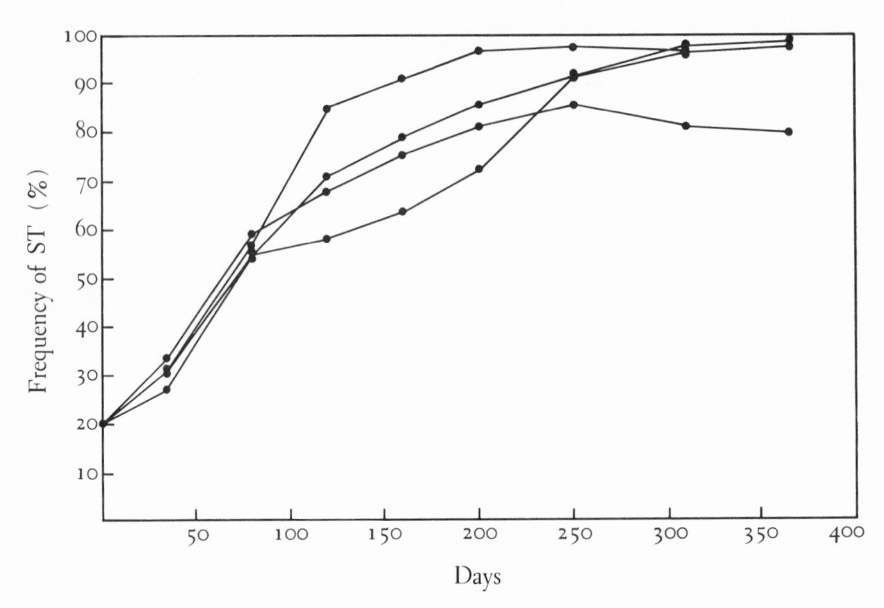

FIGURE 18-1. Frequency of *ST* chromosomes in four replicate populations of *D. pseudoobscura* that were started initially with 20 percent *ST* chromosomes from Piñon Flats, California, and 80 percent *CH* chromosomes from Chihuahua, Mexico. This figure should be compared with Figure 11-1. (After Dobzhansky and Pavlovsky, 1953.)

TABLE 18-2

CHANGES IN THE FREQUENCY OF *ST* CHROMOSOMES IN LABORATORY POPULATIONS CONTAINING *ST* CHROMOSOMES FROM PIÑON FLATS, CALIFORNIA, AND *CH* CHROMOSOMES FROM CHIHUAHUA, MEXICO.

(Dobzhansky and Pavlovsky, 1953.)

	% *ST*			
DAYS	1	2	3	4
0	20	20	20	20
35	33	31	31	27
80	59	55	57	55
120	68	58	85	71
160	76	63	91	79
200	81	72	97	86
250	86	92	98	92
310	81	98	97	97
365	80	99	—	98

geographically mixed populations, the frequency of *ST* approaches 100 percent. The four intralocality populations followed a common pattern of change represented by the average frequency of *ST* of each sample listed in Table 11-3. In the third and subsequent samples of interlocality populations, frequencies of *ST* varied so considerably from population to population that an average value would be misleading; each population followed its own pattern of change.

The results of a number of geographically mixed populations have been brought together in Table 18-3. Because changes in chromosome fre-

TABLE 18-3

RELATIVE FITNESSES OF INDIVIDUALS HOMOZYGOUS AND HETEROZYGOUS FOR VARIOUS GENE ARRANGEMENTS IN LABORATORY POPULATIONS OF D. PSEUDOOBSCURA.

Unlike those listed in Table 18-1, these are geographically mixed populations; the superscript identifies the geographic origin of each gene arrangement as Mather, Piñon, or Mexico. (Dobzhansky, 1950.)

	ARRANGEMENT		FITNESSES		
LOCALITIES	1	2	1/1	1/2	2/2
Mather versus Piñon					
45	AR^M	CH^P	1.28	1.00	0.47
46	ST^P	AR^M	0.63	1.00	1.51
47	ST^M	AR^P	1.31	1.00	0.57
48	ST^M	CH^P	1.18	1.00	0.48
49	ST^P	CH^M	1.38	1.00	0.43
Mexico versus Piñon					
55	ST^P	CH^{Mex}	1.26	1.00	0.87
56	AR^P	CH^{Mex}	1.53	1.00	1.16

quencies in mixed locality populations are erratic, estimates of fitness in these populations cannot be obtained from observed frequency changes. Instead, the estimates of fitness are based upon deviations from the 1 : 2 : 1 ratio expected in the F_2 offspring of interpopulation hybrid parents. Flies for this purpose are raised, of course, under population cage conditions. "One-generation" tests of this sort measure the differential survival of larvae, one component of fitness. One-generation tests of intralocality populations give results comparable to but not identical with those reported in Table 18-1. The results listed in Table 18-3, however, are altogether different. In every experimental population without exception, one homozygote (in one case, both homozygotes) exceeds the heterozygote in fitness. We can understand, then, why in the interlocality populations one gene arrangement frequently displaces the other and attains a frequency of 100 percent. Stable intermediate frequencies are established for the most part as the result of heterozygote superiority.

The results of these tests led Dobzhansky (1948, 1950) to suggest that chromosomes with different gene arrangements which occur together in the same geographic area are *coadapted* through natural selection so that heterozygotes exhibit high fitness. Gene arrangements from remote localities cannot be subjected to selection for coadaptation and hence the heterozygotes of mixed locality laboratory populations have fitnesses intermediate (generally) to those of the two homozygotes.

An alternative "explanation" can be suggested to account for the observations described in the preceding paragraphs. One can imagine that dissimilarity in origin decreases the probability that the two alleles at a given locus are identical (Sturtevant and Mather, 1938; Merrell, 1963). Alleles found at various loci in a given Standard chromosome from Piñon Flats, for example, are more likely to be identical to those found in a second Standard chromosome from Piñon Flats than in one from Mather, a locality several hundred miles away. Similarly, alleles found at loci in a given Standard chromosome are more likely to be identical to those in a second Standard chromosome from the same locality than to those in an Arrowhead chromosome from that locality. This argument applies particularly to loci within the inverted segment where gene exchange between the two arrangements is excluded.

According to the alternative explanation, gene arrangements A and B should show the following sequence of fitnesses in intralocality populations: $A/B > A/A \geqq B/B$. This sequence is that which is actually observed, although the differences in fitness appear to be much too great to be accounted for by the mere covering up of deleterious mutations.

The results of the mixed locality populations can be written as follows: $A/A > A/B > B/B$. The alternative explanation (covering up of deleterious mutations) now encounters difficulties. To retain it, we must make the far-fetched assumption that arrangement B of one locality is more likely to have alleles identical to those of arrangement A from a different locality than are two different A's from the same locality. No reasonable explanation can be advanced in defense of such a notion.

However, there is a third experiment, not as well known as those already described, which eliminates "greater hybridity" as an alternative explanation. Through the use of an ingenious series of crosses, Dobzhansky (1950) compared the fitnesses of

$$ST^M/ST^P : \quad \begin{array}{c} AR^P/ST^P \\ + \\ AR^M/ST^M \end{array} \quad : AR^M/AR^P$$

The superscripts *M* and *P* refer to the two localities Mather and Piñon Flats. It can be seen that the two structural homozygotes are both interlocality heterozygotes, but the structural heterozygotes carry chromosomes obtained from the same locality. The fitnesses estimated for these flies were 0.87, 1.00, and 0.46. The intrapopulation structural heterozygotes exceeded the two structural homozygotes in fitness despite the hybridity of the latter arising from their remote geographic origins. The results of this test, together with those obtained by studying single and mixed locality populations, are enough to dispose of the "greater hybridity" explanation for the observed coadaptation of gene arrangements coexisting within given geographic localities.

Coadaptation at the Gene Level

In Dobzhansky's original use of "coadaptation," the term referred specifically to chromosomal polymorphisms and to their maintenance in local populations owing to the superior fitness of structural heterozygotes. The term has implications that go beyond the maintenance of inversion polymorphisms, however. Any adjustment of the frequencies of alleles at one locus in response to changes of those at another may be regarded as an instance of coadaptation. In this sense, coadaptation is a mutual give and take among genes that find themselves coexisting in a local population. Using the platyfish, *Platypoecilus maculatus,* as an example, we can guess that black markings are needed by fish of this species. A gene that controls the existence of macromelanophores supplies the answer to this need. Macromelanophores serve their purpose only if they are restricted to certain rather small areas of the body. This restriction is supplied by genes at various other loci. The entire system of melanophore production and regulation is a coadapted system in the sense that the details and the proper working of the system depend upon sets of genes each of which has been developed within a single population. Interpopulation hybridization, as we have seen, can destroy the regulation of melanophore distribution.

Kettlewell (1958) has described for the melanic heterozygotes of *Biston betularia* a change in viability that resembles both Dobzhansky's coadaptation of inverted gene arrangements and Gordon's local control of color patterns in the platyfish. Industrial melanism among moths is an evolutionary phenomenon induced by human interference with the natural habitat; the interference can be traced to the industrial revolution. Melanic forms of the moth, *B. betularia,* were first collected in England about 1850. At various times different persons have made experimental crosses while studying the genetics of melanism in these moths. A summary of results obtained during two periods nearly a half century apart is given in Table 18-4. The data suggest that the relative viability of melanic heterozygotes improved markedly during this period.

TABLE 18-4

APPARENT CHANGE IN THE RELATIVE VIABILITY OF MELANIC (HETEROZYGOUS) AND TYPE (HOMOZYGOUS) BACKCROSS INDIVIDUALS IN BISTON BETULARIA.

(Kettlewell, 1958.)

Tests made between 1900 and 1905

MELANIC	TYPE	TOTAL	
109	123	232	
47	57	104	
11	18	29	
50	57	107	
217	255	472	46.6% melanics

Summary of results obtained in 1953–1956

MELANIC	TYPE	TOTAL	
22	14	36	
10	7	17	
30	28	58	
39	14	53	
5	1	6	
2	1	3	
108	65	173	62.4% melanics

The relative viability (and, presumably, the relative fitness) of melanic heterozygotes has apparently improved with time. If so, the evolving system resembles that of the melanophores in the platyfish very closely. For the moth, the blackening of tree trunks by soot, together with the killing of lichens that normally grow on unblackened trunks, results in an advantage for dark-winged moths. The retention and, indeed, the increase of the melanic allele in a population sets in motion selection for viability modifiers as well. At the end of fifty years, the viability of melanic heterozygotes has apparently improved so that the gene for melanism no longer impairs viability; on the contrary, it enhances the viability of these individuals.

COADAPTATION AS THE INTEGRATION OF GENE POOLS

In discussing both melanophores of fish and melanic coloration in moths, we restricted our attention to one locus and its relation to others; the main locus had a marked effect on the phenotype, the others a less well-defined, regulatory effect on the alleles at the main locus. The argument built up around the coadaptation of a "main locus" and "modifiers" need not be restricted to systems having only this format. Any locus can be taken as the main locus in which case the remainder become modifiers. For a trait such as fitness or viability, a trait subject to modification by genes at virtually all loci, one can argue that coadaptation at the level of the local population should be the rule. Some tests have confirmed this argument (in addition to the work discussed below, see Mourad, 1965); others have failed to do so (McFarquhar and Robertson, 1963; Richardson and Kojima, 1965).

One of the earliest tests of "the integration of gene pools" was that of Brncic (1954). Brncic restricted his study to the Arrowhead (*AR*) gene arrangement of *D. pseudoobscura.* The flies upon which he experimented had come from a number of localities in several western states. The account of these experiments given below is a summary of a very large study. We are interested in presenting the general results only; those interested in the subtleties and details of the different interpopulation crosses must read the original account.

The mating scheme used by Brncic is shown in Figure 3-8. The purpose of the entire scheme is to obtain wildtype flies of known genotypes in respect to the geographic origins of the wildtype chromosomes and to measure the viabilities of these flies using others that carry a genetically marked chromosome as the standard in each case. Examination of the figure shows that the wildtype flies tested were of four sorts: (1) Some carried two intact chromosomes from one locality; (2) others carried two intact chromosomes, one from each of two localities; (3) still others carried one intact chromosome and another recombinant chromosome containing blocks of genes from two different localities; and (4) some flies carried two chromosomes both of which contained blocks of genes derived from two different localities. (There are no formal terms that describe chromosomes of the sort we are talking about. "Intact" means simply that the chromosome has come from a strain of flies derived from a single locality; the allele that is found at each locus in an intact chromosome is drawn from the same locality as all others. A "recombinant" chromosome is one obtained from an F_1 interlocality hybrid female. At a given locus on such a recombinant chromosome, the allele comes from either one or the other of the two localities; among all loci alleles from both populations are probably present.)

The results obtained by Brncic are summarized in Table 18-5. The wildtype flies with highest relative viability are the F_1 interlocality hybrids; the average frequency of these flies observed in all tests has been adjusted to 1.00. The second highest average frequency of wildtype flies is that observed in the "parental" crosses; these flies carry two chromosomes obtained from flies of a single geographic locality. The third

TABLE 18-5

EFFECT OF RECOMBINATION BETWEEN WILDTYPE CHROMOSOMES OF DIFFERENT GEOGRAPHIC ORIGINS ON VIABILITY IN D. PSEUDOOBSCURA.

(Brncic, 1954.)

First experiment	
Parental	0.94
Hybrid (F_1)	1.00
Second experiment	
Parental	0.92–0.93
Hybrid (F_1)	1.00
Hybrid (F_2)	0.91
Hybrid (F_3A)	0.91
Hybrid (F_3B)	0.86

highest frequency is that observed in cultures in which the wildtype flies carry one interpopulation recombinant chromosome. Finally, the lowest frequency is observed in those cultures in which both chromosomes of the wildtype flies are interlocality recombinants.

Since the relative frequencies of the wildtype flies are used as measures of viability, several technical points should be made concerning the mating scheme shown in Figure 3-8. Our claim is that interpopulation recombinant chromosomes tend to lower the viability of their carriers. Since such recombinant chromosomes are carried in certain genetically marked flies that serve as a standard for measuring viability, their effect on the calculated relative frequencies of wildtype flies must be examined. Otherwise, varying standards may distort the experimental results and lead to erroneous conclusions. A reexamination of Figure 3-8 shows that the relative frequencies of "parental" and F_1 hybrid combinations are directly comparable; the genetically marked flies of the two types of crosses carry "intact" wildtype chromosomes. The same argument applies to flies labeled F_3A in the figure. In the case of F_3B, both third chromosomes of the wildtype flies are recombinant, but the standard also carries a recombinant wildtype chromosome. Thus the frequency of these wildtype flies, the lowest of the four categories, should in fact be still lower because the standard against which they were measured is ostensibly a standard with poor viability. The F_2 wildtype flies should have had higher frequencies than those of F_3A; instead, the two frequencies are equal. Nevertheless, recalling that there are chance errors associated with each of these observed frequencies, I believe the results as a whole are remarkably consistent.

What bearing have these results on the problem of coadaptation and the integration of gene pools? As far as the data from the recombinant chromosomes carrying blocks of genes from two remote localities are concerned, the evidence is quite clear: The breaking up of gene combinations through interpopulation recombination lowers viability; consequently, the gene combinations present in local populations must have been selected for their characteristically high viability.

If the above conclusion is true, then how does one explain the high viability of interpopulation F_1 hybrids? Why are the frequencies of parental flies not as great or greater than those of the F_1 hybrids? The answer to these questions is unclear, but it seems as if "greater hybridity" might account in part for the observed improvement in viability. Interlocality crosses result in greater average proportions of heterozygosity than do intralocality crosses; the F_1 hybrids reveal the effects of fewer deleterious genes. How, then, does one explain the poorer viability of flies carrying one "intact" chromosome and a "recombinant" homologous chromosome containing blocks of genes obtained from two other localities? Here we are forced to postulate that recombination brings together badly coordinated alleles the products (or timings) of which interact imperfectly. It would seem that the reactions affecting viability must be taking place in the immediate neighborhood of the individual chromosomes; otherwise, the F_1 interpopulation hybrids should suffer to the same extent as "three-way" hybrids.

By crossing strains of flies obtained from the same and from different geographic localities, Vetukhiv (1953, 1954, 1956) studied the survival of *D. pseudoobscura* and *D. willistoni* larvae under crowded conditions. Each type of cross—intra- and interlocality—was carried into the F_2 generation to determine the effect of recombination on survival. In addition to larval survival, Vetukhiv studied two other components of fitness: egg production and longevity.

The results of Vetukhiv's studies have been summarized in Table 18-6. In each case, the observations for the intrapopulation F_1 and F_2 have been set equal to 1.00 to serve as a standard for comparison. The results of the interpopulation crosses are consistent and agree with those obtained by Brncic. The F_1 interlocality hybrids survive better as larvae, lay more eggs, or live longer than do the intralocality flies. This increase is lost, however, following recombination; in each instance the interlocality F_2 flies perform worse than do the intralocality ones. These experiments illustrate in a very elegant way that populations inhabiting different localities come to rest on different peaks of the adaptive landscape. They also show that the F_1 hybrids are not part of the landscape; they do not represent equilibrium populations and, in a sense, hover above the landscape. The F_2 interpopulation hybrids lose all the advantage enjoyed by the F_1 hybrids and more; the F_2 hybrids show that the peaks occupied by different local populations are indeed separated by valleys.

TABLE 18-6

COMPARISON OF THE VIABILITY, FECUNDITY, AND LONGEVITY OF INTRA- AND INTERPOPULATION HYBRIDS OF D. PSEUDOOBSCURA AND D. WILLISTONI.

(Vetukhiv, 1953, 1954, 1956.)

	F_1	F_2
Survival		
D. pseudoobscura		
Intrapopulation	1.00	1.00
Interpopulation	1.18	0.83
D. willistoni		
Intrapopulation	1.00	1.00
Interpopulation	1.14	0.90
Eggs/♀/day		
D. pseudoobscura		
Intrapopulation	1.00	1.00
Interpopulation	1.27	0.94
Longevity		
D. pseudoobscura		
Intrapopulation	1.00	1.00
Interpopulation (16°)	1.25	0.94
Interpopulation (25°)	1.13	0.78–0.95

Experiments of the same general nature as those of Brncic and Vetukhiv were carried out also by Wallace (1955; see, too, Wallace and Vetukhiv, 1955) with essentially similar results. The two highest viabilities of all sorts of flies tested (*D. melanogaster*) were those representing single localities and the F_1 interlocality hybrids (highest of all). In these experiments, three-way and four-way (doublecross) hybrids were clearly inferior to the F_1 hybrids in larval survival. The average "hybridity" of these three kinds of interpopulation hybrids was identical, so it is difficult to account for the sequence of viabilities by the covering up of deleterious mutations to different degrees in the different kinds of hybrids. Rather, as in the case of Brncic's experiments, the destruction of gene combinations built up within local populations seems to offer the best explanation for the observed differences in larval survival.

OBSERVATIONS ON THE ORIGIN OF COADAPTATION WITHIN LABORATORY POPULATIONS

Coadaptation obviously requires time to develop; mutual adjustments of alleles at a number of different loci cannot be perfected in a single instant. Nevertheless, observations related to the development of harmonious intractions of various parts of the genome within relatively short periods have been made.

One series of observations, a series involving a number of experimental populations, was made by Dobzhansky (1950; Dobzhansky and Levene, 1951). These observations deal with coadaptation in its original sense—the origin of the superior fitness of inversion heterozygotes.

The observations begin with experimental population 55, a population within which Dobzhansky attempted to determine the relative viabilities of inversion homo- and heterozygotes by means of a one-generation experiment. The gene arrangements involved were Chiricahua from Mexico (CH^{Mex}) and Standard from Piñon Flats (ST^{P}). The one-generation experiment was performed as follows: ST^{P}/ST^{P} flies were crossed to others that were CH^{Mex}/CH^{Mex}. Large numbers of F_1 hybrids (ST^{P}/CH^{Mex}) were placed in cage 55; the zygotes produced by these hybrids would include 1 *ST/ST*, 2 *ST/CH*, and 1 *CH/CH*. The larvae hatching from these eggs were allowed to develop under the crowded conditions characteristic of food cups in a population cage. Young F_2 adults that finally emerged were mated individually with *ST/ST* flies; the genotypes of the F_2 flies were determined by the cytological examination of salivary chromosomes of their offspring.

Two hundred and sixty F_2 flies were tested for population 55. Of these 57 were *CH/CH*, 122 *CH/ST*, and 81 were *ST/ST*. The expected numbers of the F_2 genotypes were 65:130:65; the observed deviations fall just short of those generally accepted as significant (chi-square of 5.5 with 2 degrees of freedom; p equals 0.06). The relative fitnesses (based on larval survival) of the three genotypes are *CH/CH*, 0.94; *CH/ST*, 1.00; and *ST/ST*, 1.32. The results of this test suggest that *ST* would eventually replace *CH* in a freely breeding population involving chromosomes from Mexico and California.

Following the one-generation experiment, a new population (#66) was started. The flies used to start this population, as in the case of #55, were ST^{P}/CH^{Mex} hybrids. The one-generation experiment had revealed that *ST/ST* individuals had highest fitness while that of *ST/CH* was intermediate, so it seemed that *ST* would quickly eliminate *CH* from population 66. This expectation was not borne out. After an initial period during which the frequency of *ST* reached 75 to 80 percent, population 66 settled down at equilibrium with the frequency of *ST* lying between 65 and 70 percent.

Although a number of explanations for the origin of a stable gene frequency equilibrium are possible, the most likely explanation seems to be the superior fitness of heterozygous individuals. To determine whether heterozygous individuals did possess higher fitness than either homozygote in population 66, Dobzhansky reextracted *ST* and *CH* chromosomes from this population and performed a second one-generation experiment (population 75). Once more *ST/CH* heterozygotes were used as parents so that the expected frequencies of *CH/CH, ST/CH,* and *ST/ST* were 1:2:1. Four hundred young F_2 adults were tested. The observed numbers were 69 *CH/CH,* 234 *ST/CH,* and 97 *ST/ST*. These numbers differ considerably from the 100:200:100 expected. The adaptive values for the three genotypes calculated from this test are *CH/CH,* 0.59; *ST/CH,* 1.00; and *ST/ST,* 0.83. These values, although not precisely those needed to explain the observed equilibrium frequencies, do suggest that both gene arrangements would be retained in the population at equilibrium.

These observations illustrate the origin of coadaptation as Dobzhansky originally defined the term. At the outset, inversion heterozygotes did not possess superior fitness; at the end of the experiment they did. And in agreement with the second observation, equilibrium inversion frequencies were established within the population.

We shall discuss the above observations again in a later chapter; it is only fair, however, that some statement be made here on the origin of heterozygote superiority in population 66. Two alternatives suggest themselves: (1) The particular chromosomes that confer high fitness on heterozygotes were present in the strains of flies that were used in setting up the population, or (2) the chromosomes that conferred high fitness on the structural heterozygotes were not present as such in the original material but arose through recombination within the population and were then preserved by natural selection. If the first alternative were true, the result observed in population 66 should be repeatable; populations set up with the same initial material should lead to the same end result. If the second explanation were true, the proper recombination products need not arise in every population that is started with the same material because so many different recombinations are possible. The eight populations whose results are presented in Tables 11-3 and 18-2 (Figures 11-1 and 18-1) were set up to decide between these two alternatives; the populations of mixed-locality origins gave nonrepeatable results, as we have seen.

One more example of the origin of coadaptation within laboratory populations is provided by the origin of DDT resistance in laboratory populations of *D. melanogaster* (King, 1955). These populations were exposed each generation to aerosols containing DDT. Each population was maintained by using the survivors of the treatments as parents. Resistance to DDT was measured by the exposure time (in minutes) required to kill one half the population (LD_{50}).

If populations that were resistant to DDT were maintained without further exposure to DDT, the level of resistance would remain constant for five generations or so before dropping detectably. However, if crosses were made between different selected populations or between a selected population and the control (untreated) population, this decrease could be detected in the F_2 generation. Data on eleven interpopulation crosses are listed in Table 18-7. In each case, the LD_{50} for the F_1 generation is

TABLE 18-7

DDT RESISTANCE OF F_1 AND F_2 HYBRIDS OF D. MELANOGASTER; THE PARENTAL FLIES WERE EITHER FROM RESISTANT LABORATORY POPULATIONS OR FROM A RESISTANT POPULATION AND ITS CONTROL.

Figures represent minutes exposure needed to kill one half the exposed individuals (LD_{50}). (King, 1955.)

EXPERIMENT	PARENTAL POPULATIONS		AV.	F_1	F_2
X1	10.5	2.5	6.5	6.5	3.5
X1R	2.5	10.5	6.5	6.8	3.5
X2	8.8	10.8	9.8	10.9	5.1
X2R	10.8	8.8	9.8	10.1	5.5
X3	13.3	12.6	13.0	13.6	7.4
X3R	12.6	13.3	13.0	12.5	7.1
X4	21.3	13.5	17.4	20.0	11.6
X5	17.6	2.5	10.1	7.8	4.8
X5R	2.5	17.6	10.1	7.4	4.8
X6	(13.0)	2.5	7.8	6.4	2.6
X8	(19.0)	(18.0)	18.5	19.8	13.0

almost precisely that obtained by averaging those of the two strains involved. In each of the eleven intercrosses, the resistance of the F_2 is distinctly less than that of the F_1. Incidentally, among the eleven tests, there are eight that represent four reciprocal crosses; the repeatability of these results is very good indeed.

Coadaptation as illustrated by the data on DDT resistance (and by the results of Vetukhiv and Brncic discussed earlier) depends upon gene combinations that can be destroyed by recombination. These must be combinations of alleles at different loci and, therefore, the responsible interactions must be epistatic in nature. Crow (1956) has argued that clonal organisms should exhibit epistatic coadaptations of this sort to a greater extent than cross-fertilizing organisms. To support this argument, he cites an experiment in which two strains of *Escherichia coli* were selected for resistance to the antibiotic chloramphenicol. This selection was performed on asexually reproducing cultures. Following the development of resistance in the selected cultures, the resistant strains were allowed to undergo sexual crosses. The recombinant strains obtained had lost most of the resistance. Presumably, in the absence of recombination, gene combinations conferring high resistance as *combinations* were selected for as such; occasional mutants that improved the action of these combinations were incorporated by selection. Only after a sexual cross could these combinations be destroyed by recombination. Crow is probably correct when he argues that such complementary epistatic interactions ("coadaptation") should be especially important in clonal organisms; the occurrence of coadaptation in *Drosophila* populations shows, however, that local populations of cross-fertilizing species also avail themselves of such interactions.

19

THE EVOLUTION OF DOMINANCE

Dominance is an attribute of gene expression. In many ways the genetic bases of dominance and of coadaptation are similar so that an understanding of one helps in the understanding of the other. In recent years there has been an effort to equate dominance and recessiveness to the presence or absence of a primary gene product (or a functional enzyme, for example); this effort resembles Bateson's "presence and absence" hypothesis of a much earlier era.

In the present chapter we shall comment on the "new" and the "old" uses of the term "dominance." Following that, we shall discuss two views of the evolution of dominance, describe the biochemical framework within which dominance occurs, and present data from several genetic analyses of dominance. Finally, we shall discuss linkage in relation to the selection of dominance modifiers and then conclude with a comment on dominance modification and genetic load.

Dominance: The Old and the New Concepts

Dominance is a term that refers to the phenotype, not to the gene. It refers, as Mendel used the term, to the "roundish form" of the pea or to its "yellow colouring" (Sinnott et al., 1958, p. 423). The curly wing that is caused by the mutant gene, *Cy*, is dominant to the normal, straight wing—especially at high temperatures. The lethality that is also caused by the mutant gene, *Cy*, is recessive to the nonlethality of the normal allele. Either the terms "dominant" and "recessive" apply to the phenotype rather than the responsible allele or they apply to the allele *in respect to* specified aspects of the phenotype. *Curly* (*Cy*) is a dominant allele *in respect to* wing shape but is recessive *in respect to* lethality. To refer to *Curly* as either a dominant or a recessive allele with no further information is meaningless.

Experimental procedures in genetics have become exceedingly refined in recent years. It is possible in microbial genetics to assay directly for the presence of an enzyme in a culture of genetically identical bacteria. The normal allele at a given locus usually makes an enzyme; mutant alleles frequently do not. The test determines the presence or absence of the enzyme; consequently, it is tempting to talk of the allele as being dominant or recessive. The presence of the functional enzyme is directly related to the presence of the normal allele. The presence of the functional enzyme is also directly related to the observed "phenotype" (growth in minimal medium, for example). It seems, then, that in this case dominance might safely be used in reference to the allele itself.

It might seem in these very basic analyses that dominance can be used in reference to the allele but, even here, its use in this way is erroneous. Many allelic forms of the same gene have been revealed by the electrophoretic analysis of gene-enzyme variants. Individuals heterozygous for two dissimilar alleles frequently have enzyme variants that show up on starch gels as two distinct bands; such alleles are said to be "codominant." The starch gel becomes, in this case, the medium in which the phenotype is revealed; the presence of two colored bands means that both enzymes are present and, so it seems, the genes have equal weight in determining the phenotype. Work by Schwartz (1962) shows that even at this level the use of dominance in reference to the allele itself can be misleading. One allele of an esterase-forming gene (E) is active in the embryo and growing plant but is inactive in the endosperm. Consequently, even at the gene product level, the term "codominant" demands qualification.

Crosby (1963) has recommended that "dominance" be reserved for use in reference to what is now recognized as the competition of messenger RNA molecules for attachment to ribosomes immediately prior to protein synthesis. This is an interesting suggestion for this competition has important consequences in respect to gene action. But to identify this one problem with a term that has meant something quite different for a half century or more will surely lead to confusion. The great debates on the evolution of dominance—debates that form one of the exciting chapters of evolutionary genetics—were not concerned with a single step of the many that lie between gene action and phenotypic expression. On the contrary, these debates dealt with the control of the phenotype. To a large extent they dealt not with the ability or inability to make a certain protein but with the control of time and place at which this protein would be made.

THE EVOLUTION OF DOMINANCE: FISHER'S VIEW

Fisher (1930, Chap. 3) was one of the first to attempt to give an evolutionary interpretation of the genetic facts concerning dominance. In making this attempt he clearly saw that evolutionary genetics must deal effectively with two major questions: Can evolution be explained in terms of genetic causes? Can genetical phenomena be explained in terms of known evolutionary causes? Molecular biology, through amino acid sequence analyses and the breaking of the genetic code, is making tremendous contributions toward the solution of the first of these questions; some molecular biologists, by failing to appreciate the historical aspect of life, seem to be unaware that the second question exists.

Fisher's account begins with an enumeration of respects by which one allele can be distinguished from another:

1. One allele can be rare while another is common.
2. One allele can be selectively advantageous while the other can be disadvantageous.
3. One allele can be a mutant that has arisen from the other, earlier one.
4. One allele can be dominant in respect to some aspect of the phenotype while the other is recessive.

Associations between the four categories listed above are not random. Although evolution in the long run consists of the replacement of common alleles by once rare alleles, the common allele tends in the vast majority of cases to be the advantageous one as well. In studying mutations, common alleles generally serve as the starting point and so mutant alleles (3) are also disadvantageous (2), as a rule.

Although it is not absolutely so, one allele of a series is generally dominant to all other alleles in respect to selectively important aspects of the phenotype. Of the dozen or more alleles at the *white* locus in *D. melanogaster*, for example, the wildtype allele produces heterozygotes in combination with the other ones that are indistinguishable from the wildtype homozygote; all other heterozygous combinations are intermediate to the two corresponding homozygotes. The need to restrict this point to "selectively important aspects of the phenotype" is illustrated by work of Dobzhansky and Holz (1943) on the effects of various recessive mutations on the shape of spermathecae in female flies (Table 19-1). The wildtype allele, in respect to the shape of the spermathecae, is not at

TABLE 19-1

EFFECT OF VARIOUS RECESSIVE MUTATIONS ON THE SHAPE OF THE SPERMATHECA IN FEMALE D. MELANOGASTER.

The numbers represent mean ratios obtained by dividing the height by the diameter; "average" represents the mean of the ratio observed for the mutant homozygote and 1.56, the ratio observed in homozygous wild-type females. (Dobzhansky and Holz, 1943.)

MUTATION	HOMOZYGOTE	HETEROZYGOTE	AVERAGE
white-1	1.41	1.51	1.49
white-2	1.43	1.49	1.50
white-3	1.41	1.49	1.49
white-4	1.42	1.50	1.49
yellow-1	1.41	1.49	1.49
yellow-2	1.41	1.48	1.49
yellow-3	1.42	1.49	1.49
yellow-4	1.42	1.49	1.49
yellow-5	1.42	1.49	1.49
yellow-6	1.49	1.51	1.53
yellow-7	1.47	1.51	1.52
yellow-8	1.45	1.52	1.51
yellow-9	1.47	1.51	1.52
yellow-10	1.42	1.51	1.49
forked	1.44	1.46	1.50
vermillion	1.42	1.50	1.49
dusky	1.50	1.47	1.53
ruby-1	1.50	1.49	1.53
ruby-2	1.56	1.54	1.56

all dominant; dominance, as we emphasized earlier, is an attribute of an allele in respect to a specified aspect of the phenotype.

Dominance is not determined by the mutational origin of a new allele. That one allele has arisen from another does not necessarily mean that the newly arisen one is recessive nor that the original one is dominant. The experimental evidence bearing on this point comes from studies of back mutations. Dominant wildtype alleles can be obtained as rare mutations from mutant alleles.

If evolution occurs by the substitution of new alleles for old ones within populations (Fisher argued), and if this substitution can occur at any or all loci, then we must be prepared to admit that new genes *acquire* the dominance that we associate with common, advantageous alleles. Furthermore, the dominance of new alleles must include dominance in respect to those phenotypic traits for which the earlier wildtype allele was dominant. As they are displaced from populations, old wildtype genes acquire recessiveness.

In seeking an explanation for the acquisition of dominance by the "incoming" allele and the loss of it by the "outgoing" one, Fisher turned to the modification of gene action by the background genotype. That gene action can be modified is known. Mutations that are maintained in homozygous cultures tend to revert toward normal appearance or viability (see Table 19-2); the strong expression of the mutant phenotype can generally be regained by outcrossing it to an unrelated wildtype stock and reextracting the homozygous mutant anew. Earlier in this text we discussed the improved viability of homozygous semilethal cultures of *D. pseudoobscura* (see Table 8-10); in this case the measurements

TABLE 19-2

EVIDENCE FOR THE ACCUMULATION OF MODIFIERS THAT ENHANCE THE VIABILITY OF HOMOZYGOUS MUTANTS IN THE SWEET PEA, *LATHYRUS ODORATUS.*

In F_2 crosses, older mutants approximate the expected ratio more closely than do more recently discovered ones. (Haldane, 1957.)

FOUND	MUTANT	VIABILITY	S.E.	AV.
1700–1800	g_1 *White*	1.037	0.024	1.009
	a_1 *Red*	1.021	0.017	
	b_1 *Light axil*	1.011	0.017	
	f_1 *White*	0.996	0.038	
	d_5 *Picotee*	0.996	0.022	
	a_2 *Round Pollen*	0.990	0.024	
1880–1899	b_2 *Sterile*	0.988	0.017	0.954
	a_3 *Hooded*	0.977	0.021	
	e *Cupid*	0.976	0.032	
	f_2 *Bush*	0.936	0.030	
	d_2 *Blue*	0.931	0.024	
	g_3 *Mauve*	0.917	0.031	
1900–1912	d_1 *Acacia*	0.964	0.020	0.903
	d_4 *Smooth*	0.940	0.020	
	d_2 *Copper*	0.909	0.040	
	h *Spencer*	0.897	0.030	
	b_3 *Cretin*	0.886	0.048	
	f_3 *Marbled*	0.821	0.023	

dealt directly with larval viability, one of the main components of fitness in a *Drosophila* culture.

Fisher's argument for the modification of the dominance of a gene by selection for its background genotype *in heterozygous individuals* depends, in part, on the great excess of heterozygous over homozygous mutant individuals. This preponderance we have seen in earlier chapters. If a gene is extremely rare, the ratio of $2pq/q^2$ is very nearly $2/q$. Thus, for a gene whose frequency is 0.001, there are about 2000 heterozygotes for every homozygote. More to the point, if the mutation lowers the fitness of heterozygotes by an amount hs, its frequency at equilibrium will be u/hs (see page 192). The ratio of heterozygotes to mutant homozygotes will then be $2hs/u$ to 1.00. Modification of the heterozygotes will be subject to a great deal more selection than will the modification of the mutant homozygote.

The crucial point for Fisher's argument is the extent to which selection is able to affect heterozygotes. The heterozygotes may greatly outnumber mutant homozygotes but still be but a minuscule fraction of the population generally. To evaluate the amount of selection acting on heterozygotes, we assume that the ratio of descendants of a heterozygote to those of a homozygous normal is $x:1$; the sum of these descendants is $1 + x$. The fitness of the heterozygotes has been said to equal $1 - hs$; as a rule, in the progenies of heterozygotes, one half of the individuals are heterozygous in turn. Thus we can say that $x/(1 + x)$ equals $\frac{1}{2}(1 - hs)$ or that x equals $(1 - hs)/(1 + hs)$. The proportion of heterozygotes within an equilibrium population equals $2u/hs$. The proportionate contribution of heterozygotes to some future generation, therefore, is the product of their frequency times the proportion ascribable to a single heterozygote:

$$\frac{2u(1 - hs)}{hs(1 + hs)}$$

If u equals 10^{-6} and hs equals 10^{-2}, the proportionate contribution equals 1/5000. Thus the rate of progress achieved by selection for dominance of the wildtype gene, selection acting within the heterozygous individuals alone, is about 1/5000 of that which would be achieved if the entire population consisted of heterozygotes.

With this demonstration Fisher rested his case. Using data of the sort we shall present later in this chapter, Fisher could argue that artificial selection for gene expression in heterozygous individuals is in fact capable of modifying dominance. Furthermore, the calculations we have just made show that an identical but much less intense selection operates in natural populations. Thus Fisher concluded that the dominance exhibited by wildtype alleles is achieved by the accumulation of genetic modifiers through selection acting on heterozygous individuals.

The Evolution of Dominance: Wright's View

Criticism of Fisher's theory of the evolution of dominance through the selection of modifiers that tend to make heterozygotes resemble homozygous normals centered primarily upon the "second-order" selection involved. Selection, as we usually think of it, acts upon the relative frequencies of alternative alleles in a population. Fisher's scheme postulated that the relative frequencies of alleles at one locus are modified in response to an advantage accompanying the modification of heterozygotes at a second locus—heterozygotes whose frequency in the population is determined in turn by natural selection.

The intensity of selection in respect to the modifier under Fisher's model was calculated by Wright (1934). His calculation took the following form:

Let A and a be the alleles at the primary locus. Let M be the modifier that causes Aa to approach AA in respect to fitness but which has no effect on AA. Let M be completely dominant to m. Let q_M be the frequency of M and $1 - q_M$ be the frequency of m. Finally, let $u/h''s$ be the frequency of a, where u is the mutation rate of A to a and $h''s$ is the net disadvantage of Aa relative to AA averaged over all combinations of M and m. The disadvantage of Aa relative to AA in mm individuals is hs, while that in MM and Mm individuals is $h's$. The relation between h, h', and h'' is given by

$$h'' = (1 - q_M)^2h + [2q_M(1 - q_M) + q_M{}^2]h'$$

The distribution of genotypes in respect to A and M is given by

$$\left[\frac{u}{h''s}a + \left(1 - \frac{u}{h''s}\right)A\right]^2 \left[(1 - q_M)m + q_M M\right]^2$$

The selective disadvantage of mm individuals as a group (relative to the average of AA) equals $2uh/h''$, ignoring second-order terms. The disadvantage of M- individuals is $2uh'/h''$. Thus the advantage of M- over mm individuals, s_M, equals

$$\frac{2u(h - h')}{h''}$$

The most favorable case for the selective modification of dominance exists when M makes Aa equal in fitness to AA ($h' = 0$). In this case, h'' equals $(1 - q_M)^2h$ and s_M equals $2u/(1 - q_M)^2$. The rate of change of M equals

$$\frac{s_M q_M}{(1 - q_M)^2}$$

or

$$2uq_M$$

The value $2uq_M$ is the rate of change per generation of the dominant allele, M, brought about by the favorable selection it obtains as a modifier of the dominance of the allele A. The rate of change is so small that it would be completely erased by a mutation rate of M to m equal to $2u$ (twice the rate at which A mutates to a).

The calculations given above (based on Wright, 1934) make it seem unlikely that selection of dominance modifiers through their action on rare heterozygotes is an effective factor. As an alternative possibility, Wright suggested that there may be no need for a new mutation to *acquire* dominance. There may, indeed, be largely recessive genes whose heterozygous expression can be modified by selection; there is no reason to believe, however, that these are the ones that are seized upon in the evolution of populations. On the contrary, successful alleles, according to Wright, may not differ greatly in activity from their predecessors.

The Biochemical Framework of Dominance

A free-ranging discussion of dominance tends to reach back in time to the origin of life itself. Better that some restraints be imposed. And so, in the following discussion, we might acknowledge that life consists of a skein of interacting biochemical reactions each of which requires certain enzymes (gene products) but each of which in turn is regulated by a number of feedback controls. These feedback controls can affect both the enzymatically controlled processes and the production of the enzymes themselves. If we admit these things, and if we limit our attention momentarily to haploid organisms so that dominance will not enter our discussion, we can see that alleles favored by natural selection will be those that tend to overproduce their respective products.

The tendency to overproduce is a necessary property of any gene that is regulated by feedback controls. Just as a thermostat is useless if a furnace is incapable of overheating a house, so a feedback control is useless if a gene is incapable of performing more than adequately. On the other hand, in the presence of efficient controls, it is extremely difficult for a gene to actually overproduce. It simply gets turned off. Muller (1932b) described selection for this type of gene action although he had in mind homozygous diploid organisms. By this argument he circumvented the need to rely on selection in rare heterozygotes. In our above comments we have pointed out that overproduction of gene product is a feature "built into" a gene by selection early in the evolutionary history of life—even in haploids where there is no question of dominance.

With the admission that single functional alleles do indeed have a tendency toward dominance over inactive alleles—a tendency based on enzyme kinetics and control mechanisms established by selection under conditions where dominance need not exist; we can now turn our attention to the problem that really concerns us: Given that individuals of genotype *Aa* have a phenotype intermediate between that of the wild-type (*AA*) and the homozygous mutant (*aa*), what avenues for rectification are amenable to selection? Our discussion is limited, consequently, to those cases in which complete dominance within diploid organisms is lacking.

For purposes of discussion it is assumed that the attainment of a particular phenotype (phenotype$_{AA}$ of Figure 19-1) requires the presence of a substance *E* (for end product) in a critical concentration at a given time and place within the organism. In the absence (or with insufficient amounts) of *E*, phenotype$_{AA}$ is not formed; instead, an aberrant

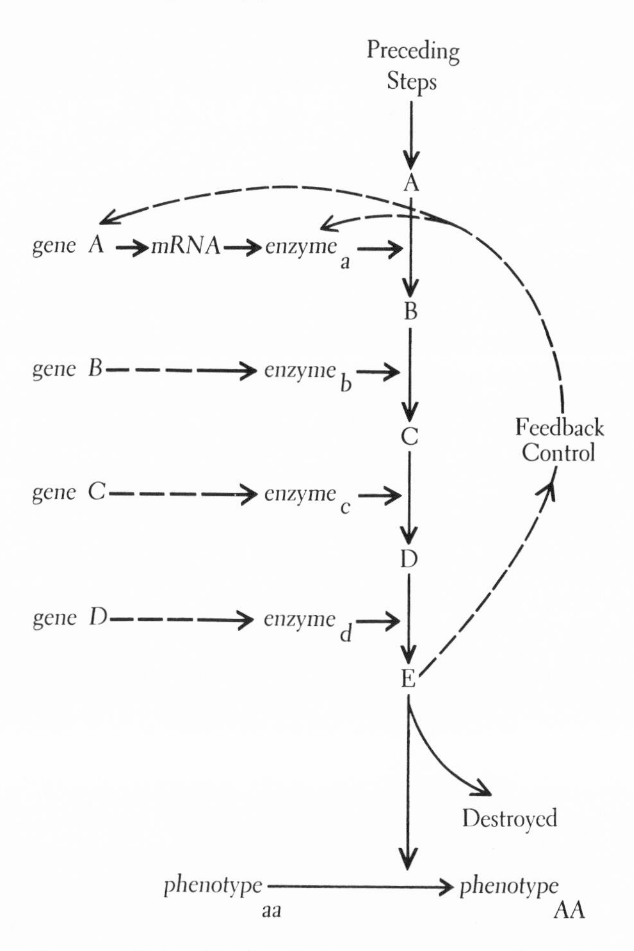

FIGURE 19-1. Biochemical steps concerned with the control of a hypothetical phenotype$_{AA}$ by a gene *A*; these steps are involved in the degree and modification of the dominance of gene *A*.

phenotype (for example, phenotype$_{aa}$ in the homozygote *aa*) results. This generalized statement covers phenotypic variation based on either quantities of phenotypic substances per cell or spatial distributions of such substances in gross patterns. Fundamentally, it merely recognizes that the difference between the appearance of one genotype and that of another is to be explained ultimately in terms of material substances.

Substance *E* is but one substance of many in a complex network of biochemical reactions. In Figure 19-1, *E* has been represented as the final step of a series leading from *A* through *B*, *C*, and *D*. As the figure shows, however, many steps precede the formation of *A*; furthermore, *E* itself is the start of a series of steps that lead to its removal or destruction. (We might note that should *E* be an inhibitor so that its removal is necessary for the development of phenotype$_{AA}$, we would merely shift our attention to a substance *E′* which is needed for the destruction of *E*. Whether the phenotype depends upon the presence or absence of a substance, the discussion remains unchanged.)

Gene *A*, shown in the figure as responsible for the synthesis of enzyme$_a$ which in turn controls the transformation of *A* to *B*, occasionally mutates to an allele *a*. Homozygous *aa* individuals are phenotypically unlike *AA* individuals, as noted above. Our interest centers on the phenotype of *Aa* individuals.

In Figure 19-1 the presence of feedback controls connecting *E* and gene$_A$ and its enzyme have been suggested by dashed lines. We have, however, agreed that these are not under discussion; such controls are essential to life and, once established, tend to confer dominance on functional alleles in diploid organisms. The interesting case for our present discussion is that in which the *Aa* heterozygote fails to develop phenotype$_{AA}$, presumably because it is unable to maintain an adequate supply of *E* at the critical time and place. What mechanisms exist for adjusting phenotype$_{Aa}$ so that it resembles phenotype$_{AA}$? Whatever these mechanisms are, they determine the nature of the evolution of dominance.

The corrective steps that are available to the organism and which when taken would be interpreted as the evolution of dominance can be inferred from an examination of Figure 19-1.

1. *The rate at which E is destroyed or otherwise removed can be lowered.* The concentration of *E* can be effectively increased by reducing the rate at which it is destroyed. Presumably this solution for dominance modification would require the modification of some gene in the destruction pathway or in still others that regulate the milieu within which this destruction is carried out. Upon genetic analysis, a corrective measure of this sort would be found to depend upon genes other than *A*—that is, upon modifier genes.

2. *The rate at which E is synthesized can be increased.* There are several means by which this can be achieved. If any one of the enzymes

b through *d* limits the overall rate of production of *E*, its quantity or its activity per molecule can be increased. Such alterations, whether caused by mutations at locus *B* (or *C* or *D*) or by still other gene changes, would represent changes in genetic modifiers of gene *A*.

If steps preceding *A* were limiting in the sense that more *E* could be produced in heterozygous individuals if more *A* were present, corrective measures could involve these "early" reactions. Once more, whatever the precise enzymatic change responsible for the adjustment, it would be looked upon as a change in a genetic modifier of gene *A*.

Finally, the transformation of *A* into *B* might be the limiting reaction in respect to the shortage of *E* in heterozygotes. In this case, a number of adjustments are possible. Upon genetic analysis, some would prove to be the work of modifiers: changes in the cellular milieu that make $enzyme_a$ more efficient, changes rendering the enzyme less sensitive to feedback control, or a reduction in the rate of destruction of $enzyme_a$. On the other hand, some of these alterations could be met by a change in the enzyme structure itself, by a change at locus *A*. Only the latter would represent an alteration of dominance by the substitution of one allele for another.

The details listed above might differ for different gene-controlled systems. Nevertheless, the main point is clear: A change in the relationship between gene dosage and phenotype is basically a change in the relationship between a given level of enzyme activity and the quantity of an often remote end product. There are a multitude of associated gene-controlled processes that influence this relationship. Consequently, when the need for change arises, the solution will often be found in the modification of physiologically associated processes that are under different primary gene control.

Feedback mechanisms promote dominance. But where sufficient control is lacking, where dominance does not exist, the corrective steps—the evolution of dominance—will frequently involve the evolution of modifier systems. Furthermore, the probability that a change in a single modifier gene will provide the entire solution to a given instance of incomplete dominance appears to be no greater than that the corrective measure will arise within the original locus. For this reason dominance will often appear to depend largely on many loci, on the "genetic milieu" of the locus in question.

Experimental Observations on Dominance

Industrial Melanism. Industrial melanism in various species of moths has given us our best opportunity to follow the evolution of dominance. "Industrial melanism" refers to the blackening of certain moths that has occurred since the onset of the industrial revolution in the early nineteenth century (see page 311). Ford (1964) has given an excellent account of this evolutionary phenomenon; in this section we shall cite only those details that are especially pertinent to the discussion of the evolution of dominance.

The natural history of industrial melanics is well known, largely through the efforts of E. B. Ford and his colleague H. B. D. Kettlewell. The moths that have been affected are those with cryptic markings; warningly colored ones or those that conceal themselves in crevices when at rest have been unaffected. The selective agent in the case of these melanics is differential predation by birds; Kettlewell and Tinbergen have excellent films showing birds of several species finding and eating moths taken at rest on tree trunks. Kettlewell has also shown that the amount of predation is considerable and that the protection offered by cryptic coloration is appreciable. Melanics are difficult to see on soot-blackened tree trunks; normals are difficult to see on the lichen-covered tree trunks of nonindustrial areas.

During the spread of melanic forms, dominance has evolved. The early heterozygotes of *Biston betularia,* while markedly different from typical wildtype specimens, had several white lines on their wings as well as pale patches; these have disappeared so that present-day heterozygotes are (with the exception of minute white spots on some) indistinguishable from homozygous melanics. The modification of the appearance of heterozygotes, an evolution of dominance that can be followed by an examination of pinned museum specimens, is important. The industrialization of Britain has resulted in the spread of melanism in over eighty species of moths in that country alone. That dominance evolves in these cases shows that fully dominant alleles are not the only ones seized upon in evolution to the exclusion of those whose heterozygotes are intermediate.

Granted that the heterozygotes of *B. betularia* have changed in appearance during the past century, can we decide whether the change came through selection for modifiers or by selection for new and more dominant alleles? At least part of the effect has been brought about by modifier genes. By transferring a dominant melanic allele from a population near Birmingham into a hybrid background of a nonmelanic population of a western, nonindustrial section of Britain, Kettlewell showed that the artificially bred heterozygotes had greater numbers of white scales than normal, Birmingham heterozygotes. Table 19-3 gives data from an analogous test of the dominance of a melanic (not *industrial* melanic, however) form of *Triphaena comes.* Experimental crosses of strains of this moth obtained from islands off the coast of Scotland (the Hebrides and the Orkneys) show that the gene for melanism is a "good" dominant; heterozygotes mated with one another produce progeny consisting of three melanics to one typical with very few hard-to-classify intermedi-

TABLE 19-3

EVIDENCE FOR THE ACCUMULATION OF UNIQUE COMBINATIONS OF DOMINANCE MODIFIERS IN ISOLATED POPULATIONS OF THE MOTH TRIPHAENA COMES.

(Ford, 1955.)

	MELANIC	INTERMEDIATE	TYPICAL	TOTAL
Barra F_2	149	3	44	196
Orkney F_2	110	2	29	141
(Barra × Orkney) F_2	99	22	39	160

ates. Matings of interisland heterozygotes, on the contrary, produce many intermediate forms; typical individuals account for one fourth of the total progeny, as before. It appears, then, that the dominance of the melanic alleles in the island populations is the result in part of modifier genes. Different modifiers have been established in the two, geographically isolated populations.

The final point to be made here in reference to industrial melanism can be expressed in the following form. Whatever allele possesses the greatest immediate advantage will spread most rapidly in a population. The greatest immediate advantage lies with the allele that confers an advantage on heterozygous individuals. These alleles may be displaced at a later time by ultimately more efficient but slower spreading ones. Nevertheless, at any moment the rapidly spreading alleles on the average will be the more common ones in the genome. This statement applies, at least in part, to the succession of melanic alleles that have been observed in moth species (see Kettlewell, 1965). More spectacularly, however, it applies to the relative proportions of partially dominant and recessive alleles used in the development of industrial melanisms. Ford (1964, p. 267) mentions a species (*Lasiocampus quercus*) in which a relatively rare but widespread and seemingly recessive allele that is responsible for a blackish genotype has been utilized for industrial melanism in each of two localities. This is the only recessive allele utilized among the more than eighty species that have developed dark forms.

The utilization of a recessive allele by only one of more than eighty species that have undergone melanization confirms our expectation concerning the more rapid spread of dominant or partially dominant alleles. Regardless of its mate, at least one half of the progeny of a rare heterozygote possess the rare phenotype if the responsible allele is dominant; rare recessive homozygotes must mate with one another (a very unlikely event) to leave progeny resembling themselves. In the case of *L. quercus*, a general darkening of both normal and dark forms has taken place in industrial areas. Assuming that industrial conditions favoring melanism in moth populations will continue, this species may yet furnish the most interesting data on the evolution of dominance. At the present time, the heterozygotes resemble the homozygous wildtype insects; it remains to be seen whether modifiers are available that will alter the phenotype of the heterozygote without altering that of typical insects to the same extent.

Dominance in Mimicry. Mimicry is a fascinating problem for population geneticists. In the case of Batesian mimicry, in which harmless organisms masquerade as others that are poisonous or otherwise harmful, the mimic must not become too common. If it does, predators may associate its appearance with "good" taste rather than with "bad." The problems raised by this need, and the answers hit upon by mimic species, underlie some of the classic studies of populations. *Ecological Genetics* by E. B. Ford (1964) has several chapters devoted to mimicry and polymorphism; these are recommended to those who are unfamiliar with the subject.

Here we are interested in the evidence mimicry presents us regarding the genetic basis of dominance. Some of this evidence is given in Table 19-4. *Papilio dardanus,* an African butterfly related to our own swallowtail, mimics a number of Danaids, butterflies related to our Monarch (an ill-tasting, ill-smelling insect that in the United States is mimicked by the Viceroy). In the area near Entebbe, on the northern edge of Lake Victoria, models are extremely common relative to mimics; in this locality, imperfect mimics are extremely rare. Some five hundred kilometers southeast near Nairobi, models are much rarer than their mimics. In this area, nearly one quarter of all mimics are imperfect in design. There is no reason to suspect that the dominant alleles controlling the mimic patterns found in the two geographic localities are different; it is much more

TABLE 19-4

DEPENDENCE OF THE FREQUENCY OF IMPERFECT MIMICS IN LOCAL POPULATIONS OF PAPILIO DARDANUS UPON THE NUMBER OF MODELS IN THE SAME LOCALITY.

(Ford, 1964, Table 15.)

	TOTAL	MIMICS	
LOCALITY	MODELS	TOTAL	% IMPERFECT
Entebbe	1949	111	4
Nairobi	32	133	32

likely that natural selection, in an area where models are rare, is not sufficiently strong to maintain the proper modifiers for the main alleles.

Dominance Modifiers of Scute *in D. pseudoobscura.* A shrewd guess about the path dominance modification will take—selection for modifiers or substitution of new alleles at the locus involved—can be made by studying the variation exhibited by partially dominant mutant genes. Helfer (1939) has made such a study using the "dominant" mutation *Scute* in *D. pseudoobscura.* The phenotypic manifestation of *Scute* is a lack of certain large bristles on the head, thorax, and scutellum of its homozygous and heterozygous carriers. Of the four dorsocentral and four scutellar bristles found on normal flies (eight bristles in all), only an average of 1.08 bristles is found on homozygous *Scute* females. The average number of bristles found on females heterozygous for *Scute* depends upon the source of the non-*Scute* chromosome ("genome" in F_1 hybrid females obtained by mating *Sc/Sc* females by wildtype males); values obtained by Helfer in his study ranged from four or more to about one. *Scute,* in other words, behaves as a complete dominant in some crosses and as a semidominant (four bristles rather than eight) in others.

The matings Helfer made in studying the variation in the dominance of *Scute* involved strains of *D. pseudoobscura* (and the sibling species *D. persimilis,* known at that time as Race B) from localities in Western United States, Mexico, and Canada. Three possibilities might explain this variation: (1) The wildtype alleles of *Scute* found in different localities may differ (isoalleles); (2) modifying genes in the rest of the genome may differ from locality to locality (dominance modifiers); or (3) both of these possibilities may be true.

The genetic analysis used by Helfer involved some rather elaborate mating schemes using genetic markers and crossover suppressors wherever needed. The essential features of his experiments are the following: Two wildtype strains (one from British Columbia and the other from Southern California) that differed considerably in their effect on the number of bristles on *Scute* heterozygotes were chosen for detailed analysis. The wildtype allele of each of these strains was transferred to a neutral genetic background. Taken out of their original backgrounds, the two alleles appear to be identical (Table 19-5). In fact, in a follow-up study based on different material, it appeared that the wildtype alleles at the *Scute* locus in the two sibling species, *pseudoobscura* and *persimilis,* are identical as well.

The analysis of background chromosomes was made by mating F_1 hybrid males (carrying genetically marked autosomes) with homozygous *Scute* females. Female flies carrying one (second, third, or fourth), two, or three major autosomes from the test locality could be identified by their phenotypes. The results of this test are also summarized in Table

TABLE 19-5

SUMMARY OF A GENETIC ANALYSIS OF THE MODIFICATION OF THE DOMINANCE OF *Scute* IN RESPECT TO BRISTLE NUMBERS IN D. PSEUDOOBSCURA.

Numbers refer to the average number of bristles (out of eight possible) possessed by the different types of flies. (Helfer, 1939.)

ORIGIN OF WILDTYPE	F_1 HYBRID
Wildtype males × *Sc/Sc* females	
Merritt, B. C.	2.5
Henshaw, California	1.2
Sc/Sc homozygotes	1.1
Wildtype allele of *Scute* in neutral background	
$_{+}$Merritt$_{/Sc}$	1.2
$_{+}$Henshaw$_{/Sc}$	1.2
$_{+}$*pseudoobscura*$_{/Sc}$	2.6
$_{+}$*persimilis*$_{/Sc}$	2.5

Amount of background from given locality:

GEOGRAPHIC ORIGIN OF WILDTYPE CHROMOSOMES	NO. OF AUTOSOMES HETEROZYGOUS			
	0	1	2	3
Merritt	1.58	1.93	2.40	3.12
Henshaw	1.61	1.28	1.07	1.21
Grand Canyon	1.52	1.74	2.03	2.82
Big Horn	1.86	1.56	1.72	1.43

19-5. Wildtype autosomes of Merritt origin carry suppressors of *Scute*; the more "Merritt" autosomes an individual carries, the greater the number of bristles. Wildtype chromosomes from Henshaw appear to enhance *Scute* relative to the "neutral" autosomes of this particular analysis. Two additional localities, Grand Canyon (Arizona) and Big Horn (Wyoming), were subjected to this chromosomal analysis; the wildtype chromosomes of one of these suppressed and the other enhanced the expression of *Scute*.

The implications of this fine analysis of dominance modification are clear. If the suppression of *Scute* were to become a major evolutionary problem in populations of *D. pseudoobscura*, the response would involve alterations in the frequencies of modifier genes at many different loci. There is no evidence that different wildtype alleles at the *Scute* locus exist; at least, if different wildtype alleles do exist, they are not markedly different in the extent to which they exhibit dominance over the mutant allele, *Scute*.

Linkage and the Selection of Dominance Modifiers

A number of studies, several of which have been cited above, have shown that the dominance exhibited by one allele over another is indeed subject to modification by alleles at other loci; the dominance of a given allele in a given locality may at times be the result of specific modifiers accumulated in that one, and to a large extent only in that one, population. On the other hand, studies such as that of Kettlewell (1965) on *Biston betularia* have shown that successive alleles, each exhibiting greater dominance than its predecessor, may sweep through populations in response to selection for dominance modification. In this section we shall consider dominance modification in algebraic terms. Following that, we shall return once more to reconsider the calculations of Fisher and Wright as given earlier in this chapter.

Polymorphism and Dominance. Suppose that a population, homozygous for gene m, is polymorphic for alleles A and a. The relative fitnesses of AA, Aa, and aa individuals are $1 - s$, 1, and $1 - t$, respectively. Consequently, the equilibrium frequencies of A and a are, in that order, $t/(s + t)$ and $s/(s + t)$.

Let the relative fitnesses listed above $(1 - s{:}1{:}1 - t)$ be based in part upon the relative partial fitnesses of AA, Aa, and aa of 1, $1 - r$, and $1 - r$, respectively. These are fitnesses ascribable to a single cause (such as predation by one particular predator) of all those that enter into and determine the selection coefficients s and t. By *partial* fitnesses 1, $1 - r$, and $1 - r$ we mean that each of the overall adaptive values $1 - s$, 1, and $1 - t$ can be written $(1 - s)(1){:}(1 + r)(1 - r)$ (approximately): $\left[\frac{1-t}{1-r}\right](1 - r)$. (It is immaterial whether the partial coefficient r is associated with a pleiotropic effect of a or with an allele at a neighboring locus that has been caught up by accident in the polymorphism at the A locus.)

Finally, let gene M, a modifier that increases the partial fitness of Aa individuals from $1 - r$ to 1.00 (that is, equal in this one respect to AA individuals), arise within the otherwise homozygous mm population. What will be the selective advantage of M over the earlier allele, m?

We can restrict our calculations to the case where M is very rare and, consequently, is found only in Mm heterozygotes. Because M is rare, the average fitness of mm homozygotes is very nearly the same as that of the entire population, $1 - st/(s + t)$. The fitness of *Mm* individuals can be calculated as the average of the following:

GENOTYPE	FREQUENCY	FITNESS
AA Mm	$t^2/(s+t)^2$	$1 - s$
Aa Mm	$2st/(s+t)^2$	$1 + r$ (very nearly)
aa Mm	$s^2/(s+t)^2$	$1 - t$

The average fitness of *Mm* individuals is $1 - st/(s + t) + 2rst/(s + t)^2$. The increase in fitness of *Mm* individuals relative to the average fitness of the population (and hence of *mm* individuals) is $2rst/(s + t)(s + t - st)$. Since only one half of the gametes of *Mm* individuals carry the gene *M*, the selective advantage of this gene relative to its allele *m* is $rst/(s + t)(s + t - st)$. This, in turn, can be written as the product of an equilibrium gene frequency and a second term:

$$\left(\frac{s}{s+t}\right)\left(\frac{rt}{s+t-st}\right) \quad \text{or} \quad \left(\frac{t}{s+t}\right)\left(\frac{rs}{s+t-st}\right)$$

In the case just described, the alleles at the *m* locus were combined at random with those at the *a* locus. If, on the contrary, one assumes that *M* is linked with either *A* or *a*, calculations similar to those given above yield the following advantages of *M* over *m*:

Linked with *A*	$rs/(s+t-st)$
Linked with *a*	$rt/(s+t-st)$

Since both $s/(s + t)$ and $t/(s + t)$ are less than 1.00, it is obvious that linkage of *M* with either *A* or *a* confers a greater advantage on *M* than this allele enjoys when its locus is independent of the *a* locus. Furthermore, the advantage of *M* is greatest when it is linked to the rarer of the *a* alleles. This point is self-evident because linkage with the rare allele virtually guarantees that *M* will be found in the heterozygous *Aa* individuals whose fitness it modifies; Crosby (1963) refers to this as the evolution of recessiveness.

A numerical example, using values of *s* and *t* commonly encountered in the case of inversion polymorphisms in *D. pseudoobscura*, shows that the selective advantage of *M* over *m* need not be negligible. Let *s* equal 0.20, *t* equal 0.70, and *r* equal 0.05. The advantages of *M* over *m* under the three linkage possibilities are as follows:

M independent of *A*	0.010
M linked to *A*	0.013
M linked to *a*	0.046

These values, we must admit, apply to extremely favorable assumptions regarding dominance modification. If, for example, the existence of *Aa* individuals were to depend upon recurrent mutation alone as in the calculations made by Fisher and Wright that were given earlier (page 323ff.), the advantage of M over m under otherwise equally favorable assumptions would be tens of thousands times smaller. Selection for dominance modifiers need not be negligible if heterozygotes are at all common.

Another Look at Fisher's and Wright's Calculations. We have made not only theoretical calculations showing that dominance modifiers may have appreciable selective advantages over their nonmodifying alleles, but we have also given examples that seem to prove that dominance is in fact under the control of modifier genes. Our biochemical model also suggested that among those events which might modify dominance, most would prove upon genetic analysis to be modifiers at various gene loci. What was the purpose, then, of repeating the initial calculations of Wright and of Fisher? Was Wright mistaken when he claimed that selection for dominance modifiers would be about equal in magnitude to mutation itself?

No, Wright was not mistaken; there were no errors in his calculations. The contrast between his results and the observations we have cited lies entirely with the frequency of heterozygotes in the population. Wright restricted his attention to the selection for dominance in rare heterozygotes—heterozygotes kept in the population by recurrent mutation and whose frequency equals that arrived at in our earlier (page 192) calculation, u/hs. Compare this assumption with the examples we cited. One example concerned the dominance of alleles that were partially dominant when first observed and that were favored by selection from the outset. Heterozygotes for these alleles, those concerned with industrial melanism, have been a substantial part of their populations for many, many years. The other alleles that we discussed, both in algebraic terms and in reference to mimicry, were involved in balance polymorphisms; heterozygotes in these cases, too, are extremely common. Here, then, is the contrast between theory and observation.

In addition to the effect of the frequency of heterozygotes, we discussed the effect of linkage on the magnitude of selection for dominance modifiers. Linkage is extremely important. An allele that is tightly linked to a rare allele and that makes it behave as a complete recessive (that is, confers dominance on the alternative allele), possesses a selection coefficient much greater than that of an unlinked allele that has precisely the same modifier effect. In fact, the new "alleles" described by Kettlewell (1965) for *Biston betularia* might easily prove to be the old allele associated with a "new," closely linked modifier.

In the calculations of Wright cited early in this chapter, the modifier, M, was allowed to segregate independently of the alleles at the a locus.

What would have been the effect if instead he had postulated tight linkage (coupling) of M and a so that, even though they were rare, all heterozygotes resembled the homozygous normals? In the original calculations, the allele M was distributed at random between AA, Aa, and aa individuals; each M carried by an AA or aa individual was wasted as far as its special "talents" (modification of the fitness of Aa individuals) were concerned. Linked to a, each M allele at the outset must be carried by an Aa individual [if $(u/hs)^2$ can be neglected]. Thus the efficiency of selection would be enormously increased; this increase amounts to approximately $hs/2u$-fold. Thus, for a lethal gene a ($s = 1.00$) that is slightly dominant ($h = 0.02$) and arises by mutation in one gamete of each million ($u = 10^{-6}$), selection for a modifier M *that is linked to the lethal allele itself* would be 10,000 times as great as for an unlinked modifier. If heterozygotes are rare, consequently, a premium is placed upon the linkage of modifiers to the rare allele. And, with linkage of this sort, selective pressures—although low—need not be negligible.

The Evolution of Dominance and the Genetic Load

The evolution of dominance offers an excellent opportunity to demonstrate that the genetic load of a population is not identical with the loss of individuals the population suffers as the result of natural selection. Further, it offers us an opportunity to demonstrate that "genetic deaths" and "nongenetic deaths" (environmentally caused deaths, ostensibly) are not independent compartments in an evolving population; as a matter of mathematical convenience they can be treated separately but only under rigidly defined, static conditions.

In the section that dealt with polymorphism and the evolution of dominance, we described an allele M that corrected a slight flaw in what were generally superior Aa heterozygotes. The fitnesses of AA, Aa, and aa individuals before M arose and after it had spread in the population can be tabulated as follows:

GENOTYPE	ORIGINAL (mm) BACKGROUND	NEW (MM)	NEW (MM) BACKGROUND (ADJUSTED)
AA	$1 - s$	$1 - s$	$1 - S$
Aa	1	$1 + r$	1
aa	$1 - t$	$1 - t$	$1 - T$

The values $1 - S$ and $1 - T$ are obtained by dividing $1 - s$ and $1 - t$ by $1 + r$. S and T are obviously greater than s and t. The genetic load, which has changed from $st/(s + t)$ to $ST/(S + T)$, has become larger. But the advantage conferred by M upon Aa individuals according to the earlier discussion was the result of *decreased* predation. We find ourselves, then, claiming that lessened predation has increased the genetic load. This seeming paradox has a simple explanation. Predation in the earlier population would have been ascribed to environmental causes; the population was uniformly *mm*, so no genetic deaths could be ascribed to that locus. The postulated advent of the allele *M* has changed all this. Deaths have been retrieved from those formerly regarded as accidental and have been relabeled "genetic deaths"; following the origin of *M*, these deaths have become part of the genetic load. If we recall that an average of only one daughter can survive per mother (Figure 1-1), we can see how artificial this apportionment of deaths can be. The allele *M* does not affect progeny size in our example; therefore, the number of offspring that must die before reaching maturity is precisely the same after the origin of *M* as before. The enlarged genetic load brought on by the allele *M* is an enlarged load in name only; some other load borne by the population has diminished correspondingly.

20

NONREPEATABLE CHANGES IN POPULATIONS

Repeatable experiments are the backbone of science. Not unexpectedly, then, when Dobzhansky and Pavlovsky (1953) published a paper entitled "Indeterminate Outcome of Certain Experiments on *Drosophila* Populations," someone should say that population genetics was no longer a science. This person was, of course, wrong. He had no appreciation for probable as opposed to assured outcomes of laboratory experiments. Chemical and physical experiments, at least the majority of them, are deterministic. Identical reactants mixed in a series of identical flasks under identical conditions will, to the best of our observational ability, yield identical results. Not all experiments are of this sort, however. A steel ball dropped into a well-constructed "pinball" board will bounce erratically from pin to pin until it comes to rest in one of a series of compartments at the lower edge. A second ball dropped in a seemingly identical manner will generally take a different path and fall into a different compartment. The path taken by an individual ball in this case is indeterminate. By observing the course taken by large numbers of balls, one learns to make accurate probability statements concerning the compartment in which a ball will come to rest. Although the fate of an individual ball is indeterminate, the distribution of probable outcomes of many repeated trials can be accurately described.

By showing that the outcomes of certain experiments on *Drosophila* populations were indeterminate, Dobzhansky and Pavlovsky did not show that population genetics does not deserve to be considered a science. They did show, however, that an "experiment" involving a year-long study of gene frequencies in a laboratory population of *Drosophila* need not be an experiment in itself. Under certain circumstances, a study of this sort is equivalent to dropping a single ball into a pinball board. Only the population takes a year to come to rest! Under circumstances such as these, a meaningful *experiment* must consist of many replicated populations; one population is but a single *observation.*

Nonrepeatable changes in populations are those most likely to affect individual natural populations periodically during their existence. Consequently, although we may not be able to predict for individual populations the outcomes of such changes in advance, we can appreciate why they are indeterminate and attempt to understand what controls the range of possible outcomes available to the population. In this chapter we shall first describe a series of three experiments performed by Dobzhansky and his colleagues. Second, we shall examine the outcomes of these experiments in terms of adaptive landscapes. Third, we shall discuss the nature of populations that develop from migrant individuals ("founders"; Mayr, 1942). Finally, we shall conclude the chapter by referring to a topic that was popular several decades ago (and still is in some branches of biology), the irreversibility of evolution.

On the Indeterminate Outcome of Experiments on Drosophila Populations

Initial Observations. The starting point of our discussion of Dobzhansky's experiments on nonrepeatable results is the series of three laboratory populations—#s, 55, 66, and 75—which were involved in the discussion of coadaptation (page 317). We can refresh our memory with the following summary: A "one-generation" experiment revealed that the larval survival of *ST/ST, ST/CH,* and *CH/CH* flies in one particular population (#55) were in the order listed; as a result, *ST* would be expected to eliminate *CH* if the population were to be continued as a free-breeding one. In a second population (#66) containing *ST* and *CH* chromosomes of the same sources, the frequencies of the two gene arrangements (after some initial irregular fluctuations) stabilized at equilibrium; *ST* did not reach 100 percent as expected. Finally, a test of larval survival utilizing *ST* and *CH* chromosomes isolated from population 66 revealed that *ST/CH* hybrid larvae exceeded both homozygotes.

The *ST* and *CH* chromosomes used in setting up population 55 and 66 were from different geographic localities; the *CH* chromosomes were from Mexico, while the *ST* chromosomes were from California. A total of twelve strains of *CH* and fifteen strains of *ST* flies were used in these experiments.

Two explanations can be advanced to explain the change in relative survival of *ST/CH* larvae from an intermediate one in population 55 (and, hence, at the outset in population 66) to a superior position within fifteen months in population 66. It is possible that one or two *CH* strains contained chromosomes that gave superior heterozygotes with those carried by one or two of the *ST* strains. The improvement of the relative survival of *ST/CH* larvae in this case would come about through selection favoring the retention of these particular *CH* and *ST* chromosomes and through the gradual elimination of the others.

Alternatively, the *CH* and *ST* chromosomes that possessed high fitness (measured in terms of larval survival) at the termination of population 66 may have arisen within the population through recombination. These selectively advantageous chromosomes may not have been present as such in the original strains.

If the first of these alternative explanations were true, replicate populations started with the same two dozen or so chromosomes should go through the same sequence of changes and reach the same final equilibrium frequency as the favored chromosomes increased in frequency and the others were eliminated. On the other hand, if the second explanation were the true one, a repetition of the sequence of changes observed in population 66 would depend upon the likelihood that the favorable combinations (or, more exactly, the *same* favorable combinations) were to reappear once more.

From data presented earlier (Tables 11-3 and 18-2 and Figures 11-1 and 18-1), we have seen that replicate populations containing the two gene arrangements obtained from Mexico and California underwent markedly different sequences of gene changes. Populations containing the same two gene arrangements both of which were taken from the same locality, on the contrary, gave highly reproducible results.

Chance and Selection. The *ultimate* source of all gene variation is mutation. But once variation exists at individual loci, recombination becomes the source of virtually endless combinations of alleles. The numbers of genotypes that can be formed by different combinations of alleles at various numbers of loci are listed (in part) in Table 20-1. Two alleles at each of 10 loci can form nearly 60,000 combinations; two alleles at each of 100 loci can give rise to 3^{100} combinations—a number involving some 47 digits. One hundred loci is an exceedingly modest number; it would not be unreasonable to base the above calculations on the basis of a thousand or more segregating loci.

TABLE 20-1

RELATION BETWEEN THE NUMBER OF GENE LOCI, THE NUMBER OF ALLELES PER LOCUS, AND THE NUMBER OF POSSIBLE GENOTYPES.

NO. OF	NO. OF LOCI					
ALLELES	1	2	. . .	10	. . .	n
1	1	1	. . .	1	. . .	1
2	3	9	. . .	59,000	. . .	3^n
.	.	.		.		.
.	.	.		.		.
.	.	.		.		.
k	$\frac{k(k+1)}{2}$	$\left[\frac{k(k+1)}{2}\right]^2$	. . .	$\left[\frac{k(k+1)}{2}\right]^{10}$	. . .	$\left[\frac{k(k+1)}{2}\right]^n$

While the number of genotypes formed by recombination is tremendously high, the number of adult flies in an experimental population is generally less than 15,000—quite often less than 5000 in laboratory populations of *D. pseudoobscura.* Thus only a minute fraction of all possible genotypes can actually arise in any one generation of any one population. The fraction is so minute and the number of genotypes so immense that it is unlikely that any two populations are alike. No! Worse than that. With such a large variety to choose from, two populations will almost certainly differ in large numbers of genotypes they happen to contain. And, to whatever extent these dissimilarities influence natural selection, the two populations as they evolve cannot possibly follow similar paths.

In a mathematical sense, the chance events we have described immediately above are related to those discussed in Chapter 6. In a biological sense they are markedly different. The earlier discussion centered on fluctuations of gene frequencies caused by repeated samples of finite size. Whenever a specific example was to be cited, the sample became quite small (ten beans or sixteen flies in the two examples used) so that fluctuations would be reasonably large. We are now discussing the inclusion or exclusion *by chance* of certain genotypes out of an immense number of possible genotypes in a population of several thousand individuals. From any single population in a single generation, most of all possible genotypes are missing entirely. Hence the chance inclusion of certain genotypes and the chance exclusion of the rest determine what future course natural selection must take. It is this combination of chance and selection that makes the historical background of a population so important in its evolution. Evolution is a historical process. It proceeds only on the basis of what is at hand at the moment, and that is determined, in turn, by what was on hand yesterday.

Population Size and Repeatability. Our guess has been that the erratic results obtained from geographically mixed populations originate in the immense number of gene combinations that can be generated in such populations. Intralocality populations of single geographic origin—populations containing much less genetic variation from the very outset—are unable to generate a comparable array. And hence one population of several thousand flies comes to possess an array of genotypes very much like those of other such populations. At least, very much like those of the others in respect to important gene combinations; it is still unlikely that any two populations are entirely alike genetically.

If this guess is correct, the apparent indeterminacy of experimental results can be increased or decreased at will. The crux of the matter lies in the inability of a finite population to contain more than a small fraction of the almost unlimited number of genotypes generated by recombination. Small numbers of individuals must represent even smaller proportions of this array of recombinant genotypes than large numbers. Consequently, a series of replicate populations each of which started with very few parents should show a greater array of end results than a series of populations each of which started with large numbers of parental flies.

The above prediction has been confirmed experimentally (Dobzhansky and Pavlovsky, 1957). The initial material consisted of Pikes Peak (*PP*)-bearing strains of *D. pseudoobscura* from Austin, Texas, and *AR*-bearing strains from Piñon Flats, California. The strains were crossed and from the *AR*/*PP* F_1 hybrids an F_2 generation was obtained. Four thousand F_2 flies produced the eggs, large numbers of eggs, from which ten populations were started. These were called "large" populations because 4000 is a relatively large number of *Drosophila.* The populations were as similar as heterogeneous populations can be because the same 4000 flies were the parents of each of the ten populations.

While the above procedure was being carried out, ten groups of 10 F_2 males and 10 F_2 females were used as parents in the production of F_3 flies in culture bottles. Each of the ten sets of F_3 progenies was transferred to a population cage. These were called "small" populations because of the bottleneck of 20 F_2 flies used in setting them up; the numbers of adult flies within the population cages were the same for the "large" and "small" populations.

The results of this study are summarized in Table 20-2 and Figure 20-1. Repeatability is measured in terms of variation in the frequency of the *PP* gene arrangement *from population to population.* If, at the time chromosomes were sampled for analysis, the ten populations were

TABLE 20-2

FREQUENCIES (PERCENT) OF *PP* CHROMOSOMES, THEIR MEAN, AND BETWEEN POPULATION VARIANCE FOR TWO SERIES OF POPULATIONS; 10 LARGE AND 10 SMALL.

(Dobzhansky and Pavlovsky, 1957.)

	LARGE		SMALL	
CAGE	OCT. 1955	NOV. 1956	OCT. 1955	NOV. 1956
1	39	32	38	18
2	42	29	31	32
3	29	35	31	46
4	38	34	32	47
5	33	23	34	33
6	36	20	42	47
7	40	32	37	16
8	41	22	25	34
9	37	26	38	32
10	42	22	25	22
Average	37.85	27.44	33.33	32.73
Variance	15.30	26.96	26.73	118.91

identical, the variation between populations would be no greater than that expected from sampling errors alone. To whatever extent the frequencies of *PP* in the different populations differ, repeatability suffers. The table and figure show that the populations do differ from one another. The variance expected from analyzing 300 chromosomes from each population is about 7.00; even the smallest observed variance is more than twice that expected if the populations were identical. About four months after the start of the experiment, the variation between small populations seemed to be larger than that between large ones but the difference between the two series is not significant. After one more year, however, the interpopulation variance in the case of small populations was clearly greater than that in the case of the large ones. The more limited the sample of genotypes taken from the initial array of recombinant genotypes, the less is our ability to predict the ultimate outcome of selection acting on the initial sample.

Initial Variability and Repeatability. On the basis of the determinate outcomes of intralocality populations and the relatively indeterminate results of geographically mixed populations, we guessed that the number of genotypes generated by interpopulation recombination was much too large to be accommodated by the limited number (several thousand)

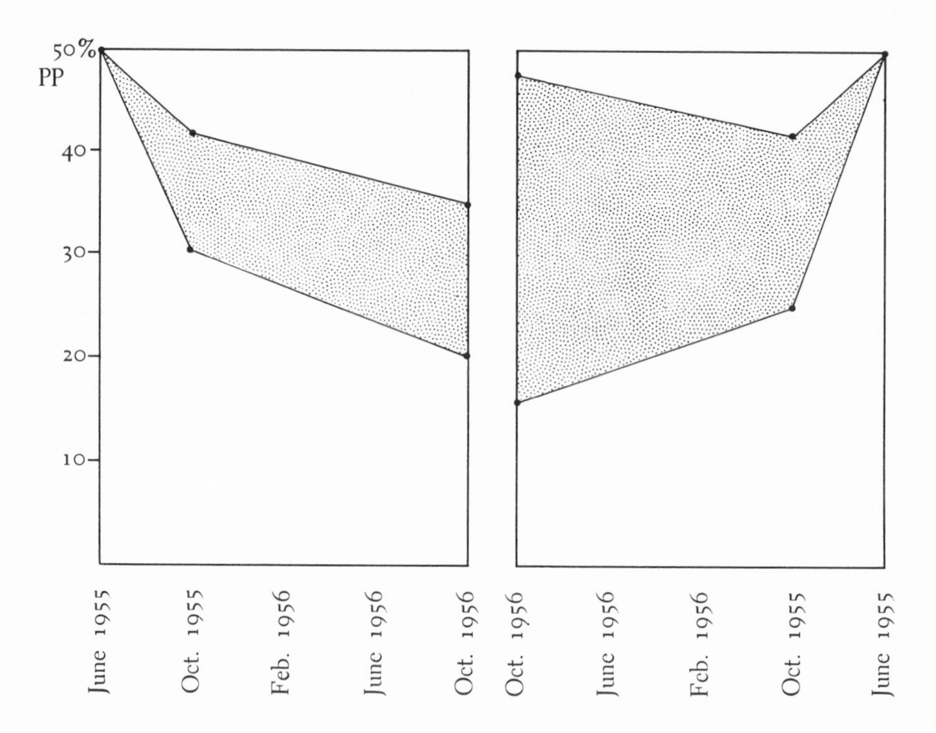

FIGURE 20-1. Interpopulation variance in the frequency of *PP* chromosomes among 10 large (left) and 10 small (right) mixed-locality populations of *D. pseudoobscura.* The first test of inversion frequencies was made 4 months after the populations were started; the second test was made 1 year later. (After Dobzhansky and Pavlovsky, 1957.)

individuals in a laboratory population and hence that each population got what was essentially a different set of genotypes with which to start its existence. In each subsequent generation, selection would act on the genotypes generated in turn by the preceding one and, as a result, the initial differences set each population off on a course of its own.

In the section immediately preceding, we described the outcome of one test of this hypothesis. We say that small populations (small in numbers of the initial founders) taken from an array of recombinants follow more divergent paths (measured in terms of the relative frequencies of cytologically distinguishable gene arrangements) than do large populations taken from the same initial array.

The second test of the original hypothesis (Dobzhansky and Spassky, 1962a) involved the study of populations of identical size taken from two arrays of genotypes, one of which was known to be larger than the other. In a sense, this experiment is a repeat of the original one; the difference is that the two arrays of genotypes were more accurately controlled in the later experiment. The larger array of genotypes ("multichromosomal") was obtained by pairwise matings of ten strains of *PP* from Texas with ten strains of *AR* from California. An F_2 was raised for each cross and five groups of 20 flies—one male and one female from each of the ten F_2's—were chosen to start five populations. The smaller array of genotypes ("bichromosomal") was obtained by crossing a single *PP* strain from Texas to an *AR* strain from California. The F_2 generation was obtained and from it five groups of 20 flies (10 males and 10 females) were chosen to start five populations.

To exaggerate the effect of selection in these populations, a sample of twenty flies were drawn from each population every four months; these twenty were used to start the freely breeding population once more. Except for these periodic bottlenecks, the populations consisted of several thousand individuals.

The outcome of these experiments is shown in Table 20-3 and Figure 20-2. The table shows the extreme frequencies of *AR* observed in the five populations at the end of each 4- or 5-month cycle. In the case of the multichromosomal populations, there is a steady increase in the range between the population with the highest and that with the lowest frequency of *AR*. In the figure, the interpopulation variance has been plotted; this variance increases steadily. The bichromosomal populations behave quite differently; neither the range nor the interpopulation variance increases regularly; the two merely fluctuate around fairly low values. Neither one of two replicate experiments, each involving five bichromosomal populations and lasting for five cycles each, underwent a change in variability (range or variance) comparable to that of the multichromosomal series.

The conclusions to be drawn from these experiments agree with those suggested by the earlier ones. Two strains, one from each of two populations, did not give rise to enough variation to seriously affect the system that was kept under observation—the relative frequencies of *AR* and *PP* gene arrangements. The multichromosomal populations, however, did have sufficient variability so that repeated small samples of it (twenty flies every four or five months) appreciably altered the nature of the selection acting on the *AR-PP* system.

Repeatability, Nonrepeatability, and Adaptive Landscapes

The landscape analogy that was described in Chapter 16 gives us a convenient basis for visualizing the events that have been described in the preceding sections. In the three figures (20-3, 20-4, and 20-5) the same rather complex landscape has been drawn; it consists of a fair number of peaks of various heights isolated from one another by "valleys." Our rule in respect to a landscape such as this is that a population will

TABLE 20-3

RANGE IN THE FREQUENCIES OF *AR* CHROMOSOMES IN DIFFERENT POPULATIONS OF D. PSEUDOOBSCURA.

Multichromosomal populations contained initially 10 different *AR* and 10 different *PP* chromosomes; bichromosomal populations contained a single *AR* and a single *PP* chromosome. (Dobzhansky and Spassky, 1962a.)

CYCLE (DATE)	*AR* RANGE OF FIVE POPULATIONS, %		
	Low	*High*	*Range*
Multichromosomal			
1. Oct. 1957	70.3	78.3	8.0
2. Mar. 1958	69.3	85.0	15.7
3. Aug. 1958	68.3	86.7	18.4
4. Jan. 1959	72.0	93.3	21.3
5. May 1959	66.7	92.0	25.3
6. Sept. 1959	62.7	91.7	29.0
. . .	. . .	. . .	. . .
9. Sept. 1960	63.0	93.7	23.7
Bichromosomal (first experiment)			
1. Oct. 1957	46.0	53.3	7.3
2. Mar. 1958	40.7	52.3	11.6
3. Aug. 1958	32.0	42.3	10.3
4. Jan. 1959	34.7	43.3	8.6
Bichromosomal (second experiment)			
1. May 1959	56.7	67.0	10.3
2. Oct. 1959	53.3	74.0	20.7
. . .	. . .	. . .	. . .
4. June 1960	54.7	65.3	10.6
5. Sept. 1960	58.3	75.7	18.4

always climb up hill, never down. If the collective genotype of a population initially embraces several hills, the population will climb the highest of them all; if the population initially embraces only one hill or if it lies on the slope of but one hill, it will climb that one whatever its height.

The initial experiment on intra- and interlocality populations is represented in Figure 20-3. The total area available to the geographically mixed recombinant genotypes is much larger than that of the intralocality ones. Consequently, the four interlocality hybrid populations have a

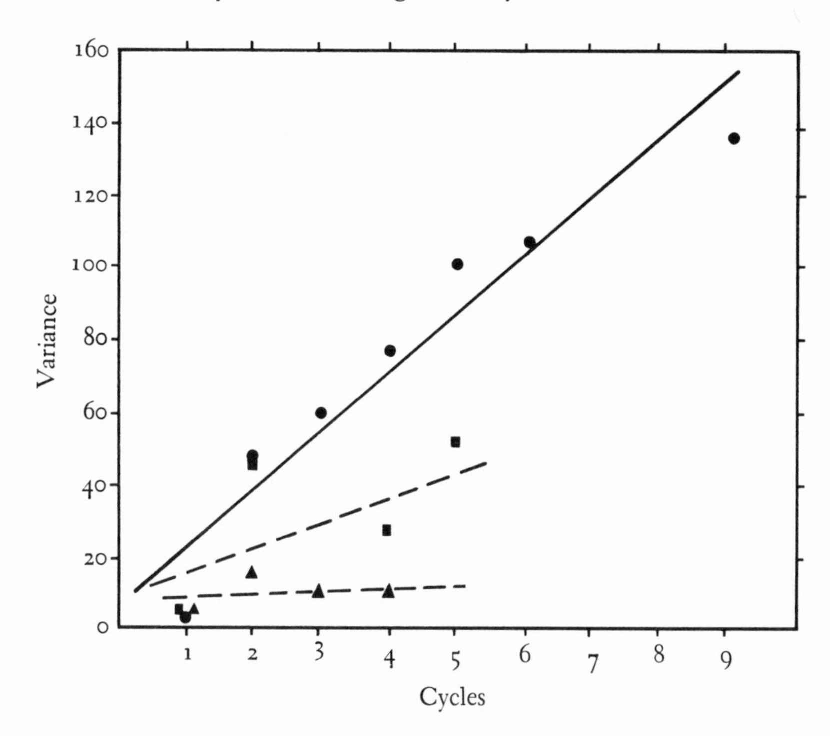

FIGURE 20-2. Increasing variance of inversion frequencies in multichromosomal (solid line) and in bichromosomal (dashed lines) populations of *D. pseudoobscura.* (After Dobzhansky and Spassky, 1962a.)

greater area over which to scatter and, as shown in the figure, come to lie on slopes leading to different hills. The intralocality populations are confined to a much smaller total range. As a result, all are found on the slope of a single hill that each promptly climbs (see Figure 11-1).

The scheme representing the experiment involving large and small populations is shown in Figure 20-4. In this figure the total array of recombinant genotypes was fixed. However, the ten small populations included such small portions of the total range that they lay scattered here and there—frequently on slopes leading to separate hills. The large populations, on the contrary, frequently overlapped on the slope of one particular hill and, as a result, many behaved in a similar manner. There was enough scatter, however, that even an occasional large population climbed its own hill.

The final series of bichromosomal and multichromosomal populations has been represented as a landscape in Figure 20-5. As in Figure 20-3, the difference between the two sets of populations lies in the amount of terrain covered by the initial array of recombinant F_2 genotypes. In the case of the multichromosomal populations this area was large, and

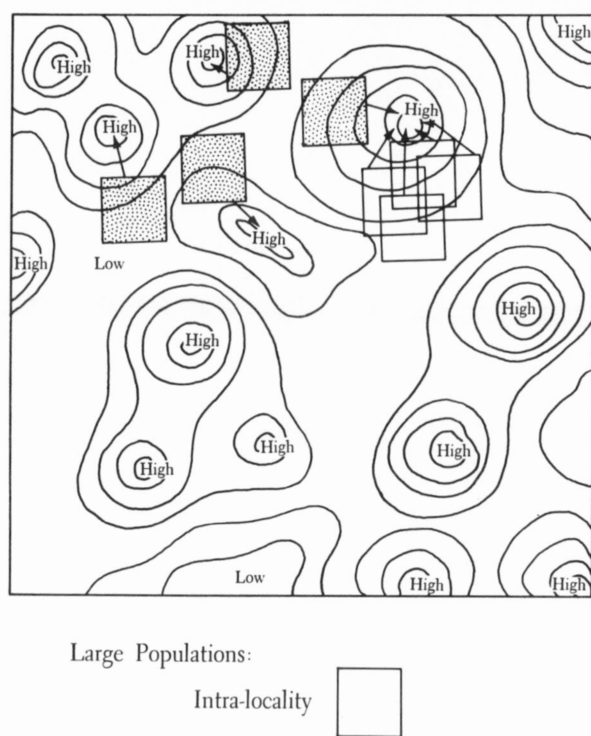

Large Populations:

Intra-locality

Inter-locality

FIGURE 20-3. Use of an adaptive landscape to illustrate the diverse outcomes of mixed locality (inter-locality) experimental populations and the uniform results of intralocality laboratory populations. Compare this figure with Figures 11-1 and 18-1.

so the small populations (twenty flies) came to lie on different hills. The repeated cycles simply opened up new vistas each unique to the hill upon which a given population lay. Consequently, the divergence between populations continued with each cycle. The bichromosomal populations were so restricted by the field of recombinant genotypes that they all lay on the slope of but one hill.

The experiments we have been discussing, then, form a reasonable extension to the work of Lewontin and White (1960). In the present case, however, we observe the "hill-climbing" activity of populations not by calculating fitnesses associated with the gene arrangements but by interpreting changes in these gene frequencies as signs of hidden turbulences affecting genes at many loci. Our indicator is not a sensitive

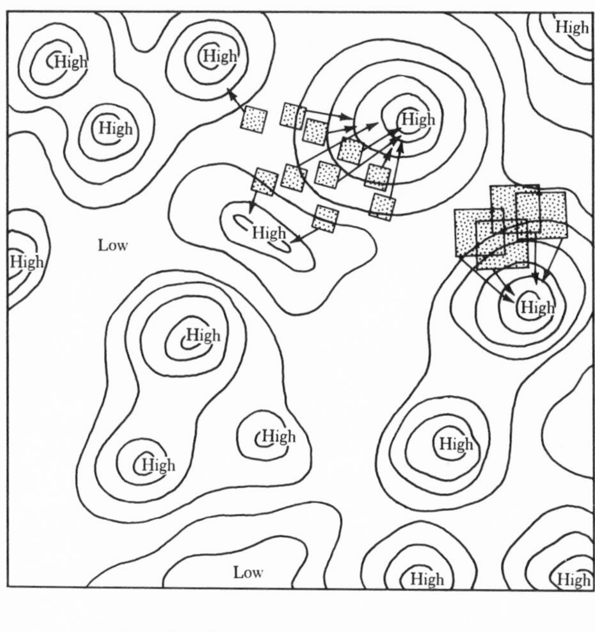

FIGURE 20-4. Use of an adaptive landscape to illustrate the diverse outcomes of large and small populations derived from an initial interlocality mixture. Compare this figure with Figure 20-1.

one but it is the best at hand. The use of such an insensitive detection system recalls a point made earlier in reference to Figure 13-2. Had we merely observed the frequency of lethal chromosomes in population 5 during the period shown in the figure, we might or might not have detected the slight increase in this frequency. Even if we had detected it, however, we would not have suspected that this reflected the replacement of nearly one half of all lethals in the population by one particular (a highly favored) lethal.

THE FOUNDER PRINCIPLE

Occasionally natural populations must arise through the accidental transport of single gravid females or other small numbers of individuals to unoccupied territories. Oceanic islands that are really the tips of volcanic

Inter-locality (all small populations):
20 Chromosomes
2 Chromosomes

FIGURE 20-5. Use of an adaptive landscape to illustrate the diverse outcomes of multi- and bichromosomal experimental populations. Compare this figure with Figure 20-2.

peaks rising from the ocean floor must be barren of terrestrial life when they first emerge from the sea. These islands are subsequently settled by immigrants that arrive in varying numbers—presumably multitudes of spores and small, windborne seeds, smaller numbers of water borne seeds and living plants, and even smaller numbers of various small and medium-sized animal species. Mayr (1942, p. 237) referred to the single gravid females and, in general, small groups of migrant individuals that succeed in establishing populations as "founders." At that time, he pointed out that these few individuals could not carry all the genetic variability of a population to the new location. Four founders, for example, can carry at best eight alleles at each gene locus. These would most likely be of a common sort. If, on the other hand, a rare allele were to be included among them, its frequency would be considerably greater (12.5 percent is the lowest frequency possible among eight alleles) than in the parental population.

Mayr's original discussion of founders dealt with the genetic uniformity of populations arising from them. In 1954 he published a further account in which his interest shifted to the variation that would exist *between different populations* that were established by small groups of founders drawn from the same large parental population. This emphasis coincides precisely with the emphasis that we have placed on the replicate populations discussed in this chapter.

Mayr's pictorial representation of the basis for the genetic divergence of isolated populations started initially by small numbers of founders is slightly different from those we have used; his is shown in Figure 20-6. Shown in this figure are the relative fitnesses (measured on an artificially expanded scale extending from 0 through 9) of alleles a_1 and a_2 in twenty background genotypes. No effort has been made to introduce the two homozygotes and the heterozygote into this model; the alleles are treated as if the organism were haploid. The fitnesses of the two alleles in each background genotype have been obtained by referring to a table of random digits; the fitnesses were chosen, however, so that each of the ten possible values was used twice in the entire series of twenty backgrounds. Consequently, the average fitnesses (4.5) of the two alleles in the series of twenty backgrounds are identical.

Now, suppose that small groups of founders are drawn from the larger group. These smaller groups are shown in the figure as the first and last groups of five background genotypes. If these represent the genotypes transported to new localities by five founders each, what about the relative fitnesses of a_1 and a_2? In one of the groups chosen, the average fitness of a_1 exceeds that of a_2, while the reverse is true in the second. According to this scheme, then, the two populations founded by these two groups

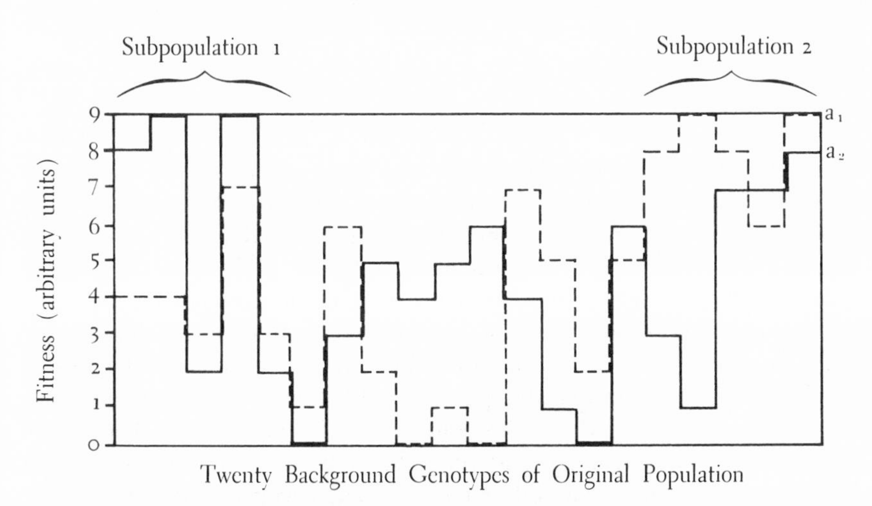

FIGURE 20-6. Model illustrating that the average fitnesses of two alleles, a_1 and a_2, need not be identical in two small subpopulations even though they were identical in the original population from which the two subpopulations were drawn. (After Mayr, 1954.)

of migrants would come to differ in composition in respect to the *a* alleles. This in turn would set off further compensatory changes, since a similar figure could have been drawn for the *b, c, d,* and other loci as well. Thus Mayr has given us an explanation for the genetic divergence of isolated populations founded by small numbers of individuals that supplements the adaptive landscape shown in Figure 20-4.

In developing his thesis concerning the founder principle, Mayr cited as an example the tremendous variation that exists among populations of kingfishers of the *Tanysiptera hydrocharis-galatea* group inhabiting the small islands surrounding New Guinea. Although there is no reason to suspect that forces other than genetic differences associated with small samples of founders are involved in Mayr's example, one must keep in mind that problems of communication also impinge upon morphological variation. Signals which when exaggerated are known to elicit exaggerated (but not pathological) responses must of necessity be tempered in populations by other, opposing considerations. Upon arriving in a new locality, migrants and their descendants may find themselves removed from these extraneous but important restraints; selection may then lead to the exaggerated condition. A flash of color, for example, may serve as a mating signal within a species of birds; the same flash may be the signal that alerts an eavesdropping predator. The counterselection of predation, then, may keep this flash at a modest level. In a remote locality, however, the predator may be absent. Under these circumstances, the flash may develop into a totally exaggerated form in a migrant colony. In many respects the evolution of communication parallels the evolution of genetic systems; some of these problems will reappear in a later chapter.

The Irreversibility of Evolution

Present-day students in biology are to a large extent spared the multitude of "laws" and "rules" that confronted students of several decades ago. Today's rules, it seems, tend to be operational ones that simplify the student's approach to experimental problems. DNA is said to be the sole carrier of genetic information for most organisms, not because this statement is necessarily true but because we have advanced our knowledge so well under this assumption that there is no compelling reason at the moment to complicate our lives with any other. Generalizations of past decades tended, on the contrary, to be descriptive. They were generalities that obviously had some basis in fact. At the same time, they were so vaguely worded or had so many exceptions that they were taught as often to be denied as to help the student organize facts. One of these rules was that "ontogeny recapitulates phylogeny." Another was Dollo's law, a law under which it was claimed (but not by Dollo, a reasonable man who actually made a much more modest statement than the one ascribed to him) that evolution is irreversible. The points that we have made in this chapter on the unpredictable changes that occur in some populations are pertinent to the question of irreversibility. The last serious genetic discussion of irreversibility seems to have been one by Muller (1939); Needham (1938) has given an account of irreversibility of evolutionary physiological changes.

From what we have said about the nonrepeatability of genetic changes in isolated populations, it follows that populations are not at all likely to retrace their steps in the face of reversed selection pressures. This point is illustrated in Figure 20-7. The discussion of nonrepeatability has shown that a population at *O* has available to it a number of paths that lead to

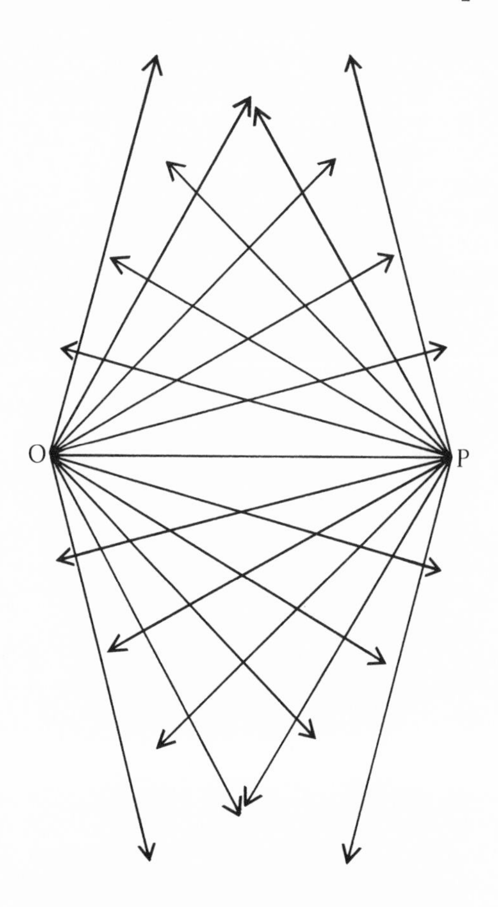

FIGURE 20-7. Irreversibility of evolution. If a population can evolve in many different directions from either point *O* or point *P*, it is highly unlikely that, having evolved from *O* to *P*, it will subsequently retrace its steps from *P* to *O*.

the right. When we say that the results of these changes are nonrepeatable, we mean that although one population may arrive at *P*, it is unlikely that any other population starting at the point *O* would also do so because there are so many alternative paths that might be taken. Consequently, should a population at *P* embark upon a path that leads to the left, it is equally unlikely that it will take precisely that path that leads to *O*. Again, there are too many alternative paths available to it. Thus the irreversibility of evolution, when expressed in relation to genetic phenomena, becomes no more than a special case of the more general phenomenon of nonrepeatability, of the indeterminate outcome of some genetic changes as described by Dobzhansky and Pavlovsky.

21

MENDELIAN POPULATIONS

Local populations in nature together with their laboratory equivalents—experimental populations—are the lowest members of a hierarchy of *Mendelian populations.* At least this is so in the case of sexually reproducing diploid individuals. The hierarchy extends from the local population to the species. Dobzhansky (1951a) defined a Mendelian population as a reproductive community of sexual and cross-fertilizing individuals that share a common gene pool. The largest Mendelian population is the biological species; a species is the largest group of individuals that share—or would share except for extraneous reasons—a common gene pool.

In describing local populations we used the bed-of-nails analogy; the "nails" represented the local populations while the "bed" represented the connecting members of the community, the migrants. Common gene pools larger than those of local populations are created and maintained by migrant individuals.

Some organisms (although their number dwindles as our information about various sexual processes increases) do not exchange genes. These are organisms that reproduce asexually—by division, by parthenogenesis, or by vegetative shoots. Such organisms cannot form Mendelian populations; the term "species," in fact, does not have the same meaning for such organisms as for sexually reproducing ones. Nor is the term "Mendelian population" applicable to self-fertilizing organisms; obligatory self-fertilization is still another technique by which an exchange of genes is prevented.

In this chapter we shall spend more time than we have in earlier ones on the various correlations that exist between local populations of the same geographical areas. Differences are found, of course, but correlations exist as well. First, we shall consider the role sex plays in evolution for sex is the very basis of the notion of Mendelian populations. Next, we shall examine a sample of various sized Mendelian populations including geographic races. Finally, we shall consider some especially important local populations, those that occur at the margins of species distributions; there are a number of reasons for believing that these border or peripheral populations frequently serve as the origins of new species.

The Role of Sex in Evolution

In 1958, at one of the national scientific meetings, Muller presented a short paper entitled "How Much Is Evolution Accelerated by Sexual Reproduction?" This rhetorical question was inadvertently answered by the program itself, for following the title was the parenthetic phrase, (10 min.). In reality, sex has been much more influential than this in permitting and shaping evolutionary change. It represents one of the greatest ideas hit upon by living things.

Sex, in a genetic sense, means the exchange and recombination of genetic material once carried by two different individuals. Recombination can take place only if the DNA of two individuals comes to lie in the same cell. Sex, in a nongenetic sense, involves the myriad of procedures by which the exchange is accomplished. In the case of bacteriophage, recombination follows the infection of one bacterium by two or more phage particles. Bacteria conjugate in carrying out their sexual process or, alternatively, rely upon phagelike particles to carry their genetic material from one individual to another. Higher plants have cycles of sexual and asexual generations that are diagrammed in nearly every genetics text and are well known to all biology students; animals have reduced the haploid phase of their life cycles to the germ cells themselves.

The advantage conferred by sex upon a species lies in the ease with which advantageous combinations of particular alleles can arise. In asexual organisms (see Figure 21-1a), the new mutations must arise one after the other within the same line of descent if the desired combination is to be formed. Mutations that are potentially useful but which arise outside a given line of descent cannot be utilized. The reason for this can be seen by looking back in time from the vantage point of a single individual; in each past generation of asexual individuals, one and only one was an ancestor of the chosen individual.

In sexually reproducing organisms, it is not necessary that mutations which make advantageous combinations arise in sequence (see Figure 21-1b). The proper alleles can arise anywhere within the species and, provided that interbreeding is not prevented by barriers of some sort, still have an opportunity to meet in the desirable combination. Again, we can choose an individual and from his position look back through time. Where has his genetic endowment come from? We get a hint when we hear someone speak of Aunt Jennie's nose and Uncle Silas's hair. In the

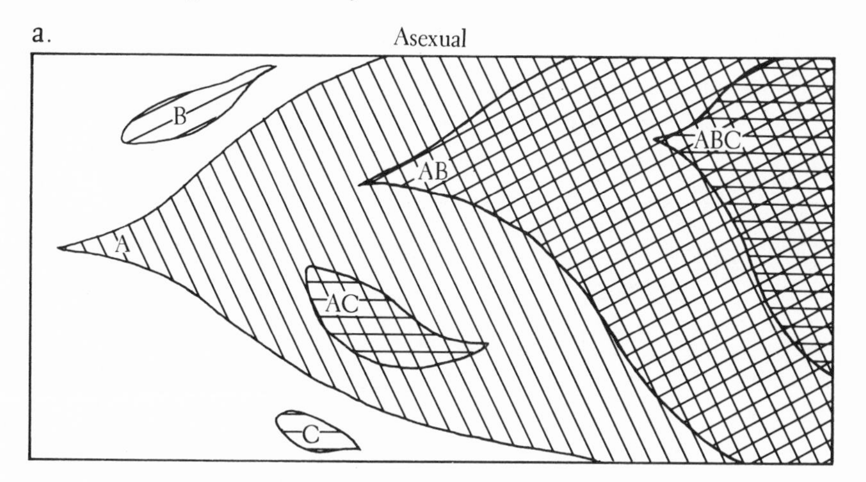

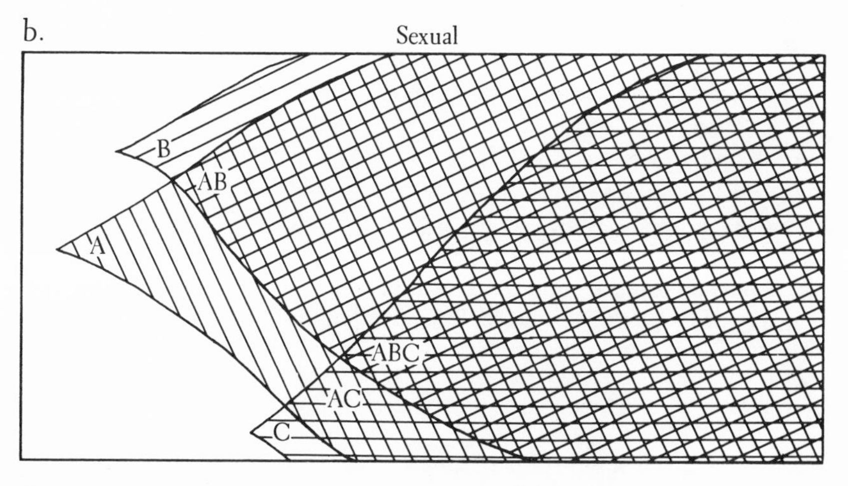

FIGURE 21-1. Advantage of sexual reproduction in giving rise to advantageous gene combinations in large populations. (After Muller, 1932a; Crow and Kimura, 1965.)

case of sexually reproducing organisms, an ancestral tree spreads out its branches through past generations. An individual does not have a single ancestor in each past generation; he has many. The genes he and his contemporaries possess are, presumably, among the best combinations that could be put together from the entire lot that was available to his ancestral predecessors.

Ratios of the rates of incorporation of new advantageous mutations into populations with and without sexual recombination have been calculated by Crow and Kimura (1965); Table 21-1 has been taken from their paper. In this table we see that small populations do not benefit from recombination to the same extent as do large ones. The reason is inbreeding; in small populations an individual possesses only a limited number of ancestors in any past generation. To spread successfully, mutations must arise sequentially in small populations to nearly the same degree as in populations of asexually reproducing organisms. On the other hand, within populations of a million or so individuals the incorporation of favorable mutations can be thousands of times more rapid than in self-fertilizing or asexually reproducing ones.

TABLE 21-1

RELATIVE RATE OF INCORPORATION OF NEW MUTATIONS INTO THE POPULATION WITH AND WITHOUT RECOMBINATION.

(Crow and Kimura, 1965.)

	POPULATION SIZE					
U/s	10^3	10^4	10^5	10^6	10^7	10^8
10^{-7}	1.0007	1.01	1.12	2.38	16.7	162
10^{-6}	1.007	1.09	2.15	14.4	139	1.4×10^3
10^{-5}	1.07	1.92	12.1	116	1.2×10^3	1.2×10^4
10^{-4}	1.69	9.75	92.6	922	9.2×10^3	9.2×10^4
10^{-3}	7.50	69.6	691	6.9×10^3	6.9×10^4	6.9×10^5
10^{-2}	46.7	462	4.6×10^3	4.6×10^4	4.6×10^5	4.6×10^6
10^{-1}	240	2.4×10^3	2.4×10^4	2.4×10^5	2.4×10^6	2.4×10^7

A HIERARCHY OF MENDELIAN POPULATIONS

Very early in our discussion of populations we stressed the "very local nature of local populations" (page 29); it was the pattern of dispersal of *Drosophila* that led us to this description. We might recall here that the composition of local populations of *Drosophila* is dependent upon the dispersal of individuals and that roughly speaking, the number of individuals that migrates a given distance falls off exponentially with the square root of distance. Most individuals scarcely move; those that do move may travel considerable distances.

Our view of the composition of local populations of *Drosophila* was borne out by the relationship between the allelism of lethals and distance. The probability that two lethals carried by two different flies are identical because of common descent seems to decrease exponentially with the square root of distance. The frequency of allelism, consequently, is much higher for lethals carried by flies captured together at one spot than for those carried by flies captured at 30, 60, or 90 meters distance.

We might note that in man the frequency with which marriage partners are born at various distances from one another reflects a dispersal pattern for man that is similar to that of flies. The logarithm of the frequencies of marriages decreases linearly with the square root of distance between the marriage partners (Cavalli-Sforza, 1959). Is there any evidence in man that would correspond to the data on the allelism of lethals mentioned above? It appears so. Schull and Neel (1965), citing Hamilton et al. (1961) have written: "More recently, with the development of methods permitting a rapid survey for the carriers of the rare recessive gene responsible for *acatalasemia,* it has been possible to demonstrate on a large scale a fact which in genetic theory comes as no surprise: even in long-settled populations, rare recessive genes are not uniformly distributed throughout the country (Japan). In the particular case under discussion, there were ten-fold differences between regions."

That populations in a geographic region tend to be subdivided to a considerable extent into local populations does not erase the similarities between these local populations. Some characteristics are advantageous under a given set of environmental conditions and will be found in the populations of a given area as a consequence. Migrants, whatever their dispersal pattern, are still more likely to come from nearby populations than from those at considerable distances. Furthermore, what is carried from one population to another by migrant individuals is more likely to succeed in nearby than in distant populations. As a result, our treatment of the hierarchy of populations must center upon differences and similarities between populations. Differences have already been emphasized; similarities, too, were summarized in Table 4-2. An appreciation of population structure is an appreciation of these contrasting facets of interpopulation relationships.

Snail Populations. Two species, more than any others, have been subjected to thoroughgoing analyses in respect to their population structures and in respect to characteristics relevant to various parameters of population genetics: man (Muller, 1950a) and the snail *Cepaea nemoralis*

(Lamotte, 1951). We want to look at just one of Lamotte's analyses, one that shows the differentiation of local populations within small geographic areas.

C. nemoralis is a widespread snail of Europe. It is a polymorphic species. Individuals of local populations differ in respect to color (brown, pink, and yellow) and the presence or absence of a series of five bands. The genetics of these polymorphisms has been studied extensively both by Lamotte (1951) and Cain and Sheppard (1954); an excellent account of these studies has been given by Ford (1964).

The absence of bands in *C. nemoralis* is governed by an allele (b^+) that is dominant to the other alleles at this locus, each of which governs its own pattern of bands. The analysis that interests us at the moment concerns the frequency of the allele b^+ in fifty-four contiguous areas each of which was ten meters square. The average frequency of b^+ over all fifty-four squares was 28.5 percent.

Lamotte has calculated the variance in the frequency of b^+ in areas that were adjacent (regarded as 10 meters distant), adjacent but one (20 meters), and so forth. These variances are listed in Table 21-2; the

TABLE 21-2

VARIANCE IN THE FREQUENCY OF b^+ BETWEEN SAMPLES OF THE WOODLAND SNAIL, C. NEMORALIS, COLLECTED AT CERTAIN DISTANCES FROM ONE ANOTHER.

(Lamotte, 1951.)

DISTANCE, METERS	NO. OF SAMPLES	VARIANCE ($\times 10^4$)
10	88	16.8
20	262	18.9
30	185	19.8
40	244	21.6
50	156	28.7
60	154	34.9
70	110	41.6
80	88	48.9
90	81	47.4

relation between variance and distance is shown graphically in Figure 21-2. The interesting feature for our purposes is the observed increase in variance that accompanies the increase in distance between the samples of snails. The increase begins ostensibly (although it is not necessarily statistically significant) even upon going from ten to twenty meters. By the time the collections have been taken from distances of fifty or sixty

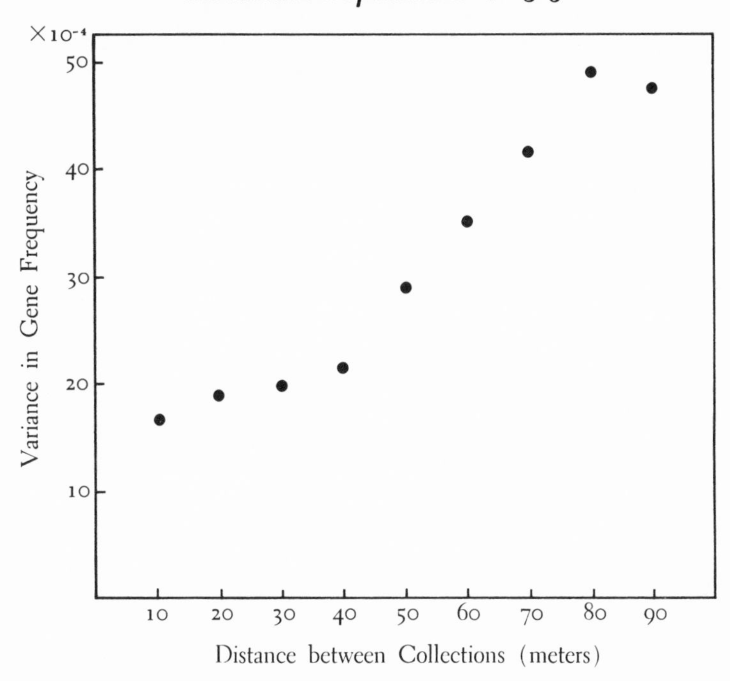

FIGURE 21-2. Variance in the frequency of the gene b^+ in collections of *Cepaea nemoralis* made at various distances from one another. (After Lamotte, 1951.)

meters, the variances in gene frequency between samples have increased markedly.

The above data show very clearly that within restricted territories there are local populations such that individuals inhabiting very small areas are differentiated from others. Nevertheless, the individuals living in adjoining areas are more alike than those living in more remote (but still nearby) areas. The data show this point clearly; they do not tell us the basis for the increasing difference. The similarity of neighboring populations may be the result of immigration, of a mere exchange of individuals. On the other hand, this similarity may conceivably result from natural selection in a patchy environment within which neighboring populations endure somewhat the same patterns of selection. Whether migrants or selection or a combination of the two is responsible, the orderly change in inter-area variance tells us that the snails studied were not distributed helter-skelter over the entire region which was only some 80 meters by 200 meters in all.

Drosophila Populations. In foregoing chapters we have discussed the frequencies of gene arrangements in various *Drosophila* species in detail.

We saw that the frequencies of these gene arrangements became progressively more similar as the distance between samples diminished. However, large samples taken from localities no farther apart than several hundred meters reveal that these local populations differ (Table 4-2). The allelism of lethals yielded data of the same sort; the allelism is higher if the lethals come from flies captured at a single site than if they come from flies captured at sites more remote (30, 60, or 90 meters) from one another (page 86).

Body size (measured as wing length, wing width, length of tibia, or in other ways) in *Drosophila* has a considerable genetic component; artificial selection for large and small size generally succeeds with no difficulty. Consequently, if the offspring (or grandchildren) of wildtype flies caught at different places differ in size even though they are raised under standard conditions of food and temperature, it is reasonable to assume that the original wild populations differed in respect to genes controlling size. Sokoloff (1965) has made a study of the size of the descendants of *D. pseudoobscura* captured at various localities throughout the western United States and Mexico. A portion of his results is listed in Table 21-3. Three of the localities listed in the table are from Mt. San Jacinto in Southern California (Figure 4-6). Even though these localities are from only six to twenty miles distant from one another, they differ in respect to genes controlling body size. Two collections were made at the same locality near the Grand Canyon at two different times during the summer (July 4, 1958, and August 16, 1958); the descendants of these flies, too, differed considerably in size when raised under standard conditions. Populations in different localities differ; populations change with time in one locality. *D. pseudoobscura* seems to differ from some other *Drosophila* species; *D. robusta* (Stalker and Carson, 1947) and *D. subobscura* (Prevosti, 1955) have been found to possess definite geographic gradients in respect to size while *D. pseudoobscura*, judging from data cited above, seemingly has none.

Gene Arrangements in D. pseudoobscura and D. persimilis. The different gene arrangements found in *Drosophila* species are excellent for reconstructing the relationships of natural populations. In Table 4-2 we used gene arrangements to show that populations of *D. pseudoobscura* inhabiting areas less than two miles apart may have different genetic compositions. In the same table, we showed that populations living thousands of miles from one another may differ in the presence or absence of certain gene arrangements, while those separated by tens or hundreds of miles may differ only in the relative frequencies within what is basically the same array of gene arrangements.

These distribution patterns deserve more attention than we have given them. They are especially useful since—unlike alleles that arise by recur-

TABLE 21-3

VARIATION IN THE SIZE OF D. PSEUDOOBSCURA COLLECTED AT VARIOUS LOCALITIES IN WESTERN UNITED STATES.

The pair of figures given in each case represents the 95 percent confidence interval; asterisks indicate measurements that are clearly different from other, comparable ones. (Sokoloff, 1965.)

	WING LENGTH	WING WIDTH	TIBIA LENGTH
Male			
Vandeventor	54.90–55.70	28.59–29.03	21.23–21.55*
Piñon Flat	55.05–55.61	28.28–28.68	21.63–21.95
Keen Camp	52.02–52.78*	27.17–27.53*	20.62–20.82**
Grand Canyon-1	56.38–56.86	29.21–29.49	21.41–21.65
Grand Canyon-2	57.01–57.37*	29.86–30.10*	21.75–21.87*
Female			
Vandeventor	61.25–61.85	31.16–31.60	22.48–22.76*
Piñon Flat	61.13–61.57	30.77–31.21*	22.70–22.98
Keen Camp	57.90–58.90*	29.59–29.95**	22.08–22.28**
Grand Canyon-1	62.26–62.74	31.49–31.93	22.58–22.78
Grand Canyon-2	62.98–63.30*	32.15–32.35*	22.92–22.98*

rent mutation—each gene arrangement has probably arisen only once in the history of the species. Therefore, the Standard (*ST*) gene arrangements found in regions as far removed as Puget Sound and Lower California, for example, are all descendants of a single chromosome, the first to bear the Standard sequence, that probably arose from a now extinct gene arrangement known as Hypothetical. Similarly, the Arrowhead (*AR*) gene arrangement that is found throughout an area extending from Canada to Mexico and from the Pacific Ocean to Nebraska arose as a single *AR* chromosome by a simple two-break inversion within a chromosome of the Standard gene arrangement. Each of these now widespread gene arrangements arose as a single event in one cell of a fly, was transmitted to a single surviving offspring, and only then had an opportunity to spread and to become established within the species. The spread from that initial local population to the distribution patterns we see today occurred in what we have referred to above as the hierarchy of Mendelian populations that constitute the species *D. pseudoobscura.*

The spread of gene arrangements within a species is not uniform. The environment is not uniform nor is the distribution of flies themselves. Each local population may accept or reject each new gene arrangement as it arrives. If selection favors the retention of a given gene arrangement in a given population, neighboring ones have increased chances to test it because migrant individuals bearing it will arrive ever more frequently. And, to whatever extent selective forces acting in neighboring localities are similar, the new arrangement is just that more likely to be adopted by the nearby populations.

Figures 21-3 and 21-4 show the distribution patterns of some gene arrangements in *D. pseudoobscura.* In Figure 21-3, the sampled populations extend from Puget Sound to a point well within Lower California. The numbers of chromosomes examined in these collections are not enormous (most are one hundred or more), but they are sufficient to show that *ST, AR,* and *CH* are common throughout the Pacific Coast region while Santa Cruz (*SC*) is found generally south of San Francisco. The figure shows, too, that Tree Line (*TL*) occurs sporadically throughout the area.

Figure 21-4 shows a distribution pattern which differs somewhat from that described above. The figure shows the replacement of one gene arrangement by another. Standard and Arrowhead, together with Chiricahua, are common gene arrangements in the coastal area of Southern California. As one proceeds toward Arizona, however, one finds that Standard becomes less frequent and finally disappears. Within Arizona and New Mexico the Pikes Peak (*PP*) arrangement appears; *PP* becomes very common as one proceeds through Texas toward the Gulf of Mexico. Chiricahua (*CH*) and Tree Line (*TL*) occur at low frequencies throughout most of the area shown in the figure. Arrowhead (*AR*), too, occurs throughout; in fact, where *ST* and *PP* are both infrequent, *AR* is the predominant gene arrangement with frequencies that at times approach 100 percent. [There have been some recent—and poorly understood—changes in the distribution patterns of these gene arrangements. *PP* is now found in low but appreciable frequencies throughout California and in Washington and Oregon as well. During the time that *PP* has invaded the Far West, the frequency of *CH* has declined. For an account of these recent changes, see Dobzhansky et al. (1966).]

Figures 21-5 and 21-6 show the distribution patterns of gene arrangements in *D. persimilis,* a species that is very closely related to *D. pseudoobscura.* In Figure 21-5 we see that Klamath (*KL*) is the predominant gene arrangement along the Pacific Coast from Washington through much of California. Two other gene arrangements, Mendocino and Standard (cytologically identical to the Standard of *D. pseudoobscura*) are also common in California.

The second distribution pattern illustrated for *D. persimilis* (Figure 21-6) shows again that at some points frequencies may change abruptly. The area illustrated in Figure 21-6 is farther inland than that illustrated in Figure 21-5. Now we see that Klamath is the predominant gene arrangement in the north but, as we move into California, Klamath is replaced by two other arrangements, Standard and Whitney. Whitney was scarcely represented in the data illustrated in Figure 21-5.

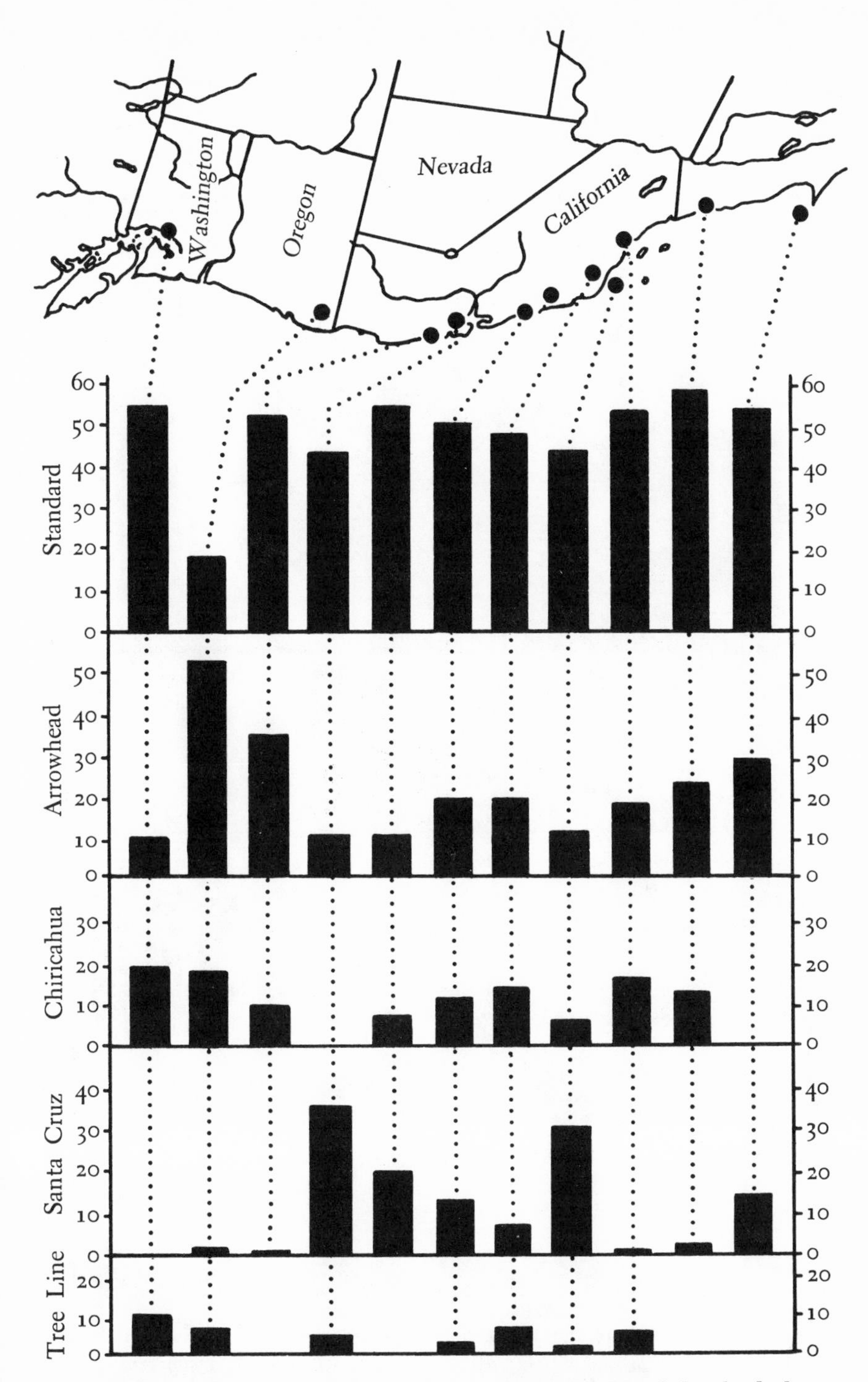

FIGURE 21-3. Frequencies of several gene arrangements of the third chromosome in populations of *D. pseudoobscura* inhabiting various regions along the Pacific Coast. (After Dobzhansky and Epling, 1944.)

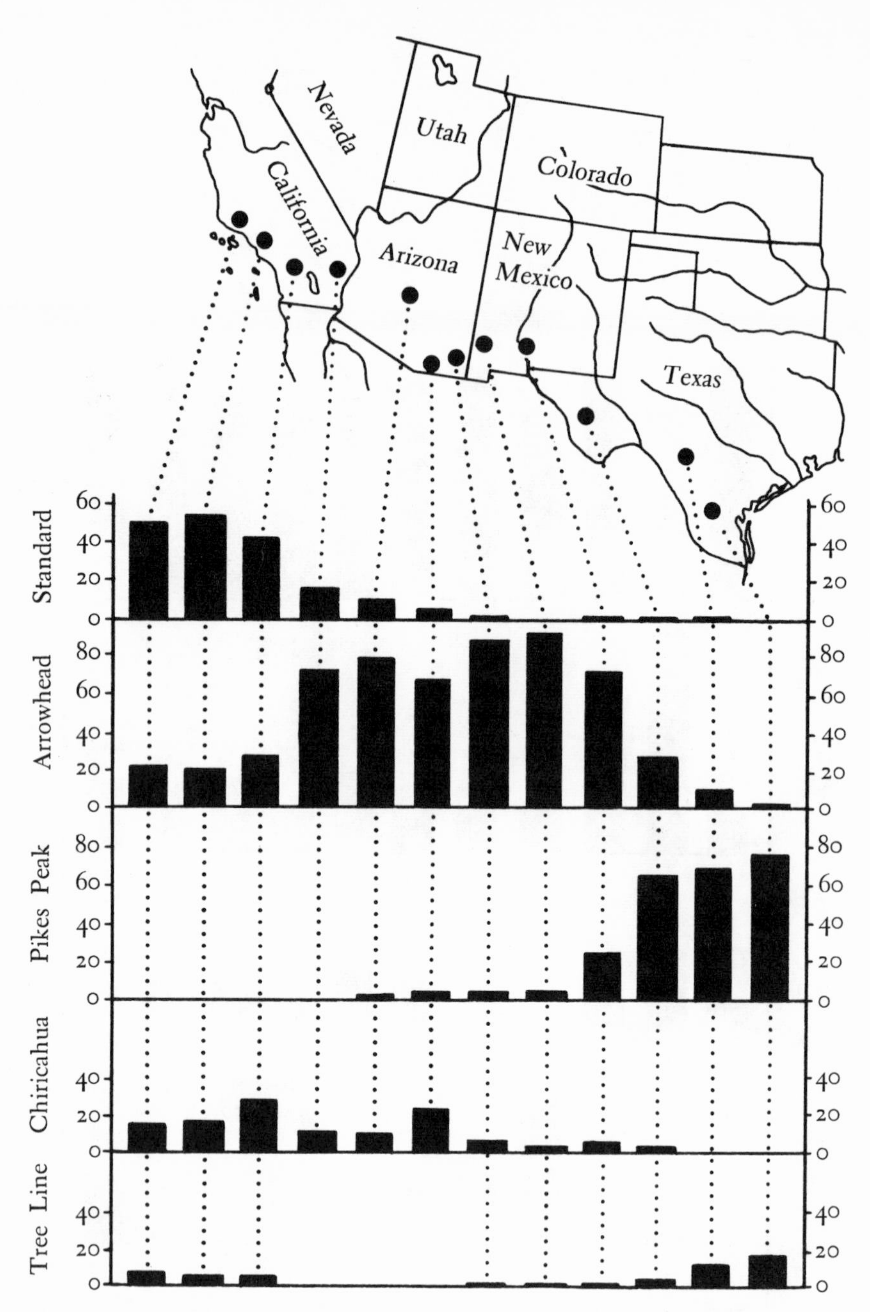

FIGURE 21-4. Frequencies of several gene arrangements of the third chromosome in populations of *D. pseudoobscura* inhabiting regions along an east-west transect in the southwestern United States. (After Dobzhansky and Epling, 1944.)

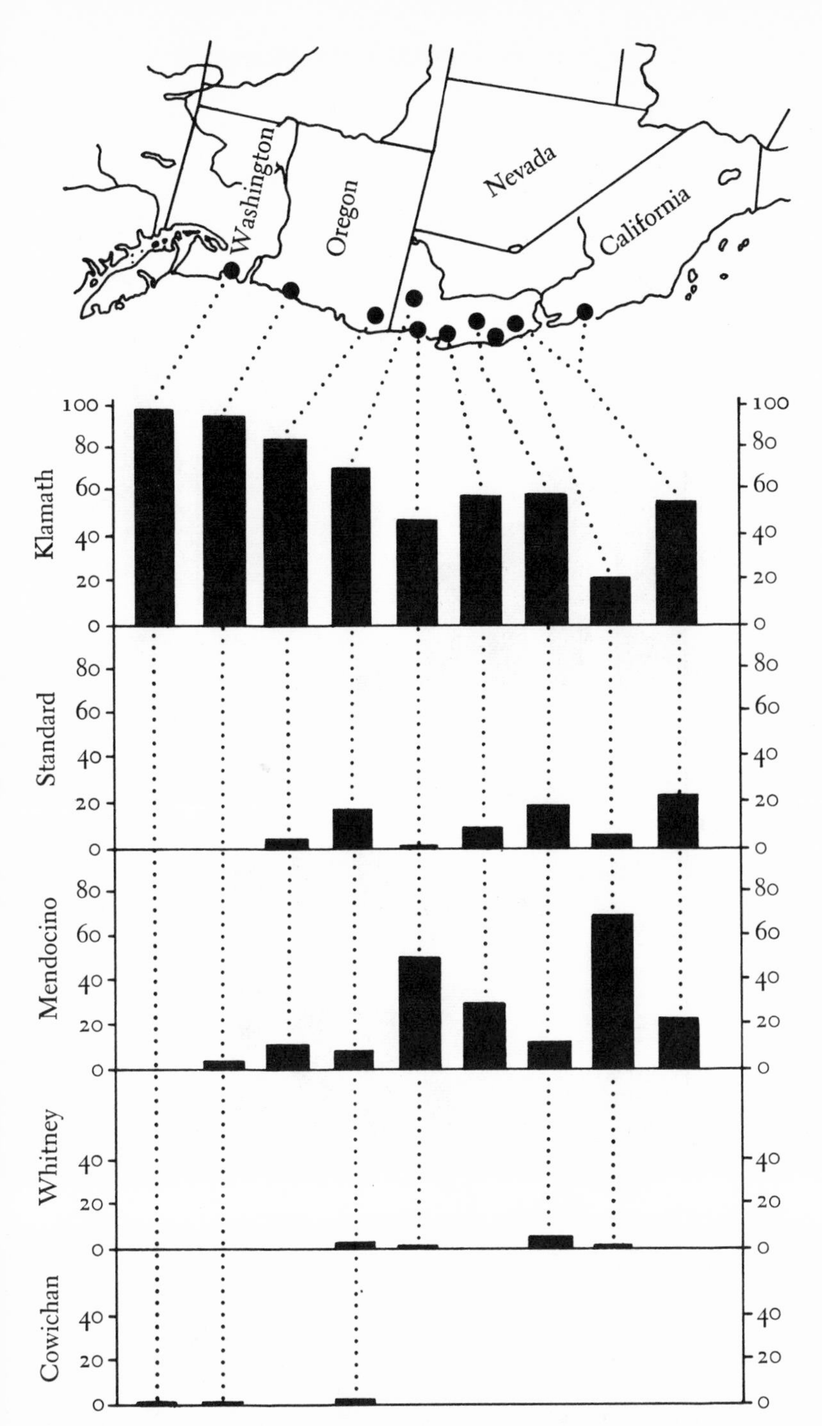

FIGURE 21-5. Frequencies of several gene arrangements of the third chromosome in populations of *D. persimilis* inhabiting various regions along the Pacific Coast. (After Dobzhansky and Epling, 1944.)

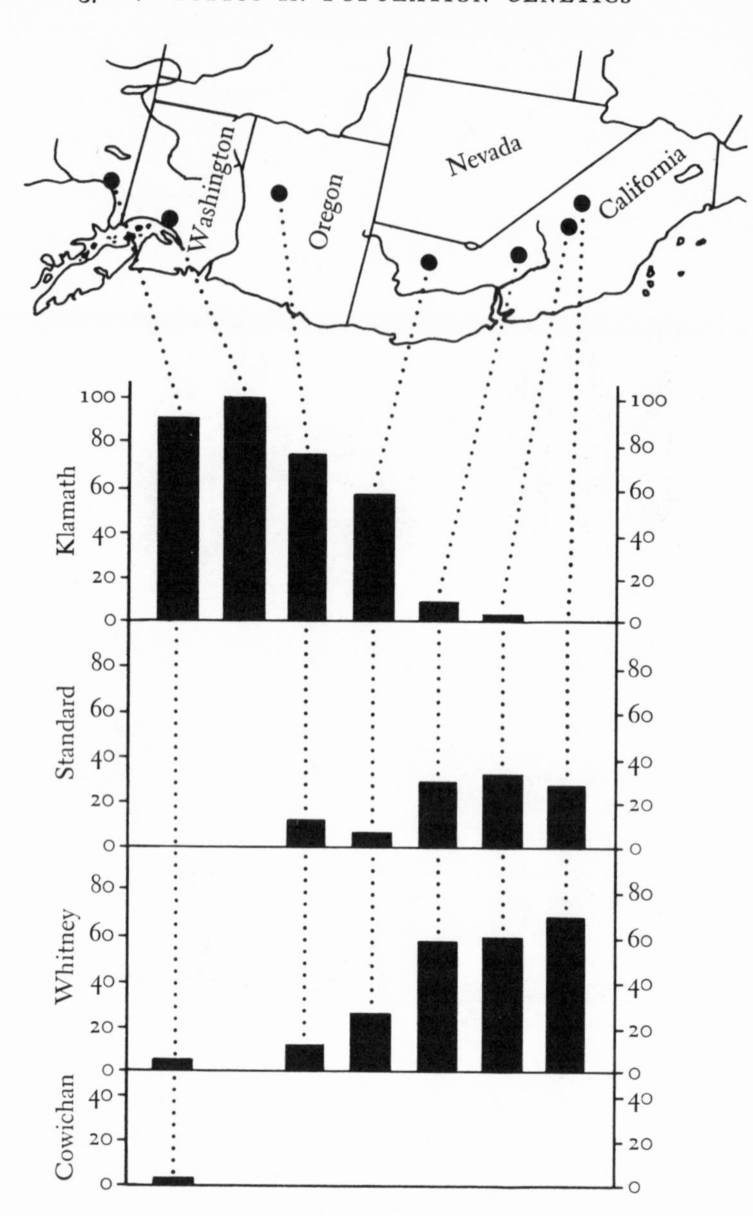

FIGURE 21-6. Frequencies of several gene arrangements of the third chromosome in populations of *D. persimilis* inhabiting regions along the Cascade–Sierra Nevada mountain chain. (After Dobzhansky and Epling, 1944.)

The data on the geographical distribution of gene arrangements upon which Figures 21-3 to 21-6 are based help us understand the concept of "race," one of the hierarchy of Mendelian populations. A race is a collection of local populations occupying a certain geographic region that differ genetically from those of other regions. Race is a *population* concept. Individuals, in the strict sense, cannot be said to represent a "race"; each individual, on the contrary, possesses a genotype. The race is a collection of genotypes generated by cross-fertilizing individuals where each genotype occurs with a frequency dependent upon gene frequencies and the actual breeding structure of the population(s).

No ambiguity exists in determining whether two populations (or collections of populations within circumscribed areas) differ genetically; following a series of observations, there are accepted statistical procedures that lead inexorably to a conclusion that two populations do (or do not) differ. On the other hand, it is impossible to enumerate all the races of a species. The reason for this is that almost any two populations can be shown to differ in gene frequencies for alleles found at once locus or another. As a result, races are named as a matter of convenience; discontinuities in gene-frequency distributions are usually at the bottom of "convenient" classifications. It is difficult to subdivide the Pacific Coast populations of *D. pseudoobscura* into races on the basis of data summarized in Figure 21-3 because the different samples have similar, although not identical, gene frequencies. The only "convenient" grouping that I can see is that based on the abrupt change in the frequency of Santa Cruz; on the basis of this, one could conceivably speak of northern and southern Pacific Coast races. Discontinuities in distribution patterns such as those illustrated in Figures 21-4 and 21-6 could be used to identify still other local races.

It is worthwhile stressing once more that *race is a population concept.* There is no Arrowhead race of *D. pseudoobscura,* for example. Arrowhead is a particular gene arrangement that may or may not be included in the gene pool of any of a number of races. Homozygous *AR/AR* individuals are an essential part of the array of genotypes generated within local populations of a great many races of *D. pseudoobscura*; there is no racial affinity between *AR/AR* individuals of different geographic localities. There is another perfectly adequate expression to describe what *AR/AR* individuals have in common: in respect to this one feature, they have identical genotypes.

Sexual Isolation in D. miranda. The divergence of local populations and of groups of local populations from one another, if it is to play any role in evolution, must have a genetic basis. To whet our interest, however, the observations themselves need not be restricted to genetic material. Thus no one need remain idle while lamenting that only some

Drosophila species, some butterflies, and a few plants are well known genetically. Although it is desirable to have for study genetic traits such as b^+ in *C. nemoralis* or the various gene arrangements in *D. pseudoobscura* or *D. persimilis,* it is not necessary to study only such traits. The size of *Drosophila,* as we have seen earlier, has been used by a number of workers to reveal genetic differences between local populations of these flies. As long as we know that the trait studied has a genetic component, the results of the study serve to tell us about the genetic similarity or dissimilarity of populations. There are many studies of this sort that could have been cited other than the few mentioned above. It would, however, be unfair to fail to mention the classical studies of Clausen, Keck, and Heisey (1940, 1948) on the climatic races (ecotypes) of plants living in varying environments; it is simply impossible to do justice to these studies in a necessarily brief space.

Physical proportions or growth habits (in the case of plants) are not the only respects in which isolated populations or races may differ. Behavior, too, is subject to genetic control. Furthermore, within local populations success in mating is an important component of fitness. The inability to secure a mate is as devastating to the genes carried by an individual as would be his death or sterility. Thus mating behavior should respond to local variations in selective pressures just as any other aspect of a group of individuals.

In the sense "mating behavior" is used here, I include various ancillary features by which members of a species recognize one another and distinguish members of other species, as well as those features of behavior that are immediately concerned in courtship and mating. Thus, in contrast to Simpson (1953, p. 170), who sees irregularities in the geographic distribution of color patterns as an indication of their nonadaptive nature, I suspect that *differences* between the coloration of similar groups in *single* localities is evidence that color patterns are indeed adaptive. The adaptive nature, however, lies in the *differences*; geographic irregularities of the colors themselves may reflect selection for optimizing local differences.

Dobzhansky and Koller (1938) have presented evidence for *D. miranda* that show not only the divergence of geographically isolated populations but also the origins of "local dialects" in mating behavior. *D. miranda,* a species closely related to *D. pseudoobscura* and *D. persimilis,* is found in two widely separated localities: the region around Puget Sound in the northwestern corner of the State of Washington and the region near Mt. Whitney in the Sierra Nevada Mountains. Flies of this species that are captured in these two areas can be mated and produce, as far as one can tell, normal offspring. However, if males from one locality are offered a choice of females from both (a choice that extended for four days in the experiments of Dobzhansky and Koller), successful matings are found to involve couples from the same locality more often than those of different ones. The results of an examination of more than 2000 females are listed in Table 21-4; the relatively greater success of homogamic (Olympic × Olympic and Whitney × Whitney) matings over heterogamic ones (Whitney × Olympic and the reverse) is clearly visible in these results. Were these flies to be left together for periods considerably longer than four days, the proportion of fertilized females would approach 100 percent in all four cells of the table.

Border Populations

The divergence of local populations is the first step along the path of evolutionary change. The correlations that unite local populations in a given geographic locality are the unifying racial characteristics of those populations, characteristics that serve to distinguish one set of local populations from another. But of all local populations the most interesting perhaps are those that occur at the border of the species distribution.

Species of plants and animals are not distributed uniformly over land areas or throughout the sea; some cosmopolitan species may appear to be distributed in this manner but most have rather well-defined distribution ranges. It was no accident that I referred to the distribution of *AR* within *D. pseudoobscura* as extending from the Pacific Coast to Nebraska; these flies are not found east of a line running roughly north through Lincoln, Nebraska. For a number of years there was a colony on Long Island, New York, founded by individuals I once released there but, to the best of my knowledge, this population is now extinct. The American Elm, *Ulmus americana*, occupies the eastern half of the United States but does not occur in the western half of Texas, Oklahoma, and

TABLE 21-4

Results of a "Multiple Choice" Test of Mating Preference Using D. miranda from the Olympic National Park (Washington) and Mt. Whitney (California).

(Dobzhansky and Koller, 1938.)

	Females					
	Olympic			Whitney		
Males	Fertilized	Total	%	Fertilized	Total	%
Olympic	474	539	87.9	400	546	73.3
Whitney	436	558	78.1	503	599	84.0

Kansas or in Wyoming and Montana (see Carson, 1955, for the distribution map). One of the most impressive sights to be seen on a transcontinental trip is the abrupt termination of the distribution of one species and its replacement by another—sometimes within an extremely short distance.

To an ecologist it is tempting to look upon a species as a group of individuals capable of sustaining themselves under certain environmental conditions. Under this view, the species border represents the position in space at which the mean number of offspring per female is precisely two (for cross-fertilizing species). Within the species range, the individuals of one generation leave ample numbers of offspring to represent the species in the next; beyond the species border, populations are not self-sustaining and hence become extinct. Only a few migrant individuals doomed to leave offspring fewer in number on the average than they themselves, according to this view, are to be found beyond the species border.

Mayr (1954), on the other hand, has described in extremely clear terms the dynamic nature of the species border. The spatial distribution of a species is not to be compared to the surface area of a lake that fills an irregularly shaped valley. A species does not merely have certain requirements for existence that are met in certain areas but not in others. A species consists of a number of populations, and these populations evolve. Local populations adapt to local conditions. If populations in the past have adapted to one geographic region after another in approaching the margin of the species distribution, why are they incapable of making the further adaptation that will enable the species to extend its range over the next few miles? And the next? And the ones after that?

Mayr emphasizes the dynamic fluidity of the species border. It represents an equilibrium established by the adaptive response to border conditions and the destruction of this response through genetic "contamination" by migrant individuals. The species border, then, is not a border in the sense of the shoreline of a lake but rather in the sense of the upper limit of a vertical jet of water. Although a fountain may reach a well-defined height (a large one in Geneva, Switzerland reaches 100 meters), this height is not defined by static factors; it represents a well-defined equilibrium between the initial velocity of the water and gravity.

Carson (1955) has given an excellent picture of the species border that helps clarify Mayr's notions. According to Mayr's account of the border, it is the arrival of migrant individuals from well within the species range that prevents successful adaptation to border conditions. Carson enlarges upon this by describing the ebb and flow of the species border in time. He likens the border to the water's edge on a large tidal flat. At high tide the water flows over and covers the entire flat. During the ebb tide it recedes, leaving behind tidal pools and small interconnecting channels. These are swamped by the next high tide. The pools left behind following a receding tide do not contain the same water as they did following the previous one. In fact, the pools themselves may be different from the earlier ones; only certain well-defined pools remain more or less constant over a long period of rising and falling tides.

At a given moment the border of a species may be surprisingly sharp but there will be pockets of individuals beyond it, nevertheless, where local conditions have enabled them to survive and reproduce. With time climatic conditions may change so that the species distribution expands and sweeps over these pockets or shrinks once more to re-expose them. The individuals left in the pockets after each periodic surge, however, are not necessarliy genetically similar to those that were there before; the earlier ones are swamped by interbreeding with the new arrivals. Consequently, after the contraction of the species range, the individuals left in the refuge pockets are no more able to spread out and thrive beyond these pockets than their predecessors had been during the earlier period of retraction.

The interest in border populations lies in the opportunity these possess to be the first to expand the species range by colonizing hitherto unoccupied territory. Of all individuals that might be adapted for conditions that exist at and just beyond the border, those that live on the border itself or in small pockets beyond the border are the most nearly adapted. To perfect these adaptations, they need a respite from the intermixture with the (less well adapted) remainder of the species.

Under Carson's scheme, a border population in isolation from the bulk of the species occasionally becomes structurally monomorphic (in contrast, for example, to the widespread inversion polymorphisms of many *Drosophila* species), reorganizes its gene pool with the result that hybrids with the main population possess poor fitness, develops a reluctance to mate with the migrant individuals from the main body of the species, and, freed in this manner from old restraints, spreads out into the area beyond the old species border. The dark "balloon" attached to the left of the main species range in Figure 21-7a symbolizes Carson's view. The new species, seen in this light, has been a new species since its inception. The early reproductive isolation between the border population and its mother species that develops while the new species is still no more than an isolated pocket is an important feature of Carson's scheme.

Fundamentally, Carson seems to have given a reasonable account of the origin of new species. It contains the answer to a question once raised by Moore (1957): If two species occupy adjoining areas, why should members of the two species from all parts of their respective distributional areas exhibit sexual isolation? One would expect for reasons that will be clearer in a later chapter that many isolating mechanisms would be selected for only in the actual zone of contact. Carson

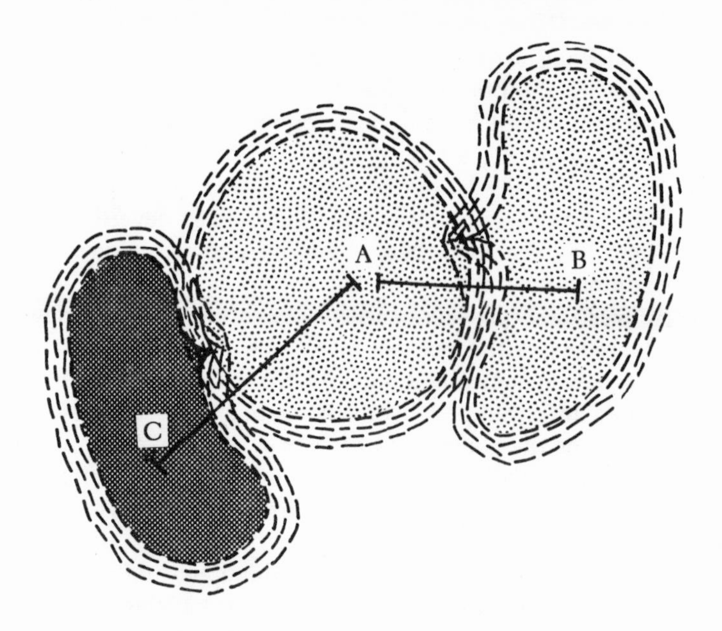

FIGURE 21-7. Diagrammatic representation of the extension of a species range (B) or of the origin of a daughter species (C) from a peripheral population of the original species distribution (A).

answers this question by having one of the two species arise from a small population of the other; the small population, in preserving its genetic structure, develops isolating mechanisms that prevent it from mating with the older species. These isolating mechanisms, then, are part of the fabric of the new species even when it has grown in numbers so that it occupies what is largely its own territory.

The available evidence is not overwhelming, but it seems unlikely that the marginal population that successfully breaks through the species border need be structurally monomorphic. *D. paulistorum*, a polymorphic species that is divided into a number of groups that could easily be regarded as species (with the present *D. paulistorum* becoming the "*paulistorum* group"), gives no evidence of having been monomorphic before splitting into its subgroups (Kastritsis, 1966).

Wallace (1966d) has shown that chromosomes which are monomorphic within *D. pseudoobscura* are nevertheless capable of exhibiting fitness heterosis and hence do not differ in a qualitative way from the third chromosome which is highly polymorphic. Nevertheless, when two related *Drosophila* species are polymorphic for a single chromosome each, this chromosome is frequently the same one in both species. If the newer species arose from a monomorphic population, and if all chromosomes exhibit fitness heterosis, the utilization of the same chromosome by both the new and the old species would be a matter of chance. The facts are otherwise, however, for the observed coincidence is too great to be a chance phenomenon. For example, ten different species of the *mulleri* group of *Drosophila* are polymorphic; in nine of these chromosome 2 is the only one that is polymorphic while in the tenth both chromosomes 2 and 3 are polymorphic (Wasserman, 1962, Table 1). The species of this group possess four long autosomal arms each, and so the observed distribution of polymorphisms among these arms has about 1 chance in 100,000 of being a chance accident.

As an alternative to the monomorphism suggested by Carson (1959), Wallace (1953, 1959b) has suggested that certain combinations of three or more inversions are incompatible within local *Drosophila* populations. They are incompatible because, in the presence of all, no two of them can utilize the hybrid vigor of inversion heterozygotes to the fullest extent. On the contrary, the recombination within inverted blocks of genes that is permitted by the incompatible combinations of inversions might be deleterious in the same manner as the interpopulation recombination studied by Brncic (page 313) and Vetukhiv (page 315). If this suggestion is correct (but it seems not to be universally true among *Drosophila* species; see, for example, Levitan et al., 1954), only two of the three or more incompatible arrangements should remain in any one population. The suggestion goes on to say that the third gene arrangement upon arising in a border population offers the means by which this population can break through the old restraints and expand into a new area.

According to the suggestion just made, the expanding population need not be a new species although there is no reason why it should not be. Thus, in Figure 21-7, we have shown on the right of the main species range a balloon that has arisen initially within a single border population. The line *AB* represents a transect through the old and newly occupied areas. The scheme just proposed would lead us to expect to see gradients of gene arrangements along this transect such as that shown in Figure 21-4 for *D. pseudoobscura* (where *ST-AR-PP* form an incompatible combination of gene arrangements) and in Figure 21-6 for *D. persimilis* (where *KL-ST-WH* form a second incompatible series). The gradient along the transect *AC* of Figure 21-7 has not been illustrated, because it passes from one species into a second which may or may not have retained its clearly distinct geographic range. However, an examination of Figures 21-3 and 21-5 reveals that a very common pair of gene arrangements in *D. pseudoobscura* populations is *ST-AR*. On the other hand, Klamath (*KL*) is an exceedingly common gene arrangement in *D. persimilis*; the combination *AR-ST-KL* is one that would be classified as incompatible if the three arrangements existed in a single species.

The heart of the above discussion is that the discontinuities seen in the geographic distributions of gene arrangements in *Drosophila* might represent the shattering of former species borders by small, marginal populations. And further, that this is true whether the discontinuity falls within the species (as the *ST-AR, AR-PP* discontinuity of Figure 21-4 or the *KL, ST-WH* discontinuity of Figure 21-6) or spans two closely related species (such as the *AR-ST, KL* discontinuity observed between *D. pseudoobscura* and *D. persimilis*).

The account given here of peripheral populations and their role in the spread of species or in the origin of new species, I believe, is a reasonably conservative view. It is difficult, though, to superimpose this view onto one concerned with the question as to whether species arise sympatrically or allopatrically; both periodic contact and some degree of physical isolation is needed under the "border population" model. Nevertheless, I see very little conflict between this scheme and allopatric speciation. On the other hand, I find it difficult to visualize a mechanism for sympatric speciation that foregoes the periods of isolation that are essential to the scheme presented above.

What seems to be essential for the spread of border populations is that a barrier be erected between them and the rest of the species just as the breeder erects pens about his experimental animals or establishes isolation plots for his plants. It seems to me (Wallace, 1959b) that a number of genetic barriers that serve this purpose can be identified: self-fertilization, apomixis, parthenogenesis, polyploidy, certain combinations of translocations and inversions, and extrachromosomes. The geographic distribution of organisms carrying these built-in barriers is often precisely that which would be expected had the barrier first arisen in a border population.

22

STABILIZING AND DISRUPTIVE SELECTION

Natural selection has been frequently mentioned in this book. Although some evolutionary changes must be determined by chance, the majority are adaptive and are brought about by selection. Our primary concern to this point has been with the genetic consequences of selection, consequences expressed in terms of changes in or the stabilization of gene frequencies. In the present chapter we shall discuss selection primarily from the phenotypic aspect. In the final analysis, selection operates on individuals by virtue of their phenotypes, not genotypes. As a rule, of course, phenotype and genotype are correlated so that individuals of a given phenotype are not a random sample of all genotypes found in a population. (Homozygous inbred lines are an exception, because these individuals are genetically identical; their phenotypic variation is entirely environmental in origin.) However, the genotypic bases of quantitative aspects of the phenotype such as height or weight are not usually single-gene or even two-gene systems. On the contrary, they involve genes at many loci. Hence, we shall make little effort to discuss specific genetic responses to selection in this chapter; instead, we shall stay very close to phenotypes.

Types of Selection

There are three main types of selection: directional, stabilizing, and disruptive (Figure 22-1). Directional selection is that which concerned Darwin and which concerns most plant and animal breeders today. Texts on quantitative genetics deal almost exclusively with directional selection; the reader might consult books such as those by Falconer (1960), Lerner (1954, 1958), Rasmuson (1961), and W. B. Mather (1964).

Crow (1966) has pointed out that directional selection is not common within natural populations. At least, directional selection over extended periods of time is not common. The bulk of selection is concerned with maintaining conditions as they are. This involves the elimination of deviants of all sorts. Evidence for the lack of directional selection in man, for example, can be obtained from an analysis of his height. Ignoring the past century, man has not changed much in height in a million years. Now, if selection had been effective in increasing man's height an average of 0.001 inch per generation (an amount that would have escaped detection in any reasonable comparison of the mean heights of parents and offspring or even of grandparents and grandchildren), the mean height of man would have increased roughly 40 inches in 40,000 generations. Certainly, the average effect of selection in respect to height over this extended period of time has not been directional.

Stabilizing Selection

Stabilizing selection is selection that favors an intermediate optimum phenotype (Figure 22-1a). The individuals that are removed by death or infertility from a population are those that represent the tails of a distribution curve. Generally, the farther an individual deviates from

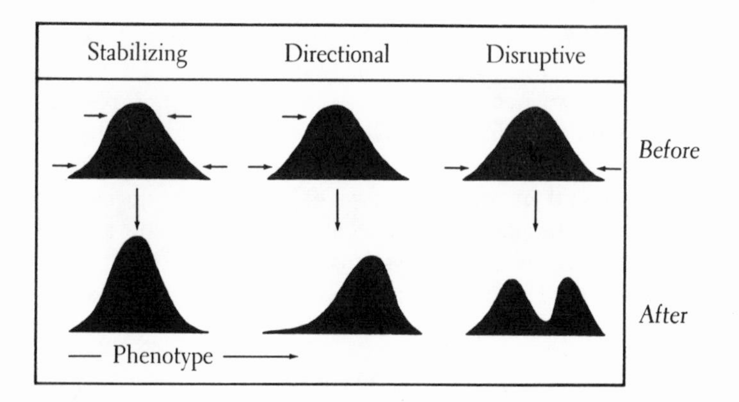

FIGURE 22-1. Stabilizing, directional, and disruptive selection. (After Mather, from Scharloo, 1965.)

the optimum phenotype, the more intense is selection against him and the less is his genetic contribution to the following generation.

A number of examples of selection for an intermediate optimum have been given by Haldane (1954a) together with a measure of the intensity of selection involved. He cites a study by Rendel (1943), for example, which revealed that the variance of the weight of duck eggs is significantly lower for those that hatch than for the entire initial sample of eggs. In the case of human births, too, there is an intermediate optimum weight; babies that are too heavy or too light are more likely to die during the first month of life than are those of approximately average size.

The existence of an intermediate optimum for sternopleural bristle number in *D. melanogaster* has been demonstrated by Scossiroli (1959). In fact, two earlier observations suggest that these bristles are related to fitness: (1) The average numbers of sternopleurals in different stocks of flies are very similar, as are the average numbers in successive generations of an unselected strain of flies; and (2) genetic variation in respect to bristle number exists because one can quite easily select lines with either higher or lower numbers than that of the initial unselected strain. These observations suggest either that bristle number is unrelated to fitness or that selection operates to keep the number at the observed average. The latter is the most plausible (on the basis of what we have just said); otherwise unselected strains of different sources should differ considerably.

Scossiroli obtained estimates of fitness for flies with different numbers of bristles in two artificially selected lines of flies. His results are shown in Figure 22-2. The parabolic curves in this figure have been fitted to the estimates of fitness; it is quite clear that there are intermediate optima in these two lines. Furthermore, as one would expect in lines subjected to artificial selection, the mean numbers of bristles on the flies in these selected cultures lie beyond the number corresponding to maximum fitness. If it were otherwise, there would be no need to practice artificial selection for higher bristle numbers; natural selection, itself, would increase bristle numbers. If artificial selection were to be stopped, the two lines would tend to revert toward their earlier low numbers of bristles. Finally, however, we should note that the two lines do not have identical optima. During artificial selection for higher bristle number there is an unavoidable selection, too, for shifting the number of bristles associated with optimum fitness. The optimum number does not increase as rapidly, however, as the mean bristle number and so almost all selected lines tend to reach a plateau that is not surpassed by directional selection. Furthermore, as we mentioned above, nearly all selected lines tend to revert somewhat once selection is relaxed.

The disruption of the initial pattern of stabilizing selection by a program of artificial directional selection has a great deal to do with the

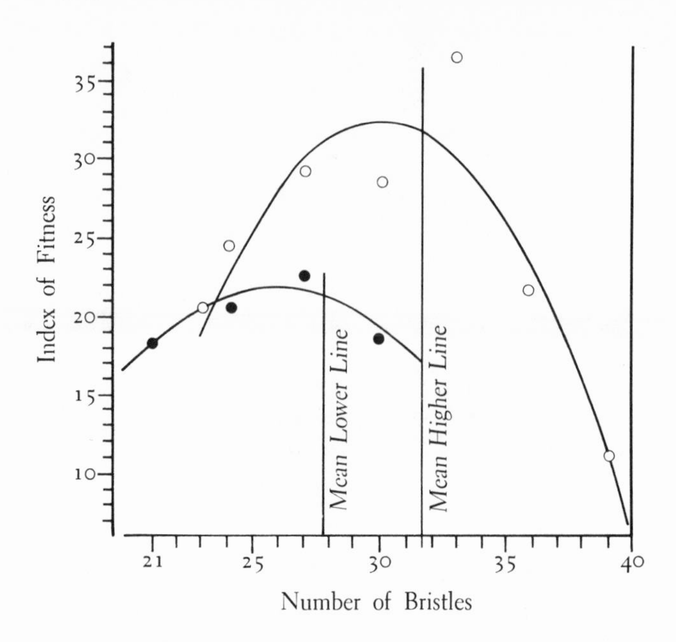

FIGURE 22-2. Correlation between index of fitness and sternopleural bristle number in two selected lines of *D. melanogaster*. (After Scossiroli, 1959.)

ultimate success or failure of the selection program. For example, artificial selection might eventually lead to homozygosis for the genes involved in determining the selected trait. At this point, progress should cease. Occasionally this actually is the outcome of selection, but certainly other outcomes are possible. Figure 22-3 shows the progress made in one of Scossiroli's selected lines of flies; the mean number of bristles increased to 32 and then remained constant. As the average number of bristles approached this plateau, however, the estimates of *heritability* steadily *increased*! Heritability is a measure of the fraction of all variance that is caused by segregating loci. If selection had caused the selected line to become homozygous, the estimate of heritability would have become progressively smaller; when the plateau was reached, heritability should have equaled zero. Rather than reaching a plateau because of homozygosis, this particular line has become stabilized because artificial selection is counterbalanced by an opposing natural selection that favors lower bristles.

DISRUPTIVE SELECTION

Within natural populations it is surely true that a good deal of selection is involved in maintaining the status quo by eliminating deviant individuals and in preserving the optimal phenotype (and its responsible

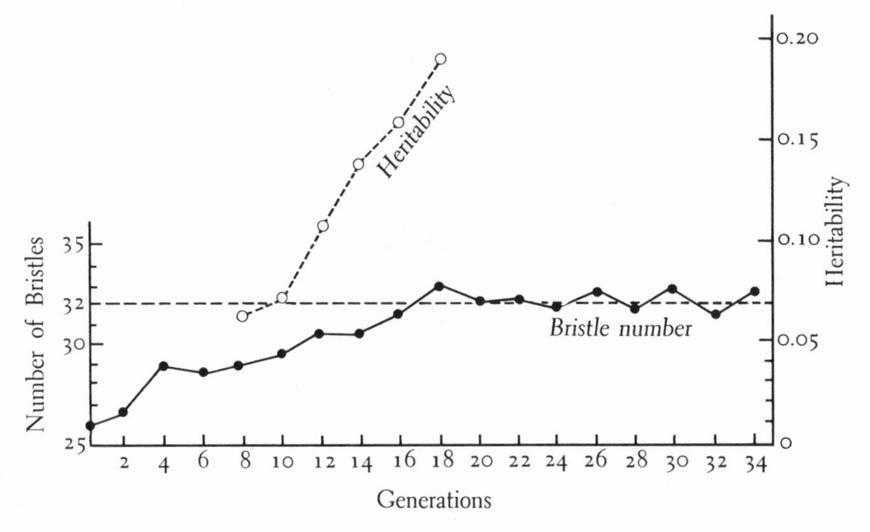

FIGURE 22-3. Curves showing the progress of artificial selection for high sternopleural bristle number in *D. melanogaster* together with the estimates of heritability. The existence of a plateau in this example does not mean the exhaustion of genetic variation. (After Scossiroli, 1959.)

genotypes). A great deal of selection is stabilizing selection. On the other hand, the environment is not uniform—either in space or in time. A great deal of the selection that takes place in the local population, or in neighboring populations, must be somewhat self-contradictory. Although fast larval development may be favored in one locality even at the expense of the size of adult flies, slow development and large adult body size may be favored in another. At times, dissimilar responses to the same environmental challenge may be developed by partially isolated populations (see, for example, Table 18-7); an exchange of migrant individuals then would confront each population with conflicting alternatives for future selection. Conflicts of this sort, in fact, must underlie the origins of coadaptation; when the conflict is extreme, it gives rise to disruptive selection. The most extreme representation is that shown in Figure 22-1c. In this case, the tails of the initial normal distribution have been designated as "optimal"; a successful response to the opposing selections would be the formation of a bimodal population in which each mode would represent one of the optimal phenotypes.

Selection for Sexual Dimorphism. In many animals, the two sexes are phenotypically different. A great many of the differences between males and females can be grouped under the heading "secondary sexual characteristics." These characteristics, however, are not immutable. In some species of birds, the color patterns of males play an important role during courtship and mating; females of these same species may possess drab, cryptic patterns. On isolated islands where confusion during courtship is reduced by the presence of few or no related species, it sometimes happens that both males and females have similarly drab patterns (Mayr, 1963, p. 317). The possession of a distinctive pattern in these species of birds is not an intrinsic property of maleness; it is a necessity developed under the influence of selection for sexual dimorphism.

As further examples of sexual dimorphism, we might cite the sex-influenced dominance of certain genes such as that for baldness in man or for the presence of horns in cattle. The modification of dominance by sex can produce a crude sexual dimorphism; in fact, the maximum frequency of chance encounters of horned (for example) males and nonhorned females in a population of cattle would occur if the frequencies of the two responsible alleles were equal. Should a need for further change arise, we know that dominance is readily modified by selection.

How easily can a sexual dimorphism be developed in a population should it become advantageous? This question led to a series of selection experiments (Wallace and Madden, unpublished) in which an effort was made to develop strains of *D. melanogaster* having high numbers of sternopleural bristles in one sex and low numbers in the other. Ordinarily the average numbers of bristles on the two sexes of a given strain are very close. In the experiments on sexual dimorphism, hybrids were obtained by intercrossing flies from Syosset, New York, and experimental population 1 (see Table 13-1); four culture bottles were set up from which hybrid flies could be obtained. For flies from each of the four cultures, the number of sternopleural bristles (one side only) was counted for one hundred males and one hundred females. Two cultures were used to start experiments AHM (A, high males) and BHM (B, high males) in which the ten males with the highest numbers of bristles were mated with the ten females with the lowest number. The other two cultures were used to start the converse experiments, ALM (A, low males) and BLM (B, low males), in which the ten males with the lowest numbers of bristles were mated with the ten females with the highest numbers. Each generation after that, bristles on one hundred flies of each sex were counted of which the ten highest and the ten lowest (of the proper sex for the particular experiment) were chosen as parents.

The results of this series of experiments suggest that sexual dimorphism is rather difficult to establish by selection. The raw materials from which a genetic system can be composed in which males will have high (or low) bristle numbers while females will have low (or high) numbers are not available in every population. Three of the four experimental population showed no systematic response whatsoever to the selection for sexual dimorphism. The fourth (ALM) did respond in the manner illustrated in Figure 22-4; each sex came to have a bimodal distribution

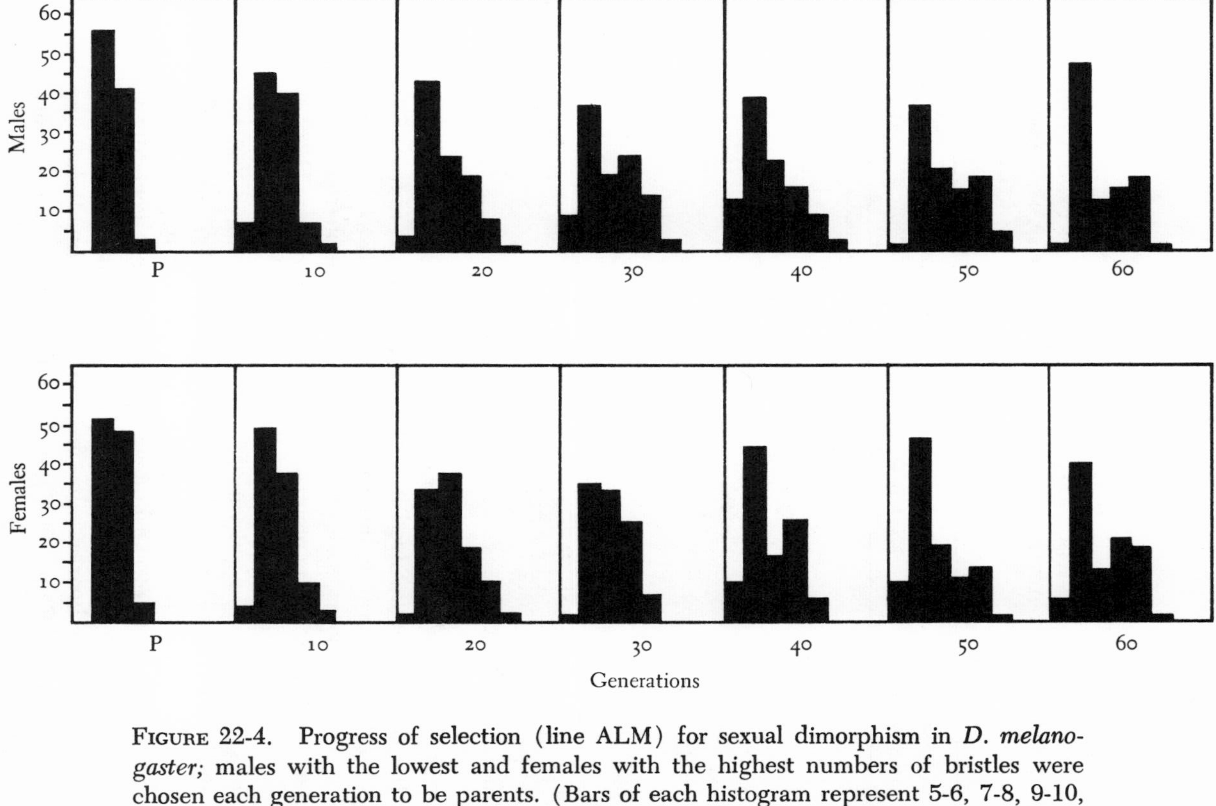

FIGURE 22-4. Progress of selection (line ALM) for sexual dimorphism in *D. melanogaster;* males with the lowest and females with the highest numbers of bristles were chosen each generation to be parents. (Bars of each histogram represent 5-6, 7-8, 9-10, 11-12, 13-14, 15-16, 17-18, and 19+ sternopleural bristles.) (Wallace, unpublished.)

of bristle numbers. The population was unable, however, to rid either sex of the unwanted portion of its distribution curve.

Further information about line ALM is given in Table 22-1 and Fig-

TABLE 22-1

STERILITY OF MALES ASSOCIATED WITH HIGH NUMBERS OF STERNOPLEURAL BRISTLES IN AN ARTIFICIALLY SELECTED LINE OF D. MELANOGASTER.

(Wallace, unpublished.)

NO. OF BRISTLES ON MALE	NO. OF FEMALES EXAMINED	NO. OF FEMALES INSEMINATED	% INSEMINATED
2	27	4	14.8
5	67	24	35.8
6	293	101	34.5
7	284	112	39.4
8	292	113	38.7
9	280	65	23.2
10	287	9	3.1
11	282	3	1.1
12	287	0	0
13	271	0	0
14	233	1	0.4
15	52	1	1.9

ure 22-5. Males with high bristle numbers in this line were largely sterile. In three tests that involved flies of all bristle numbers from 2 to 15 (no 3s or 4s) in which individual males were exposed to ten virgin females for 4, 8, and 16 hours, males with 5, 6, 7, or 8 sternopleural bristles inseminated some 35 to 40 percent of the available females. Those with nine bristles inseminated only one half as many, while those with ten or more bristles inseminated very few. It should be noted that this pattern of sterility (and subsequent dissection of males showed that those with high numbers of bristles were indeed sterile) is unique to ALM; males with high numbers of sternopleural bristles are not generally sterile.

The data summarized in Table 22-1 and Figure 22-5 illustrate three points. First, they bear out the observation on optimal phenotypes made by Scossiroli (see Figure 22-2); males with only two sternopleural bristles appear to be less successful at inseminating females than those with somewhat higher (but still low) numbers. Second, they emphasize the extent to which adaptive responses by a population fit the framework within which challenges are posed. In the selected line ALM, males with high numbers of bristles were never used as parents. Because of the scheme by which line ALM was maintained, the sterility of males with

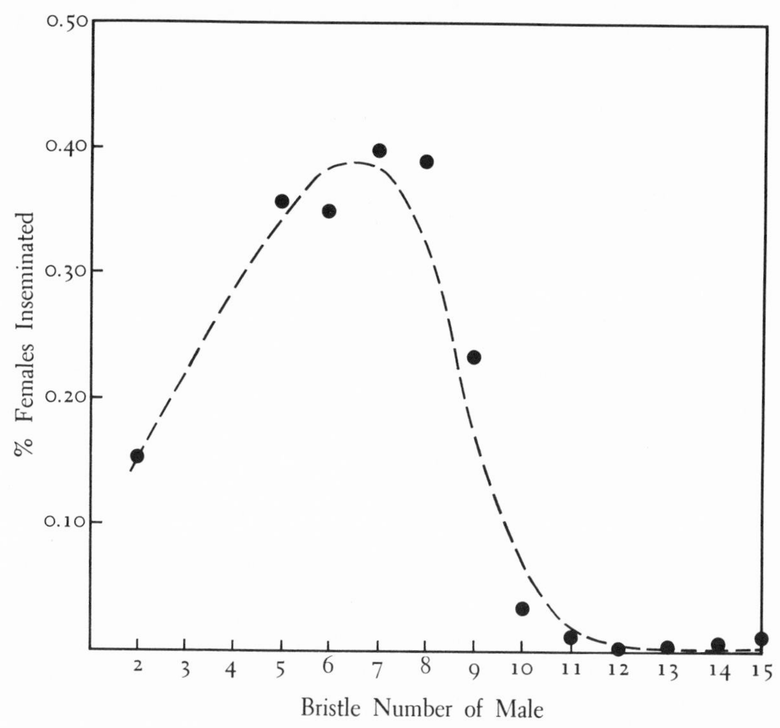

FIGURE 22-5. Relation between the number of sternopleural bristles of males in line ALM and the ability of the males to fertilize females. (Wallace, unpublished.)

high numbers of bristles was a matter of complete indifference. These males never served as parents; consequently, a response to selection that involved their sterility was an acceptable response. Finally, the data emphasize the need for care in interpreting the results of experiments on disruptive selection. Suppose that instead of being sterile, males of ALM with high numbers of bristles had died during early development. The distribution curves of bristle numbers in this case would have suggested that selection for sexual dimorphism was quite successful in this line. At least, the males would have had uniformly low numbers. However, this solution to the problem of devising a successful response to selection for sexual dimorphism would have been a false solution. It would have been based upon a loss of approximately one half of all male zygotes; alleles that cause the death of their carriers are deleterious alleles. They tend to be eliminated from populations according to rules that were discussed in Chapters 9 and 10.

Selection for Contrasting Genotypes. Some of the most exquisite examples of successful responses to disruptive selection—both that involving sexual dimorphism and that involving two or more forms of a single sex—are to be found among butterflies that mimic other species. The genetics of mimicry in *Papilio dardanus* (see page 334) has been studied extensively by Clarke and Sheppard; their work has been reviewed by Ford (1964). In this species, the male is nonmimetic; it is monomorphic —yellow and black—with swallow tails "typical" of the genus. Females, except in Madagascar where they resemble the nonmimetic males, mimic various noxious species of butterflies. The various mimic forms are controlled by single "supergenes" that not only alter the color pattern of the females to match that of the appropriate model but also produce tailless hind wings and, in some instances, noticeably affect the behavior of the mimic so that its habits also resemble those of the model.

This entire genetic complex so beautifully analyzed, as we said above, by Clarke and Sheppard represents one of the expected outcomes of a successful program of disruptive selection. In effect, the species has developed a series of alleles A, A', A'', and so on. In males, these alleles are phenotypically equivalent: A equals A' equals A''; these alleles form the basis of the nonmimetic, monomorphic phenotype. In females, these same alleles exhibit dominance such that AA equals AA' equals AA'' while $A'A'$ equals $A'A''$. $A''A''$ is the bottom recessive in this illustration. The three phenotypes corresponding to A-, A'-, and $A''A''$ form (within the proper genetic background; see page 334) three mimic forms.

For nearly ten years, Thoday and his co-workers have carried on an intensive study of the effects of disruptive selection on *D. melanogaster* (see Gibson and Thoday, 1964, for references). One of their populations, D+, is especially interesting for our discussion. This experimental population (Thoday, 1958; Thoday and Boam, 1959) was maintained by disruptive selection with assortative mating. The population consisted of four cultures numbered 1 through 4 (following Thoday and Boam's description), two of which (1 and 2) were designated H for high and the other two (3 and 4) were designated L for low. The mating scheme can be presented in tabular form:

PARENT IN GENERATION	CULTURE (= MALE SUBLINE) 1(H) ♀	1(H) ♂	2(H) ♀	2(H) ♂	3(L) ♀	3(L) ♂	4(L) ♀	4(L) ♂
n	H3	H1	H4	H2	L1	L3	L2	L4
$n+1$	H4	H1	H3	H2	L2	L3	L1	L4
$n+2$	H3	H1	H4	H2	L1	L3	L2	L4
$n+3$	H4	H1	H3	H2	L2	L3	L1	L4

In this mating scheme, the males carry the designation of the culture; females come from elsewhere. Females labeled H3 and H4 are the females with the highest numbers of sternopleural bristles that hatched in cultures 3(L) and 4(L). Similarly, females designated L1 and L2 are females with the lowest numbers of bristles among those hatching in cultures 1(H) and 2(H). The numbers of bristles were counted on twenty flies of each sex; the matings shown in the table are single pair matings involving the highest and the lowest flies of the appropriate sex.

The mating scheme is somewhat unusual and needs to be carefully examined to be understood. First, brother-sister matings are not involved. Second, no crosses involve males and females from cultures 1 and 2 or from cultures 3 and 4; the selection scheme, consequently, is not a simple directional one. Finally, the matings do involve the highest females of the low lines times the highest males of the high lines, and the lowest females of the high lines and the lowest males of the low lines. If "high" and "low" are thought of as two very demanding environmental conditions, we see that males that meet these conditions always mate with "migrant" females from the opposite environment but females that meet as well as possible the requirements of their new, foster environment. Simple directional selection would involve matings of males and females that meet the demands of their proper environment but with no exchange ("contamination" or "migration") between environments.

What was the outcome of this mating scheme, a scheme designed as if to tax the "ingenuity" of the selected population? The cultures within the selected population diverged. The high cultures (1 and 2) gave rise to flies with high bristle numbers, while the lows (3 and 4) gave rise to flies with low numbers (Figure 22-6). Thus it seems that the scheme of selection imposed on this population has succeeded—that the population has responded by developing a polymorphism that satisfies both extremes of selection.

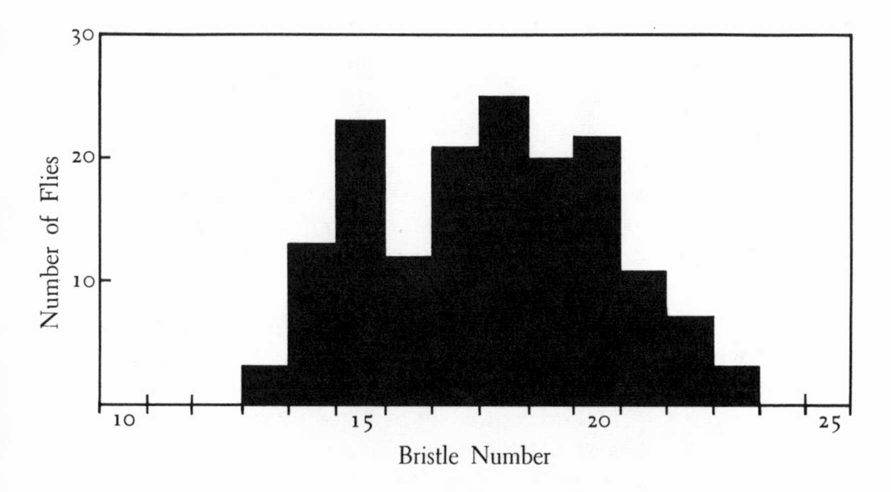

FIGURE 22-6. Distribution of sternopleural bristle numbers in a line (D^+) of *D. melanogaster* subjected to disruptive selection. (After Thoday and Boam, 1959.)

Genetic analyses show that the population has indeed hit upon a solution but, in some respects at least, it resembles what was earlier called a "false" solution. Two loci are involved; one is at 27.5 and the other at 47.5 on the second chromosome. Flies with low bristle numbers are of genotype (+ −)/(− −) or (− +)/(− −); those with the higher number are (− +)/(+ −), (− +)/(− +), or (+ −)/(+ −). The (− −) combination of alleles is a recessive lethal. About ten percent of the gametes of (+ −)/(− +) females carry (− −) chromosomes, while another ten percent carry the complementary chromosome (+ +). The latter acts as a dominant lethal and is never recovered (Gibson and Thoday, 1959, 1962a, 1962b).

The selection program has succeeded because low × low matings give both (+ −)/(− −) and (− +)/(− −) (low) and (+ −)/(+ −), (− +)/(− +), and (− +)/(+ −) (high) genotypes. Similarly, high × high matings give (− +)/(− +), (+ −)/(+ −), and (− +)/(+ −) (high) and (+ −)/(− −) and (− +)/(− −) (low) genotypes. The low segregants from the high crosses are genotypically like those from the low crosses; on the other hand, the high segregants from the low crosses are genetically the same as those from the high crosses.

The cost of this polymorphism is high. One fourth of the progeny of low × low are homozygous lethal (− −)/(− −). One tenth of the progeny of many high × high matings carry a chromosome (+ +) that behaves as a dominant lethal. This cost is quite separate from any decline in fitness that may have accompanied the selection program for other reasons. The system is obviously not as efficient as that governing mimicry in *Papilio dardanus*; the switch genes responsible for the different genotypes of *Papilio* shunt the phenotype into certain channels with, as far as is known, no lethality as a result of the genetic switch mechanism itself. Of course, *Papilio* has had considerably more time to develop an efficient system in nature than have the experimental lines of Thoday and his colleagues. We might also recall the balanced lethal system found in some populations of *D. tropicalis*; not all genetic solutions hit upon in response to natural selection have to be highly efficient.

The Development of Sexual Isolation through Disruptive Selection. In the experiments on disruptive selection described above, the male and female that served as parents in a given culture were taken from both high and low lines. Whether the female in a given cross was the highest of the lows or the lowest of the highs, the experiment continued only by the forced intercrossing of progeny from high and low cultures.

With a slightly modified experimental procedure, one can visualize still another outcome of successful disruptive selection—reproductive isolation. Suppose that two environments are designated "high" and "low" and that only high males and females survive in one while only low males and females survive in the other. Suppose further that the intermediate offspring obtained by the intercrossing of high with low survive in neither environment. One of the possible solutions available to the population is its subdivision into two reproductively isolated populations with the highs adapted to the high environment and the lows to the low. In effect, this is what speciation is all about.

The experiments dealing with the role of disruptive selection in the development of sexual isolation within populations have also been done largely by Thoday and his colleagues (Thoday and Gibson, 1962; Thoday, 1964). In this series of experiments, the flies (*D. melanogaster*) have an opportunity to choose their own mates. A typical experiment is started by counting the numbers of sternopleural bristles on eighty males and eighty females and selecting from them the eight males and the eight virgin females with the highest numbers of bristles together with the eight males and eight virgin females with the lowest numbers. These thirty-two flies are placed together in a small vial for twenty-four hours. After this period, males are discarded, while the females are separated into "highs" and "lows" once more. The females are then placed in culture bottles. When the progeny flies hatch, forty males and forty virgin females are collected from the "high" cultures while similar numbers are collected from the "low" ones. The eight highest males and females of the high line and the eight lowest males and females of the low line are again placed together in the small mating vial for twenty-four hours. Males are again discarded; the females are separated into highs and lows and are placed in culture bottles. The cycle is repeated each generation.

The results obtained from an experiment such as that just described were published by Thoday and Gibson (1962) in a paper entitled "Isolation by Disruptive Selection." Their results are shown in Figures 22-7 and 22-8. Figure 22-7 shows the divergence of the arrays of bristle numbers in the high and low lines. Disruptive selection practiced in the manner described above successfully divided the population into two portions, a high-bristle-number portion and a low-bristle-number portion.

How does the disruptive selection bring about the division of the population in these experiments? We must recall that all thirty-two flies (8 high males, 8 high females, 8 low males, and 8 low females) were placed together for twenty-four hours in a small mating vial. We should also recall that in selecting the highest offspring of high females and the lowest offspring of low females, we are almost certainly selecting offspring from high $\times$ high and low $\times$ low matings. Consequently, if there is any genetic basis for sexual preference between flies with similar numbers of sternopleural bristles, selection for flies with high and low bristle numbers will also represent selection for sexual isolation. Females

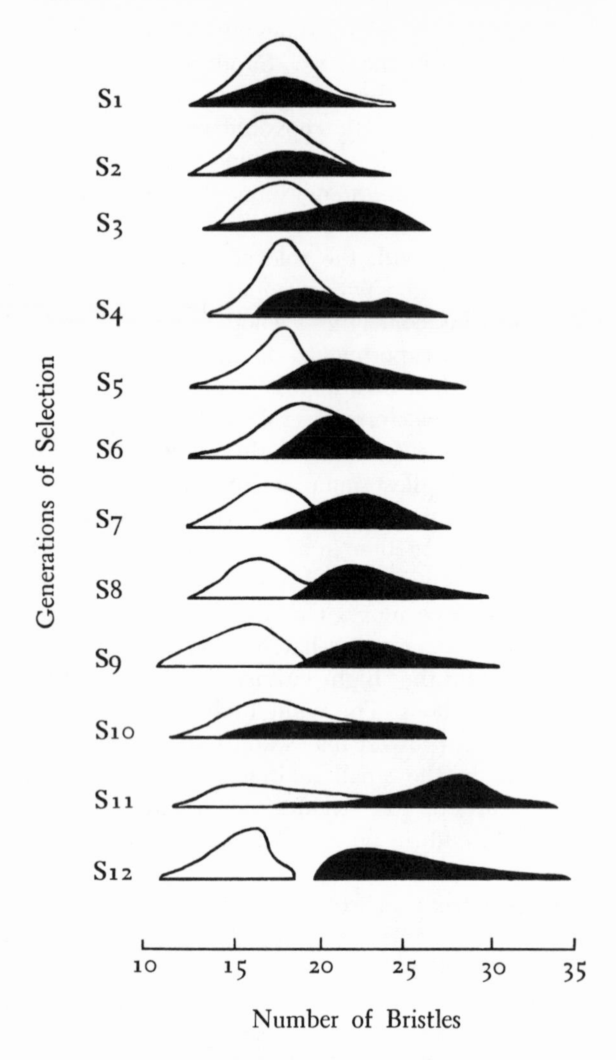

FIGURE 22-7. Divergence of "high" and "low" lines of *D. melanogaster* that was obtained despite the opportunity for random mating to occur provided by placing males and females with high and low bristle numbers together as virgins in a single mating chamber. (After Thoday and Gibson, *Nature, 193*: 1164–1166, 1962.)

with high bristle numbers, for example, that accept males with low numbers cannot contribute to the selected progeny. Nor can any other mating of dissimilar types contribute to the progeny selected to be the parents of the following generation.

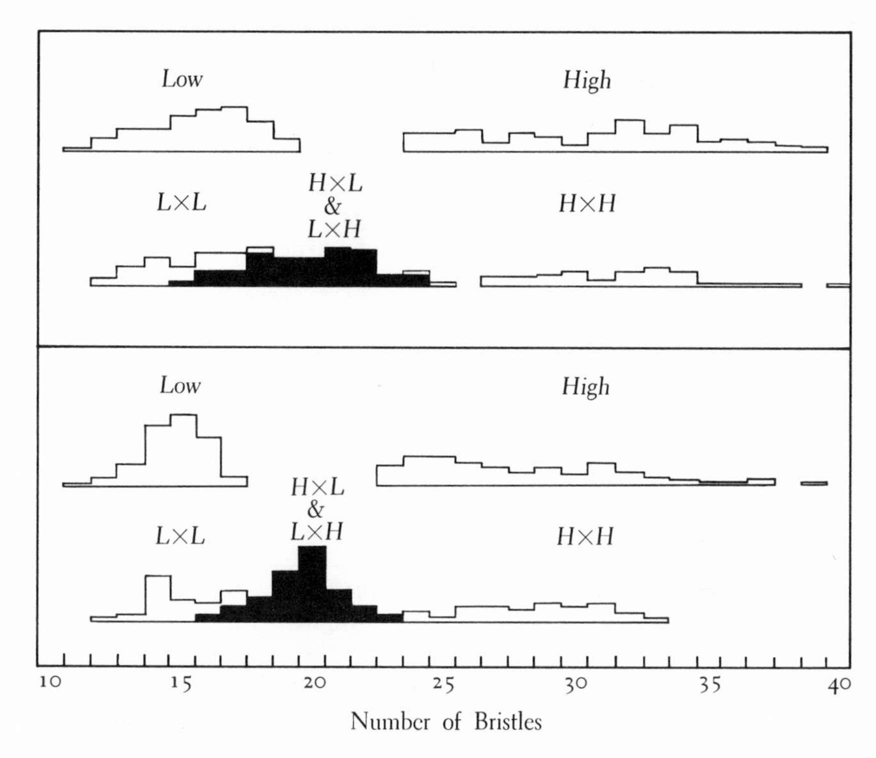

FIGURE 22-8. Evidence for sexual isolation between lines kept by disruptive selection (see Figure 22-7). The upper curves represent the bristle numbers observed among the progeny of low and high females of the two selected lines. The lower curves are those of progeny of high × high, high × low, and low × low matings. Few, if any, offspring of high × low matings are found within the selected lines even though the opportunity for such matings was provided. (After Thoday, *Insect Reproduction*, Symp. 2, Royal Ent. Soc. London, pp. 108–119, 1964.)

The notion that the disruptive selection has operated through the development of sexual isolation has been tested; the results are shown in Figure 22-8 and Table 22-2. The figure shows the arrays of bristle numbers seen in the high and low lines during the twentieth and twenty-first generations. Together with the actual arrays are those obtained by mating low × low, low × high, high × low, and high × high flies. Clearly, the low × low and high × high portions of the reconstructed distribution corresponds closely to the high and low lines themselves. The arrays obtained by mating high × low and low × high do not have a counterpart in the high and low lines. Evidence of this sort was obtained during the sixth and eleventh generations; the evidence obtained during the twentieth and twenty-first generations is very clear indeed.

The data listed in Table 22-2 show that sexual preferences are in fact involved in the absence of intermediate bristle numbers from the selected lines. By simply determining the type of male that successfully mated with the two types of females in experiments corresponding to the regular mating procedure, Thoday was able to show that there was a decided preference of high males for high females and of low males for low females. Relatively few of the matings involve flies with dissimilar (high versus low) numbers of sternopleural bristles.

These are remarkable results. As Waddington and Robertson (1966) have pointed out in reference to them, selection has succeeded in establishing sexual isolation in this experiment where it has failed in other, even more stringent attempts. In my own laboratory (Wallace, 1954), seventy-three generations of selection for sexual isolation succeeded in altering mating preferences only slightly. Knight et al. (1956) were no more successful in their attempts to develop sexual isolation by selecting for it directly. Furthermore, not every experiment involving disruptive selection for sternopleural bristle numbers has been successful; Mrs. Chabora (Alice Schwartz) did not succeeed in obtaining lasting divergence in any line out of eight tested. Scharloo (1964) was also unable to obtain evidence for sexual isolation in a line subjected to disruptive selection on the expression of *cubitus interruptus,* a wing-vein mutation. Thus it seems that the outcome observed by Thoday and his colleagues is one that occurs infrequently at best. But if it occurs at all, that is what matters.

TABLE 22-2

EVIDENCE FOR PREFERENTIAL MATING OF MALES WITH HIGH AND LOW NUMBERS OF STERNOPLEURAL BRISTLES FOLLOWING A NUMBER OF GENERATIONS OF DISRUPTIVE SELECTION.

(Thoday, 1964.)

Female:	HIGH		LOW	
Successful male: / Generation	HIGH	LOW	HIGH	LOW
7	12	3	4	12
8	14	2	6	10
9	10	4	6	7
10	8	4	3	13
19	27	2	8	20
	71	15	27	62

Disruptive Selection and Sympatric Speciation

The real importance of the work done by Thoday and his colleagues is the impetus it has given to the once-beleaguered notion of sympatric speciation. Until the results of this work became known, the idea that one species could split into two without an intervening period of physical (nongenetic) isolation was all but abandoned in many circles; in at least one modern textbook on evolution (Stebbins, 1966), the term "sympatric speciation" is not included in the index nor, presumably, is it discussed in the text.

Just what is sympatric speciation? How does it differ from geographic or allopatric speciation? And why are these two patterns of speciation regarded as opposites when, as we saw in the preceding chapter, the origin of species from (species) border populations includes elements of both? The remainder of this chapter will be devoted to a discussion of these and related questions and to an effort to fit disruptive selection into the scheme of species origins.

Stripped of all nonessential qualifications, the difference between the allopatric and sympatric origin of species involves the postulated need or lack of need for a period of spatial isolation, nongenetic in origin, during which two populations can diverge genetically. By far the most common basis for this period of isolation is geographic isolation, the occurrence of populations of a given species in geographically remote areas. What is geographically "remote" for one species need not, of course, be remote for another.

Genetic divergence can occur in isolation, but it can be speeded up if urged on by selection. This fact, as far as speciation is concerned, is one of the chief advantages possessed by populations found at or near species borders. There are local selective forces that tend to adapt these populations to prevailing border conditions. Basically, these must be unfavorable conditions. There is an obvious inability of the species at large to exploit these conditions fully, otherwise the border would not be a border at all. And there are periods of contact alternating with periods of isolation which serve to speed along any incipient genetically determined isolating mechanism.

Simple geographic isolation, the isolation of large segments of a species in hospitable but disjunct areas, does not furnish a strong basis for divergence. The selection for sexual isolation in *D. melanogaster* mentioned above serves as an illustration. The selection experiment (Wallace, 1954) did not give rise to isolation even after seventy-three generations of selection; still, mating preferences were altered. At the same time, however, tests of sexual isolation were made using flies from experimental populations of *D. melanogaster* (the same populations as those described in

Table 13-1) that had also been isolated for seventy generations (nearly three years). Tests of these populations involving an examination of nearly 2000 females failed to reveal any evidence for intrapopulation mating preference. As we suggested above, selection can bring about events rather quickly that would otherwise happen very slowly, if at all.

In its purest form, sympatric speciation would require no spatial isolation during the period when a population adapts to two different environments and, simultaneously, erects reproductive barriers that serve to isolate the two differently adapted portions of the population. To some proponents of sympatric speciation, sympatry means that the isolating distance between the two segments of the population be small in the judgment of the observer. This, I believe, is wrong. Sympatry occupies the analogous position in respect to speciation processes that the null hypothesis occupies in respect to statistical analyses. In analyzing data we claim that our observations do not differ from those expected under a given model (this claim is known as the "null hypothesis"); we then determine whether there is reason to reject this claim. We cannot prove the null hypothesis because there is always the possibility that a larger set of observations than that tested might have succeeded in revealing a significant departure of observed from expected results.

By analogy, allopatric speciation is speciation aided by a period of geographic isolation. Sympatric speciation, then, is speciation without the need of isolation—the null hypothesis against which observations must be tested. It is the failure to recognize the unique ("unique" in the formal or mathematical sense) position required of proponents of sympatric speciation that enables them at times to cite isolation by small distances or over short periods of time as evidence in support of their views.

What is the nature of the experiments that have been made in an effort to bring about sexual isolation in the laboratory? The first of these was made by Koopman (1950) using *D. pseudoobscura* and *D. persimilis,* two closely related species that are rather difficult to cross in the laboratory —except at low temperatures. Koopman began each generation of his experiment with fairly large numbers of males and virgin females of each species, suitably marked with recessive mutations so that offspring of both intra- and interspecific matings could be identified. Each generation the wildtype hybrid offspring were discarded while males and virgin females of the two species were collected to serve as parents for the next generation. The gametes of promiscuous individuals who mated with members of the wrong species were thrown out with the wildtype hybrids. And to the extent that mating discrimination has a genetic basis, Koopman selected for greater discrimination by (1) using only members of the pure species as parents but (2) by offering them an opportunity to err each generation. Within very few generations, a marked isolation between the two species developed in Koopman's cages.

The experiments of Wallace (1954) and Knight et al. (1956) used procedures that were basically the same as those used by Koopman. The difference lay in the attempt in these experiments to build up sexual isolation between two strains of the same species (*D. melanogaster*) rather than between two species. In the experiments with *D. melanogaster,* the two strains were marked with recessive mutations so that wildtype "hybrids" could be recognized and discarded. At the start of each generation, males and virgin females of both mutant strains were placed together in a mating chamber so that there would be an opportunity for both intra- and interstrain matings to occur. Selection in favor of intrastrain matings arose through the elimination of interstrain (wildtype) hybrid offspring.

The experiments on *D. melanogaster* described above failed to give rise to sexual isolation as quickly as those of Koopman; indeed, isolation was never obtained, only an alteration in mating preference. But what if they had given rise to isolation? The crux of the matter lies in the destruction of the hybrids. The rationale was that the act of discarding wildtype hybrid progeny was *a laboratory procedure that took the place of a prolonged period of geographic isolation* during which the strains were said to have diverged enough to exhibit "hybrid breakdown," to produce inviable or infertile hybrids. In the study involving *D. pseudoobscura* and *D. persimilis,* this breakdown was already a fact because *pseudoobscura* × *persimilis* hybrid males are sterile; in the study of *D. melanogaster* the strains gave perfectly "good" hybrids. Human life, though, is too short for an experimenter to wait for genetic divergence leading to inviable hybrids to occur between isolated strains of a single species. And so the discarding of interstrain hybrids was introduced as a substitute procedure. In this way, the events that might follow the coming together of two formerly isolated populations can be studied experimentally.

The studies on disruptive selection, especially that which gave rise to sexually isolated strains with widely differing numbers of sternopleural bristles, have been cited as evidence that sympatric speciation is possible. In these studies, as we have pointed out earlier, flies with the highest bristle numbers are the progeny of high × high matings while those with the lowest numbers are progeny of low × low matings. The experimental procedure used in the study of disruptive selection, then, does not differ from that used in the earlier studies. In my own experiments, for example, *sepia* flies were the offspring of *sepia* × *sepia* matings, while *straw* flies were the progeny of *straw* × *straw* matings. Hence one view of the experiments on disruptive selection is that the act of discarding the carriers of intermediate numbers of bristles was an act that substituted for a period of isolation leading to hybrid breakdown. That was the interpretation of the experimental procedure in the earlier studies. And because the effect of disruptive selection is to duplicate these earlier procedures, this is an acceptable interpretation for the experiments on disruptive selection. Under this interpretation, of course, disruptive selection has no bearing whatsoever on the matter of sympatric speciation, no more than Koopman's experiments had.

The alternative interpretation of the experimental procedure is that which we have used several times in this chapter; this is the interpretation that leads us to say that there are "high" and "low" environments that permit only flies with high and low members of sternopleural bristles to survive. Given that there may be such environments and that the optimal bristle numbers for each are above and below the extremes of the present-day distribution, then the experiments on disruptive selection do bear on the matter of sympatric speciation. In the same way, however, the earlier ones did, too. To accommodate this view, we merely assume that there are two environments—a "*sepia*" and a "*straw*" environment—which permit only *sepia* and *straw* flies to survive. If these two environments are distributed as a fine mosaic so that adult flies have an opportunity to mate at random, will selection intervene and give rise to sexual isolation? In the case of *sepia* and *straw* the answer was, No, not within seventy-three generations. In the case of sternopleural bristles, the answer in one case was, Yes, and within a very short time.

Having established that the experiments of Thoday and of Knight, Robertson, Waddington, and myself are fundamentally similar, it remains to ask why I, for example, had not regarded my own experiment as a test of the plausibility of sympatric speciation. The reason is the artificial nature of the postulated environments, of their demands, and of their spatial distribution. It must be extremely rare, indeed, to have a patchwork of environments in which only alternative forms can survive and in both of which intermediate forms perish. In general, the demands of one environment extend beyond but are not in direct conflict with those of a neighboring one. It is even rarer that these environments should favor the tails of what is presently a normal-shaped, intermediate (and hence dead) population. Despite an intensive search, pesticide experts have failed to develop a pair of poisons such that resistance to one confers an obligatory sensitivity to the other. I do not think that adjacent environments pose problems that conflict to the same extent as those used in studying sexual isolation and disruptive selection.

If natural conditions are not as stringent as those set up in laboratory experiments, they take on properties identical to those discussed in Chapter 11, in which we dealt with selection in favor of heterozygotes. If the genes affected by selection are alternate alleles at a given locus and if the elimination of intermediates (heterozygotes) is not complete, then a balanced polymorphism may be established; this polymorphism would not differ from the multitude of other known instances of polymorphisms. The only requirement for establishing a polymorphic system under these conditions is that the heterozygotes have higher fitness (survive in larger proportion, for example) than the geometric mean of the two homozygotes. (If strains of *Drosophila* were found that developed sexual isolation rather quickly under artificial selection, the above expectation could be tested rather elegantly.)

In considering the possibility that sympatric speciation may occur, it is worthwhile remembering the distinction between what is merely imaginable and what is actually probable. There are many things that are possible in life. New gene arrangements can arise from preexisting ones by three or more simultaneous breaks. It is *possible* for them to arise in this manner. Chromosomal breaks are normally so infrequent, however, that even two-break inversions arise only rarely; the frequency with which three or more breaks arise spontaneously in the same chromosome arm must be vanishingly small. It is *possible* that one enzyme might dismantle and reconstruct organic molecules in a most thorough manner. Enzymes do not work this way, however; each controls a single step in a biochemical pathway. Devising an enzyme capable of controlling one step is difficult; devising one that can carry on an entire battery of steps is next to impossible.

Finally, the statement is frequently made that no matter how improbable sympatric speciation may be, life has been evolving for a long time and so this type of speciation probably has taken place. What *can* happen, in other words, *will* happen. Now, this argument is valid for chance events. There is a remote chance, for example, that one gene arrangement will arise from another spontaneously (without the aid of mutator genes, radiation, or the like) through the simultaneous occurrence of three or more breaks in a single chromosome arm. Such things must occur with their given probabilities.

The argument is not valid if there is competition among those events that might take place. Sympatric speciation may be one possible outcome of a given (albeit highly unlikely) pattern of selection. If there are alternative outcomes, if these are more likely than sympatric speciation, and if having occurred they *exclude* sympatric speciation as a possible outcome, then sympatric speciation will be an extremely rare event indeed. (Suppose one event occurs with a probability of 1/100 while another occurs with a probability of 1/10. One would expect the second event to occur ten times while the first occurs once. According to the Poisson distribution, there is only 1 chance in 20,000 that the first event will occur without being preceded by at least one of the second.) In this connection, it seems that polymorphism is a much more likely response to the often conflicting demands of selection than is reproductive isolation. Having developed polymorphic systems, populations do not split because of them; on the contrary, they seem to go to great lengths to preserve these systems.

23

FUSION, INTROGRESSION, OR ISOLATION?

Suppose that two populations have been isolated from each other for some time. Suppose, too, that during this period each has been subjected to different environments to which it has responded genetically owing to natural selection. The environments in which the populations were found may have differed at least slightly in temperature (both in mean temperature and in daily and annual extremes), humidity, rainfall, duration and intensity of sunlight, altitude, soil conditions, and in many other respects. Furthermore, the environments may have differed as well because of association with different neighboring organisms. Each population was surrounded by and interspersed among other plants or animals. Some of these were "enemies" (predators or grazers), others "friends," still others indifferent. Some cast deep shadows while others offered physical support as a means for escaping shadows. Some neighbors contributed to the richness of the soil while others tended to deplete it.

The suppositions that follow from the initial one that two populations are physically isolated are endless. We cannot possibly enumerate all the different influences to which such isolated populations are subjected. Despite this inability, however, we want now to suppose that the two populations find themselves in contact once more. What will happen? What new patterns of natural selection will arise from the intermingling of the individual members of these two populations?

The remainder of the chapter will be devoted to an outline of possible outcomes together with examples of each. With this chapter we will have completed our discussion of evolutionary changes occurring within populations—changes that cause populations to become dissimilar, changes that lead to the origin of new species.

After Geographic Overlap: Some Possibilities

Two formerly isolated populations that make contact once more can follow any one of three paths:

1. The two can fuse into one common population with subsequent events taking place according to rules discussed in the first half of this book.
2. The two can exchange genes at a low but more or less continuous rate.
3. The two can develop isolating mechanisms that keep them apart—behaviorally or physically.

Which of these three options will be chosen is determined by the fitness relationships of the parental populations, their F_1 hybrids, and the F_2 or backcross generations. Figure 23-1 is an attempt to show some of these relationships diagrammatically. As with most such diagrams, this one is grossly oversimplified. Nevertheless, it helps us see the bases underlying the three broad paths described above.

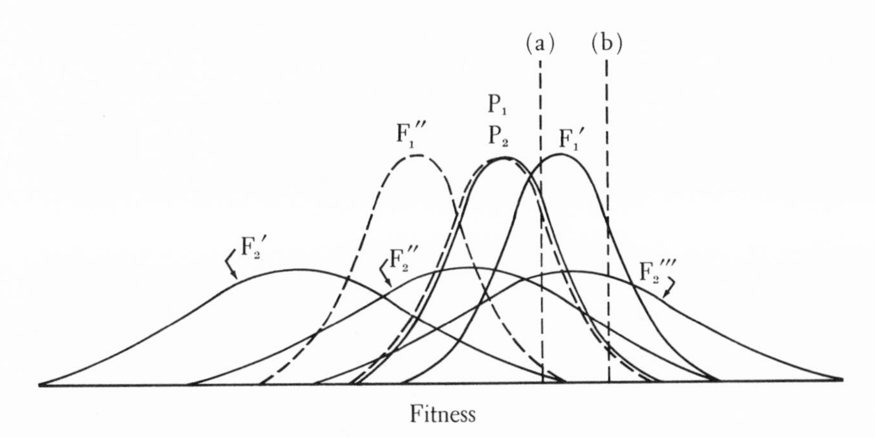

Figure 23-1. Some of the possible relationships between the mean fitness and the variance in fitness of two populations (P) and their F_1 and F_2 hybrids. Further details in the text.

The two populations that have come together are represented by the heavy curves (one solid and one dashed) labeled P_1 and P_2. The two have come together after following different routes, different both geographically and evolutionally. Having reached the same locality, however, we assume that they have devised about equally efficient, but not identical, genetic systems for coping with the local environment. Therefore, the curves are superimposed. Except that the case under discussion is much more complicated, the populations resemble those studied by King (see Table 18-7). In developing nearly identical levels of DDT resistance, we may recall, isolated populations of *D. melanogaster* took different but equally effective genetic routes.

The F_1 hybrids formed by the intercrossing of individuals of the two parental populations may exhibit heterosis (F_1'), may be sterile or inviable (F_1''), or may be about as fit as the parental populations. If the hybrids are sterile or inviable, selection will tend to favor any characteristic of either or both of the two species that promotes intrapopulation matings and discourages interpopulation gene exchange. On the other hand, if the F_1 hybrids are fertile and possess normal (or better) viability (F_1'), these individuals become part of the breeding population. The dashed line (a) in Figure 23-1 represents a fitness level below which individuals are not successful in leaving offspring; F_1'' falls entirely to the left of this arbitrary level of success while that of the F_1' falls largely to the right of it; only a relatively small fraction [one half in man according to an estimate by Penrose (1957)] of the total array of individuals of any population actually serve as parents of the next generation.

If the F_1 hybrids enter the ranks of successfully reproducing individuals, the following generation will contain F_2 and backcross hybrids (shown only as F_2 in the diagram). Once more, a number of possibilities exist. Frequently the recombinations that occur in F_1 hybrids are disharmonious; the offspring of these hybrids may suffer from a high mortality or sterility (F_2'). The backcross offspring of *D. pseudoobscura* × *D. persimilis* hybrids are of this sort. If this "breakdown" is severe, F_1 hybrids are effectively sterile. Sterility, of course, reduces the fitness of these hybrids to zero despite their apparent individual viabilities or fecundities. On the other hand, the second generation hybrids may have an average fitness equal to that of the F_1 hybrids (F_2''') or just slightly less than that of the parental populations (F_2'') (see Table 18-7, for example); in either of these cases, large numbers of F_2 or backcross hybrids may become successful breeders. Whether the two populations fuse together or merely exchange genes through these hybrid individuals depends upon the ultimate fate of the "migrant" genes themselves.

The sequence of events that precedes and then follows from the curves shown in Figure 23-1 can be outlined as follows:

1. Geographic isolation (or any other isolation that is sufficiently prolonged but is not based on genetic differences).

2. Genetic divergence. This divergence is based on genetic changes that occur within the isolated populations as the result of different patterns of natural selection. As an accidental concomitant of this divergence, the members of the two populations may fail to give rise to hybrids when they regain the opportunity to do so. On the other hand, if hybrids are produced, they may be inviable, sterile, or the F_2 or backcross progeny may be inviable or sterile. Finally (and most frequently), the two populations may simply merge into one new and highly variable population.

3. The reinforcement of isolation by natural selection. Two populations, upon meeting again after a period of isolation, may give rise to hybrid (F_1 or later) progeny that are unable for one reason or another to be included among the successful breeders of the population. In this case there will be an active selection for the cessation of hybridization. This isolation generally consists of the reinforcement or exaggeration of slight average differences that are already present; the characters involved can frequently be seen and measured, so the exaggerated differences are detected as "character displacements" (Brown and Wilson, 1956). The reinforcement of isolation may involve sexual, ecological, seasonal, or mechanical isolating mechanisms.

The Fusion of Once-Isolated Populations

By far the most common outcome of the reunion of two formerly isolated portions of a species is their complete intermingling and the formation of a new, more variable, population. Certainly, if individuals from any two or more of the experimental populations of *D. melanogaster* described in Table 13-1 had been brought together in a single population cage, the outcome would have been a single population containing the genetic elements of the earlier ones. This is why it is difficult to determine which of two populations of a single species is better adapted to a particular set of environmental conditions; if the test involves placing the populations together in a given locality, they merely fuse into one new population.

With the exception of human populations, relatively few studies exist that deal with the outcome that follows the fusion of once-isolated populations. In the case of human populations, there has been an interest in populations that arise from the intermingling of racial groups as in Brazil, the Caribbean Islands, Hawaii, and the United States (see Schull and Neel, 1963, p. 350, for recent references). In the case of plants or other animals, the experimental data on the consequences of merging two populations come largely from breeding records.

Forbes Robertson (1955) has given a rather complete account of changes wrought in several experimental populations of *D. melanogaster* by selection, together with a further account of the progress of selection following the hybridization of these populations. Altogether he described the outcome of selection and hybridization in four populations, three from Scotland and one from Italy. The results obtained by a study of one pair of populations (Crianlarich, Scotland, and Ischia, Italy) have been reproduced schematically in Figure 23-2.

The two lines of flies were subjected to directional selection for large and small body size. The curves in the figure show that response in both directions occurred fairly readily in both lines. It appears that the strain from Scotland was somewhat more amenable to selection for large size and less so to selection for small size than was the strain from Ischia. At any rate, all four selected lines eventually reached plateaus with no (or very little) further progress. At that time, the selected lines were intercrossed. The F_1 hybrids tended to retain or, in the case of the small lines, to regain their large size. The F_2 individauls were smaller, on the average, than the F_1 hybrids in both intercrosses. Selection was resumed in the F_2 generation and, in each case, progress was resumed. In the case of the large lines, the renewed selection program exceeded the earlier plateaus; each line evidently contributed genes for large body size that were not carried by the other. In the particular case illustrated in the figure, the renewed selection for small thorax did not transgress the lower of the two original plateaus.

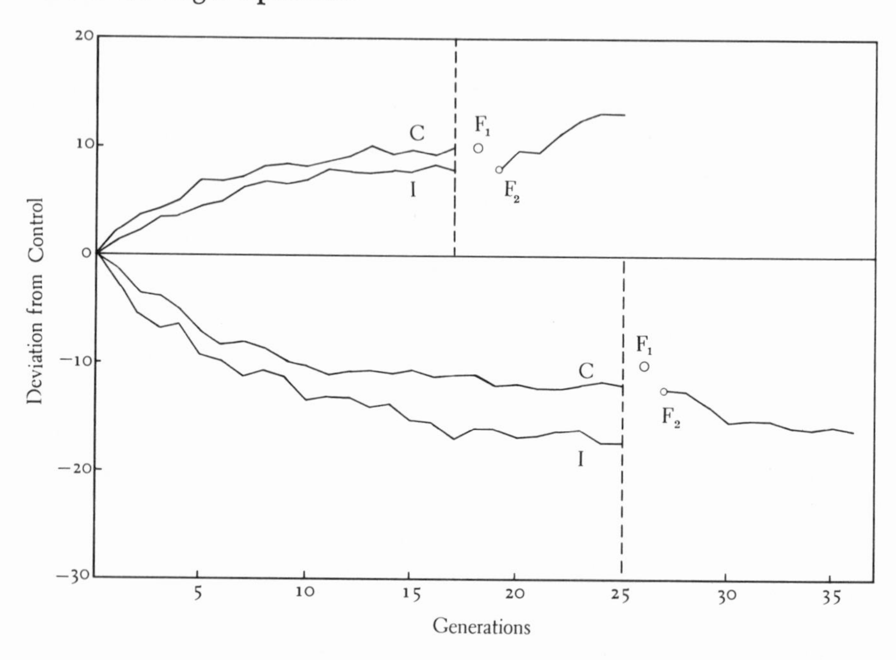

Figure 23-2. Results of artificial selection for long and short thoraxes in two geographic strains of *D. melanogaster* followed by intercrossing the two high lines and the two low lines and a resumption of selection. (After F. Robertson, 1955.)

If we were to relate the results of this selection experiment to the diagram shown in Figure 23-1, we would say that it resembles the collection of curves labeled P_1, P_2, and F_2''. Selection was not applied to the F_1 generation and so it had an effectively high fitness by default. On the other hand, had the parents for the F_2 generation of the low lines been selected from a mixture of P_1 (Crianlarich), P_2 (Ischia), and F_1 hybrids between these two, there would have been a rather intense selection in favor of Ischia alone; both Crianlarich and the F_1 hybrids would have been at a disadvantage.

Experiments by Koopman (1950) in which sexual isolation between *D. pseudoobscura* and *D. persimilis* was enhanced in the laboratory were described in Chapter 22 (page 000). These two species and their F_1 and backcross hybrids (F_1 males are sterile in this case so there can be no F_2 generation) have fitness relationships such as those of curves P_1, P_2, F_1'', and F_2' of Figure 23-1. The backcross progeny in the case of these two species are sufficiently inviable that isolation could probably have arisen experimentally without the need for actually eliminating the hybrids. The backcross hybrids would have genetically "discarded" themselves.

More precisely, the hybrids in Koopman's experiment would have discarded themselves had he been able to dispense with the need for mutant strains of the two species. Without the mutant phenotypes, however, it would have been impossible to recognize the two species or to distinguish them from their hybrids. Consequently, the procedure followed by Koopman was a necessary one if his experiment were to be carried out.

More recently Van Valen (1963) has shown that the presence of debilitating mutations in the "pure" strains of *D. pseudoobscura* and *D. persimilis* can lead to a fusion of these two species into a single hybrid population. The mutants lower the fitnesses of the parental populations. Consequently, these and their F_1 and backcross hybrids have fitnesses corresponding to the curves P_1, P_2, F_1', and F_2'' of Figure 23-1 but with the arbitrary cutoff point moved from position (a) to position (b). The enormous egg-laying capacity of the wildtype F_1 hybrid females now counts heavily in favor of interbreeding; nor are all the backcross progeny seriously inviable in comparison with the mutant parental species themselves.

The above statements are borne out by experiments based on populations containing initially *orange* (eye color) *persimilis* and *glass* (eye morphology) *pseudoobscura* males and females in equal numbers. The results of one experiment are listed in Table 23-1 and are shown graphically in Figure 23-3. The mutant homozygotes give way rapidly to wildtype individuals that must be interspecific hybrids of one sort or another. The hybrid nature of these individuals can be verified cytologi-

TABLE 23-1

RELATIVE PROPORTIONS (PERCENT) OF *orange*, WILDTYPE, AND *glass* FLIES IN A POPULATION STARTED INITIALLY WITH ORANGE-EYED D. PERSIMILIS AND GLASS-EYED D. PSEUDOOBSCURA.

(Van Valen, 1963.)

DAYS	n	*orange*	WILDTYPE	*glass*
9	513	39	1	60
73	358	0	67	33
114	112	0	70	30
134	559	0	76	24
154	139	0	84	16
175	242	0	78	22
196	178	0	88	12
237	391	0	96	4
256	520	0	91	9
275	1227	0	92	8
288	360	0	95	5

cally; gene arrangements typical of both *pseudoobscura* and *persimilis* are found for both the X and second chromosomes in the hybrid population.

Mettler (1957) has also shown that two different species of *Drosophila* can, under laboratory conditions, merge their gene pools into one containing portions of both. The two species in this case are *D. mojavensis* and *D. arizonensis,* two closely related species of the "*mulleri*" group. As far as is known, these species occur allopatrically in southwestern United States and northern Mexico.

Female *mojavensis* when crossed with *arizonensis* males produce fertile male and female hybrids. The reciprocal cross (*arizonensis* females × *mojavensis* males) produces fertile females but sterile males.

Table 23-2 gives the outcome of six artificial populations studied by Mettler. The data listed in this table are "raw" data. Nevertheless, it can be seen from the data as presented that after even as long as two years, hybrid combinations of gene arrangements for two different autosomes are found in the hybrid populations living in these cages. Only the *arizonensis* X chromosome seems to be eliminated systematically from these artificial populations; in population 6, however, where *arizonensis* males lived to inseminate the F_1 hybrid females, *arizonensis* chromosomes (both X and autosomal) fared considerably better than in the other populations.

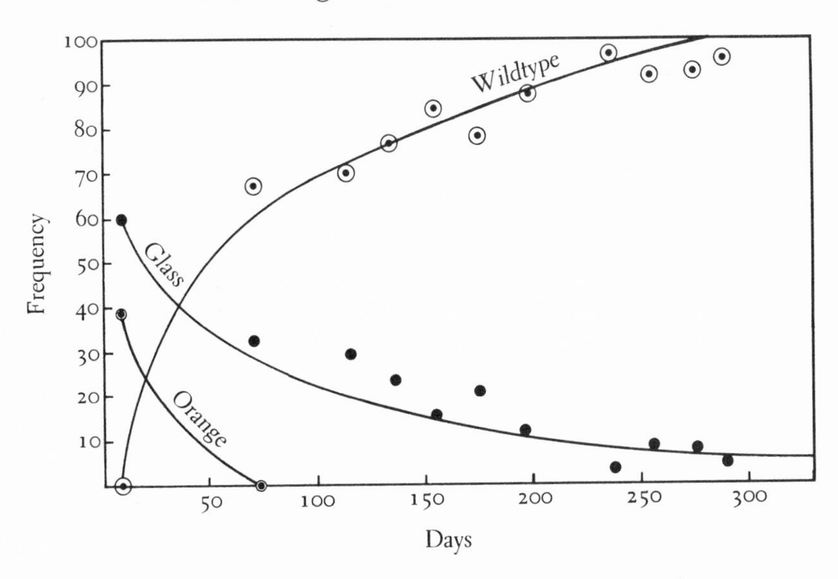

FIGURE 23-3. Formation of a wildtype hybrid swarm by the intercrossing of mutant strains of *D. pseudoobscura* and *D. persimilis* in a laboratory population. (After Van Valen, 1963.)

INTROGRESSION

If the flow of genes between two populations is restricted in any way, the two populations may retain a great deal of their individual characteristics for extended periods of time while showing the consequences of the intermixture. Should ecological preferences keep two populations physically isolated even though both have invaded the same geographic region, introgression is brought about by migrant individuals.

Experimental studies of introgression between laboratory populations have been carried out. Streams and Pimentel (1961), for example, found that progress under artificial selection is slowed down but not destroyed entirely by "contaminating" selected lines of *D. melanogaster* with individuals from unselected control stocks. A more extensive study has been carried out by Dobzhansky and Spassky (1962b and unpublished) on lines of *D. pseudoobscura* selected for geotaxis. Selection has been carried out both with and without the introgression of genes from one population to another. The results of some of these studies have been given in Table 23-3 and are shown in Figures 23-4 and 23-5.

These experiments involve (at least initially) lines of flies polymorphic for the Arrowhead and Chiricahua gene arrangements. The initial experiment consisted of directional selection for positive and negative geotaxis. In such an experiment, flies are introduced into one side of a ver-

TABLE 23-2

CHROMOSOMAL COMPOSITION OF FLIES FOUND IN LABORATORY POPULATIONS STARTED INITIALLY WITH MIXTURES OF D. MOJAVENSIS AND D. ARIZONENSIS.

X, 2, and 3 refer to the X, second, and third chromosomes; A, H, and M refer to *arizonensis*, hybrid, and *mojavensis*. (Mettler, 1957.)

			X:	A	A	A	A	A	A	A	A	A	H	H	H	H	H	H	H	H	H	M	M	M	M	M	M	M	M	M
			2:	A	A	A	H	H	H	M	M	M	A	A	A	H	H	H	M	M	M	A	A	A	H	H	H	M	M	M
CAGE	♀♀	♂♂	TIME, DAYS 3:	A	H	M	A	H	M	A	H	M	A	H	M	A	H	M	A	H	M	A	H	M	A	H	M	A	H	M
1	M, A	M, A	500–700	—	—	—	—	—	—	—	—	—	—	—	—	—	—	—	—	—	—	—	17	31	—	97	248	—	70	137
4	M, A	M, A	100–190	1	—	—	—	—	—	—	—	—	2	8	9	5	20	21	—	18	15	—	12	13	4	52	98	—	54	68
2	A	M	550–700	—	—	—	—	—	—	—	—	—	—	—	—	—	—	—	—	—	—	—	1	12	—	27	130	—	42	188
5	A	M	100–190	1	—	—	—	2	—	—	—	—	1	9	2	4	23	15	—	5	11	2	27	20	1	78	104	8	39	48
3	M	A	—	contaminated																										
6	M	A	100–190	55	34	—	30	17	2	3	1	—	32	28	7	24	41	25	4	8	7	9	8	2	9	29	12	3	9	2
7	H	H	105–195	—	1	—	1	—	—	—	—	—	7	4	1	8	19	10	2	13	2	6	13	3	22	66	47	8	45	21

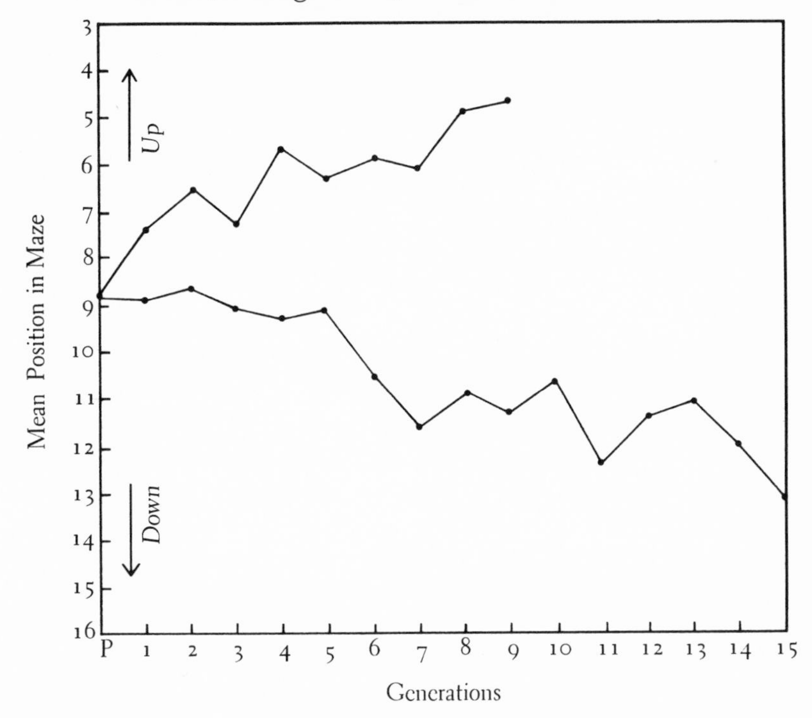

FIGURE 23-4. Artificial selection for positive (down) and negative (up) geotaxis in *D. pseudoobscura.* (After Dobzhansky and Spassky, 1962b.)

tical maze. Once in the apparatus, they can reach the opposite side only by making a series of decisions that involve walking up or down (possibly *falling* down). Should a fly choose to walk upward at every choice, he would emerge in the uppermost tube (#1); such a fly would be said to exhibit a strong negative geotaxis. On the other hand, a consistent choice to turn downward at every opportunity would lead the fly to tube 16; flies making such choices are said to exhibit positive geotaxis. Unselected flies are generally indifferent; the mean tube number occupied by a group of unselected flies that have passed through the maze is usually between 8 and 9. To arrive in these tubes, the flies must go upward and downward about an equal number of times.

Figure 23-4 (see the leftmost columns of Table 23-3) shows that geotaxis has a genetic basis. Selection in either direction goes relatively smoothly. Selection in an upward direction has been found to lead in at least two instances (one of which is illustrated here) to monomorphism for the Arrowhead (*AR*) gene arrangement; monomorphism for *AR* is not an invariable outcome of upward selection, however.

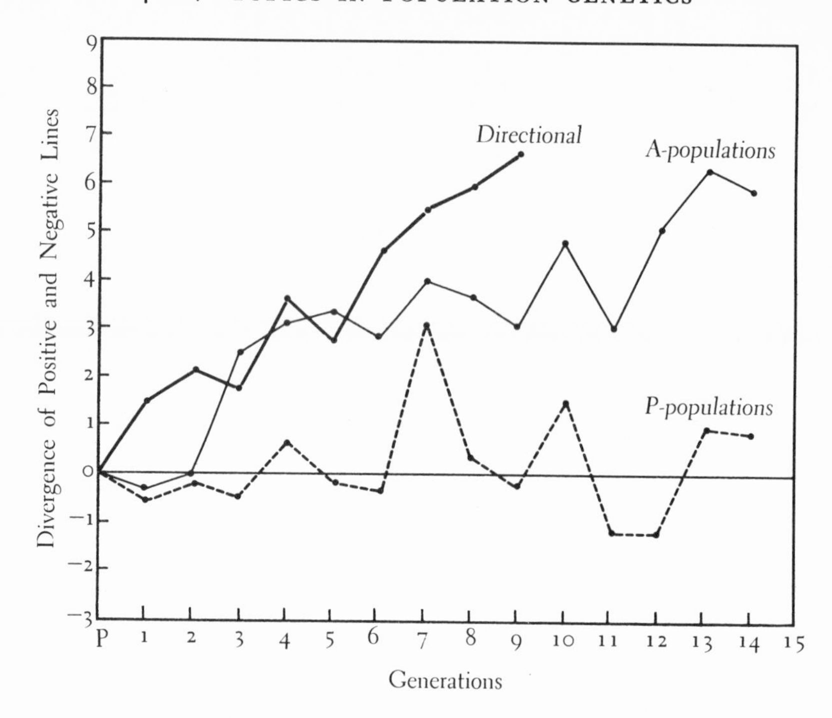

FIGURE 23-5. Divergence of lines of *D. pseudoobscura* selected for positive and negative geotaxis. Solid, heavy line: no introgression (same as Figure 23-4); solid, light line: large amount of introgression from weakly selected lines; dashed line: little introgression from strongly selected lines. (After Dobzhansky and Spassky, 1962b and unpublished.)

The results shown in Figure 23-4 can also be represented in terms of the divergence between the two selected lines. This has been done in Figure 23-5, where the most rapidly ascending curve is the distance between the two curves of Figure 23-4.

To study the effects of migration on introgression, Dobzhansky and Spassky set up four new populations. Two of these (the so-called A populations) underwent rather intense positive and negative selection (20 individuals of each sex selected from 300) for geotaxis; to each selected line, however, five flies (20 percent of the total number of selected parents) were added from so-called P populations that were being selected less intensely in the same direction. The P populations underwent weak selection (245 selected individuals chosen as parents from 300) for positive and negative geotaxis. To the selected 245 individuals, however, were added five individuals from the corresponding but more intensely selected A population.

The continued introgression of genes from the A population to the

TABLE 23-3

RESULTS OF ARTIFICIAL SELECTION FOR POSITIVE AND NEGATIVE GEOTAXIS IN D. PSEUDOOBSCURA.

(See the text for explanation.) (Dobzhansky and Spassky, 1962b and unpublished.)

	DIRECTIONAL			A'S			P'S		
GENERATION	UP	DOWN	DIFFERENCE	UP	DOWN	DIFFERENCE	UP	DOWN	DIFFERENCE
0	8.8	8.8	0	8.1	8.1	0	13.7	13.7	0
1	7.4	8.9	1.5	9.2	8.9	—0.3	13.8	13.3	—0.5
2	6.6	8.7	2.1	9.7	9.7	0	13.2	13.0	—0.2
3	7.3	9.1	1.8	8.2	10.7	2.5	11.9	11.4	—0.5
4	5.7	9.3	3.6	8.3	11.4	3.1	10.1	10.7	0.6
5	6.3	9.1	2.8	8.8	12.2	3.4	10.1	10.0	—0.1
6	5.9	10.5	4.6	9.3	12.2	2.9	12.3	11.9	—0.4
7	6.1	11.6	5.5	8.0	12.0	4.0	8.9	12.0	3.1
8	4.9	10.9	6.0	9.1	12.8	3.7	12.1	12.4	0.3
9	4.7	11.4	6.7	9.5	12.6	3.1	11.8	11.7	—0.1
10	—	10.6	—	8.3	13.2	4.9	11.3	12.8	1.5
11	—	12.4	—	8.6	11.7	3.1	12.4	11.2	—1.2
12	—	11.4	—	7.9	12.9	5.1	11.6	10.4	—1.2
13	—	11.1	—	7.1	13.5	6.4	11.0	12.0	1.0
14	—	12.0	—	7.3	13.3	6.0	10.5	11.4	0.9
15	—	13.1	—	7.3	—	—	—	11.9	—

P's and back again to the A's would be expected to act as a drag on the progress of the intensively selected populations (the A populations); conversely, the admixture of individuals from the A populations would be expected to speed up the progress in the P populations. The first of these expectations is borne out by the experimental results (Table 23-3 and Figure 23-5); the A populations do diverge but more slowly than the corresponding (but not simultaneous) populations subjected to simple directional selection. On the contrary, no consistent divergence was observed between the P populations despite the slight (2½ percent of all selected individuals) contribution of the more intensely selected A populations.

A second trait was followed during the course of these experiments: the P populations were not only selected (weakly, to be sure) for positive and negative geotaxis but were also forced to reproduce on medium containing ordinary table salt (NaCl), a substance toxic in high concentrations to *Drosophila* larvae. Contrary to the selection for geotaxis, that for salt resistance was quite intense. The P populations responded to selection for salt resistance despite the small amount of "contamination" from the unselected A lines. Furthermore, the continual inflow of genes for resistance to salt from the P to the A populations brought about an intermediate level of resistance in these populations even though no table salt was added to the medium on which these flies were raised.

It is appropriate at this point in our discussion of introgression to refer once more to the populations of *D. melanogaster* studied by Carson (1958), which were the basis for Figures 5-3 and 5-4. The point to be made is that an introduction of as little as a single genome from one population into another can have very appreciable consequences. Carson carefully studied the number of flies and the total weight of flies in a number of experimental populations. Two of these populations contained flies homozygous for the mutations *se ss k* e^s and *ro* (sepia eye color, spineless, short bristles, ebony body color, and rough eye), the flies in two others were wildtype. As shown in Table 23-4, the average num-

TABLE 23-4

EFFECT OF ADDING A SINGLE WILDTYPE GENOME ON THE POPULATION SIZE (NUMBER OF FLIES AND TOTAL WEIGHT) OF LABORATORY POPULATIONS OF MUTANT D. MELANOGASTER.

(Carson, 1958.)

	NO.	WEIGHT, MG	WEIGHT PER FLY, MG
1. Oregon-R wildtype population	355	208	0.59
2. *se ss k* e^s *ro* population	157	89	0.53
3. *se ss k* e^s *ro* population after introducing one Oregon-k genome	480	306	0.64

ber of flies found in the mutant populations during twenty weekly counts was approximately 157; the average weight of all flies counted each week (the population's "biomass") was nearly 90 milligrams. Similar measurements for the Oregon-R wildtype populations gave some 355 flies and a total weight of about 210 milligrams. Now, at a given moment, *one* hybrid female (*se ss k* e^s *ro*/Oregon-R) was introduced into each of the two mutant populations. The populations, starting about three generations after introducing these single wildtype genomes, literally exploded in size. The average number of flies found during the weekly counts in these populations (following the sharp increase) was 480; the average weight of the flies counted each week was about 305 milligrams. As the populations grew in size, the mutations dropped in frequency but, as we have seen in Figures 5-3 and 5-4, they were not completely eliminated. If introgression is looked upon as the rather restricted effect of genes from one population upon the characteristics of a second population, it does not follow that a restricted introduction of genes will lead to introgression. The populations studied by Carson more nearly "fused" following the introduction of but a single genome.

When discussing introgression, persons do not think usually of populations that would fuse into a single hybrid population if given the chance; generally they have in mind populations that are somewhat reluctant to accept alleles from one another. Possibly the hybrid vigor accompanying relatively large blocks of genes favors their retention in early (F_1, F_2, or early backcross) hybrids; with continued recombination, these same genes may eventually lose their advantage and pass out of the population.

A typical pattern of introgression between divergent forms has been illustrated in Figures 5-5 and 5-6; the hybridization of the two species of Towhees in a narrow geographic zone led to populations with intermediate phenotypes (Sibley, 1954).

An example similar to that involving Towhees occurs in the case of two species of tree frogs living in southern United States, *Hyla cinerea* and *H. gratiosa*. These species differ in a number of respects: *H. gratiosa* is large, stocky, and heavily spotted; *H. cinerea* on the other hand is small, slender, unspotted, and with a light stripe running along its side. A hybrid index can be constructed on the basis of these differences. Members of the two species found in Florida and in Alabama have index distributions such as those shown in Figure 23-6a. At Auburn, Alabama, though, in a region where the natural habitat has been upset by the construction of numerous artificial ponds, the individuals have index values distributed as shown in Figure 23-6b. In the Auburn, Alabama, population there is a sizable group of individuals possessing values almost exactly intermediate to those of the "pure" species; these are probably F_1 hybrids. There seem to be other values, too, that might belong to backcross or even F_2 individuals. Most interesting, however, are the modes that correspond to the pure species; each has been shifted noticeably toward the center of the range of values. Introgression between these species, over a period of some twelve years apparently, has caused a detectable modification of the phenotype of each.

The Development of Reproductive Isolation (Character Displacement)

There are very few evolutionary events that satisfy the student of natural populations as much as the divergence that occurs between populations living in the same area, a divergence that subdivides the environment on a smaller scale than it was subdivided previously, a divergence that squeezes a little more life out of our inanimate surroundings than before.

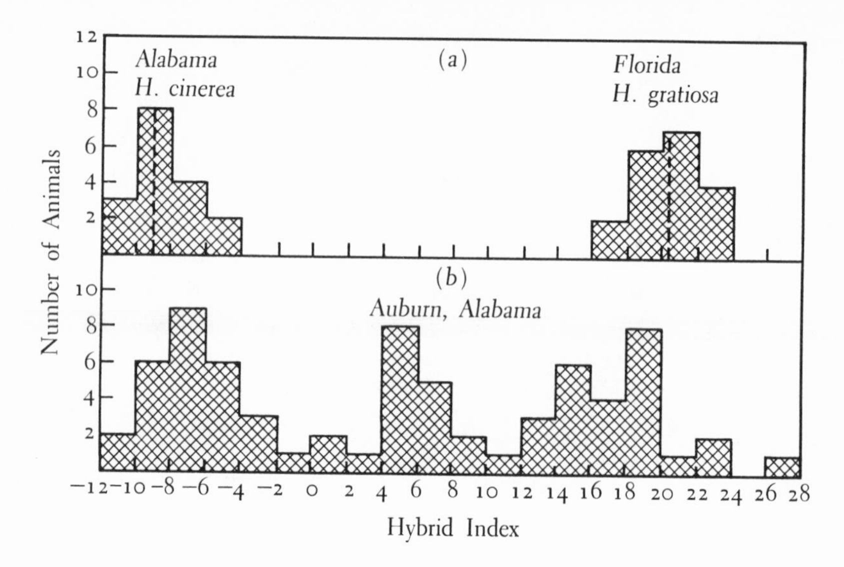

FIGURE 23-6. Introgression between two species of frogs in a disturbed habitat in Alabama. (*a*) Hybrid indices of populations *Hyla cinerea* and *H. gratiosa* as those species are found in Alabama and Florida. (*b*) Hybrid indices of the same two species in a disturbed area near Auburn, Alabama. (After Mecham, 1960.)

There are many reasons why plant and animal species that at one time were quite similar come eventually to be dissimilar. Some of this dissimilarity takes place in isolation under long-term selection and evolutionary change. Despite their distribution today, the racial characteristics of man were originally characteristics of geographically isolated (or largely isolated) groups. Even the "English" sparrow, which was introduced into the United States about 1850, has developed a number of fairly distinct geographic races (Johnston and Selander, 1964).

More interesting than the divergence of geographically isolated groups is the divergence that occurs between two groups in sympatry. If the groups involved are distinct species—or are well on their way to becoming distinct species—at the time they come in contact, their divergence is analogous to the deformation of hollow rubber balls that are forced into a restricted space.

Wallace and Srb (1964) have presented a schematic illustration of ecological niches (Figure 23-7) that includes the divergence of sympatric populations. According to this view, a niche can be represented as a collection of points, each of which refers to a specific attribute of the environment. (Some attributes—temperature, for example—are measured on a continuous scale; to represent these as points, the values have to be grouped on some appropriate scale.) Now, a given species can

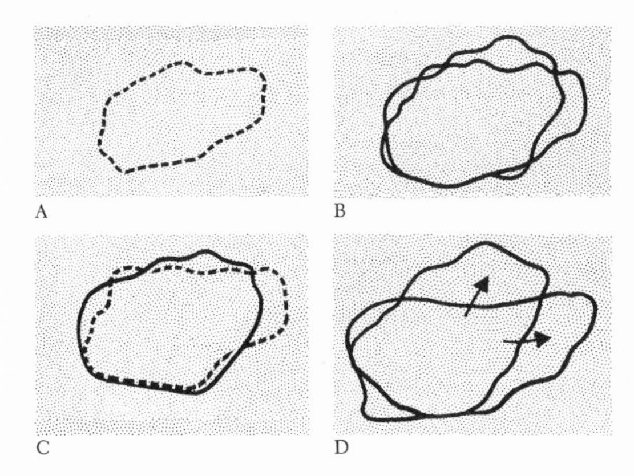

FIGURE 23-7. Representation of an ecological niche as a set of points (A), the fate of two species whose niches were largely identical (B and C), and of two species whose niches differed enough for both to exist and for selection for divergence to occur (D). (After Wallace and Srb, 1964.)

successfully occupy a series of environments in a given geographical region; the niche that this species occupies can be represented by surrounding the appropriate set of points by a line. A second species may occupy almost precisely the same set of points. According to Gause (1934) such species pairs cannot coexist indefinitely; one will eventually displace the other (Figure 23-7b and c).

On the other hand, the second species may occupy a niche with many points in common with the first but one with many unique points of its own and one that leaves many points uniquely to the first (Figure 23-7b and d). If the unique points are capable of sustaining the respective populations long enough for natural selection to be effective, the two populations will diverge as shown in Figure 23-7d; each will come to use more (and more effectively) the unique portions of its niche while relying less on the common points. These are the evolutionary changes that give rise to the different adaptations of sympatric species, changes that exaggerate earlier, preexisting differences. As a consequence, we say that the original differences were reinforced or, alternatively, that certain physical or behavioral characteristics have been "displaced."

The displacement of characters or the reinforcement of early, slight differences comes about very often in response to selection for sexual isolation. When two populations come together and interbreed despite the sterility or inviability of their hybrid progeny, selection will act to prevent this interbreeding, this wastage of gametes. The basis of natural selection in this case is the effective sterility of those individuals that accept mates from the wrong population. If there are genetic differences between those individuals that do and those that do not accept such mates, the genes responsible for intrapopulation mating are those favored by natural selection. More and more individuals will come to possess these genes.

The remainder of the chapter will consist of four examples in which species in sympatry have seemingly reinforced their isolating mechanisms under selection for more or less complete isolation.

The first example (Figure 23-8) shows the geographic variation in one aspect of the mating calls of two species of chorus frogs, *Pseudacris* (Mecham, 1961). The striking feature of these distribution patterns is the divergence that occurs in the calls of *P. triseriata* and *P. nigrita* in the area where these species overlap. The difference that is established in the zone of overlap compares favorably in magnitude with that between *P. triseriata* and *P. clarki* or *P. brachyphona* in other geographic areas. The impression that one gains from this figure is that the discrimination of these frogs in respect to pulses per second is rather crude; to borrow a term from television, I would suggest that there are not a large number of channels available to these species in this particular method of communication.

The second example (Waldron, 1964) concerns two species of *Drosophila* with which we have some familiarity, *D. pseudoobscura* and *D. persimilis*. These species, as we have learned, can hybridize in the labo-

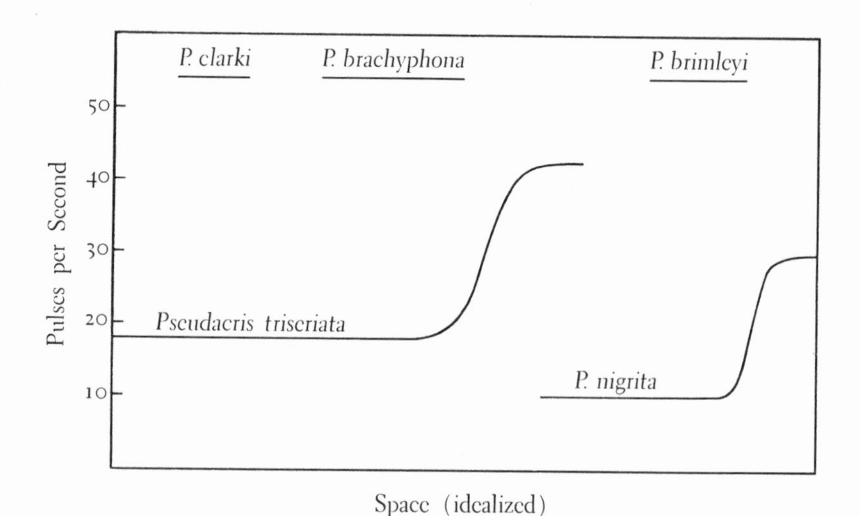

Figure 23-8. Alterations in the mating calls (recorded as pulses per second) of two species of frogs within a geographic area where they occur sympatrically. (After Mecham, 1961.)

ratory; in fact, we learned earlier in this chapter that they can be forced by appropriate means to establish a hybrid population.

Despite their ability to mate and produce hybrids in the laboratory, these species do not hybridize in nature. Their isolation is achieved not by one complete isolating mechanism but by the summation of a series of mechanisms each of which is only partially effective. The broad ecological preferences of the species differ: *D. pseudoobscura* tends to occur in more southerly drier climates than does *D. persimilis.* Where they do occur together geographically, *pseudoobscura* tends to occur at lower elevations than does *persimilis.* Where the two overlap in the same locality at the same elevation, *persimilis* favors cooler, more humid microenvironments than does *pseudoobscura*; their times of peak activity are slightly displaced as a consequence. Finally, as Figure 23-9 suggests, there are physiological differences associated with mating behavior. Waldron has succeeded in recording the vibrations produced by males during courtship; the species differ in the frequency of the underlying vibration, in the time between "pulses" during which the amplitude of the basic vibration is greatly increased, and in the duration of the pulse. The role these vibration and pulse patterns play in mating—indeed, if any—is unknown.

The third example (Figure 23-10) is one based on data given by Lack (1961, Tables 23 to 26). It shows the frequency distributions (based on ranges given by Lack) of beak depths for species belonging to three genera of Darwin's Finches found on four islands of the Galapagos—Culpepper, a small isolated island lying northwest of the main islands of the archipelago; Abingdon and Bindloe, two neighboring islands just

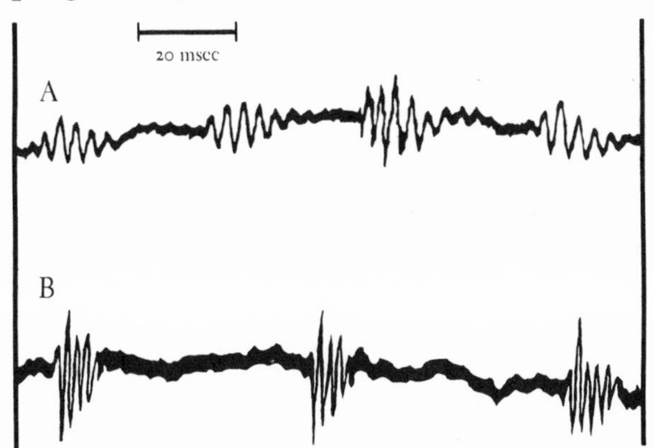

FIGURE 23-9. "Noises" made by males of *D. pseudoobscura* (A) and *D. persimilis* (B) while courting. (After Waldron, *Science,* Fig. 1, Vol. 144, p. 191, 10 April 1964. Copyright 1964 by the American Association for the Advancement of Science.)

Tone 1 

Certhidea olivacea

Camarhynchus parvulus

Geospiza fuliginosa

Tone 2 

Camarhyncus pallidus

Geospiza difficilis

Tone 3 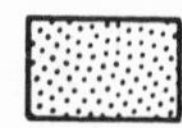

Camarhyncus psittacula

Geospiza scandens (+conirostris)

Tone 4 

Camarhyncus crassirostris

Geospiza fortis

Tone 5 

Geospiza magnirostris

north of the main islands; and James, a large island surrounded by other equally large or larger islands. The data reveal that the number of species on an island decreases with its size and with increasing isolation.

More interesting for our discussion is the sequence of beak depths observed for species of a given genus. On each island the total range of all possible depths is subdivided so that each species has its own characteristic depth range. It appears in at least one instance that the relative standings of two species are reversed in two different localities (James as opposed to Abingdon and Bindloe); this reversal is excellent evidence that a *difference* in beak depth, not a given beak depth, is the basis upon which selection acts in these populations of finches. Finally, it is quite clear that the orderly subdivision of the total range of depths of bills is accomplished only for species of the same genus; species of different genera living on the same island frequently possess bills of the same average depth.

The final example (Fulton, 1952) concerns the average time of emergence of individuals belonging to four "races" of field crickets in North

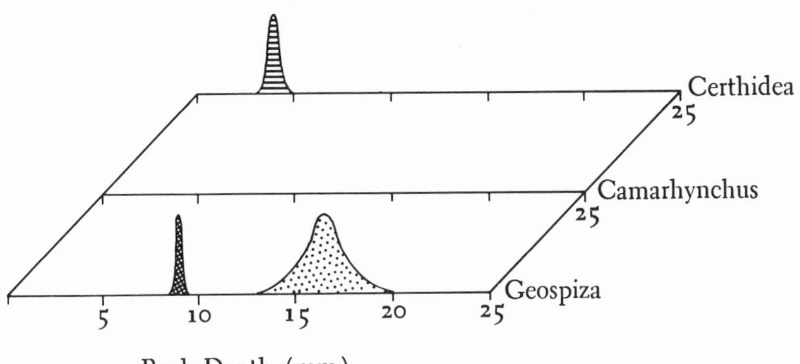

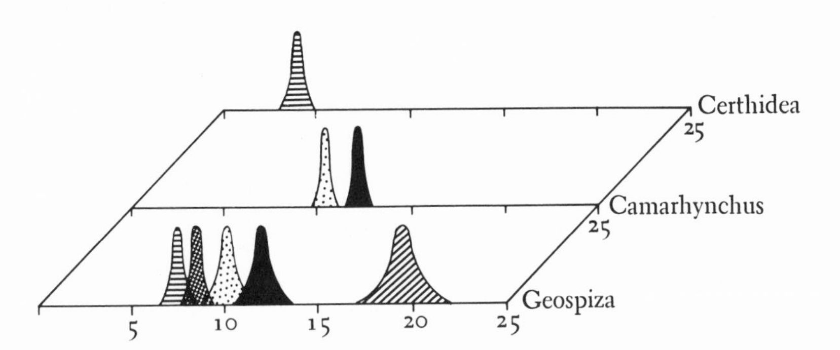

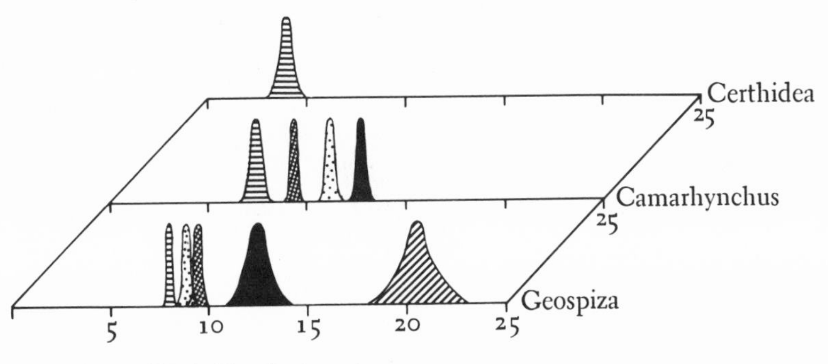

FIGURE 23-10. Beak depths of different species of Darwin's finches on various islands of the Galapagos archipelago. (After Lack, 1961, Tables 23–26.)

Carolina (Figure 23-11). Crickets emerge at all times from March until November. It is clear, however, that the times of peak emergence (marked by the white vertical bars) for the different races occur at regularly spaced intervals throughout the spring and summer months. These "races" differ, too, in habitat preference, song, and morphological characters. Because offspring are not produced by interracial crosses, it appears that these "races" are good biological species.

The perfection of various isolating mechanisms, a perfection that takes place very often sympatrically in response to what can be an intense selection, adds the final touches to the long series of changes that make up the process of speciation. These final touches are important; they do not mean, as many students believe, that speciation occurs sympatrically. The events that confer the selective advantage on the perfection of these last details have already taken place during the preceding isolation.

Future events are hard to predict with assurance. And so, when we observe hybridization occurring between species of tree frogs in Alabama (Figure 23-6), we cannot say what the future will bring: fusion, introgression, or isolation. And since we do not know, we disagree on terminology. Some biologists may argue that various organisms now referred to as species are not really species; others claim that groups not now recognized as species should be recognized as such. These arguments, seemingly important to some, seem trivial compared to an understanding and an appreciation of the nature and causes of the evolutionary changes that are occurring in living organisms all about us.

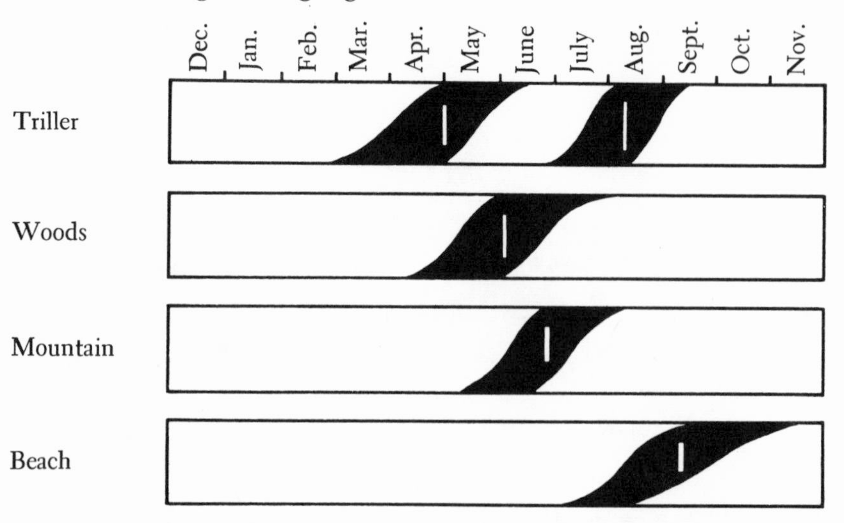

FIGURE 23-11. Emergence of adults of four "races" of field crickets in North Carolina. All races are differentiated by habitat preference, song, and morphological characters. No offspring are produced by interracial crosses. (After Fulton, 1952.)

24

THE ECOLOGICAL ASPECTS OF POPULATION GENETICS

> The ecological problem of populations has to do with the number of animals and what determines these numbers. The genetical problem of populations has to do with the kind or kinds of animals and what determines kind. These two disciplines meet when the questions are asked, how does the kind of animal . . . influence the numbers and how does the number of animals influence the kind. . . . These questions are as much ecological as they are genetical.
>
> *The Genetic Factor in Population Ecology,* Birch (1960)

In the foregoing chapters we have been concerned almost exclusively with intrapopulational events. The early chapters dealt with the Hardy-Weinberg equilibrium and factors such as mutation, selection, migration, and chance that qualify the alleged constancy of gene frequencies or the expected distribution of alleles among genotypes. The later chapters dealt with more complex problems arising from rather large-scale processes such as coadaptation or the evolution of dominance that take place within populations. The last of these chapters dealt with processes that are intimately related to the splitting of single species into two or more derived ones.

Although the events occurring within populations are complex indeed, they are fairly well understood in principle. The simpler ones could, it would seem, have been grasped at any time subsequent to the discovery of Mendelian inheritance although, in reality, they have been understood only since the 1920s and the early 1930s following the mathematical analyses of Chetverikov, Fisher, Haldane, and Wright. As Lewontin (1964) has described it, subsequent mathematical geneticists have been polishing with finer and finer grades of jeweller's rouge those colossal monuments of population genetics: *The Genetical Theory of Natural Selection, The Causes of Evolution,* and *Evolution in Mendelian Populations.* And, as Lewontin adds, "it is due mainly to man's infinite capacity to make more and more out of less and less, that [population geneticists] are not currently among the unemployed."

In the present chapter we shall discuss the relationship of population genetics and ecology. This relationship, as Birch has said, is concerned with kinds and numbers of individuals. It also involves the notion of fitness—population fitness. In contrast to the relative fitnesses of individuals within populations (adaptive values or relative Darwinian fitnesses), the fitness of a population has not been defined adequately. Perhaps, since the ideal definition should include a prediction of future events, population fitness never will be satisfactorily defined. Later on in this chapter, we shall claim that the ratio of adult daughters to mothers, because it is a measure of their adaptive values, is also a measure of the Darwinian fitnesses for genotypes within a population; this ratio has an average value of 1.00 for stable populations. We shall argue, then, that this use of Darwinian fitness permits us to say something about the more elusive concept of population fitness. In addition, starting with an example that deals with intrapopulation Darwinian fitness, we shall proceed gradually to the question, what bearing has genetic load on the number of individuals in a population—that is, on the population's existence?

Intrapopulation Fitness

Although it may be difficult to measure experimentally, the concept of the relative fitnesses of individuals in a population is not difficult to grasp. The fitness or adaptive value of an individual is measured by his ability to leave adult, fertile offspring. The relative Darwinian fitness of a given genotype, its adaptive value, is the average fitness of individuals of that genotype. Experimental difficulties arise because some components of fitness are difficult to measure while others are not. And so it is tempting, for example, to count survivors but avoid the more laborious task of measuring fertility. Or to score fertility and sterility as clear-cut alternatives without bothering to measure slight quantitative differences of fertility. Or to score carefully the fertility of fertilized females but to neglect the varying efficiencies with which these females obtain mates.

Despite the multitude of components of fitness that might be measured, physical limitations force the investigator to choose not only those that are important but, of these, only those that are also easily measured. He then hopes that the unmeasured components of fitness will be correlated with those he has observed. Thus, in studying the morphological polymorphism of a marine copepod, *Tisbe reticulata*, Battaglia (1958) determined the deviations of V^VV^V, V^VV^M, and V^MV^M individuals from the 1 : 2 : 1 ratio expected in F_2 cultures and used those deviations to estimate the relative fitnesses (adaptive values) of the three genotypes (Table 11-2). This is an estimate of Darwinian fitness based upon events taking place between fertilization and adulthood; it neglects the remainder of the life cycle.

In a later study (Battaglia, 1965), the tolerances of three types of individuals to varying salinities of the culture medium were determined (Figure 24-1). In this test, the number of surviving eggs (number of living offspring) per female was determined. As the figure shows, each form has an optimal salinity that is different from those of the others.

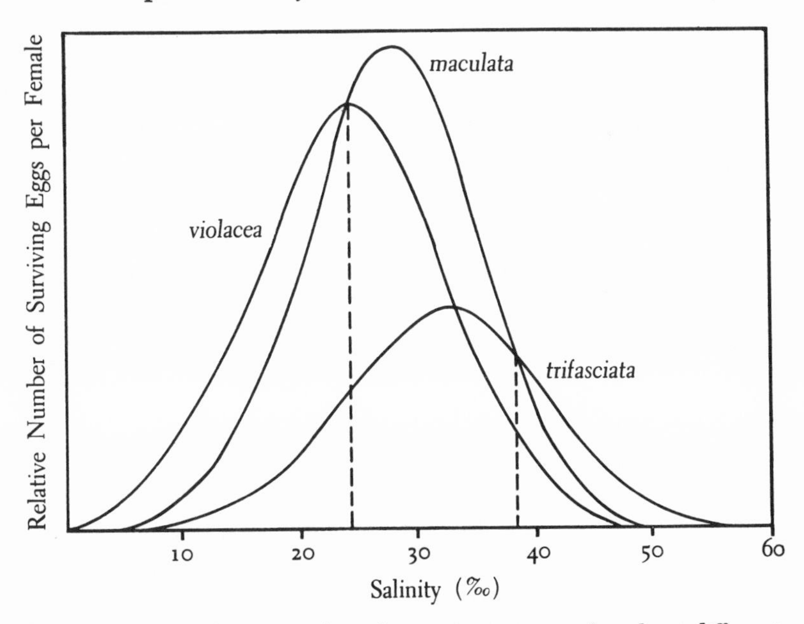

Figure 24-1. Relative number of surviving eggs per female at different salinities for three genotypes of the marine copepod *Tisbe reticulata*. (After Battaglia, 1965.)

Using these measures of Darwinian fitnesses, one can assign from the curves adaptive values at any given level of salinity. Thus *violacea* is superior to the other two forms at salinities below 25 $^0/_{00}$ (25 parts per thousand), *maculata* is superior between this level and one that is nearly 40 $^0/_{00}$, while *trifasciata* is superior above 40 $^0/_{00}$. On the basis of this type of information, together with further information on the resistance of these forms to various temperatures, Battaglia was able to make inferences concerning the relative frequencies of different genes in marine populations and in populations inhabiting the lagoon of Venice. By using the estimates of fitness revealed by laboratory experiments, he has attempted to understand the relative proportions of individuals observed in collections from various natural populations.

Relative Survival of Different Species

Data analogous to those collected by Battaglia for various genotypes of *Tisbe reticulata* can be gathered for members of different species. Tantawy and Mallah (1961), for example, collected *D. melanogaster* and *D. simulans,* two closely related species, at various localities in Lebanon, Egypt (U.A.R.), and Uganda. Tests were made on the proportion of eggs that gave rise to viable adults at 10, 15, 18, 22, 25, 28, 30, and 31°. The results of these tests are summarized in Table 24-1. In the table the survival at various temperatures has been converted into total area beneath the survival curve over the entire range from 10 to 31°. This has been accomplished by summing the areas of small segments of the total area. For the entire temperature range, *melanogaster* has a higher average survival than *simulans*. For the part of this range from 18 to 25°, however, *simulans* has the generally higher survival. *D. simulans,* according to this information, might be expected to outnumber *melanogaster* in regions where temperature extremes are not encountered; in geographic regions that are subjected to rather wide temperature fluctuations, on the other hand, one might expect *melanogaster* to displace *simulans*. As a matter of fact, within the United States this expectation is borne out rather well. *D. simulans* is the common species of the two in the southern United States where the climate is equitable throughout the year. *D. melanogaster* is more common in the northern United States where winters are severe and wide daily fluctuations in temperature occur.

The type of information obtained by Tantawy and Mallah is obviously not sufficient in itself to predict which of the two sibling species will be the dominant one in a given region. But it is one kind of information that would be needed to make such a prediction. Furthermore, it emphasizes the limitation of studies limited to the adaptive values of genotypes that are found within a species. A study of the relative fitnesses of several genotypes within a species is unable to tell us whether the

TABLE 24-1

MEASURES OF THE PERCENT EMERGENCE OF D. MELANOGASTER AND D. SIMULANS OVER TWO TEMPERATURE RANGES CALCULATED AS THE RELATIVE AREAS (IN ARBITRARY UNITS) UNDER THE DISTRIBUTION CURVES.

(After Tantawy and Mallah, 1961.)

SOURCE	10–31°C *D. melanogaster*	*D. simulans*
Lebanon	1379	1161
Luxor, Egypt*	1287	1078
Alexandria, Egypt	1213	1089
Wadi-el-Natroon, Egypt	1097	1070
Uganda	1027	1053
	18–25°C	
Lebanon	529	520
Luxor, Egypt*	488	490
Alexandria, Egypt	459	484
Wadi-el-Natroon, Egypt	413	478
Uganda	396	473

* *D. simulans* from Beni-Swef, Egypt.

species will exist or not. Even the results of Battaglia shown in Figure 24-1 do not reveal a level of reproduction below which the species *T. reticulata* would be eliminated from a locality. It is entirely possible, for instance, that no population of *T. reticulata* homozygous for *vv* (*trifasciata*) can exist in nature; the data in Figure 24-1 do not tell us whether this genotype can sustain itself in the presence of other organisms.

"HARD" AND "SOFT" SELECTION

Under equilibrium or near equilibrium conditions only a small fraction of all zygotes of any generation actually survives to become the breeding portion of the population (Figure 24-2). As a reasonable approximation, the number of breeding individuals should remain fairly constant from generation to generation; hence the fraction of "successful" individuals is roughly twice the reciprocal of the average progeny size.

I use the terms "hard" and "soft" in describing the basis by which natural selection determines which individuals are to be excluded from the ranks of successful breeders. One possibility is that the "cutoff point" is determined on an invariate fitness scale or by unconditional selective factors. Consequently, as the distribution of fitnesses that characterize a population fluctuates relative to the constant cutoff point (Figure

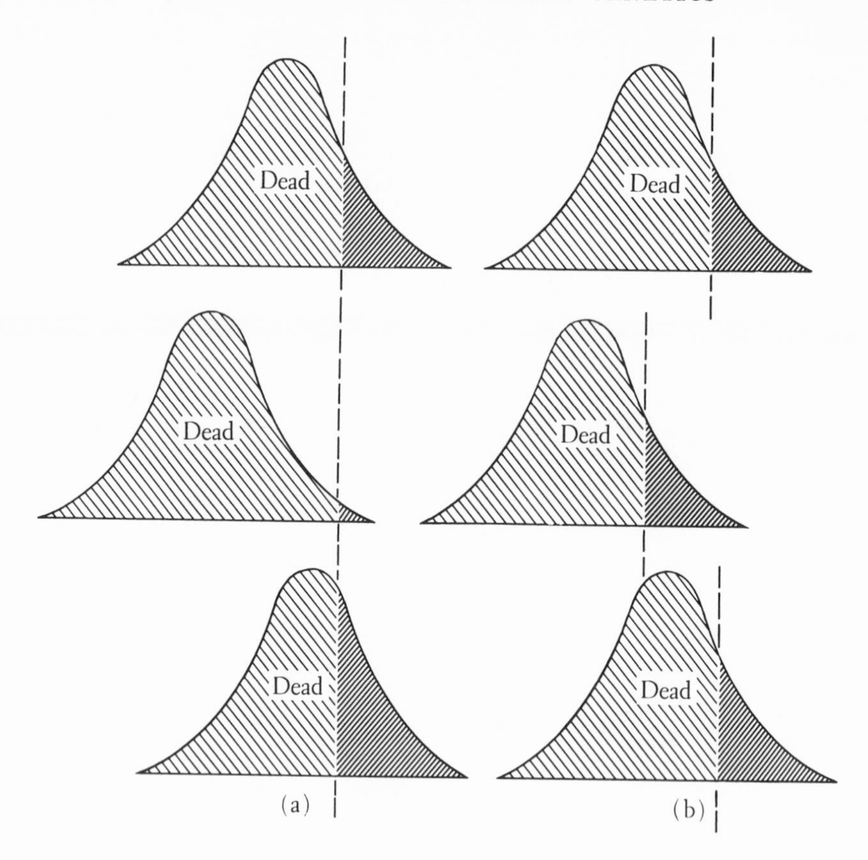

FIGURE 24-2. Two types of selective forces acting on populations. "Hard" selection (*a*) eliminates all individuals except those that meet rigid requirements (such as not being homozygous for a lethal mutation); "soft" selection (*b*) permits a certain, relatively constant, proportion of the population to survive and reproduce.

24-2a), the number of individuals leaving progeny also fluctuates. Another possibility, though, is that the cutoff point is not a constant determined according to some fixed fitness scale. Under this possibility, the number of parents may remain relatively constant from generation to generation despite fluctuations in the distribution of fitnesses within the population (Figure 24-2b). These two possibilities are not mutually exclusive; the following examples will show that both play a role in determining the number of adult individuals that constitute a breeding population.

Hard Selection. An obvious situation under which the fitness of individuals of one species is measured according to relatively unyielding standards is that in which survival is measured relative to that of a second (standard) species under conditions such that only a fraction of either survives. Neglecting for the moment specific inhibitions or cross-feeding of the sort studied by Weisbrot (1966), we can use Timofeeff-Ressovsky's (1935a) studies to illustrate this point.

The technique used by Timofeeff-Ressovsky consisted of allowing groups of 300 *Drosophila* larvae to develop in vials that, on the basis of previous experience, were known to support only about one half that number (Figure 24-3); this technique is very much like that used later by Vetukhiv (see Table 18-6) in his studies. Timofeeff-Ressovsky studied the survival of different geographic strains of *D. funebris* under three different temperatures: 15, 22, and 29°. He measured the survival of *D. funebris* relative to that of a standard strain of *D. melanogaster*. Into

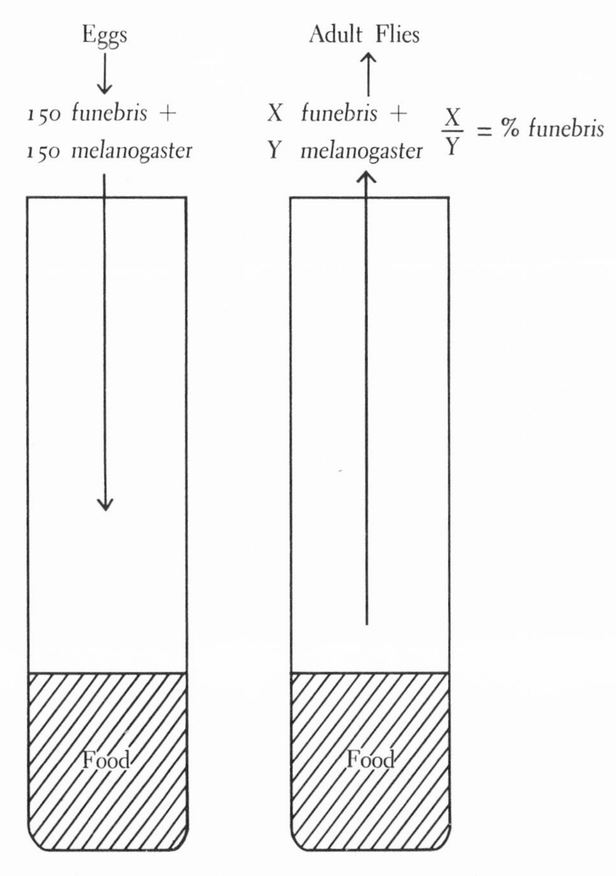

FIGURE 24-3. Experimental procedure that imposes relatively "hard" selection upon competing individuals. (After Timofeeff-Ressovsky, 1935a.)

each vial he placed 150 eggs of *D. funebris* and 150 eggs of *D. melanogaster*. The adults that survived were counted and the survival of *funebris* was expressed as a fraction of the survival of *D. melanogaster*. Some of the results obtained in this study are listed in Table 24-2. The essential point to be gained from the table is that different geographic strains of *D. funebris* differ in their ability to survive under the experimental conditions. Strains of *funebris* from Moscow, Perm, and Tomsk survive in numbers nearly as great as those of *D. melanogaster* at the low temperature; strains of *funebris* from the Mediterranean region survive rather poorly at low temperatures. At 29° *D. funebris* survives poorly as a rule but the strains from Egypt, Tripoli, and Spain survive somewhat better than do strains from more northern regions. The results of these experiments are consistent with the known origins of the *funebris* strains. The estimates of fitness have been obtained under what I have called hard selection because the available food was capable of supporting only a limited number of flies; among those that survived, the relative proportions were determined by the fitness distributions of the two species relative to the prevailing temperature and culture conditions. *D. melanogaster*, in this case, served as a standard for revealing different survivals of genetically different strains of *D. funebris*; a standard *funebris* strain could have served the same role in respect to different geographic strains of *D. melanogaster*.

Soft Selection. The alternative possibility by which the number of successfully breeding individuals might be limited is by means of factors

TABLE 24-2

SURVIVAL OF DIFFERENT GEOGRAPHIC STRAINS OF D. FUNEBRIS AT DIFFERENT TEMPERATURES EXPRESSED AS A PERCENTAGE OF THE SURVIVAL OF A STANDARD STRAIN OF D. MELANOGASTER.

(After Timofeeff-Ressovsky, from Dobzhansky, 1937.)

	TEMPERATURE, °C		
STRAIN	15	22	29
Egypt	68	46	30
Tripoli	64	47	31
Spain	69	48	30
France	80	44	25
Sweden	88	40	21
Moscow	101	43	28
Perm	98	41	26
Tomsk	96	42	28

that govern numbers of survivors virtually independently of genotype. The number of successfully reproducing individuals under this possibility would be nearly constant despite fluctuations in the fitness distribution of the population.

The idea that factors controlling the number of breeding individuals may indeed be "soft" can be illustrated by some simple observations on *Drosophila* cultures. During a short stay at the Genetics Institute of the University of Groningen, Netherlands, I was introduced to a medium consisting of agar and brewer's yeast. This medium has a tendency to liquify as the larvae develop.

The natural history of flies in vials with the agar-brewer's yeast food is interesting. The larvae, for instance, tend to pupate at a fixed level within the vials. Figure 24-4 shows the proportion of pupae that formed at various heights above the surface of the food in six vials. In five of these (the solid line), 80 percent of all larvae pupated in a band some two or three centimeters above the food. Not all strains behave alike. The dashed line shows the sixth vial (one similar to the other five but contaminated with wildtype flies of unknown origin), in which 80 percent of the larvae pupated in the first two centimeters above the surface of the culture medium.

To keep a daily record of pupae (as they formed) and to avoid counting the same pupae twice, the position of each one was marked on the outside of the vial by a small drop of India ink. Within a day or two it became clear that pupae did not always remain where they had first been seen, for upon reexamination of the vials it was found that some ink spots no longer marked the position of pupae. Simultaneously, drowned pupae began to appear in the food. When vials were fifteen days old, a count was made of all pupae on the wall of the vial (including those from which adult flies had already emerged) and of those pupae that had been dislodged and drowned in the medium. Nineteen vials representing five additional strains were studied as well. The results are listed in Table 24-3; from one tenth to one third of all pupae formed in these vials were dislodged and drowned by larvae arriving later at the preferred pupation site a few centimeters above the surface of the food.

The interesting feature of the data for the point we are making concerning soft selection is brought out by the chi-square analysis of the variation in numbers of pupae between vials of the same strain. From the actual numbers and the expected number (the strain average) one can compute a chi-square that, together with its degrees of freedom, leads to an estimate of p, the probability of obtaining by chance alone deviations as large or larger than those observed. One can also consolidate the series of observations by adding chi-square values and degrees of freedom.

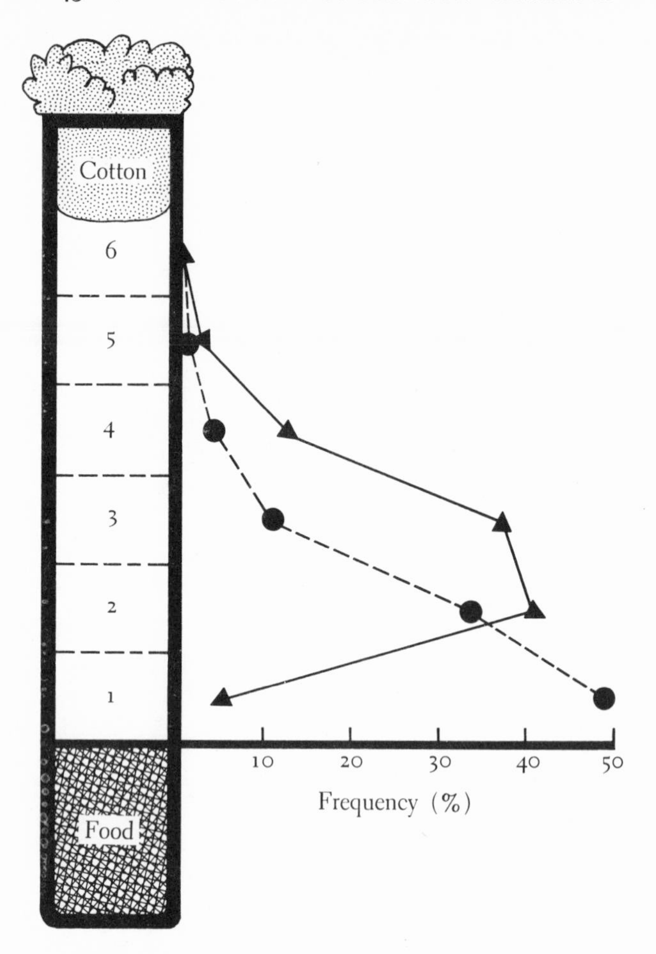

FIGURE 24-4. Distribution of pupae on the wall of culture vials for two strains of *D. melanogaster*. (Wallace, unpublished.)

The results of this analysis are listed in Table 24-4. In the case of total pupae (both those found on the wall of the vial and those that were drowned in the medium), the total chi-square with 14 degrees of freedom is 27.32. The probability that deviations as large as those seen among the vials would arise from chance events alone is less than 0.02. This result is not unexpected. The amount of food in the vials was only approximately the same; furthermore, different females do lay different numbers of eggs. One might expect, then, that the total number of pupae should vary from vial to vial more than chance events alone can explain.

A corresponding chi-square test made of pupae found on the walls of the vials gives quite different results. The total chi-square with 14

TABLE 24-3

PROPORTION OF DROWNED PUPAE IN VIAL CULTURES OF D. MELANOGASTER FROM DIFFERENT GEOGRAPHIC LOCALITIES.

(Wallace, unpublished.)

STRAIN	*n*	DROWNED	% DROWNED
Pyrenees	509	115	22.6
Madeira Wine Association	747	193	25.8
Madeira, Reid's Hotel	574	99	17.2
Madeira, Public Market	261	32	12.3
$bw^D/+$ (Riverside, California)	923	137	14.8
$bw^D/+$ plus $+/+$ contaminant	121	14	11.6
Riverside, California	883	274	31.0
Total	4018	864	21.5

degrees of freedom is only 5.89; the probability that deviations as large or larger than those seen would arise by chance is about 0.97. The usual interpretation of a probability of this size is that the data are "not significant." Significance, however, depends upon the question one asks of his data. A probability value of 0.97 means that the probability of obtaining through chance alone deviations as small or smaller than those observed is only 0.03. Although the variation in the total number of pupae from vial to vial was significantly *greater* than one would expect by chance, the number on the walls (undrowned) was significantly *less* so. The conclusion is clear: There is room in the preferred zone for a given number of pupae and so the total number found in that zone tends to be a constant rather than a random variable. Not always, however, does the race go to the fleet. There is a more or less constant turnover of pupae as late arriving larvae dislodge those formed earlier and take their places on the vial wall.

The result of the events described above is identical to that illustrated in Figure 24-2b. Despite fluctuations in the total number of pupae formed, the number of individuals surviving remained virtually constant. This same pattern of constant numbers of survivors should arise in the case of all species in which individuals stake out territories and for which the number of suitable territories, rather than the number of progeny produced, determines the number of breeding individuals (or pairs).

Before leaving this section, some comment on hard and soft selection in determining the fate of a species should be made. Both kinds of factors are at work in nature. There is evidence (Ives, 1954) that *D. simulans* has largely displaced *D. melanogaster* in the southern regions of

TABLE 24-4

SUMMARY OF CHI-SQUARE TESTS FOR HOMOGENEITY OF SURVIVING PUPAE AND TOTAL PUPAE IN REPLICATED CULTURES OF D. MELANOGASTER OF VARIOUS GEOGRAPHIC ORIGINS.

(Wallace, unpublished.)

SURVIVING PUPAE			TOTAL PUPAE	
CHI-SQUARE	DEGREES OF FREEDOM	STRAIN*	CHI-SQUARE	DEGREES OF FREEDOM
0.16	2	Pyrenees	0.36	2
0.60	4	Wine Assn.	9.86	4
0.35	1	Public Market	0.46	1
2.83	4	$bw^D/+$	8.41	4
1.95	3	California	8.23	3
5.89	14		27.32	14

* Omitted from the above analysis: Madeira, Reid's Hotel; surviving pupae chi-square (4 degrees of freedom) 49.43, total pupae chi-square (4 degrees of freedom) 47.90.

the United States. English sparrows and starlings through sheer aggression have displaced many once-common bird species. The Norway rat has successfully displaced the black rat in many areas of the world. In each of these examples it appears that there was not room for both species involved; every retreat on the part of one species was seized upon and converted into an advance by the other in a manner reminiscent of Timofeeff-Ressovsky's experiments.

Instances such as those described above, in which one form has displaced another, might give the impression that nature is an enormous, overcrowded vial of food similar (except for size and complexity of its contents) to those studied by Timofeeff-Ressovsky. I prefer to believe, however, that soft selection generally outweighs hard selection in controlling populations. Within stable communities, for example, I believe that each species is sufficiently well adapted to the requirements of its niche that "superfit" individuals are not required for maintaining the population at a constant level. Furthermore, the spread of a species into a niche that happens to be vacant at the moment and its efficient adaptation to the intricately adjusted requirements of that niche are not rapid or easy processes according to my view. Thus, if someone armed with an enormous vacuum sweeper could remove every individual of the species *D. repleta* from all the dark and musty spaces under and behind the wine vats on Madeira, I believe this niche would remain unoccupied. And for many, many years, almost any wildtype strain of *D. repleta*—despite its apparent fitness relative to other wildtype strains of the same species—if reintroduced into Madeira would in my opinion be able to reoccupy the old "*repleta*" niche.

The concept of genetic load tends, I believe, to be preoccupied with what I have called hard selection. That is, the relative fitnesses of various genotypes are given fixed values $1 - s$, $1 - t$, . . . Many lethal mutations do, indeed, express their lethal effects in all known environments. The concept of "lethal equivalents" (page 51) is completely wedded to the notion that lesser causes of inviability are also rigidly determined. These fitness values can be regarded as fixed only if soft selection is unimportant relative to hard selection. This is not to say that hard selection, as we use the term, reflects unremittingly difficult selection while soft selection refers to the lax selection of an equitable environment. As shown in Figure 24-2, the proportion of surviving individuals can be larger under hard than under soft selection. The "softness" of soft selection lies in the flexibility with which it allows a number of individuals to survive and reproduce; the remainder of the population is eliminated as surely as if it had failed to meet the requirements of hard selection.

To illustrate the effect soft selection has on fitness calculations, we can cite the following example. Suppose that the fitness of one genotype in *D. melanogaster* has been found under certain culture conditions to exceed that of another genotype 1000-fold. A population consisting of one of the high-fitness larvae and one hundred of low-fitness ones would possess, according to current genetic load theory, a mean fitness of approximately 0.011, or a genetic load of 0.989. Imagine now that groups of these 101 larvae are grown in vials for which the average number of pupae tends to be twenty-six. We might grant the single high-fitness larva a successful life in every one of these vials. Now, the twenty-five other surviving pupae will be individuals of the low-fitness genotype. The probability of survival for the low-fitness larvae, then, is 0.25. Hence, despite our statement that the high fitness was 1000 times greater than the low fitness, only a 4-fold difference in the probability of survival can be demonstrated in these particular vials. The true mean fitness in these vials is 0.2574 rather than 0.011, and the genetic load is only 0.7426. Sved et al. (1967) have suggested that discussions of fitness ratios greater than ten may be meaningless; the existence of soft selection encourages this view.

Darwinian Fitness and the Self-Regulation of Populations

A model describing the stabilization of populations at certain sizes was described in Chapter 1. According to this model, the average number of adult daughters produced per mother declines as population size increases. When populations are small in size, the average number of adult daughters left by each mother exceeds 1.00 and so the population grows (Figure 24-5). On the other hand, when populations exceed a certain size, the average number of daughters per mother is less than 1.00 and, consequently, the population tends to become smaller. These opposing tendencies lead to an equilibrium size at which the average number of daughters per mother is *exactly* 1.00. This is a stable equilibrium in the sense that small displacements in either direction tend to be corrected so that the equilibrium size is arrived at once more.

Two comments concerning the above model may be made at this point. Haldane (1953) has presented a more inclusive model than that illustrated here. In his model, the number of adult daughters per mother also drops to zero when the population size falls below a certain level; thus his model includes the possibility of extinction from too great a reduction in numbers of individuals.

The second point is that the horizontal axis need not represent size at all. It could just as well represent density. Or, to pick still another variable, it could represent the accumulation of waste products (see Weisbrot, 1966, for example) which, in turn, might be only imperfectly correlated with numbers of individuals. Nevertheless, numbers of individuals is the variable most easily grasped and so "size," set off by quotation marks when necessary, will continue as the basis for discussion. Furthermore, since "adaptive value" is a measure of the relative number of offspring produced by individuals of different genotypes, the ratio of daughters to mothers will also be referred to as "fitness" (see Penrose, 1949). Under this use of fitness, *population "size" is stabilized when the mean Darwinian fitness of a population equals 1.00.*

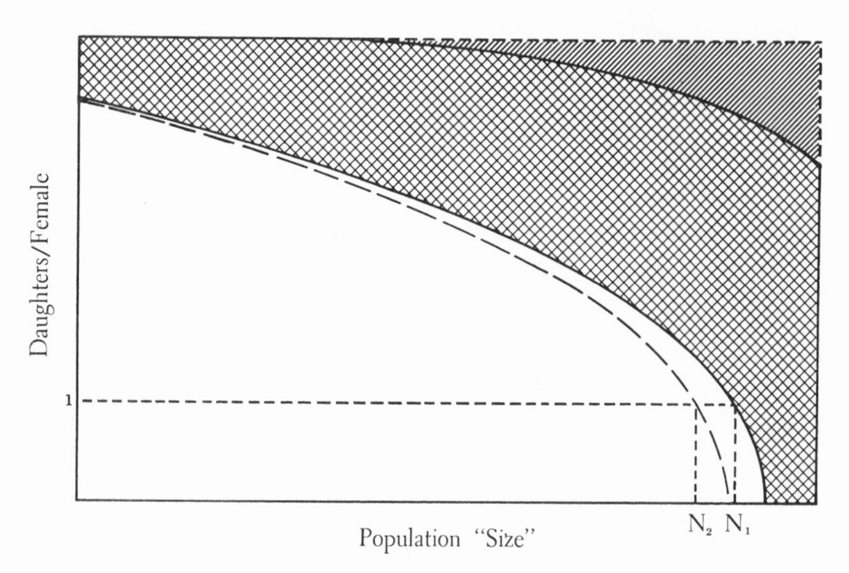

FIGURE 24-5. Conditions that lead to stable population size (see Figure 1-1). This diagram permits us to say that *at equilibrium* the mean Darwinian fitness of a population is 1.00. (Modified from Haldane, 1953.)

The relation between mean Darwinian fitness and population size shown in Figure 24-5 is a weighted average based on all types and frequencies of individuals within a population. There is no need to assume that individuals of different genotypes respond in precisely the same manner to changes in population size. If it is granted that individuals of dissimilar genotypes respond differently, though, it must also be granted (1) that relative fitnesses are not constant and (2) that calculations based exclusively on hard selection to the neglect of soft selection are inadequate for an understanding of populations.

Data capable of illustrating the claims made immediately above are not plentiful. Sokal and his colleagues (see for example, Sokal and Sullivan, 1963; Sullivan and Sokal, 1963, 1965; Sokal and Huber, 1963; and Bhalla and Sokal, 1964) have studied the relation between population density and survival for various mutant and wildtype strains of the housefly and of the flour beetle. There have been studies, too, on the productivity and biomass of monomorphic and polymorphic populations of *D. pseudoobscura* (see Beardmore et al. 1960; Strickberger, 1963; and Dobzhansky et all. 1964). These studies, however, do not supply the data needed to illustrate the point we have in mind. To make this point we have constructed the curves shown in Figure 24-6. The curves represent the survival of a number of homozygous strains of a hypothetical organism as a function of population size; the survival of hybrids of these strains is also shown. For our purpose, we assume that the curves representing the pure strains represent single alleles and, furthermore, that they are also suitable for representing the homozygous strains in mixed populations together with heterozygotes. This assumption is almost certainly wrong, but at the moment the correct relationship for any set of genotypes could be arrived at only empirically.

The relationship between survival and population size shown in Figure 24-6 is not adequate for identifying equilibrium population sizes. Some level of survival must be chosen to represent that which on the average allows one daughter to survive per mother. In the figure a survival of 0.35 has been chosen arbitrarily to represent a fitness of 1.00, where "fitness" is a measure of the average number of daughters per mother.

Figure 24-6 tells us that populations which are monomorphic for either *A* or *B* can actually exist but that no population monomorphic for *C* can maintain itself. Given a population monomorphic for *A* which had established a constant size, the introduction of either *B* or *C* would result in an increased size because the heterozygous females would leave more than an average of one daughter each. As the new population increased in size, however, the fitness of the homozygous females would decrease. Finally, a new equilibrium size would be established. At this new size, the two "alleles" would behave as lethals. The population would now give the impression of consisting of a balanced lethal system. At the new equilibrium, one half of the population would have zero fitness; therefore, the number of daughters per heterozygous female required to maintain the population at a constant size (an average fitness for the population of 1.00) would be 2.00.

Should all three alleles be present in the same population, the population would (according to the curves shown in Figure 24-6) expand slightly more and re-establish an equilibrium when the mean fitness (adult daughters/mother) was reduced once more to 1.00. Since, with three alleles, each generation would consist of inviable homozygotes (one third of all zygotes) and viable heterozygotes (the remaining two thirds), the viable heterozygous mothers would have to produce an average of 1.5 daughters each to maintain a stable population. Hence the population size shown in the figure as the stable one for $A + B + C$ is the one for which heterozygotes have a Darwinian fitness of 1.5.

Figure 24-6 illustrates the pitfalls that accompany a rigid adherence to the concept of hard selection. Within the polymorphic populations, "alleles" *A* and *B* appear to be lethals. It may seem, then, that populations monomorphic for *A* or *B* could not exist. This, of course, is wrong. The monomorphic populations can exist, presumably as excellent populations. What the figure shows is that population size can help absorb the genetic load of a population. The load need not be contained entirely within the average progeny size! The zero fitness of *B* individuals in polymorphic populations does not tell us that these individuals are incapable of producing progenies large enough to sustain a monomorphic population; the anticipated genetic load of the monomorphic population is absorbed not by the production of infinite numbers of offspring but by increased survival in a population somewhat diminished in size.

When one computes a genetic load, one obtains a number that applies to a population under a given set of conditions. One cannot safely say that a given load is too large for a population to sustain; indeed, one can point to other populations that successfully bear what appear by identical calculations to be larger loads. One cannot say that a given load will lead to extinction for extinction itself is a dynamic process, one that leads to new fitness values. As a result, the concept of genetic load, as it has been defined in rigorous mathematical terms, does not help us understand the ecological aspects (including those related to long-term evolutionary changes) of population genetics.

Population Fitness

The concept of population fitness, as we pointed out early in this chapter, has never been defined adequately. Nor is it likely to be so defined because of the uncertainty of future events.

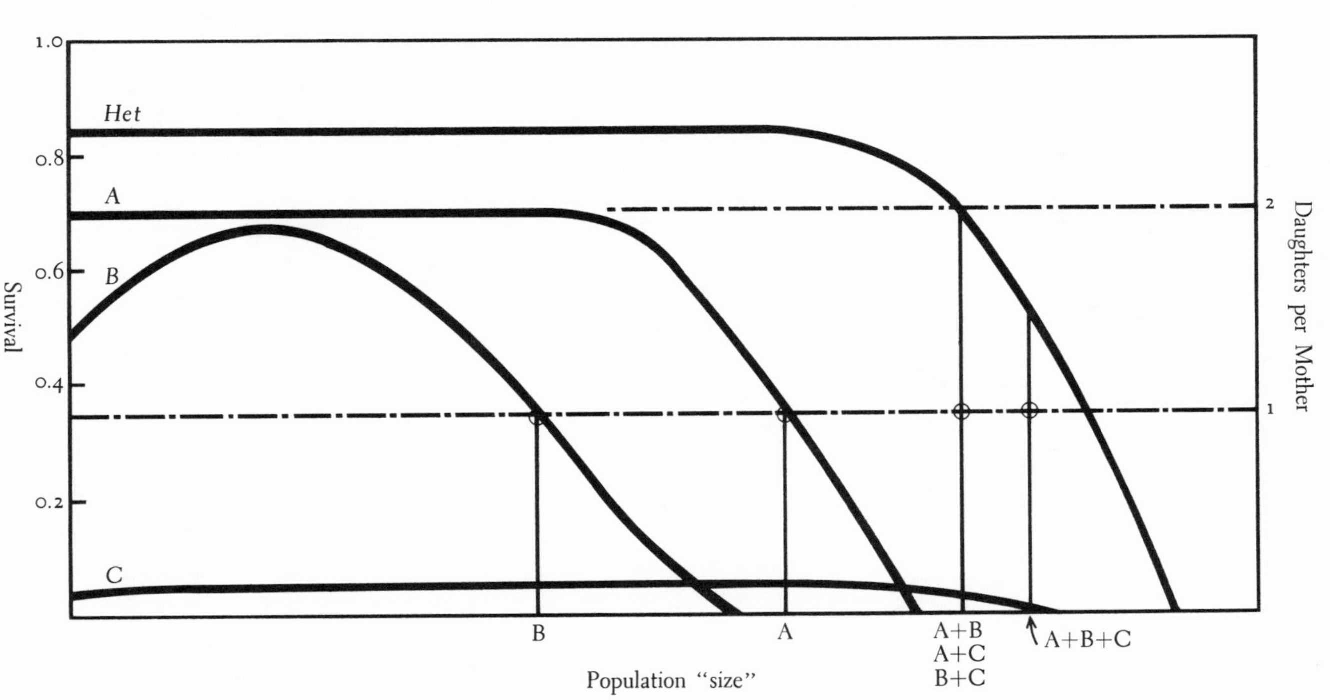

Figure 24-6. Egg to adult survival of homozygous individuals in three (hypothetical) monomorphic populations (A, B, and C) and that of their heterozygotes. Survival of 0.35 has been chosen arbitrarily to represent a mean Darwinian fitness (ratio of daughters to mothers) of 1.00. The relative sizes of equilibrium monomorphic and polymorphic populations can be determined from the curves.

Nevertheless, other things being equal, the aspect of the population that we have called "size" seems to be related to what we would intuitively regard as population fitness. Under moderate environmental conditions, for example, the size of a population increases; this is not a change that would be regarded normally as harmful to the population. On the contrary, we would say that a population is much more likely to continue its existence under moderate than under severe environmental stresses. The extreme examples, perhaps, are the mutant strains of many organisms that are maintained by investigators, nurserymen, and fanciers; the care expended on these individuals increases the fitnesses of these mutants far above what they would be in natural populations.

By the same argument, a genetic change that enables a population to cope with its surroundings in a more efficient manner and hence to increase its equilibrium size would probably be looked upon as a change that improves the fitness of the population. Thus in Figure 24-6 the monomorphic *A* population would be looked upon as "more fit" than the monomorphic *B* population and the polymorphic population as more fit than either. But the genetic load in the polymorphic population is considerably greater than that (no load) in either monomorphic population. From the relationships discussed for Figure 24-6, it is clear that population fitness is unrelated to genetic load; the change from monomorphism to two-allele polymorphism introduces a genetic load of 0.5 despite an increase in population size (and hence in population fitness). The further change to a three-allele polymorphism increases population size still more but, in this case, with a somewhat reduced genetic load.

A Generalized Fitness: Frequency Diagram

The standard relationship between gene frequency and the average fitness of a population in the case of a two-allele balanced polymorphism was shown in Figure 16-2. In the meantime we have learned that this figure illustrates the relationship between fitness and gene frequency under hard selection where population size can be ignored. We have also learned that the fitnesses of individuals of different genotypes within a polymorphic population change with population size. Consequently, the simple parabolic curve of Figure 16-2 is unable to convey the full relationship between fitness and gene frequency; nonequilibrium populations can deviate from equilibrium conditions in two dimensions: size and gene frequency.

An attempt has been made in Figure 24-7 to illustrate the approach of a polymorphic population to equilibrium under conditions where population size affects the individual adaptive values. On the left of the figure is a chart illustrating the fitnesses of three genotypes—*AA*, *Aa*, and *aa*—in relation to population size. These line segments have been drawn

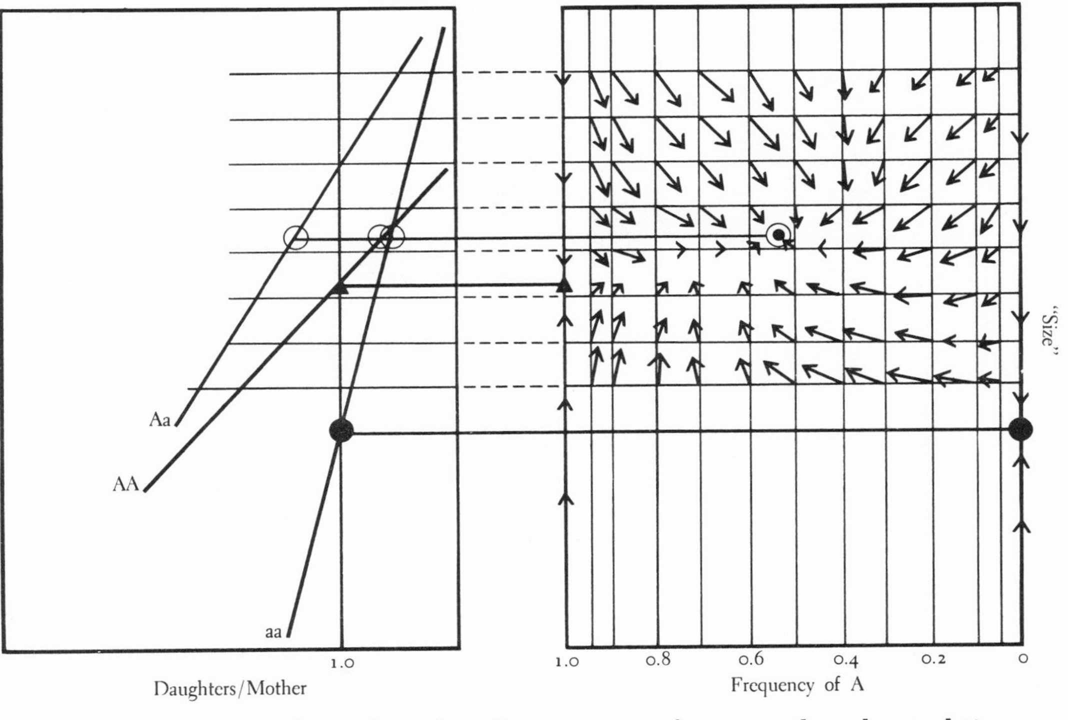

Figure 24-7. Two charts that relate fitness to gene frequency through population size (the common axis) as the two must be related in freely breeding, equilibrium populations.

straight for simplicity; they are not unlike the rightmost portions of the curves representing *A, B,* and the heterozygotes in Figure 24-6. There are three equilibrium sizes: (1) that of the monomorphic *aa* population, (2) that of the monomorphic *AA* population, and (3) that of the polymorphic population. The relative fitnesses of the three genotypes in the polymorphic population enable one to calculate equilibrium gene frequencies for this population.

The chart on the right of the figure represents gene frequency and population size. The three equilibrium frequencies and sizes are known from the chart on the left; these are the two monomorphic populations (100 and 0 percent *A*) and the polymorphic population whose equilibrium frequencies of *A* and *a* can be calculated (about 53 percent *A*). Now, nonequilibrium populations, because they represent artificial situations, can be constructed on the basis of any combination of size and frequency. How does each of these unstable populations evolve? Simple calculations on a desk calculator give the results indicated by arrows in the right-hand chart. There is no simple parabola leading to the equilibrium gene frequency in the polymorphic population. Although some paths lead rather directly to the equilibrium point, many others approach it in a roundabout fashion. In general, populations that are too large contract rapidly while converging on the equilibrium point; those starting with 40 to 50 percent *A* tend to lose *A* momentarily before regaining it. Populations that are below equilibrium size approach the equilibrium point by relatively large gains in size or in frequency of *A* depending upon their starting gene frequencies.

The dependence of the adaptive values of the different genotypes upon population size calls for the replacement of the simple parabolic curve of Figure 16-2 by a curved surface. The reader is left to contemplate the notion that the two charts of Figure 24-7 represent the floor and one side of a three-dimensional figure (with fitness forming the vertical axis). In this case, the three lines shown in the chart on the left could in turn be represented as surfaces extending back into the box; consequently, fitnesses that depend upon gene frequencies (see page 289) as well as upon population size could be incorporated into the generalized diagram.

25

POPULATION GENETICS AND DEVELOPMENTAL GENETICS

Our final topic concerns the relation between population genetics and developmental studies. In Chapter 24 we saw that population genetics (especially that part of population genetics concerned with evolutionary problems) cannot remain aloof for long from the actual populations of individuals with which it is concerned. Population genetics, in a word, must eventually come to include ecological genetics. We outlined a simple procedure for uniting ecological and genetic concepts, for relating kinds and numbers of individuals.

It is equally true that population genetics cannot proceed indefinitely on abstract symbols (A and a, a_1, a_2, . . .) alone, with no concern as to the meaning of these symbols for the development of the individual. The mathematical manipulation of symbols, it seems to me, has proceeded about as far as possible without recourse to developmental as well as ecological concepts.

Through their use of the term "gene action," quantitative geneticists sometimes infer that the techniques of population genetics can shed light on the mechanism of gene action, the role of the gene in development.

It should be realized that an analysis of phenotypic frequencies does not lead to an understanding of gene action in the sense—what do genes do?

That a statistical analysis of phenotypic proportions cannot reveal the basic causes of these phenotypes can be illustrated by an analogy based on the resistance of microorganisms to antibiotics. Suppose, as we have done in Figure 25-1, that persons developing certain symptoms (fever, nausea, or similar symptoms) report to their family doctors and are treated by an injection of penicillin, streptomycin, or a mixture of the two. Suppose, too, that after a week or so, the symptoms recur in some persons and, despite a resumption of the original treatment, death ensues.

An examination of the (hypothetical) records of many such cases reveals that 37 percent of those treated with penicillin or streptomycin

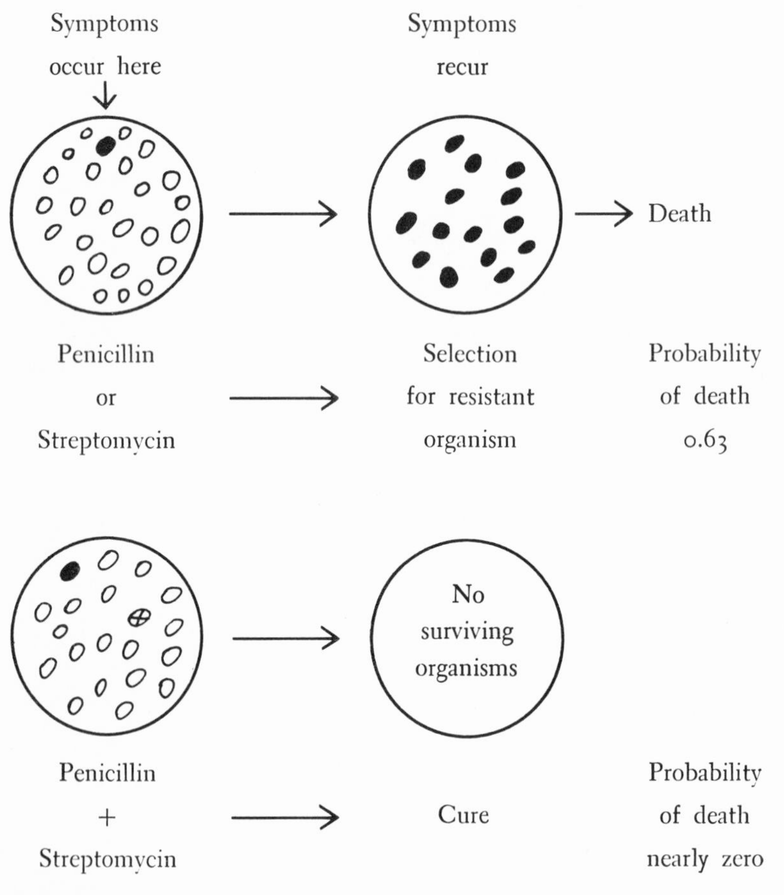

FIGURE 25-1. Analogy based on antibiotic-resistant mutants in bacteria illustrating how independent, randomly distributed events (mutations) can appear to interact in producing their final effect: life or death of the infected individual.

are cured while 63 percent develop the symptoms anew. Those individuals treated simultaneously with both penicillin and streptomycin are, as far as the data show, invariably cured; no relapses and deaths occur among these persons.

An analysis of these data will reveal that penicillin reacts *synergistically* with streptomycin in bringing about the cure of individuals receiving the double treatment. We say that the two antibiotics interact synergistically because if they did not we would expect that 0.63×0.63, or 0.40, of the individuals treated with both penicillin and streptomycin should develop once more the now-fatal symptoms.

The "interaction" that is observed at the "phenotypic" level in this case has nothing to do with the conditions under which symptoms develop or with the action of antibiotics on the causative organism. The situation described above is based, in fact, on the *independent* origin of penicillin- and streptomycin-resistant organisms and the *independent* action of the two antibiotics. To obtain the events we described, we merely assume that symptoms develop when the population of organisms in the body reaches a certain number, say 10^9. We assume, too, that in a bacterial population of this size there is an average of one penicillin-resistant and one streptomycin-resistant individual. The persons that are cured by the initial treatment are those in which no resistant organisms exist; the proportion of these individuals (for either antibiotic) is e^{-1}, or 37 percent. The remainder have one or more resistant organisms and, consequently, these individuals eventually develop symptoms anew and eventually die. Finally, it is highly unlikely that a single bacterium resistant to both penicillin and streptomycin will arise by chance (this chance would be approximately 10^{-18}); hence the double treatment is always effective.

To obtain a "nonsynergistic" effect of double treatment in our example, it would have been necessary to change from treatment by one antibiotic to treatment by the other at the time when symptoms recurred. The proportion of successful cures under this use of antibiotics would have been 37 percent by the first treatment and 23 percent (0.37×0.63) by the second; about 40 percent of all treated individuals would not be cured.

As the action of antibiotics in this example, the action of genes in cells is so far removed from the groupings of individuals into gross phenotypic classes that such groupings cannot tell us about gene action in any biochemical or physiological sense.

If population genetics can tell us little or nothing about developmental genetics, then what can its role be as far as developmental geneticists are concerned? Its role, as I see it, is to reveal problems that might otherwise go unnoticed. This population genetics can do in a thoroughly effective way. A case in point is the mounting evidence based on population studies that each of many gene loci is occupied by a series of different alleles rather than by a single "normal" gene. This evidence has culminated in the calculation (Lewontin and Hubby, 1966) that some 30 to 40 percent of all gene loci are polymorphic in any one population. These calculations and the results of other experiments such as those of Wallace (1958), Mukai and Yoshikawa (1964), and Vann (1966) raise questions about the role of hybridity in development that most developmental biologists have failed to recognize.

Not only can the studies of populations at times raise questions for developmental biologists but population concepts can also aid in pinpointing reasonable explanations. We can continue with the discussion of the high proportion of loci exhibiting gene-enzyme (or gene-protein) variation. One possible explanation for this variation is that all the observed variants are selectively neutral (Shaw, 1965). Selectively neutral alleles, however, are lost rapidly from populations (Chapter 6), and hence neutrality should lead to a homogeneous rather than to a heterogeneous population. So many supposedly "neutral" polymorphisms have been found in the past to be maintained by natural selection that neutrality seems to be the least attractive explanation for the preservation of polymorphisms.

A second possible explanation is one based on high mutation rates. This is a testable (but unlikely) notion; it was once proposed (Neel, 1950) as an explanation for the high frequency of sickle-cell anemia in Africa. One can calculate the mutation rate needed to preserve a polymorphism in the absence of selection. One can then collect data to show whether mutations do in fact occur at the postulated rate.

Finally, of course, there is the possibility that individuals possessing two forms of a given enzyme perform more efficiently and, therefore, have higher fitnesses than those possessing only one or the other of the two forms. The evidence that this type of advantage exists for a great many polymorphic loci now, and presumably for still others in the recent history of the species, must come eventually from biochemical and developmental studies. It may be necessary, however, to know more about the site and mode of action of enzymes within cells to appreciate fully the virtues of the molecular diversity of enzyme structure. In any case, studies that concentrate wholly on the active site and the kinetics of enzyme activity in solution scarcely seem capable of dealing with the problem of polymorphism realistically.

This discussion of the role of heterozygosity in development can be concluded with an examination (and modification) of a model proposed originally by R. A. Fisher (1930, Chap. 1). According to this model (see Figure 25-2), no organism is perfect in respect to a particular adaptation. Consequently, if *0* (in the figure) represents the optimum adaptation, the organism will be found somewhere on a circle whose radius measures the imperfect nature of the organism. Now, suppose the phenotype of that individual were to be altered slightly. The position of the altered phenotype in respect to this particular adaptation can be represented by a circle drawn about the original phenotype. The radius of the second circle indicates the degree to which the phenotype has been altered.

An examination of the figure shows that, according to Fisher's model, the smaller the alteration (the smaller the radius of the second circle), the more nearly the probability that it will be an advantageous one approaches one half (that is, nearly one half the circumference of the small circle lies within the larger circle). Large alterations, on the other hand, might never be beneficial; specifically, if the radius of the second circle were to exceed the diameter of the original one, the altered phenotype could never be closer to point *0* than the original phenotype.

It is easy to regard Fisher's model as an indication that small *mutations* (mutations with small phenotypic effects) may have a 50 : 50 chance of being beneficial. In a strict sense, however, his model applies to the two homozygotes, *aa* and *a′a′*, as shown in the lower part of Figure 25-2. If the allele that possesses the greater beneficial effect (whose homozygote lies closer to point *0*) in respect to the adaptation of the organism were to be dominant to any appreciable degree over the less favorable one, the possible positions of the heterozygote, *aa′*, relative to optimal adaptation (point *0*) are represented by the stippled area of the figure. A great deal more than one half of the stippled area lies within the original circle. The modified model suggests, then, that mutations with small effects may improve their heterozygous carriers more frequently than not. This, of course, is reminiscent of the results observed by both Wallace (1958) and Mukai and Yoshikawa (1964).

Limitations Imposed on Evolution by Development

In commenting upon the papers presented by Haldane and himself at the Sixth International Congress of Genetics, Fisher (1932) made the following point: Haldane (1932b) was asking in his talk whether genetic facts can explain evolution while he (Fisher) was asking in his whether evolutionary facts could explain genetics. It was Fisher's genius to see that genetic phenomena do not stand aside, directing but untouched by evolutionary events. Rather, in being but one aspect of life, these phenomena too have had an evolutionary history and are the product of natural selection.

We can express an analogous view about evolution and development. We can ask, what modifications of developmental processes have occurred through the action of evolution? We can also ask, what limitations have developmental processes placed upon evolutionary events? It is the second of these questions that concerns us now.

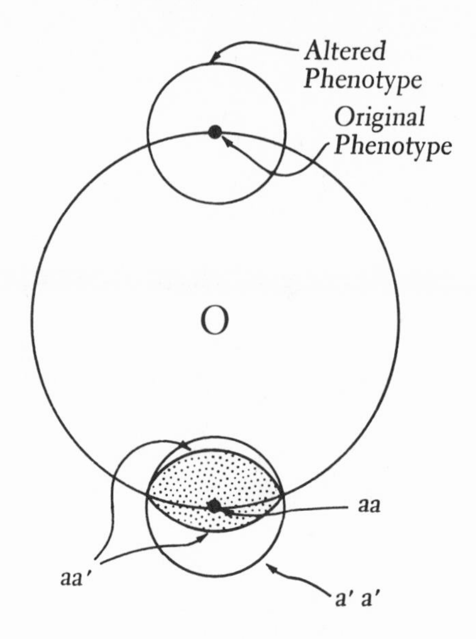

FIGURE 25-2. Model upon which Fisher based his argument that small phenotypic changes have nearly a 50 : 50 chance of improving the adaptation of the individual. The modified model implies that, with dominance in respect to advantageous aspects of the phenotype, mutations with small effects have greater than a 50 : 50 chance of being beneficial.

The constraints within which the evolution of life must occur are those arising from the physical limitations of living material, those that are inherent in the finite size of organisms, and those imposed by the limited time within which changes must take place if they are to be appropriate for a given need.

Examples that illustrate the limitations imposed upon evolutionary change are numerous. There is, for instance, an upper limit to the size that a land animal can obtain. There are even more stringent restrictions upon the size of flying organisms. Warm-blooded animals cannot exist if they are below a critical size because, if they are too small, heat is lost more rapidly than the metabolic furnace can be stoked.

Life continues to exist in the face of the conflicting demands of a rather ruthless environment only by making a series of compromises. Baldwin (1949), for example, cites examples of compromises made in response to freshwater as opposed to marine life. Eggs laid by freshwater organisms must have an impermeable coat that prevents the diffusion of water molecules into the egg's interior. When equipped with an impermeable coat, however, an egg must contain enough material of all sorts to sustain the development of the embryo; consequently, freshwater forms must lay larger eggs than their marine counterparts. And, finally, because the total amount of material that can be converted into "egg stuff" by an individual is not unlimited, freshwater forms should produce fewer eggs than marine forms. This seems to be true. Baldwin cites a shrimp, a whelk, and a mussel for which the numbers of eggs laid per year by marine and comparable freshwater forms are 320 : 25, 12,000 : 100, and 2,000,000 : 18,000.

More spectacular examples of the physical limitations imposed by the materials of life can now be cited from the "messages" carried by DNA molecules. Three base pairs of the DNA molecule are required to specify one amino acid. Thus a very precise relationship exists between the amount of DNA carried by organisms and the total number and size of the protein molecules they can make. Because the number of enzymes limits the number of biochemical reactions that can take place, very small organisms simply cannot carry out complicated reactions. Cells, that is, probably have a minimum size (Steward, 1966, p. 13).

The Modification of Development by Evolutionary Changes

That evolution has led to a diversity of living things, each with its characteristic pattern of development, is not our concern in this section. Rather, we want to discuss three examples that illustrate different aspects of the development-evolution relationship. The first of these deals with phenocopies, noninherited changes in the phenotype that mimic the effects of mutant genes. The second concerns the developmental aspects of a successful response to artificial selection. Finally, the third considers the evolutionary significance of embryological sequences; does, in fact, ontogeny recapitulate phylogeny?

Phenocopies. One of the experimental techniques used two or three decades ago (before the present elegant techniques of molecular biology were possible) in the study of gene action consisted of duplicating the phenotypic effects of mutant genes by environmental agents. The abnormal phenotypes produced were known as phenocopies. Heat shocks given to pupae of *D. melanogaster* at the proper moment can lead to incomplete crossveins in the wing of the adult; the mutant gene *cross-veinless* has a similar effect. The addition of silver nitrate to the culture medium produces pale-bodied adult flies resembling those homozygous for *straw*, a color mutant (Rapoport, 1939).

Phenocopies were defined as nonhereditary changes induced by environmental agents that simulated genetically caused ones. Largely as the result of a series of studies carried out by Waddington (1953, 1956), it is now known that this definition is not entirely adequate. The morphological change that is induced by the environmental agent is not a heritable change, to be sure; the environmental agent does not alter the genic content of the affected individual's germ cells. The ability to respond to the environmental agent and to produce the phenocopy, however, is a function of the genotype of the affected individual. Consequently, one can select artificially for an increased proportion of phenocopies. Having succeeded in this selection program, one may find that the "phenocopy" is expressed even in the absence of the once-necessary environmental agent. Waddington (1953) referred to this process as "the genetic assimilation of an acquired character."

The nature of Waddington's "genetic assimilation" experiment has been illustrated in Figure 25-3 by means of a hypothetical example based on selection for tail length in mice. Suppose we have been directed to establish a short-tailed strain of mice by artificial selection. Our director, however, has seen fit to force us to proceed with only a limited view of the mice with which we are to work. Each generation he lines the mice along the top of a box in which there is a narrow window (*A* in Figure 25-3). The tail of each mouse hangs past the window so that it is im-

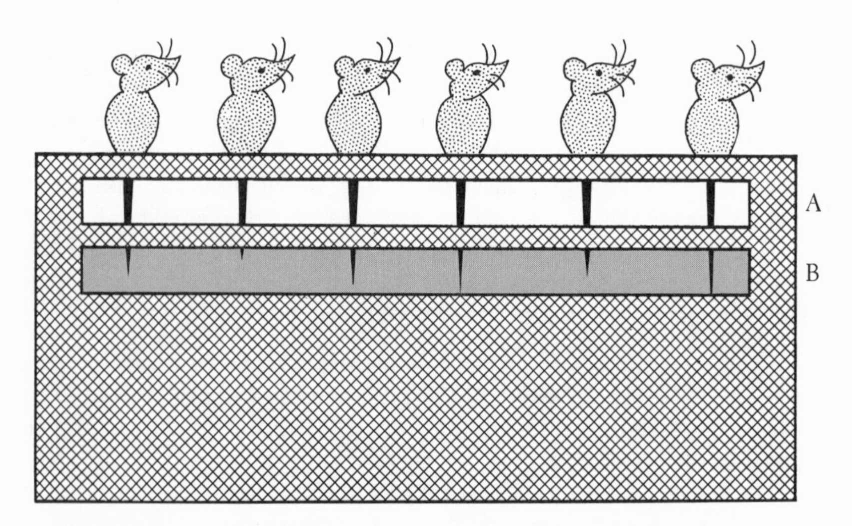

Figure 25-3. Hypothetical experiment involving selection for short tails in mice. Window *A* does not allow the experimenter to see the ends of the tails; window *B* does.

possible to recognize short-tailed and long-tailed individuals. Obviously, then, we cannot carry out the selection program. With no knowledge about the lengths of tails on different individuals, we can at best make what are actually random matings among the mice. The result of these matings will be mice whose tails resemble those of the parental generation.

Suppose now, however, that we shift the window downward from position *A* to position *B* as shown in the figure. The ends of some tails are now visible. And now our job is very simple. Short-tailed mice can be identified and used as parents. A short-tailed strain is indeed possible to make. Eventually, some mice will have tails too short to be visible in window *B*. As the proportion of such mice increases, the ends of the tails of the shortest-tailed mice may become visible in window *A*. From that moment, window *B* is no longer necessary because we can now select directly from short tails that are visible in the top window.

To summarize this example, a selection program that depended initially upon a window at *B* can, in time, be carried out in its absence. Compare that summary with the statement made earlier: One can select for an increased proportion of phenocopies; having succeeded in this selection, one may find that the "phenocopy" is expressed even in the absence of the once-necessary environmental agent.

There are two ingredients necessary for the successful "assimilation" of a phenocopy. The first is genetic variation in respect to the morphological variant produced as the phenocopy. The second is an environmental agent that produces the phenocopy—a means for shifting the window.

Dun and Fraser (1958, 1959) have reported on the selection for high and low vibrissae ("whisker") number in mice despite the almost invariate number of these hairs on normal individuals. The second window in this case was provided by a semidominant, sex-linked mutant gene, *Tabby*. *Tabby* males have an average number of 8.7 vibrissae while heterozygous *Tabby* females have an average of 15.1. Normal mice have 19 vibrissae. While the variance in the number of vibrissae in normal mice is virtually zero (nearly all individuals have exactly 19 bristles), considerable variation exists in the case of *Tabby* males and heterozygous *Tabby* females.

Dun and Fraser used the variation among *Ta*/+ females to start lines of mice in which the mothers were chosen for low ("low" lines) and high ("high" lines) numbers of vibrissae. On alternate generations, the mating involved *Tabby* males and heterozygous *Tabby* females. This cross produces wildtype males. By the sixth generation of selection for high and low bristle numbers in *Ta*/+ females, wildtype males were found that were aberrant: Vibrissae were missing in the low lines, extra vibrissae were present in the high lines. Thus, by shifting the window through which a given trait was viewed, it was possible to see variation that was not visible before. This variation proved to be genetic, at least in part, so that selection was effective in changing vibrissae number. The effectiveness of selection extended to an alteration of vibrissae number even as viewed through the window provided by normal mice. (As an important corollary, we learn from this experiment that a highly uniform phenotype cannot be accepted as evidence for the absence of genetic variation. This we learned, too, in Chapter 3.)

Response to Selection. Except for experiments such as those of Maynard Smith and his colleagues (see Sondhi, 1961, 1963) on the modification of pattern, the usual selection program is concerned with increasing or decreasing a given quantitative trait: high milk production in cattle, high egg production in poultry, high or low oil content in corn, high or low bristle numbers in *Drosophila,* or resistance or sensitivity to insecticides for houseflies and mosquitoes.

Because the response to artificial selection in quantitative selection programs can be described as high-or-low or more-or-less, one thinks of the various genotypes as consisting of combinations of plus and minus genes; these can be located on particular chromosomes by standard genetic tests.

What is it, however, that plus and minus alleles do to bring about the corresponding increase or decrease in the phenotypic trait? Obviously, these genes act through developmental processes just as do other genes. In a selection program where only the final phenotype is examined to determine which individuals will be chosen to serve as parents for the next generation, any developmental alteration that leads to the desired phenotype and does not interfere with the use of the affected individual as a parent will be compatible with the selection program (see Table 22-1 and Figures 22-4 and 22-5). Thus, an increase or decrease in cell number as well as an increase or decrease in cell size will affect the final size of an individual or one of his organs. In some cases cell number may be modified by postponing or advancing the time at which an ancestral cell is committed to a given developmental fate. Cell size, in turn, can be controlled in some instances by alterations in the physiological control of the mitotic cycle.

Some of the hidden genetic variation underlying the response of *D. melanogaster* to selection has been revealed in a study by Spickett (1963). The initial material in this study consisted of two strains of flies that differed by an average of more than five sternopleural bristles on each side: the strain with the higher number had been obtained by artificial selection.

By the use of mutant markers, Spickett was able to identify three chromosomal regions in the selected line that had pronounced effects on bristle number (Table 25-1 and Figure 25-4). The action of these three chromosomal regions (the analysis was not fine enough to permit one

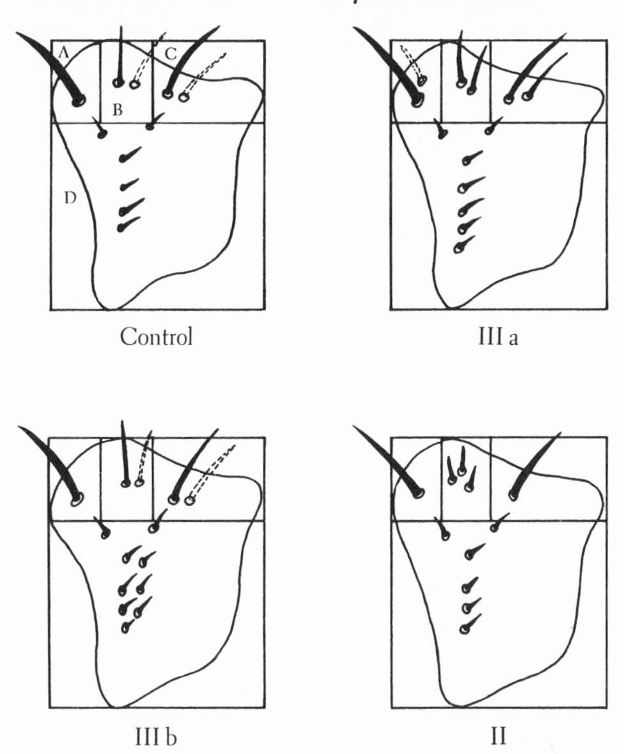

FIGURE 25-4. Effect of certain chromosomal regions of a high (sternopleural) bristle line of *D. melanogaster* on the development of sternopleural bristles. (After Spickett, *Nature, 199:* 870–873, 1963.)

to say "locus") differed considerably. One (IIIa) increased the size of the fly and its sternopleurae and, in this way, brought about an increase in the number of bristles in all four areas shown in Figure 25-4. A second chromosomal region (IIIb) had its effect in area D of the sternopleura; it accomplished this by tending to produce two rows of hairs rather than a single row. A third region (II) led to the suppression of the large bristle usually found in area B; in its place there developed two or three smaller, later-appearing bristles. The developmental effects of these three chromosomal regions were compatible with one another and so the high bristle number of the selected line came from the combination of all three.

The analysis by Spickett makes one point abundantly clear. Developmental responses required to bring about seemingly trivial alterations in the phenotype are not necessarily trivial in themselves. They frequently call for substantial alterations in the control of developmental processes. As a result, selection programs, as we have seen in Chapter 22, frequently reach plateaus where no further progress is possible. Such

TABLE 25-1

EFFECT OF THREE CHROMOSOMAL REGIONS ON THE NUMBERS OF STERNOPLEURAL BRISTLES IN D. MELANOGASTER.

(Spickett, 1963.)

	REGION OF THE STERNOPLEURA				
	A	B	C	D	TOTAL
Control	1.02	1.61	1.43	5.99	10.05
IIIa	1.12	1.88	1.72	7.03	11.74
IIIb	1.04	1.72	1.53	7.58	11.82
II	1.06	2.37	1.42	6.02	10.87
IIIa + IIIb	1.10	1.83	1.82	8.57	13.46
IIIa + II	1.14	3.81	1.75	7.06	13.76
IIIb + II	1.07	2.49	1.48	7.58	12.61
IIIa + IIIb + II (= high line)	1.14	3.73	1.78	8.59	15.24

plateaus are often caused by natural selection acting within the selected lines. In these cases, the individuals that most closely meet the demands of selection suffer from severe developmental disorders that leave them physically weak or partially sterile. Figure 25-5 shows the tremendous effect on fitness brought about by artificial selection for increased shank length in chickens. The actual increase in shank length brought about by this selection program seems modest indeed in comparison to the effect it has had on fitness.

Ontogeny Recapitulates Phylogeny. Earlier in this book we examined the "law" of the irreversibility of evolution. An understanding of the many paths open to an evolving population lead, in turn, to an appreciation of the improbability that a population will follow any one path. To reverse its past evolution, a population must take precisely that path it has just followed—but in the reverse direction. As in the case of any specified path, it is highly improbable that the population will take this one either.

Another "law," one originally stated by Haeckel, which we shall now discuss is that which says, "Ontogeny recapitulates phylogeny." This law, like the one on irreversibility, is taught to students more as something to be criticized than as something to be taken seriously. It cannot be taken *literally*, of course, because embryos do not pass through developmental forms that represent those of their adult phylogenetic ancestors. Embryos do not exhibit a series of antiquated adult characteristics.

There is a fundamental truth to the statement that ontogeny recapitu-

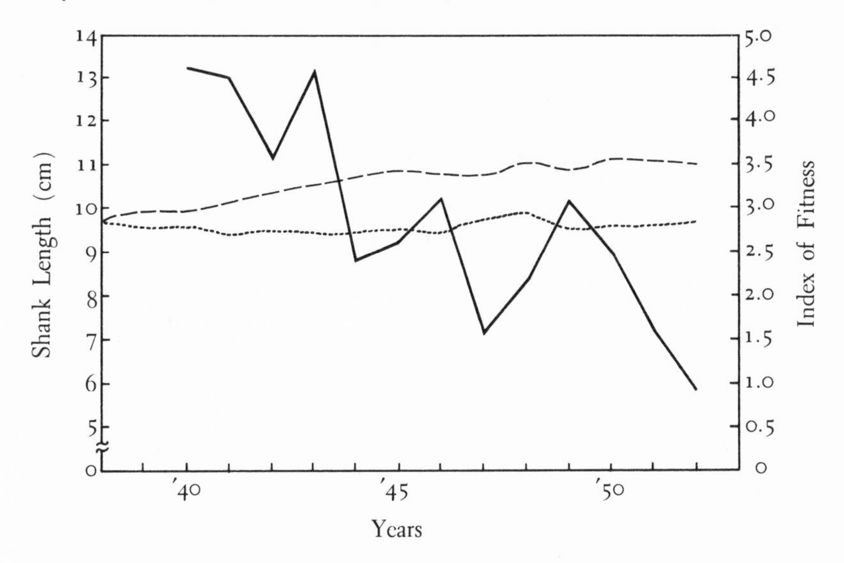

FIGURE 25-5. Selection for increased shank length in chickens and its effect on fitness (offspring raised to maturity per dam). Dotted line: control shank length; dashed line: selected shank length; solid line: index of fitness. (After Lerner, 1954, Table 1.)

lates phylogeny, however, that should not be overlooked. If there exist on other planets a number of evolving systems of life such as that found on earth, I suspect that each one conforms to a descriptive statement such as that made originally by Haeckel.

Three reasons exist for making the above claim. First, evolution proceeds by the modification of existing structures, not by the invention of preperfected structures. Second, the development of each individual must be accomplished within a finite period of time. Third, speeding up the rate at which a series of interrelated processes takes place is a rather simple evolutionary problem.

Throughout the discussions in this book we have emphasized that evolution is a historical process. Every evolutionary change occurs on the basis of and in a manner compatible with conditions as they exist at the moment. A wing, for example, is not *created*. It represents a modified arm in the case of both bats and birds. A paddle in the case of the penguin is a modified wing, whereas that of the seal is a modified leg.

Now, if evolution proceeds by the attachment of one modification after another upon structures and processes that have already been modified many times in the past, the time required for development would be immense. The nature of this problem has been outlined in Figure 25-6. The figure starts with the ontogenetic development of a primitive individual through the stages A, A′, and A″; for convenience, we shall

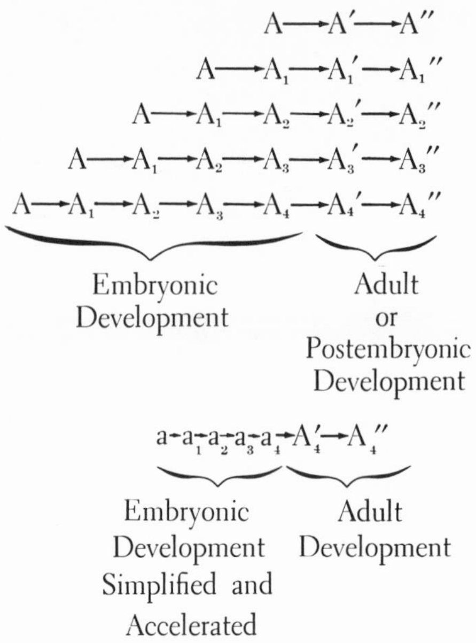

$$A \to A' \to A''$$
$$A \to A_1 \to A_1' \to A_1''$$
$$A \to A_1 \to A_2 \to A_2' \to A_2''$$
$$A \to A_1 \to A_2 \to A_3 \to A_3' \to A_3''$$
$$A \to A_1 \to A_2 \to A_3 \to A_4 \to A_4' \to A_4''$$

Embryonic Development — Adult or Postembryonic Development

$$a \to a_1 \to a_2 \to a_3 \to a_4 \to A_4' \to A_4''$$

Embryonic Development Simplified and Accelerated — Adult Development

FIGURE 25-6. Schematic account of ontogeny recapitulating phylogeny. Further details in the text.

let A′ and A″ represent adult stages of development while A represents an embryonic stage.

An evolutionary modification of A occurs by forcing A into a different path such that it reaches A_1, which then goes on to develop into the later stages A_1' and A_1''. Still another modification might lead from A to A_1 to A_2, with the latter developing through the stages A_2' and A_2'' (where the primes designate adult stages).

The point brought out in the figure is that the list of letters that symbolize various developmental stages becomes longer and longer. The time available for an organism to develop does not grow in a comparable manner. More precisely, individuals that succeed in shortening the time required to go from A to A_4'', for example, have a selective advantage over those that do not. And it is a simpler evolutionary problem to speed up the rate at which one passes through A_1, A_2, and A_3 than it is to bypass them. As a result, the development of an individual takes on the form shown in the lower part of the figure where small letters represent stages passed through quickly.

The consolidation of the developmental processes of the embryo entails more than the mere bypassing of adult forms of these processes (A′, A″, A_1', A_1'', A_2', . . .); superfluous embryonic aspects are sloughed off as well. What is left finally is a series of stages, reduced in time and reduced in complexity, each of which is essential (for the moment) in controlling or permitting the proper development of the stage that follows. These embryonic stages, then, which resemble relics of the individual's phylogenetic past are exactly that. They are, in a sense, comparable to the ducklike noises that come from a tape recorder as one goes rapidly from one part of the tape to another. We must beware, however, of thinking of these relic stages as manifestations for corresponding relic genes ("fish" genes in a mammal, for example); the phenomena of coadaptation, hybrid vigor, and hybrid breakdown have already shown us that the genetic bases of physiological and developmental processes are more transitory than the processes themselves.

Our survey of population genetics is over. We started by learning to calculate gene and zygotic frequencies. We have ended by saying a few words in support of Haeckel's biogenetic law as we understand it in the light of evolutionary genetics. In between, we have dealt with a variety of problems concerned with populations and the individuals they contain.

If I were to express but one hope for this book, it would be that I have raised problems with which the reader was unaware. This, after all, is what education is all about. If I were permitted a second hope, it would be that the reader has learned that many evolutionary problems can be expressed quantitatively and, as a result, have logical solutions. And, if I were allowed still a third, it would be that among my readers are some young ones possessing skills that will enable them to solve problems which without my help might have escaped their attention.

References

Allard, R. W., and C. Wehrhahn. 1964. A theory which predicts stable equilibrium for inversion polymorphisms in the grasshopper *Moraba scurra*. Evolution *18:* 129–130.

Allison, A. C. 1961. Abnormal haemoglobins and erythrocyte enzyme-deficiency traits. In *Genetical Variation in Human Populations* (Editor, G. A. Harrison). Pergamon Press, London.

Anderson, E. 1949. *Introgressive Hybridization*. Wiley, New York.

Andrewartha, H. G. 1963. *Introduction to the Study of Animal Populations*. Univ. Chicago Press, Chicago.

Ashton, G. C. 1965. Serum amylase (thread protein) polymorphism in cattle. Genetics *51:* 431–437.

Ayala, F. J. 1966. Evolution of fitness. I. Improvement in the productivity and size of irradiated populations of *Drosophila serrata* and *Drosophila birchii*. Genetics *53:* 883–895.

Baldwin, E. 1949. *An Introduction to Comparative Biochemistry*. Cambridge Univ. Press, New York.

Bateman, A. J. 1947a. Contamination of seed crops. I. Insect pollination. J. Genet. *48:* 257–275.

Bateman, A. J. 1947b. Contamination of seed crops. II. Wind pollination. Heredity *1:* 235–246.

Bateman, A. J. 1947c. Contamination of seed crops. III. Relation with isolation distance. Heredity *1:* 303–336.

Bateman, A. J. 1950. Is gene dispersion normal? Heredity *4:* 353–363.

Battaglia, B. 1958. Balanced polymorphism in *Tisbe reticulata*, a marine copepod. Evolution *12:* 358–364.

Battaglia, B. 1965. Advances and problems of ecological genetics in marine animals. Proc. 11th Intern. Congr. Genet. *2:* 451–463.

Beardmore, J. A., Th. Dobzhansky, and O. Pavlovsky. 1960. An attempt to compare the fitness of polymorphic and monomorphic experimental populations of *Drosophila pseudoobscura*. Heredity *14:* 19–33.

Bernstein, F. 1925. Zusammenfassende Betrachtungen über die erblichen Blutstrukturen des Menschen. Z. induktive Abstammungs-Vererbungslehre, *37:* 237–270.

Bhalla, S. C., and R. R. Sokal. 1964. Competition among genotypes in the housefly at varying densities and proportions (the *green* strain). Evolution *18:* 312–330.

Birch, L. C. 1960. The genetic factor in population ecology. Am. Naturalist *94:* 5–24.

Bodmer, W. F., and P. A. Parsons. 1960. The initial progress of new genes with various genetic systems. Heredity *15:* 283–299.

Bonnier, G., U.-B. Jonsson, and C. Ramel. 1958. Selection pressure on irradiated populations of *Drosophila melanogaster*. Hereditas *44:* 378–406.

Brncic, D. 1954. Heterosis and the integration of the genotype in geographic populations of *Drosophila pseudoobscura*. Genetics *39:* 77–88.

Brown, W. L., Jr., and E. O. Wilson. 1956. Character displacement. Systematic Zool. *5:* 49–64.

Bryson, V., and H. J. Vogel. 1965. *Evolving Genes and Proteins*. Academic Press, New York.

Buri, P. 1956. Gene frequency in small populations of mutant *Drosophlia*. Evolution *10:* 367–402.

Burla, H., A. B. da Cunha, A. G. L. Cavalcanti, Th. Dobzhansky, and C. Pavan. 1950. Population density and dispersal rates in Brazilian *Drosophila willistoni*. Ecology *31:* 393–404.

Burns, J. M. 1966. Preferential mating versus mimicry: Disruptive selection and sex-limited dimorphism in *Papilio glaucus*. Science *153:* 551–553.

Cain, A. J., and P. M. Sheppard. 1954. Natural selection in *Cepaea*. Genetics *39:* 89–116.

Carson, H. L. 1955. The genetic characteristics of marginal populations of *Drosophila*. Cold Spring Harbor Symp. Quant. Biol. *20:* 276–287.

Carson, H. L. 1958. Increase in fitness in experimental populations resulting from heterosis. Proc. Natl. Acad. Sci. U.S. *44:* 1136–1141.

Carson, H. L. 1959. Genetic conditions which promote or retard the formation of species. Cold Spring Harbor Symp. Quant. Biol. *24:* 87–105.

Carson, H. L. 1961. Heterosis and fitness in experimental populations of *Drosophila melanogaster*. Evolution *15:* 496–509.

Cavalli-Sforza, L. L. 1959. Some data on the genetic structure of human populations. Proc. 10th Intern. Congr. Genet. *1:* 389–407.

Cesnola, A. P. di. 1904. Preliminary note on the protective value of colour in *Mantis religiosa*. Biometrika *3:* 58–59.

Chao, C. Y. 1959. Heterotic effects of a chromosome segment in maize. Genetics *44:* 657–677.

Chetverikov, S. S. 1926. On certain features of the evolutionary process from the viewpoint of modern genetics. J. Exptl. Biol. (Russian) *2:* 3–54.

Clausen, J., D. D. Keck, and W. M. Heisey. 1940. Experimental studies on the nature of species. I. Effects of varied environments on western North American plants. Carnegie Inst. Wash. Publ. *520:* 1–452.

Clausen, J., D. D. Keck, and W. M. Heisey. 1948. Experimental studies on the nature of species. III. Environmental responses of climatic races of *Achillea*. Carnegie Inst. Wash. Publ. *581:* 1–129.

Crenshaw, J. W. 1965. Radiation-induced increases in fitness in the flour beetle *Tribolium confusum*. Science *149:* 426–427.

Crew, F. A. E. 1966. *The Foundations of Genetics*. Pergamon Press, London.

Crosby, J. L. 1963. The evolution and nature of dominance. J. Theoret. Biol. *5:* 35–51.

Crow, J. F. 1956. Genetics of DDT resistance in *Drosophila*. Proc. Intern. Genet. Symp. Tokyo (Cytologia, Suppl.): 408–409.

Crow, J. F. 1957. Possible consequences of an increased mutation rate. Eugen. Quart. *4:* 67–80.

Crow, J. F. 1958. Some possibilities for measuring selection intensities in man. Human Biol. *30:* 1–13.

Crow, J. F. 1963. The concept of genetic load: A reply. Am. J. Human Genet. *15:* 310–315.

Crow, J. F. 1966. The quality of people: Human evolutionary changes. Bioscience *16:* 863–867.

Crow, J. F., and M. Kimura. 1965. Evolution in sexual and asexual populations. Am. Naturalist *99:* 439–450.

Crow, J. F., and R. G. Temin. 1964. Evidence for the partial dominance of recessive lethal genes in natural populations of *Drosophila.* Am. Naturalist *98:* 21–33.

Cunha, A. B. da. 1949. Genetic analysis of the polymorphism of color pattern in *Drosophila polymorpha.* Evolution *3:* 239–251.

Dixon, W. J., and F. J. Massey. 1951. *Introduction to Statistical Analysis.* McGraw-Hill, New York.

Dobzhansky, Th. 1937. *Genetics and the Origin of Species.* Columbia Univ. Press, New York.

Dobzhansky, Th. 1939. Experimental studies on genetics of free-living populations of *Drosophila.* Biol. Rev. *14:* 339–368.

Dobzhansky, Th. 1943. Genetics of natural populations. IX. Temporal changes in the composition of populations of *Drosophila pseudoobscura.* Genetics *28:* 162–186.

Dobzhansky, Th. 1947. A directional change in the genetic constitution of a natural population of *Drosophila pseudoobscura.* Heredity *1:* 53–64.

Dobzhansky, Th. 1948. Genetics of natural populations. XVIII. Experiments on chromosomes of *Drosophila pseudoobscura* from different geographic regions. Genetics *33:* 588–602.

Dobzhansky, Th. 1950. Genetics of natural populations. XIX. Origin of heterosis through natural selection in populations of *Drosophila pseudoobscura.* Genetics *35:* 288–302.

Dobzhansky, Th. 1951a. Mendelian populations and their evolution. In *Genetics in the 20th Century* (Editor, L. C. Dunn). Macmillan, New York.

Dobzhansky, Th. 1951b. *Genetics and the Origin of Species* (3rd edition). Columbia Univ. Press, New York.

Dobzhansky, Th. 1952a. Genetics of natural populations. XX. Changes induced by drought in *Drosophila pseudoobscura* and *Drosophila persimilis.* Evolution *6:* 234–243.

Dobzhansky, Th. 1952b. Nature and origin of heterosis. In *Heterosis* (Editor, J. W. Gowen). Iowa State College Press, Ames.

Dobzhansky, Th. 1964. How do the genetic loads affect the fitness of their carriers in *Drosophila* populations? Am. Naturalist *98:* 151–166.

Dobzhansky, Th., W. W. Anderson, and O. Pavlovsky. 1966. Genetics of natural populations. XXXVIII. Continuity and change in populations of *Drosophila pseudoobscura* in western United States. Evolution *20:* 418–427.

Dobzhansky, Th., and C. Epling. 1944. Contributions to the genetics, taxonomy, and ecology of *Drosophila pseudoobscura* and its relatives. Carnegie Inst. Wash. Publ. *554:* 1–183.

Dobzhansky, Th., and A. M. Holz. 1943. A re-examination of the problem of manifold effects of genes in *Drosophila melanogaster*. Genetics *28:* 295–303.

Dobzhansky, Th., and P. Koller. 1938. An experimental study of sexual isolation in *Drosophila*. Biol. Zentr. *58:* 589–607.

Dobzhansky, Th., and H. Levene. 1948. Genetics of natural populations. XVII. Proof of operation of natural selection in wild populations of *Drosophila pseudoobscura*. Genetics *33:* 537–547.

Dobzhansky, Th., and H. Levene. 1951. Development of heterosis through natural selection in experimental populations of *Drosophila pseudoobscura*. Am. Naturalist *85:* 247–264.

Dobzhansky, Th., R. C. Lewontin, and O. Pavlovsky. 1964. The capacity for increase in chromosomally polymorphic and monomorphic populations of *Drosophila pseudoobscura*. Heredity *19:* 597–614.

Dobzhansky, Th., and O. Pavlovsky. 1953. Indeterminate outcome of certain experiments on *Drosophila* populations. Evolution *7:* 198–210.

Dobzhansky, Th., and O. Pavlovsky. 1955. An extreme case of heterosis. Proc. Natl. Acad. Sci. U.S. *41:* 289–295.

Dobzhansky, Th., and O. Pavlovsky. 1957. An experimental study of interaction between genetic drift and natural selection. Evolution *11:* 311–319.

Dobzhansky, Th., and O. Pavlovsky. 1962. A comparative study of the chromosomes in the incipient species of the *Drosophila paulistorum* complex. Chromosoma *13:* 196–218.

Dobzhansky, Th., and M. L. Queal. 1938. Genetics of natural populations. II. Genic variation in populations of *Drosophila pseudoobscura* inhabiting isolated mountain ranges. Genetics *23:* 463–484.

Dobzhansky, Th., and M. M. Rhoades. 1938. A possible method for locating favorable genes in maize. J. Am. Soc. Agron. *30:* 668–675.

Dobzhansky, Th., and B. Spassky. 1944. Genetics of natural populations. XI. Manifestation of genetic variants in *Drosophila pseudoobscura* in different environments. Genetics *29:* 270–290.

Dobzhansky, Th., and B. Spassky. 1947. Evolutionary changes in laboratory cultures of *Drosophila pseudoobscura*. Evolution *1:* 191–216.

Dobzhansky, Th., and B. Spassky. 1953. Genetics of natural populations. XXI. Concealed variability in two sympatric species of *Drosophila*. Genetics *38:* 471–484.

Dobzhansky, Th., and B. Spassky. 1954. Genetics of natural populations. XXII. A comparison of the concealed variability in *Drosophila prosaltans* with that in other species. Genetics *39:* 472–487.

Dobzhansky, Th., and B. Spassky. 1962a. Genetic drift and natural selection in experimental populations of *Drosophila pseudoobscura*. Proc. Natl. Acad. Sci. U.S. *48:* 148–156.

Dobzhansky, Th., and B. Spassky. 1962b. Selection for geotaxis in monomorphic and polymorphic populations of *Drosophila pseudoobscura*. Proc. Natl. Acad. Sci. U.S. *48:* 1704–1712.

Dobzhansky, Th., B. Spassky, and W. Anderson. 1965. Bichromosomal synthetic semilethals in *Drosophila pseudoobscura*. Proc. Natl. Acad. Sci. U.S. *53:* 482–486.

Dobzhansky, Th., and S. Wright. 1941. Genetics of natural populations. V. Relations between mutation rate and accumulation of lethals in populations of *Drosophila pseudoobscura*. Genetics *26:* 23–51.

Dobzhansky, Th., and S. Wright. 1943. Genetics of natural populations. X. Dispersion rates in *Drosophila pseudoobscura*. Genetics *28:* 304–340.

Dobzhansky, Th., and S. Wright. 1947. Genetics of natural populations. XV. Rate of diffusion of a mutant gene through a population of *Drosophila pseudoobscura*. Genetics *32:* 303–324.

Dubinin, N. P., and G. G. Tiniakov. 1946. Inversion gradients and natural selection in ecological races of *Drosophila funebris*. Genetics *31:* 537–545.

Dun, R. B., and A. S. Fraser. 1958. Selection for an invariant character—"vibrissa number"—in the house mouse. Nature *181:* 1018–1019.

Dun, R. B., and A. S. Fraser. 1959. Selection for an invariant character, vibrissa number, in the house mouse. Australian J. Biol. Sci. *12:* 506–523.

Ehrman, L., B. Spassky, O. Pavlovsky, and Th. Dobzhansky. 1965. Sexual selection, geotaxis, and chromosomal polymorphism in experimental populations of *Drosophila pseudoobscura*. Evolution *19:* 337–346.

Emerson, S. 1939. A preliminary survey of the *Oenothera organensis* population. Genetics *24:* 524–537.

Epling, C., D. F. Mitchell, and R. H. T. Mattoni. 1953. On the role of inversion in wild populations of *Drosophila pseudoobscura*. Evolution *7:* 342–365.

Falconer, D. S. 1960. *Introduction to Quantitative Genetics*. Oliver and Boyd, Edinburgh.

Falk, R. 1961. Are induced mutations in *Drosophila* over-dominant? II. Experimental results. Genetics *46:* 737–757.

Falk, R., and N. Ben-Zeer. 1966. Viability of heterozygotes for induced mutations in *Drosophila melanogaster*. II. Mean effects in irradiated autosomes. Genetics *53:* 65–77.

Fisher, R. A. 1930. *Genetical Theory of Natural Selection.* Clarendon Press, Oxford.

Fisher, R. A. 1932. The evolutionary modification of genetic phenomena. Proc. 6th Intern. Congr. Genet. *1:* 165–172.

Fisher, R. A. 1949. *The Theory of Inbreeding*. Oliver and Boyd, Edinburgh.

Fisher, R. A. 1958. *Genetical Theory of Natural Selection* (2nd revised edition). Dover, New York.

Ford, E. B. 1940. Polymorphism and Taxonomy. In *The New Systematics* (Editor, J. Huxley). Clarendon Press, Oxford.

Ford, E. B., 1955. Polymorphism and Taxonomy. Heredity *9:* 255–264.

Ford, E. B. 1964. *Ecological Genetics.* Methuen, London.

Ford, E. B. 1965. *Genetic Polymorphism.* M.I.T. Press, Cambridge, Mass.

Fulton, B. B. 1952. Speciation in the field cricket. Evolution *6:* 283–295.

Gause, G. F. 1934. *The Struggle for Existence*. Williams and Wilkins, Baltimore.

Gershenson, S. 1928. A new sex-ratio abnormality in *Drosophila obscura.* Genetics *13:* 488–507.

Gibson, J. B., and J. M. Thoday. 1959. Recombinational lethals in a polymorphic population. Nature *184:* 1593–1594.

Gibson, J. B., and J. M. Thoday. 1962a. An apparent 20 map-unit position effect. Nature *196:* 661–662.

Gibson, J. B., and J. M. Thoday. 1962b. Effects of disruptive selection. VI. A second chromosome polymorphism. Heredity *17:* 1–26.

Gibson, J. B., and J. M. Thoday. 1964. Effects of disruptive selection. IX. Low selection intensity. Heredity *19:* 125–130.

Glass, B., and C. C. Li. 1953. The dynamics of racial intermixture—an analysis based on the American Negro. Am. J. Human Genet. *5:* 1–20.

Goldschmidt, E., and R. Falk. 1959. On the dominance of "recessive" lethals. Bull. Res. Council Israel *B8:* 1–8.

Gordon, C. 1935. An experiment on a released population of *Drosophila melanogaster*. Am. Naturalist *69:* 381.

Gordon, H., and M. Gordon. 1957. Maintenance of polymorphism by potentially injurious genes in eight natural populations of the platyfish, *Xiphophorus maculatus*. J. Genet. *55:* 1–44.

Gordon, M. 1947. Speciation in fishes. Distribution in time and space of seven dominant multiple alleles in *Platypoecilus maculatus*. Advan. Genet. *1:* 95–132.

Green, M. M. 1959. The discrimination of wild-type iso-alleles at the *white* locus of *Drosophila melanogaster*. Proc. Natl. Acad. Sci. U.S. *45:* 549–553.

Green, M. M. 1963. Unequal crossing over and the genetical organization of the *white* locus of *Drosophila melanogaster*. Z. Vererbungslehere *94:* 200–214.

Greenberg, R., and J. F. Crow. 1960. A comparison of the effect of lethal and detrimental chromosomes from *Drosophila* populations. Genetics *45:* 1153–1168.

Haldane, J. B. S. 1932a. *The Causes of Evolution*. Harper & Row, New York.

Haldane, J. B. S. 1932b. Can evolution be explained in terms of known genetical facts? Proc. 6th Intern. Congr. Genet. *1:* 185–189.

Haldane, J. B. S. 1937. The effect of variation on fitness. Am. Naturalist *71:* 337–349.

Haldane, J. B. S. 1948. The theory of a cline. J. Genet. *48:* 277–284.

Haldane, J. B. S. 1953. Animal populations and their regulation. New Biol. *15:* 9–24.

Haldane, J. B. S. 1954a. The measurement of natural selection. Proc. 9th Intern. Congr. Genet. *1:* 480–487.

Haldane, J. B. S. 1954b. The statics of evolution. In *Evolution as a Process* (Editors, J. S. Huxley, A. C. Hardy, and E. B. Ford). Allen and Unwin, London.

Haldane, J. B. S. 1956. The estimation of viabilities. J. Genet. *54:* 294–296.

Haldane, J. B. S. 1957. The cost of natural selection. J. Genet. *55:* 511–524.

Haldane, J. B. S. 1964. A defense of beanbag genetics. Perspectives Biol. Med. *7:* 343–359.

Hamilton, H. B., J. V. Neel, and K. Ozaki. 1961. The frequency in Japan of carriers of the rare "recessive" gene causing acatalasemia. J. Clin. Invest. *40:* 2199–2208.

Hardy, G. H. 1908. Mendelian proportions in a mixed population. Science *28:* 49–50.

Harlan, H. V., and M. L. Martini. 1938. The effect of natural selection on a mixture of barley varieties. J. Agr. Res. *57:* 189–199.

Heed, Wm. B. 1963. Density and distribution of *Drosophila polymorpha* and its color alleles in South America. Evolution *17:* 502–518.

Helfer, R. G. 1939. Dominance modifiers of *Scute* in *Drosophila pseudoobscura.* Genetics *24:* 278–301.

Hiraizumi, Y., and J. F. Crow. 1960. Heterozygous effects on viability, fertility, rate of development, and longevity of *Drosophila* chromosomes that are lethal when homozygous. Genetics *45:* 1071–1083.

Ives, P. T. 1945. The genetic structure of American populations of *Drosophila melanogaster.* Genetics *30:* 167–196.

Ives, P. T. 1954. Genetic changes in American populations of *Drosophila melanogaster.* Proc. Natl. Acad. Sci. U.S. *40:* 87–91.

James, A. P. 1959. The spectrum of severity of mutant effects. I. Haploid effects in yeast. Genetics *44:* 1309–1324.

Jensen, Neal. 1965. Multiline superiority in cereals. Crop Sci. *5:* 566–568.

Johnson, F. M., C. G. Kanapi, R. H. Richardson, M. R. Wheeler, and W. S. Stone. 1966. An analysis of polymorphisms among isozyme loci in dark and light *Drosophila ananassae* strains from American and Western Samoa. Proc. Natl. Acad. Sci. U.S. *56:* 119–125.

Johnston, R. F., and R. K. Selander. 1964. House sparrows: Rapid evolution of races in North America. Science *144:* 548–550.

Jukes, T. H. 1966. *Molecules and Evolution.* Columbia Univ. Press, New York.

Kastritsis, C. D. 1966. A comparative chromosome study in the incipient species of the *Drosophila paulistorum* complex. Chromosoma *19:* 208–222.

Kennedy, H. C. 1965. A generalized Hardy-Weinberg law. Bioscience *15:* 418.

Kerr, W. E., and S. Wright. 1954. Experimental studies of the distribution of gene frequencies in very small populations of *Drosophila melanogaster.* I. Forked. Evolution *8:* 172–177.

Kettlewell, H. B. D. 1958. A survey of the frequencies of *Biston betularia* (L) (Lep) and its melanic forms in Great Britain. Heredity *12:* 51–72.

Kettlewell, H. B. D. 1961a. Geographic melanism in the Lepidoptera of Shetland. Heredity *16:* 393–402.

Kettlewell, H. B. D. 1961b. Selection experiments on melanism in *Amathes glareosa* (Lepidoptera). Heredity *16:* 415–434.

Kettlewell, H. B. D. 1965. Insect survival and selection for pattern. Science *148:* 1290–1296.

Kettlewell, H. B. D., and R. J. Berry. 1961. The study of a cline. Heredity *16:* 403–414.

King, J. C. 1955. Evidence for the integration of the gene pool from studies of DDT resistance in *Drosophila.* Cold Spring Harbor Symp. Quant. Biol. *20:* 311–317.

Knight, G. R., A. Robertson, and C. H. Waddington. 1956. Selection for sexual isolation within a species. Evolution *10:* 14–22.

Kojima, K., and K. M. Yarbrough. 1967. Frequency dependent selection at the Esterase 6 locus in *Drosophila melanogaster.* Proc. Natl. Acad. Sci. U.S. *57:* 645–649.

Koopman, K. F. 1950. Natural selection for reproductive isolation between

Drosophila pseudoobscura and *Drosophila persimilis*. Evolution *4:* 135–148.

Lack, David. 1961. *Darwin's Finches*. Harper & Row, New York.

Lamotte, M. 1951. Recherches sur la structure genetique des populations naturelles de *Cepaea nemoralis* (L). Bull. Biol. France Belg. Suppl. *35:* 1–238.

Lerner, I. M. 1954. *Genetic Homeostasis*. Oliver and Boyd, Edinburgh.

Lerner, I. M. 1958. *The Genetic Basis of Selection*. Wiley, New York.

Levitan, M., H. L. Carson, and H. D. Stalker. 1954. Triads of overlapping inversions in *Drosophila robusta*. Am. Naturalist *88:* 113–114.

Lewontin, R. C. 1955. The effects of population density and composition on viability in *Drosophila melanogaster*. Evolution *9:* 27–41.

Lewontin, R. C. 1962. Interdeme selection controlling a polymorphism in the house mouse. Am. Naturalist *96:* 65–78.

Lewontin, R. C. 1964. The role of linkage in natural selection. Proc. 11th Intern. Congr. Genet. *3:* 517–525.

Lewontin, R. C. 1965. *The Theory of Inbreeding* by R. A. Fisher (review). Science *150:* 1800–1801.

Lewontin, R. C. 1966. Is nature probable or capricious? Bioscience *16:* 25–27.

Lewontin, R. C., and L. C. Dunn. 1960. The evolutionary dynamics of a polymorphism in the house mouse. Genetics *45:* 705–722.

Lewontin, R. C., and J. L. Hubby. 1966. A molecular approach to the study of genic heterozygosity in natural populations. II. Amount of variation and degree of heterozygosity in natural populations of *Drosophila pseudoobscura*. Genetics *54:* 595–609.

Lewontin, R. C., and M. J. D. White. 1960. Interaction between inversion polymorphisms of two chromosome pairs in the grasshopper, *Moraba scurra*. Evolution *14:* 116–129.

Li, C. C. 1948. *An Introduction to Population Genetics*. National Peking Univ. Press, Peking, China.

Li, C. C. 1955. *Population Genetics*. Univ. Chicago Press, Chicago.

Li, C. C. 1963. The way the load ratio works. Am. J. Human Genet. *15:* 316–321.

Lorenz, K. Z. 1958. The evolution of behavior. Sci. Am. *199* (December) 67.

Mather, W. B. 1964. *Principles of Quantitative Genetics*. Burgess, Minneapolis.

Matsunaga, E. 1955. Intra-uterine selection by the ABO incompatibility of mother and foetus. Am. J. Human Genet. *7:* 66–71.

Matsunaga, E., Y. Hiraizumi, T. Furusho, and H. Izumiyama. 1962. Studies on selection in ABO blood groups. Ann. Rept. Natl. Inst. Genet. Japan *13:* 103–106.

Mayr, E. 1942. *Systematics and the Origin of Species*. Columbia Univ. Press, New York.

Mayr, E. 1954. Change of genetic environment and evolution. In *Evolution as a Process* (Editors, J. Huxley, A. C. Hardy, and E. B. Ford). Allen and Unwin, London.

Mayr, E. 1959. Where are we? Cold Spring Harbor Symp. Quant. Biol. *24:* 1–14.

Mayr, E. 1963. *Animal Species and Evolution.* Harvard Univ. Press, Cambridge.

McFarquhar, A. M., and F. W. Robertson. 1963. The lack of evidence for co-adaptation in crosses between geographical races of *Drosophila subobscura coll.* Genet. Res. *4:* 104–131.

Mecham, J. S. 1960. Introgressive hybridization between two southeastern treefrogs. Evolution *14:* 445–457.

Mecham, J. S. 1961. Isolating mechanisms in anuran amphibians. In *Vertebrate Speciation* (Editor, W. F. Blair). Univ. Texas Press, Austin.

Merrell, D. J. 1963. "Heterosis" in *Drosophila.* Evolution *17:* 481–485.

Merrell, D. J. 1965. Lethal frequency and allelism in DDT resistant populations and their controls. Am. Naturalist *99:* 411–417.

Mettler, L. E. 1957. Studies on experimental populations of *Drosophila arizonensis* and *Drosophila mojavensis.* Texas, Univ. Publ. *5721:* 157–181.

Mettler, L. E., S. E. Moyer, and K. Kojima. 1966. Genetic loads in cage populations of *Drosophila.* Genetics *54:* 887–898.

Moore, J. A. 1957. An embryologist's view of the species concept. AAAS Publ. *50:* 325–338.

Morton, N. E., J. F. Crow, and H. J. Muller. 1956. An estimate of the mutational damage in man from data on consanguineous marriages. Proc. Natl. Acad. Sci. U.S. *42:* 855–863.

Mourad, A. E. 1965. Genetic divergence in M. Vetukhiv's experimental populations of *Drosophila pseudoobscura.* 2. Longevity. Genet. Res. *6:* 139–146.

Mukai, T., and I. Yoshikawa. 1964. Heterozygous effects of radiation-induced mutations on viability in homozygous and heterozygous genetic backgrounds in *Drosophila melanogaster.* Japan. J. Genet. *38:* 282–287.

Mukai, T., I. Yoshikawa, and K. Sano. 1966. The genetic structure of natural populations of *Drosophila melanogaster.* IV. Heterozygous effects of radiation-induced mutations on viability in various genetic backgrounds. Genetics *53:* 513–527.

Muller, C. J. 1961. A comparative study on the structure of mammalian and avian hemoglobins. Ph.D. thesis, University of Groningen, the Netherlands.

Muller, H. J. 1927. Artificial transmutation of the gene. Science *66:* 84–87.

Muller, H. J. 1930. Radiation and genetics. Am. Naturalist *64:* 220–251.

Muller, H. J. 1932a. Some genetic aspects of sex. Am. Naturalist *64:* 118–138.

Muller, H. J. 1932b. Further studies on the nature and causes of gene mutations. Proc. 6th Intern. Congr. Genet. *1:* 213–255.

Muller, H. J. 1939. Reversibility in evolution considered from the standpoint of genetics. Biol. Rev. *14:* 261–280.

Muller, H. J. 1950a. Our load of mutations. Am. J. Human Genet. *2:* 111–176.

Muller, H. J. 1950b. Radiation damage to the genetic material. I. Effects manifested mainly in descendants. II. Effects manifested mainly in exposed individuals. Am. Scientist *38:* 33–59, 399–425.

Muller, H. J. 1958a. The mutation theory re-examined. Proc. 10th Intern. Congr. Genet. *1:* 306–317.

Muller, H. J. 1958b. How much is evolution accelerated by sexual reproduction? AAAS General Program, p. 205.

Muller, H. J., and R. Falk. 1961. Are induced mutations in *Drosophila* overdominant? I. Experimental design. Genetics *46:* 727–735.

Nabours, R. K., and L. L. Kingsley. 1934. The operations of a lethal factor in *Apotettix eurycephalus* (grouse locusts) Genetics *19:* 323–328.

Needham, J. 1938. Contributions of chemical physiology to the problem of reversibility of evolution. Biol. Rev. *13:* 225–251.

Neel, J. V. 1950. The population genetics of two inherited blood dyscrasies in man. Cold Spring Harbor Symp. Quant. Biol. *15:* 141–158.

Nicholson, A. J. 1955. Density governed reaction, the counterpart of selection in evolution. Cold Spring Harbor Symp. Quant. Biol. *20:* 288–293.

Novitski, E., W. J. Peacock, and J. Engel. 1965. Cytological basis of "sex ratio" in *Drosophila pseudoobscura.* Science *148:* 516–517.

Oshima, C. 1961. The persistence of some natural lethal chromosomes of *Drosophila melanogaster* in experimental populations. Ann. Rept. Natl. Inst. Genet. Japan *12:* 11–12.

Owen, D. F. 1966. Polymorphism in pleistocene land snails. Science *152:* 71–72.

Parsons, P. A. 1965. Assortative mating for a metrical characteristic in *Drosophila.* Heredity *20:* 161–167.

Parsons, P. A., and W. F. Bodmer. 1961. The evolution of overdominance: Natural selection and heterozygote advantage. Nature *190:* 7–12.

Pearl, R., S. L. Parker, and B. M. Gonzalez. 1923. Experimental studies on the duration of life. VII. The Mendelian inheritance of duration of life in crosses of wildtype and quintuple stocks of *Drosophila melanogaster.* Am. Naturalist *57:* 153–192.

Pearson, K. 1904. On a generalized theory of alternative inheritance with special reference to Mendel's law. Phil. Trans. Roy. Soc. London *A203:* 53–86.

Pearson, K., and A. Lee. 1903. On the laws of inheritance in man. I. Inheritance of physical characters. Biometrika *2:* 357–462.

Penrose, L. S. 1949. The meaning of "fitness" in human populations. Ann. Eugen. *14:* 301–304.

Penrose, L. S. 1957. Mutation in man. In *Effect of Radiation on Human Heredity,* World Health Organization, Geneva.

Prevosti, A. 1955. Geographical variability in quantitative traits in populations of *Drosophila subobscura.* Cold Spring Harbor Symp. Quant. Biol. *20:* 294–299.

Rapoport, J. A. 1939. Specific morphoses in *Drosophila* induced by chemical compounds. Bull. Biol. Med. Exptl. URSS *7:* 415–417.

Rasmuson, M. 1961. *Genetics on the Population Level.* Scandinavian Univ. Books, Stockholm.

Reed, S. C., and E. W. Reed. 1950. Natural selection in laboratory populations of *Drosophila.* II. Competition between a white-eye gene and its wild type allele. Evolution *4:* 34–42.

Rendel, J. M. 1943. Variations in the weights of hatched and unhatched ducks' eggs. Biometrika *33:* 48–58.

Rendel, J. M. 1945. Genetics and cytology of *Drosophila subobscura.* II. Normal and selective matings in *Drosophila subobscura.* J. Genet. *46:* 287–302.

Rendel, J. M. 1951. Mating of *ebony vestigial* and wildtype *Drosophila melanogaster* in light and dark. Evolution *5:* 226–230.

Richardson, R. H., and K. Kojima. 1965. The kinds of genetic variability in relation to selection responses in *Drosophila* fecundity. Genetics *52:* 583–598.

Robertson, F. W. 1955. Selection response and the properties of genetic variation. Cold Spring Harbor Symp. Quant. Biol. *20:* 166–177.

Sanghvi, L. D. 1963. The concept of genetic load: A critique. Am. J. Human Genet. *15:* 298–309.

Sankaranarayanan, K. 1964. Genetic loads in irradiated experimental populations of *Drosophila melanogaster*. Genetics *50:* 131–150.

Sankaranarayanan, K. 1965. Further data on the genetic loads in irradiated experimental populations of *Drosophila melanogaster*. Genetics *52:* 153–164.

Sankaranarayanan, K. 1966. Some components of the genetic loads in irradiated experimental populations of *Drosophila melanogaster*. Genetics *54:* 121–130.

Scharloo, W. 1964. The effect of disruptive and stabilizing selection on the expression of a *cubitus interruptus* mutant in *Drosophila*. Genetics *50:* 553–562.

Scharloo, W. 1965. Stabiliserende en disruptieve selectie. Genen en Phaenen *10:* 8–23.

Schull, W. J., and J. V. Neel. 1963. *Genetic Selection in Man.* Univ. Michigan Press, Ann Arbor.

Schull, W. J., and J. V. Neel. 1965. *The Effects of Inbreeding on Japanese Children.* Harper and Row, New York.

Schwartz, D. 1962. Genetic studies on mutant enzymes in maize. III. Control of gene action in the synthesis of pH 7.5 esterase. Genetics *47:* 1609–1615.

Scossiroli, R. E. 1959. Selezione artificiale per un carattere quantitativo in popolazioni di *Drosophila melanogaster* irradiate con raggi X. Comitato Nazionale per le Recherche Nucleari, Divisione Biologica (CNB–4).

Shaw, C. R. 1965. Electrophoretic variation in enzymes. Science *149:* 936–943.

Sheppard, P. M. 1952. A note on non-random mating in the moth *Panaxia dominula* (L). Heredity *6:* 239–241.

Sibley, C. G. 1954. Hybridization in the red-eyed towhees of Mexico. Evolution *8:* 252–290.

Simpson, G. G. 1953. *The Major Features of Evolution.* Columbia Univ. Press, New York.

Sinnott, E., L. C. Dunn, and Th. Dobzhansky. 1958. *Principles of Genetics.* McGraw-Hill, New York.

Sokal, R. R., and I. Huber. 1963. Competition among genotypes in *Tribolium castaneum* at varying densities and gene frequencies (the *sooty* locus). Am. Naturalist *97:* 169–184.

Sokal, R. R., and R. L. Sullivan. 1963. Competition between mutant and wild-type house-fly strains at varying densities. Ecology *44:* 314–322.

Sokoloff, A. 1965. Geographic variation of quantitative characters in populations of *Drosophila pseudoobscura*. Evolution *19:* 300–310.

Sondhi, K. C. 1961. Developmental barriers in a selection experiment. Nature *189:* 249–250.

Sondhi, K. C. 1963. The biological foundations of animal patterns. Quart. Rev. Biol. *38:* 289–327.

Sonneborn, T. M. 1947. Recent advances in the genetics of Paramecium and Euplotes. Advan. Genet. *1:* 263–358.

Sonnenblick, B. P. 1940. Cytology and development of the embryos of X-rayed adult *Drosophila melanogaster.* Proc. Natl. Acad. Sci. U.S. *26:* 373–381.

Sonnenblick, B. P., and P. S. Henshaw. 1941. Influence on development of certain dominant lethals induced by X-rays in *Drosophila* germ cells. Proc. Soc. Exptl. Biol. Med. *48:* 74–79.

Spassky, B., Th. Dobzhansky, and W. Anderson. 1965. Genetics of natural populations. XXXVI. Epistatic interactions of the components of the genetic load in *Drosophlia pseudoobscura.* Genetics *52:* 653–664.

Spencer, W. P. 1947. Mutations in wild populations in *Drosophila.* Advan. Genet. *1:* 359–402.

Spencer, W. P. 1957. Genetic studies on *Drosophila mulleri.* I. The genetic analysis of a population. Texas, Univ. Publ. *5721:* 186–205.

Spickett, S. G. 1963. Genetic and developmental studies of a quantitative character. Nature *199:* 870–873.

Sprague, G. F. 1941. The location of dominant favorable genes in maize by means of an inversion. Genetics *26:* 170 (abstract).

Stalker, H. D., and H. L. Carson. 1947. Morphological variation in natural populations of *Drosophila robusta* Sturtevant. Evolution *1:* 237–248.

Stebbins, G. L. 1950. *Variation and Evolution in Plants.* Columbia Univ. Press, New York.

Stebbins, G. L. 1966. *Processes of Organic Evolution.* Prentice-Hall, Englewood Cliffs, N.J.

Stern, C. 1943. The Hardy-Weinberg law. Science *97:* 137–138.

Stern, C. 1960. *Human Genetics.* Freeman, San Francisco.

Stern, C., G. Carson, M. Kinst, E. Novitski, and D. Uphoff. 1952. The viability of heterozygotes for lethals. Genetics *37:* 413–449.

Steward, F. C. 1966. *About Plants: Topics in Plant Biology.* Addison-Wesley, Reading, Mass.

Streams, F. A., and D. Pimentel. 1961. Effects of immigration on the evolution of populations. Am. Naturalist *95:* 201–210.

Strickberger, M. W. 1963. Evolution of fitness in experimental populations of *Drosophila pseudoobscura.* Evolution *17:* 40–55.

Sturtevant, A. H. 1965. *A History of Genetics.* Harper and Row, New York.

Sturtevant, A. H., and Th. Dobzhansky. 1936a. Inversions in the third chromosome of wild races of *Drosophila pseudoobscura* and their use in the study of the history of the species. Proc. Natl. Acad. Sci. U.S. *22:* 448–450.

Sturtevant, A. H., and Th. Dobzhansky. 1936b. Geographical distribution and cytology of "sex-ratio" in *Drosophila pseudoobscura* and related species. Genetics *21:* 473–490.

Sturtevant, A. H., and K. Mather. 1938. The inter-relations of inversions, heterosis, and recombination. Am. Naturalist *72:* 447–452.

Sullivan, R. L., and R. R. Sokal. 1963. The effects of larval density on several strains of the house fly. Ecology *44:* 120–130.

Sullivan, R. L., and R. R. Sokal. 1965. Further experiments on competition between strains of house flies. Ecology *46:* 172–182.

Sumner, F. B. 1935. Evidence for the protective value of changeable coloration in fishes. Am. Naturalist *69:* 245–266.

Sved, J. A., T. E. Reed, and W. F. Bodmer. 1967. The number of balanced polymorphisms that can be maintained in a natural population. Genetics *55:* 469–481.

Tan, C. C. 1946. Genetics of sexual isolation between *Drosophila pseudoobscura* and *D. persimilis.* Genetics *31:* 558–573.

Tantawy, A. O., and G. S. Mallah. 1961. Studies on natural populations of *Drosophila.* I. Heat resistance and geographical variation in *Drosophila melanogaster* and *D. simulans.* Evolution *15:* 1–14.

Teissier, G. 1954a. Conditions d'équilibre d'un couple d'alleles et supériorité des hétérozygotes. Compt. Rend. *238:* 621–623.

Teissier, G. 1954b. Sélection naturelle et fluctuation génétique. Compt. Rend. *238:* 1929–1931.

Temin, R. G. 1966. Homozygous viability and fertility loads in *Drosophila melanogaster.* Genetics *53:* 27–46.

Thoday, J. M. 1958. Effects of disruptive selection: The experimental production of a polymorphic population. Nature *181:* 1124–1125.

Thoday, J. M. 1964. Genetics and the integration of reproductive systems. In *Insect Reproduction* (Editor, K. C. Highnam). Symp. 2, Roy. Entomol. Soc. London, pp. 108–119.

Thoday, J. M., and T. B. Boam. 1959. Effects of disruptive selection. II. Polymorphism and divergence without isolation. Heredity *13:* 205–218.

Thoday, J. M., and J. B. Gibson. 1962. Isolation by disruptive selection. Nature *193:* 1164–1166.

Timofeeff-Ressovsky, N. W. 1932. Mutations of the gene in different directions. Proc. 6th Intern. Congr. Genet. *1:* 308–330.

Timofeeff-Ressovsky, N. W. 1934. Über die Vitalität einiger Genmutationen und ihrer Kombinationen bei *Drosophila funebris* und ihre Abhängigkeit vom genotypischen und vom ausseren Milieu. Z. induktive Abstammungs.- Vererbungslehre *66:* 319–344.

Timofeeff-Ressovsky, N. W. 1935a. Über geographische Temperaturrassen bei *Drosophila funebris* F. Arch. Naturgesch. *4:* 245–257.

Timofeeff-Ressovsky, N. W. 1935b. Auslösung von Vitalitätmutationen durch Röntgenbestralung bei *Drosophila melanogaster.* Nachr. Ges. Wiss. Gottingen, Biol. N. F., *1:* 163–180.

Timofeeff-Ressovsky, N. W. 1940. Mutations and geographical variation. In *The New Systematics* (Editor, J. Huxley). Clarendon Press, Oxford.

Timofeeff-Ressovsky, N. W., and H. A. Timofeeff-Ressovsky. 1940a. Populationsgenetische Versuche an *Drosophila.* I. Zeitliche und raumliche verteilung der Individuen einiger *Drosophila*-Arten über das Gelande. Z. induktive Abstammungs.- Vererbungslehre *79:* 28–34.

Timofeeff-Ressovsky, N. W., and H. A. Timofeeff-Ressovsky. 1940b. Populationsgenetische Versuche an *Drosophila.* II. Aktionsbereiche von *D. fune-*

bris und *D. melanogaster*. *Z. induktive Abstammungs.- Vererbungslehre* *79:* 39–43.

Timofeeff-Ressovsky, N. W., and H. A. Timofeeff-Ressovsky. 1940c. Populationsgenetische Versuche an *Drosophila*. III. Quantitative Untersuchung an einigen *Drosophila*-populationen. Z. induktive Abstammungs.- Vererbungslehre *79:* 44–49.

The United Nations Scientific Committee on the Effects of Atomic Radiation, Report of, 1962. United Nations, New York.

Van Valen, L. 1963. Introgression in laboratory populations of *Drosophila persimilis* and *D. pseudoobscura*. Heredity *18:* 205–214.

Vann, E. G. 1966. The fate of X-ray induced chromosomal rearrangements introduced into laboratory populations of *Drosophila melanogaster*. Am. Naturalist *100:* 425–449.

Vetukhiv, M. 1953. Viability of hybrids between local populations of *Drosophila pseudoobscura*. Proc. Natl. Acad. Sci. U.S. *39:* 30–34.

Vetukhiv, M. 1954. Integration of the genotype in local populations of three species of *Drosophila*. Evolution *8:* 241–251.

Vetukhiv, M. 1956. Fecundity of hybrids between geographic populations of *Drosophila pseudoobscura*. Evolution *10:* 139–146.

Waddington, C. H. 1953. Genetic assimilation of an acquired character. Evolution *7:* 118–126.

Waddington, C. H. 1956. Genetic assimilation of the *bithorax* phenotype. Evolution *10:* 1–13.

Waddington, C. H., and E. Robertson. 1966. Selection for developmental canalization. Genet. Res. *7:* 303–312.

Waldron, I. 1964. Courtship sound production in two sympatric sibling *Drosophila* species. Science *144:* 191–192.

Wallace, B. 1948. Studies on "sex-ratio" in *Drosophila pseudoobscura*. I. Selection and "sex-ratio." Evolution *2:* 189–217.

Wallace, B. 1950. Allelism of second chromosome lethals in *Drosophila melanogaster*. Proc. Natl. Acad. Sci. U.S. *36:* 654–657.

Wallace, B. 1952. The estimation of adaptive values of experimental populations. Evolution *6:* 333–341.

Wallace, B. 1953. On coadaptation in *Drosophila*. Am. Naturalist *87:* 343–358.

Wallace, B. 1954. Genetic divergence of isolated populations of *Drosophila melanogaster*. Proc. 9th Intern. Congr. Genet. *1:* 761–764.

Wallace, B. 1955. Inter-population hybrids in *Drosophila melanogaster*. Evolution 9*:* 302–316.

Wallace, B. 1956. Studies on irradiated populations of *Drosophila melanogaster*. J. Genet. *54:* 280–293.

Wallace, B. 1958. The average effect of radiation-induced mutations on viability in *Drosophila melanogaster*. Evolution *12:* 532–556.

Wallace, B. 1959a. The role of heterozygosity in *Drosophila* populations. Proc. 10th Intern. Congr. Genet. *1:* 408–419.

Wallace, B. 1959b. Influence of genetic systems on geographical distribution. Cold Spring Harbor Symp. Quant. Biol. *24:* 193–204.

Wallace, B. 1962. Temporal changes in the roles of lethal and semilethal chromosomes within populations of *Drosophila melanogaster*. Am. Naturalist *96:* 247–256.

Wallace, B. 1963a. The elimination of an autosomal lethal from an experimental population of *Drosophila melanogaster*. Am. Naturalist *97:* 65–66.

Wallace, B. 1963b. Further data on the over-dominance of induced mutations. Genetics *48:* 633–651.

Wallace, B. 1964. *Population Genetics.* BSCS Pamphlet, Heath, Boston.

Wallace, B. 1965. The viability effects of spontaneous mutations in *Drosophila melanogaster*. Am. Naturalist *99:* 335–348.

Wallace, B. 1966a. On the dispersal of *Drosophila.* Am. Naturalist *100:* 551–563.

Wallace, B. 1966b. Distance and the allelism of lethals in a tropical population of *Drosophila melanogaster*. Am. Naturalist *100:* 565–578.

Wallace, B. 1966c. The fate of *sepia* in small populations of *Drosophila melanogaster*. Genetica *37:* 29–36.

Wallace, B. 1966d. Natural and radiation-induced chromosomal polymorphism in *Drosophila.* Mutation Res. *3:* 194–200.

Wallace, B., and Th. Dobzhansky. 1962. Experimental proof of balanced genetic loads in *Drosophila.* Genetics *47:* 1027–1042.

Wallace, B., and J. C. King. 1951. Genetic changes in populations under irradiation. Am. Naturalist *85:* 209–222.

Wallace, B., and J. C. King. 1952. A genetic analysis of the adaptive values of populations. Proc. Natl. Acad. Sci. U.S. *38:* 706–715.

Wallace, B., J. C. King, C. V. Madden, B. Kaufmann, and E. C. McGunnigle. 1953. An analysis of variability arising through recombination. Genetics *38:* 272–307.

Wallace, B., and C. Madden. 1953. The frequencies of sub- and supervitals in experimental populations of *Drosophila melanogaster*. Genetics *38:* 456–470.

Wallace, B., and A. M. Srb. 1964. *Adaptation,* (2nd edition). Prentice-Hall, Englewood Cliffs, N.J.

Wallace, B., and M. Vetukhiv. 1955. Adaptive organization of the gene pools of *Drosophila* populations. Cold Spring Harbor Symp. Quant. Biol. *20:* 303–310.

Wallace, B., E. Zouros, and C. Krimbas. 1966. Frequencies of second and third chromosome lethals in a tropical population of *Drosophila melanogaster*. Am. Naturalist *100:* 245–251.

Wasserman, M. 1962. Cytological studies of the *repleta* group of the genus *Drosophila.* V. The *mulleri* subgroup. Texas, Univ. Publ. *6205:* 85–117.

Weinberg, W. 1908. Über den Nachweis der Vererbung beim Menschen. Jahresh. Verein. vaterl. Naturk. Württemberg *64:* 368–382. (see Stern, C., 1943).

Weisbrot, D. R. 1966. Genotypic interactions among competing strains and species of *Drosophila.* Genetics *53:* 427–435.

Wolfenbarger, D. O. 1946. Dispersion of small organisms. Distance dispersion rates of bacteria, spores, seeds, pollen, and insects; incidence rates of diseases and injuries. Am. Midland Naturalist *35:* 1–152.

Wright, S. 1931. Evolution in Mendelian populations. Genetics *16:* 97–159.

Wright, S. 1932. The roles of mutation, inbreeding, crossbreeding, and selection in evolution. Proc. 6th Intern. Congr. Genet. *1:* 356–366.

Wright, S. 1934. Physiological and evolutionary theories of dominance. Am. Naturalist *68:* 24–53.

Wright, S. 1964. The distribution of self-incompatibility alleles in populations. Evolution *18:* 609–619.

Wright, S. 1968. Dispersion of *Drosophila pseudoobscura.* Am. Naturalist *102:* 81–84.

Wright, S., Th. Dobzhansky, and W. Hovanitz. 1942. Genetics of natural populations. VII. The allelism of lethals in the third chromosome of *Drosophila pseudoobscura.* Genetics *27:* 363–394.

Wright, S., and Th. Dobzhansky. 1946. Genetics of natural populations. XII. Experimental reproduction of some of the changes caused by natural selection in certain populations of *Drosophila pseudoobscura.* Genetics *31:* 125–156.

Index

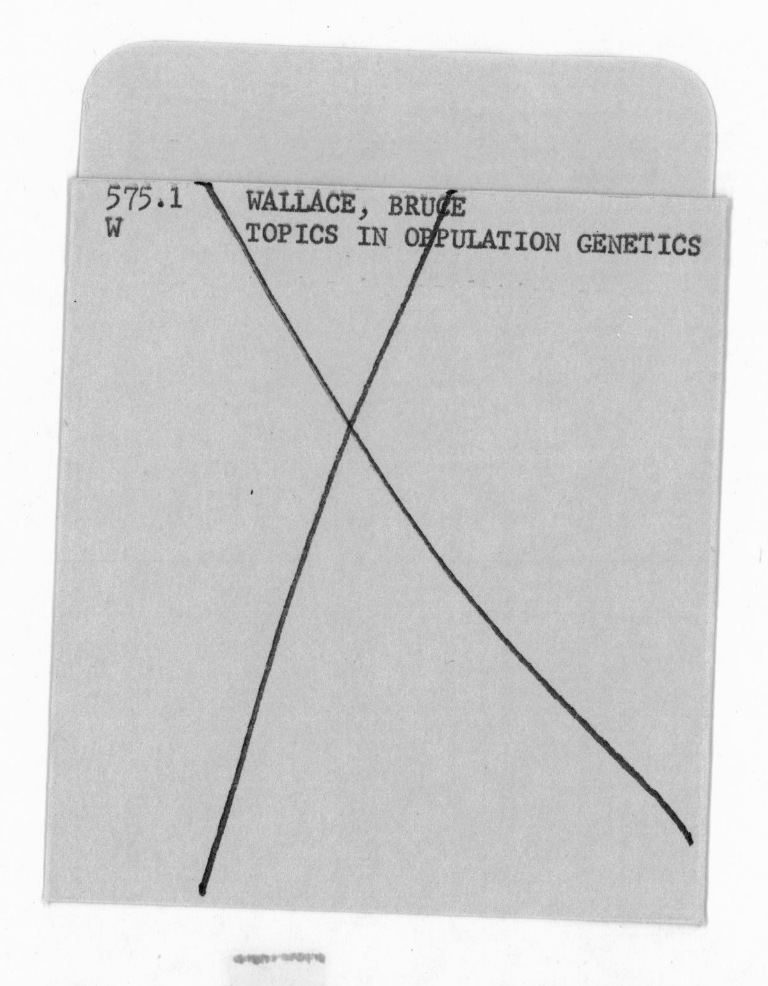